CAMBRIDGE LIBRARY COLLECTION

Books of enduring scholarly value

Mathematical Sciences

From its pre-historic roots in simple counting to the algorithms powering modern desktop computers, from the genius of Archimedes to the genius of Einstein, advances in mathematical understanding and numerical techniques have been directly responsible for creating the modern world as we know it. This series will provide a library of the most influential publications and writers on mathematics in its broadest sense. As such, it will show not only the deep roots from which modern science and technology have grown, but also the astonishing breadth of application of mathematical techniques in the humanities and social sciences, and in everyday life.

The Scientific Papers of Sir George Darwin

Sir George Darwin (1845-1912) was the second son and fifth child of Charles Darwin. After studying mathematics at Cambridge he read for the Bar, but soon returned to science and to Cambridge, where in 1883 he was appointed Plumian Professor of Astronomy and Experimental Philosophy. His family home is now the location of Darwin College. His work was concerned primarily with the effect of the sun and moon on tidal forces on Earth, and with the theoretical cosmogony which evolved from practical observation: he formulated the fission theory of the formation of the moon (that the moon was formed from still-molten matter pulled away from the Earth by solar tides). This volume of his collected papers covers periodic orbits and some miscellaneous papers, including two investigating the health statistics of the marriage of first cousins – of interest to a member of a dynasty in which such marriages were common.

Cambridge University Press has long been a pioneer in the reissuing of out-of-print titles from its own backlist, producing digital reprints of books that are still sought after by scholars and students but could not be reprinted economically using traditional technology. The Cambridge Library Collection extends this activity to a wider range of books which are still of importance to researchers and professionals, either for the source material they contain, or as landmarks in the history of their academic discipline.

Drawing from the world-renowned collections in the Cambridge University Library, and guided by the advice of experts in each subject area, Cambridge University Press is using state-of-the-art scanning machines in its own Printing House to capture the content of each book selected for inclusion. The files are processed to give a consistently clear, crisp image, and the books finished to the high quality standard for which the Press is recognised around the world. The latest print-on-demand technology ensures that the books will remain available indefinitely, and that orders for single or multiple copies can quickly be supplied.

The Cambridge Library Collection will bring back to life books of enduring scholarly value (including out-of-copyright works originally issued by other publishers) across a wide range of disciplines in the humanities and social sciences and in science and technology.

The Scientific Papers of Sir George Darwin

VOLUME 4:
PERIODIC ORBITS
AND MISCELLANEOUS PAPERS

GEORGE HOWARD DARWIN

CAMBRIDGE
UNIVERSITY PRESS

CAMBRIDGE UNIVERSITY PRESS

Cambridge New York Melbourne Madrid Cape Town Singapore São Paolo Delhi

Published in the United States of America by Cambridge University Press, New York

www.cambridge.org
Information on this title: www.cambridge.org/9781108004473

This edition first published 1911
This digitally printed version 2009

ISBN 978-1-108-00447-3

SCIENTIFIC   PAPERS

CAMBRIDGE UNIVERSITY PRESS
London: FETTER LANE, E.C.
C. F. CLAY, Manager

Edinburgh: 100, PRINCES STREET
Berlin: A. ASHER AND CO.
Leipzig: F. A. BROCKHAUS
New York: G. P. PUTNAM'S SONS
Bombay and Calcutta: MACMILLAN AND CO., Ltd.

SCIENTIFIC PAPERS

BY

SIR GEORGE HOWARD DARWIN,

K.C.B., F.R.S.

FELLOW OF TRINITY COLLEGE
PLUMIAN PROFESSOR IN THE UNIVERSITY OF CAMBRIDGE

VOLUME IV

PERIODIC ORBITS

AND

MISCELLANEOUS PAPERS

Cambridge:
at the University Press
1911

𝕮𝖆𝖒𝖇𝖗𝖎𝖉𝖌𝖊 :

PRINTED BY JOHN CLAY, M.A.
AT THE UNIVERSITY PRESS.

PREFACE

THE papers contained in this fourth and last volume are too diverse in character to admit of such a complete arrangement according to subjects, as was made in the earlier volumes. I begin, however, with three papers on Periodic Orbits, a subject which has played a very important part in the recent developments of dynamical astronomy. The middle one of the three is by Mr Hough, and it is reproduced, by his permission, as explaining the sequence of ideas which led from the first to the third paper.

The proprietors of the *Encyklopädie der mathematischen Wissenschaften* have kindly allowed me to print here the English text of the article 'Die Bewegung der Hydrosphäre.' It is the long paper entitled 'The Tides,' which forms Part II. The section on the dynamical theory of the tides is by Mr Hough, but I wrote the rest. According to my general scheme this paper would naturally have belonged to the group of papers contained in Vol. I, but it was published after the appearance of that volume. The same remark also applies to the two recent papers on tides contained in Part IV.

Part III consists of miscellaneous papers arranged in chronological order. Some of them are obviously of slight value and are reproduced merely for the sake of completeness.

The longest paper in this part is that on 'The mechanical conditions of a swarm of meteorites,' and I find it very difficult to estimate its value. If success is to be measured by the amount of comment to which a paper gives rise, it is a failure; for it has received but little notice.

There are two principal schemes of cosmogony; in the one a gaseous nebula is the parent of the solar system, and in the other the sun and the planets are formed by the accretion of myriads of meteorites. It seems to me that it was worth while to attempt to explain how two such different origins may have led to the same result.

Cosmogonists are of course compelled to begin their survey of the solar system at some arbitrary stage of its history, and they do not, in general, seek to explain how the solar nebula, whether gaseous or meteoritic, came to exist. My investigation starts from the meteoritic point of view, and I assume the meteorites to be moving indiscriminately in all directions. But the doubt naturally arises as to whether at any stage a purely chaotic motion of the individual meteorites could have existed, and whether the assumed initial condition ought not rather to have been an aggregate of flocks of meteorites moving about some central condensation in orbits which intersect one another at all sorts of angles. If this were so the chaos would not be one consisting of individual stones which generate a quasi-gas by their collisions, but it would be a chaos of orbits. But it is not very easy to form an exact picture of this supposed initial condition, and the problem thus seems to elude mathematical treatment. Then again have I succeeded in showing that a pair of meteorites in collision will be endowed with an effective elasticity? If it is held that the chaotic motion and the effective elasticity are quite imaginary, the theory collapses. It should however be remarked that an infinite gradation is possible between a chaos of individuals and a chaos of orbits, and it cannot be doubted that in most impacts the colliding stones would glance from one another. It seems to me possible, therefore, that my two fundamental assumptions may possess such a rough resemblance to truth as to produce some degree of similitude between the life-histories of gaseous and meteoritic nebulæ. If this be so the Planetesimal Hypothesis of Chamberlin and Moulton is nearer akin to the Nebular Hypothesis than the authors of the former seem disposed to admit.

Even if the whole of the theory should be condemned as futile, yet the paper contains an independent solution of the problem of Lane and Ritter; and besides the attempt to discuss the boundary of an atmosphere, where the collisions have become of vanishing rarity, may still perhaps be worth something.

Paper 20 on the Eulerian Nutation in Part III also demands a word of explanation. It was originally written in the form of a letter to the late M. Folie, who held heterodox views as to the earth's nutation, and it was presented by him with my permission to the Royal Academy of Belgium. It was written with the object of clearing my own ideas and those of M. Folie, and does not claim to contain anything novel.

Part V consists of addresses to learned societies. The scientific aspects of several of the subjects have changed considerably since the dates of their delivery, but no attempt has been made to introduce more modern views into the discussions.

An appendix at the end of the volume contains, by permission of the Royal Statistical Society, two papers on the marriage of cousins written many years ago. As, even after the lapse of 36 years, I still receive letters on the subject, it seemed advisable to include them in this collection, although they are biological in character.

The chronological list of papers is again repeated, and a complete list is given of all the errata hitherto discovered in the four volumes, some of them of course being repeated from the previous volumes. So accurate has been the work of the compositors and readers that typographical mistakes are hardly to be found, and I feel greatly indebted to them for their care.

In conclusion I desire to thank the Syndics of the Cambridge University Press for the honour they did me four years ago, when they invited me to undertake the task which is now completed.

G. H. DARWIN.

CAMBRIDGE,
May, 1911.

CONTENTS

PART V. ADDRESSES TO SOCIETIES.

APPENDIX.

CHRONOLOGICAL LIST OF PAPERS WITH REFERENCES TO THE VOLUMES IN WHICH THEY ARE CONTAINED.

YEAR	TITLE AND REFERENCE	Volume in collected papers
1875	On some proposed forms of slide-rule. London Math. Soc. Proc., 6, 1875, p. 113.	IV
1875	On two applications of Peaucellier's cell. London Math. Soc. Proc., 6, 1875, pp. 113, 114.	IV
1875	The mechanical description of equipotential lines. London Math. Soc. Proc., 6, 1875, pp. 115—117.	IV
1875	On a mechanical representation of the second elliptic integral. Messenger of Math., 4, 1875, pp. 113—115.	IV
1875	On maps of the World. Phil. Mag., 50, 1875, pp. 431—444.	IV
1876	On the influence of geological changes on the Earth's axis of rotation. Roy. Soc. Proc., 25, 1877, pp. 328—332; Phil. Trans., 167, 1877, pp. 271—312.	III
1876	On an oversight in the *Mécanique Céleste*, and on the internal densities of the planets. Astron. Soc. Month. Not., 37, 1877, pp. 77—89.	III
1877	A geometrical puzzle. Messenger of Math., 6, 1877, p. 87.	IV
1877	A geometrical illustration of the potential of a distant centre of force. Messenger of Math., 6, 1877, pp. 97—98.	IV
1877	Note on the ellipticity of the Earth's strata. Messenger of Math., 6, 1877, pp. 109—110.	III
1877	On graphical interpolation and integration. Messenger of Math., 6, 1877, pp. 134—136; or abstract Brit. Assoc. Rep., 1876, p. 13.	IV
1877	On a theorem in spherical harmonic analysis. Messenger of Math., 6, 1877, pp. 165—168.	IV
1877	On a suggested explanation of the obliquity of planets to their orbits. Phil. Mag., 3, 1877, pp. 188—192.	III
1877	On fallible measures of variable quantities, and on the treatment of meteorological observations. Phil. Mag., 4, 1877, pp. 1—14.	IV

ERRATA.

Vol. I.,   p. 275, equation (26), line 9 from foot of page should be
$$\vartheta_{2s} = 4) + 2p_s\kappa_m - \kappa_{2s}.$$

„   p. 377, line 14 from foot of page,
for 28 read 27.

Vol. II.,   p. 369, lines 5 to 8. This paragraph purports to give an outline of the results obtained previously, but it does not do so, and I cannot explain how this passage came to stand as it does. The paragraph should run :

    As the moon recedes from the earth the period of her revolution increases both absolutely and relatively to the period of the earth's rotation. At a very early stage of the history the month has increased until it has become equal to $1\tfrac{7}{11}$ days. This is the critical epoch at which the lunar orbit ceases to be circular. Subsequently the eccentricity of the orbit continually increases.

Vol. III., p. 86, second line of equation (22),

$$\text{\emph{for}} \;\; S_0' a^6 \, 2 \left\{ \frac{df}{ada} \; ... \right\} \;\; \text{\emph{read}} \;\; S_0' a^6 \left\{ 2 \, \frac{df}{ada} \; ... \right\}.$$

„   p. 87, in the first and second of equations (26),

$$\text{\emph{for}} \; = \frac{S_0'}{a^2} \{ \, ... \, \} \;\; \text{\emph{read}} \; = S_0' a^2 \{ \, ... \, \}.$$

„   „ in the middle line of equation (28),

$$\text{\emph{for}} \; +S' \left\{ ... \, -\tfrac{52}{21} ah \frac{dh}{da} \; ... \right\} \;\; \text{\emph{read}} \; +S' \left\{ ... \, -\tfrac{52}{21} h \frac{dh}{da} \; ... \right\}.$$

„   p. 103, the equation in line 5 should be

$$\frac{\rho'}{\rho_0} = \frac{\rho}{\rho_0} + \left(1 - \frac{\rho}{\rho_0} \right) \left(\frac{a}{a'} \right)^3.$$

„   p. 394, line 4 from foot of page,
for $\mathfrak{A}_1{}^1$ read $A_1{}^1$.

„   p. 402, line 12 from foot of page,
for $EF' + F'F - FF'$ *read* $EF' + E'F - FF'$.

Vol. IV., p. 74, in the formula given in the last line but one,
for Ψ_0 *read* ψ_0.

PART I

PERIODIC ORBITS

1.

PERIODIC ORBITS.

[*Acta Mathematica*, Vol. XXI., (1897), pp. 99—242, and (with omission of certain tables of results) *Mathematischen Annalen*, Vol. LI., pp. 523—583.]

§ 1. *Introduction.*

THE existing methods of treating the Problem of the three Bodies are only applicable to the determination, by approximation, of the path of the third body when the attraction of the first largely preponderates over that of the second. A general solution of the problem is accordingly not to be obtained by these methods.

In the Lunar and Planetary theories it has always been found necessary to specify the motion of the perturbed body by reference to a standard curve or intermediate orbit, of which the properties are fully known. The degree of success attained by any of these methods has always depended on the aptness of the chosen intermediate orbit for the object in view. It is probable that future efforts will resemble their precursors in the use of standard curves of reference.

Mr G. W. Hill's papers on the Lunar Theory* mark an epoch in the history of the subject. His substitution of the Variational Curve for the ellipse as the intermediate orbit is not only of primary importance in the Lunar Theory itself, but has pointed the way towards new fields of research.

The variational curve may be described as the distortion of the moon's circular orbit by the solar attraction. It is one of that class of periodic solutions of the Problem of the three Bodies which forms the subject of the present paper.

* *American Journal of Mathematics*, Vol. I., pp. 5—29, 129—147, 245—260, and *Acta Mathematica*, T. VIII., pp. 1—36.

Of these solutions M. Poincaré writes:

"Voici un fait que je n'ai pu démontrer rigoureusement, mais qui me paraît pourtant très vraisemblable.

"Etant données des équations de la forme définie dans le n° 13 et une solution particulière de ces équations, on peut toujours trouver une solution périodique (dont la période peut, il est vrai, être très longue), telle que la différence entre les deux solutions soit aussi petite qu'on le veut, pendant un temps aussi long qu'on le veut. D'ailleurs, ce qui nous rend ces solutions périodiques si précieuses, c'est qu'elles sont, pour ainsi dire, la seule brèche par où nous puissions essayer de pénétrer dans une place jusqu'ici réputée inabordable *."

He tells us that he has been led to distinguish three kinds of periodic solutions. In those of the first kind the inclinations vanish and the eccentricities are very small; in those of the second kind the inclinations vanish and the eccentricities are finite; and in those of the third kind the inclinations do not vanish †.

If I understand this classification correctly the periodic orbits, considered in this paper, belong to the first kind, for they arise when the perturbed body has infinitely small mass, and when the two others revolve about one another in circles.

M. Poincaré remarks that there is a quadruple infinity of periodic solutions, for there are four arbitrary constants, viz. the period of the infinitesimal body, the constant of energy, the moment of conjunction, and the longitude of conjunction ‡. For the purpose of the present investigation this quadruple infinity may however be reduced to a single infinity, for the moment and longitude of conjunction need not be considered; and the scale on which we draw the circular orbit of the second body round the first is immaterial. Thus we are only left with the constant of relative energy of the motion of the infinitesimal body as a single arbitrary.

Notwithstanding the great interest attaching to periodic orbits, no suggestion has, up to the present time, been made by any writer for a general method of determining them. As far as I can see, the search resolves itself into the discussion of particular cases by numerical processes, and such a search necessarily involves a prodigious amount of work. It is not for me to say whether the enormous labour I have undertaken was justifiable in the first instance; but I may remark that I have been led on, by the interest of my results, step by step, to investigate more and again more cases. Now that so much has been attained I cannot but think that the conclusions will prove of interest both to astronomers and to mathematicians.

* *Mécanique Céleste*, T. i., p. 82.
† *Mécanique Céleste*, T. i., p. 97, and *Bull. Astr.*, T. i., p. 65.
‡ *Mécanique Céleste*, T. i., p. 101.

In conducting extensive arithmetical operations, it would be natural to avail oneself of the skill of professional computers.   But unfortunately the trained computer, who is also a mathematician, is rare.   I have thus found myself compelled to forego the advantage of the rapidity and accuracy of the computer, for the higher qualities of mathematical knowledge and judgment.

In my earlier work I received the greatest assistance from Mr J. W. F. Allnutt; his early death has deprived me of a friend and of an assistant, whose zeal and care were not to be easily surpassed.   Since his death Mr J. I. Craig (of Emmanuel College) and Mr M. J. Berry (of Trinity College) have rendered and are rendering valuable help.   I have besides done a great deal of computing myself*.

The reader will see that the figures have been admirably rendered by Mr Edwin Wilson of Cambridge, and I only regret that it has not seemed expedient to give them on a larger scale.

The first part of the paper is devoted to the mathematical methods employed, the second part contains the discussion of the results, and the tables of numerical results are relegated to an Appendix.

PART I.

§ 2.   *Equations of motion.*

The particular case of the problem of the three bodies, considered in this paper, is where the mass of the third body is infinitesimal compared with that of either of the two others which revolve about one another in circles, and where the whole motion takes place in one plane.

For the sake of brevity the largest body will be called the Sun, the planet which moves round it will be called Jove, and the third body will be called the planet or the satellite, as the case may be.

Jove J, of unit mass, moves round the Sun S, of mass ν, in a circle of unit radius SJ, and the orbit to be considered is that of an infinitesimal body P moving in the plane of Jove's orbit.

Let S be the origin of rectangular axes; let SJ be the x axis, and let the y axis be such that a rotation from x to y is consentaneous with the orbital motion of J.   Let x, y be the heliocentric coordinates of P, so that $x - 1$, y are the jovicentric coordinates referred to the same x axis and a parallel y axis.

---

* About two-thirds of the expense of these computations have been met by grants from the Government Grant and Donation Funds of the Royal Society.

Let r denote SP, and θ the angle JSP; let ρ denote JP, and let the angle SJP be $180° - \psi$. Thus r, θ are the polar heliocentric coordinates, and ρ, ψ the polar jovicentric coordinates of P.

Let n denote Jove's orbital angular velocity, so that in accordance with Kepler's law

$$n^2 = \nu + 1$$

The equations of motion of a particle referred to axes rotating with angular velocity ω, under the influence of forces whose potential is U, are

$$\frac{d}{dt}\left(\frac{dX}{dt} - \omega Y\right) - \omega\left(\frac{dY}{dt} + \omega X\right) = \frac{\partial U}{\partial X}$$

$$\frac{d}{dt}\left(\frac{dY}{dt} + \omega X\right) + \omega\left(\frac{dX}{dt} - \omega Y\right) = \frac{\partial U}{\partial Y}$$

where t is the time.

Now in the present problem, if the origin be taken at the centre of inertia of the Sun and Jove with SJ for the X axis, the coordinates of P are $X = x - 1/(\nu + 1)$, $Y = y$. Also the potential function is $\nu/r + 1/\rho$. Hence the equations of motion are

$$\frac{d^2x}{dt^2} - 2n\frac{dy}{dt} - (\nu + 1)\left(x - \frac{1}{\nu + 1}\right) = \frac{\partial}{\partial x}\left(\frac{\nu}{r} + \frac{1}{\rho}\right)$$

$$\frac{d^2y}{dt^2} + 2n\frac{dx}{dt} - (\nu + 1)y = \frac{\partial}{\partial y}\left(\frac{\nu}{r} + \frac{1}{\rho}\right)$$

But $r^2 = x^2 + y^2$, $\rho^2 = (x - 1)^2 + y^2$. Hence if we put

$$2\Omega = \nu\left(r^2 + \frac{2}{r}\right) + \left(\rho^2 + \frac{2}{\rho}\right) *\quad \dots\dots\dots\dots\dots\dots(1)$$

the equations of motion may be written

$$\left.\begin{array}{l} \dfrac{d^2x}{dt^2} - 2n\dfrac{dy}{dt} = \dfrac{\partial\Omega}{\partial x} \\[3mm] \dfrac{d^2y}{dt^2} + 2n\dfrac{dx}{dt} = \dfrac{\partial\Omega}{\partial y} \end{array}\right\}\quad \dots\dots\dots\dots\dots\dots(1)$$

where $n^2 = \nu + 1$.

Let the second of (1) be multiplied by $2dx/dt$, and the third by $2dy/dt$, let the two be added together and integrated, and we have Jacobi's integral

$$V^2 = \left(\frac{dx}{dt}\right)^2 + \left(\frac{dy}{dt}\right)^2 = 2\Omega - C\quad \dots\dots\dots\dots\dots(2)$$

where C is a constant, and V denotes the velocity of P relatively to the rotating axes.

---

* It is perhaps worth noting that 2Ω may be written in the form

$$\nu(r - 1)^2\left(1 + \frac{2}{r}\right) + (\rho - 1)^2\left(1 + \frac{2}{\rho}\right) + 3(\nu + 1)$$

Let s be the arc of the planet's relative orbit measured from any fixed point, and let ϕ be the inclination to the x axis of the outward normal of the orbit. Then

$$\frac{dx}{ds} = -\sin\phi, \qquad \frac{dy}{ds} = \cos\phi$$

Hence if P be the component of inward effective force,

$$P = -\frac{\partial\Omega}{\partial x}\cos\phi - \frac{\partial\Omega}{\partial y}\sin\phi \dots\dots\dots\dots\dots\dots(3)$$

Therefore
$$PV = -\frac{\partial\Omega}{\partial x}\frac{dy}{dt} + \frac{\partial\Omega}{\partial y}\frac{dx}{dt}$$

Now if R denotes the radius of curvature at the point x, y, of the relative orbit of P,

$$\frac{1}{R} = \frac{\dfrac{d^2y}{dt^2}\dfrac{dx}{dt} - \dfrac{d^2x}{dt^2}\dfrac{dy}{dt}}{\left[\left(\dfrac{dx}{dt}\right)^2 + \left(\dfrac{dy}{dt}\right)^2\right]^{\frac{3}{2}}}$$

On substituting for the second differentials from (1), we have

$$\frac{V^3}{R} = \frac{\partial\Omega}{\partial y}\frac{dx}{dt} - \frac{\partial\Omega}{\partial x}\frac{dy}{dt} - 2n\left[\left(\frac{dx}{dt}\right)^2 + \left(\frac{dy}{dt}\right)^2\right]$$

Hence by means of (2) and (3)

$$\frac{1}{R} = \frac{P}{V^2} - \frac{2n}{V} \dots\dots\dots\dots\dots\dots\dots\dots\dots\dots\dots(4)$$

If the value of Ω in (1) be substituted in (3) we easily find

$$\left.\begin{aligned} P &= \nu\left(\frac{1}{r^2} - r\right)\cos(\phi - \theta) + \left(\frac{1}{\rho^2} - \rho\right)\cos(\phi - \psi) \\ V^2 &= \nu\left(r^2 + \frac{2}{r}\right) + \left(\rho^2 + \frac{2}{\rho}\right) - C \end{aligned}\right\} \dots\dots\dots(4)$$

and

Thus the curvature at any point of the orbit is expressible in terms of the coordinates and of the direction of the normal. If s_0, ϕ_0, x_0, y_0, t_0 be the initial values of the same quantities, it is clear that

$$\left.\begin{aligned} \phi &= \phi_0 + \int_{s_0}^{s}\frac{ds}{R} \\ x &= x_0 - \int_{s_0}^{s}\sin\phi\, ds \\ y &= y_0 + \int_{s_0}^{s}\cos\phi\, ds \\ n(t - t_0) &= \int_{s_0}^{s}\frac{n}{V}ds \end{aligned}\right\} \dots\dots\dots\dots\dots\dots\dots(5)$$

Also the polar coordinates of P relatively to axes fixed in space with heliocentric origin are r, $\theta + n(t - t_0)$, and with jovicentric origin are ρ, $\psi + n(t - t_0)$.

Hence the determination of x and y involves in each case two integrations, and another integration is necessary to find the time, and the orbit in space.

§ 3. *Partition of relative space according to the value of the relative energy* *.

It may be easily shown that the function Ω arises from three sources, and that it is the sum of the rotation potential, the potential of the Sun and the disturbing function for motion relatively to the Sun. Hence Ω is the potential of the system, inclusive of the rotation potential. Thus the equation $V^2 = 2\Omega - C$ may be called the equation of relative energy.

For a real motion of the planet V^2 must be positive, and therefore 2Ω must be greater than C. Accordingly the planet can never cross the curve represented by $2\Omega = C$, and if this curve has a closed branch with P inside, it must always remain inside; or if P be outside, it must always remain so.

This is Mr Hill's result in his celebrated memoir † on the Lunar Theory, save that the value of Ω used here has not been reduced to an approximate form.

We shall now proceed to a consideration of the family of curves $2\Omega = C$. That is to say we shall find, for a given value of C, the locus of points for which the three bodies may move for an instant as parts of a single rigid body. We are clearly at the same time finding the curves of constant velocity relatively to the moving axes for other values of C.

For any given value of ρ, the values of r are the roots of the cubic equation

$$r^2 + \frac{2}{r} = \frac{1}{\nu}\left(C - \rho^2 - \frac{2}{\rho}\right)$$

If C' be written for the value of the right hand of this equation, the cubic becomes

$$r^3 - C'r + 2 = 0$$

The solution is

$$r = 2\sqrt{\tfrac{1}{3}C'}\cos\alpha, \quad \text{where} \quad \cos 3\alpha = -C'^{-\frac{3}{2}}\sqrt{27}$$

* A somewhat similar investigation is contained in a paper by M. Bohlin, *Acta Math.*, T. x., p. 109 (1887). The author takes the Sun as a fixed centre, which is equivalent to taking the Sun's mass as very large compared with that of Jove; he thus fails to obtain the function Ω in the symmetrical form used above.
† *Amer. Journ. of Math.*, Vol. i., pp. 5—29.

In order that α may be a real angle, such a value of ρ must be assumed that C' may be greater than 3, or $\rho^2 + 2/\rho$ less than $C - 3\nu$. The limiting form of this last inequality is $\rho^2 + 2/\rho = C - 3\nu$, a cubic of the same form as before. Hence it follows that $C - 3\nu$ must be greater than 3, if there is to exist any curve of the kind under consideration. Thus the minimum value of C consistently with the existence of a curve of which the equation is $2\Omega = C$, is $3(\nu + 1)$. For smaller values of C no curve of zero velocity exists.

With C greater than $3(\nu + 1)$, let β be the smallest possible angle such that $\cos 3\beta = C'^{-\frac{3}{2}}\sqrt{27}$. Then β is clearly less than $30°$, and the three roots of the cubic are

$$2\sqrt{\tfrac{1}{3} C'} \cos(60° \pm \beta), \quad -2\sqrt{\tfrac{1}{3} C'} \cos \beta$$

The third of these roots is essentially negative, and may be omitted as not corresponding to a geometrical solution. But the first two roots are positive and will give a real geometrical meaning to the solution provided that if $\rho > 1$,

$$r < \rho + 1$$
$$> \rho - 1$$

and if $\rho < 1$,

$$r < \rho + 1$$
$$> 1 - \rho$$

In some cases there are two solutions, in others one and in others none.

By the solution of a number of cubic equations I have found a number of values of r, ρ which satisfy $2\Omega = C$, and have thus traced the curves in Fig. 1, to the consideration of which I shall return below.

Some idea of the nature of the family of curves may be derived from general considerations; for when r and ρ are small the equation approximates to $2\nu/r + 2/\rho = C$, and the curves are like the equipotentials due to two attractive particles of masses 2ν and 2.

Thus for large values of C they are closed ovals round S and J, the one round S being the larger. As C declines the ovals swell and coalesce into a figure-of-8, which then assumes the form of an hour-glass with a gradually thickening neck.

When on the other hand r and ρ are large the equation approximates to $\nu r^2 + \rho^2 = C$, and this represents an oval enclosing both S and J, which decreases in size as C decreases.

It is thus clear by general reasoning that for large values of C the curve consists of two closed branches round S and J respectively, and of a third closed branch round both S and J. The spaces within which the velocity of the planet is real are inside either of the smaller ovals, and outside the larger one. Since the larger oval shrinks and the hour-glass swells, as C declines, a stage will be reached when the two curves meet and coalesce.

This first occurs at the end of the small bulb of the hour-glass which encloses J. The curve is then shaped like a horse-shoe, but is narrow at the toe and broad at the two points.

For still smaller values of C, the horse-shoe narrows to nothing at the toe, and breaks into two elongated pieces. These elongated pieces, one on each side of SJ, then shrink quickly in length and slowly in breadth, until they contract to two points when C reaches its minimum.

This sketch of the sequence of changes shows that there are four critical stages in the history of the curves,

(α) when the internal ovals coalesce to a figure-of-8;

(β) when the small end of the hour-glass coalesces with the external oval;

(γ) when the horse-shoe breaks;

(δ) when the halves of the broken shoe shrink to points.

The points of coalescence and rupture in (α), (β), (γ) are obviously on the line SJ (produced either way), and the points in (δ) are symmetrically situated on each side of SJ.

We must now consider the physical meaning of the critical points, and show how to determine their positions.

In the first three cases the condition which enables us to find the critical point is that a certain equation derived from $2\Omega = C$ shall have equal roots.

(α) The coalescence into a figure-of-8 must occur between S and J; hence $r = 1 - \rho$, and $2\Omega = C$ becomes

$$\nu\left[(1-\rho)^2 + \frac{2}{1-\rho}\right] + \rho^2 + \frac{2}{\rho} = C \dots\dots\dots\dots\dots(6)$$

This equation must have equal roots. Accordingly by differentiation we find that ρ must satisfy,

$$-\nu(1-\rho) + \frac{\nu}{(1-\rho)^2} + \rho - \frac{1}{\rho^2} = 0$$

or $\qquad (\nu+1)\rho^5 - (3\nu+2)\rho^4 + (3\nu+1)\rho^3 - \rho^2 + 2\rho - 1 = 0$

a quintic equation from which ρ may be found.

This equation may be put in the form,

$$(3\nu+1)\rho^3 = 1 - \frac{\rho(1-\rho^3)(1-\tfrac{2}{3}\rho)}{1-\rho+\tfrac{1}{3}\rho^2}$$

When the Sun is large compared with Jove ν is large, and ρ is obviously small, and we have approximately

$$(3\nu+1)^{\frac{1}{3}}\rho = 1 - \tfrac{1}{3}\rho$$

whence $\qquad\qquad \rho = \dfrac{1}{(3\nu+1)^{\frac{1}{3}} + \tfrac{1}{3}} \dots\dots\dots\dots\dots\dots(7)$

If this value of ρ be substituted in (6) we obtain the approximate result

$$C = 3\nu + \frac{2\nu}{3\nu + 1} + 3\left(3\nu + 1\right)^{\frac{1}{3}} \quad \dots\dots\dots\dots\dots(8)$$

In this paper the value adopted for ν is 10, and the approximate formulæ (7) and (8) give

$$\rho = \cdot 28779, \quad r = \cdot 71221, \quad C = 40\cdot 0693$$

The correct results derived from the quintic equation and from the full formula for C are

$$\rho = \cdot 28249, \quad r = \cdot 71751, \quad C = 40\cdot 1821\dots\dots\dots\dots(9)$$

Thus for even so small a value of ν as 10, the approximation is near the truth, and for such cases as actually occur in the solar system it would be accurate enough for every purpose.

The formula from which ρ has been derived is equivalent to $\partial\Omega/\partial x = 0$, and since $y = 0$, we have also $\partial\Omega/\partial y = 0$. Hence the point is one of zero effective force at which the planet may revolve without motion relatively to the Sun and Jove.

This position of conjunction between the two larger bodies is obviously one of dynamical instability.

(β)   The coalescence of the hour-glass with the external oval must occur at a point in SJ produced beyond J; hence $r = 1 + \rho$, and $2\Omega = C$ becomes

$$\nu\left[(1 + \rho)^2 + \frac{2}{1 + \rho}\right] + \rho^2 + \frac{2}{\rho} = C$$

This equation must have equal roots, and ρ must satisfy

$$\nu\left(1 + \rho\right) - \frac{\nu}{(1 + \rho)^2} + \rho - \frac{1}{\rho^2} = 0$$

or $$(\nu + 1)\rho^5 + (3\nu + 2)\rho^4 + (3\nu + 1)\rho^3 - \rho^2 - 2\rho - 1 = 0$$

This quintic equation may be written in the form

$$(3\nu + 1)\rho^3 = 1 + \frac{\rho\left(1 - \rho^3\right)\left(1 + \frac{2}{3}\rho\right)}{1 + \rho + \frac{1}{3}\rho^2}$$

With the same approximation as in (α)

$$\rho = \frac{1}{(3\nu + 1)^{\frac{1}{3}} - \frac{1}{3}} \quad \dots\dots\dots\dots\dots\dots(10)$$

$$C = 3\nu - \frac{2\nu}{3\nu + 1} + 3\left(3\nu + 1\right)^{\frac{1}{3}} \quad \dots\dots\dots\dots(11)$$

When ν is 10, the approximate formulæ (10), (11) give

$$\rho = \cdot 35612, \quad r = 1\cdot 35612, \quad C = 38\cdot 7790$$

The correct results derived from the quintic equation are

$$\rho = \cdot 34700, \quad r = 1\cdot 34700, \quad C = 38\cdot 8760 \quad \dots\dots\dots(12)$$

The approximation is not so good as in (α), but in such cases as actually occur in the solar system the formulæ (10), (11) would lead to a high degree of accuracy.

This second critical point is another one at which the planet may revolve without motion relatively to the Sun and Jove, and such a motion is dynamically unstable.

(γ) The thinning of the toe of the horse-shoe to nothing must occur at a point in JS produced beyond S; hence $\rho = r + 1$, and $2\Omega = C$ becomes

$$\nu\left(r^2 + \frac{2}{r}\right) + (r+1)^2 + \frac{2}{r+1} = C$$

This equation must have equal roots, and r must satisfy

$$\nu\left(r - \frac{1}{r^2}\right) + (r+1) - \frac{1}{(r+1)^2} = 0$$

or $\qquad (\nu + 1)\, r^5 + (2\nu + 3)\, r^4 + (\nu + 3)\, r^3 - \nu\, (r^2 + 2r + 1) = 0$

a quintic for finding r.

If we put $r = 1 - \xi$, the equation becomes

$$(\nu + 1)\, \xi^5 - (7\nu + 8)\, \xi^4 + (19\nu + 25)\, \xi^3 - (24\nu + 37)\, \xi^2 + (12\nu + 26)\, \xi - 7 = 0$$

This equation may be solved by approximation, and the first approximation, which is all that I shall consider, gives

$$\xi = 1 - r = \frac{7}{12\nu + 26} \quad \dots\dots\dots\dots\dots\dots(13)$$

Thus the approximate solution is $r = 1 - \dfrac{7}{12\nu + 26}$.

We also find

$$C = \nu\,(1 - 2\xi + \xi^2 + 2 + 2\xi + 2\xi^2 \dots) + 4 - 4\xi + \xi^2 + 1 + \tfrac{1}{2}\xi + \tfrac{1}{4}\xi^2$$
$$= 3\nu + 5 - \tfrac{7}{2}\xi + (3\nu + \tfrac{9}{4})\,\xi^2 \quad \dots\dots\dots\dots\dots\dots\dots(14)$$

If we take only the term in ξ in (14), and put $\nu = 10$ the approximate result is

$$r = \cdot 95205, \quad \rho = 1\cdot 95205, \quad C = 34\cdot 9012$$

The exact solution derived from the quintic equation is

$$r = \cdot 94693, \quad \rho = 1\cdot 94693, \quad C = 34\cdot 9054 \quad \dots\dots\dots(15)$$

With large values of ν the first approximation would give nearly accurate results. This critical point is another one at which the three bodies may move round without relative motion, but as before the motion is dynamically unstable.

(δ) The fourth and last critical position occurs when C is a minimum, consistently with the existence of the curves under consideration. Now C is a minimum when $\partial C/\partial r = 0$, $\partial C/\partial \rho = 0$; whence $r = 1$, $\rho = 1$, and $C = 3\nu + 3$. We arrived above at this minimum value of C from another point of view.

If an equilateral triangle be drawn on SJ, its vertex is at this fourth critical point; and since this vertex may be on either the positive or negative side of SJ, there are two points of this kind.

It is well known that there is an exact solution of the problem of three bodies in which they stand at the corners of an equilateral triangle, which revolves with a uniform angular velocity.

Thus all the five critical points correspond with particular exact solutions of the problem, and of these solutions three are unstable. It will be proved in a postscript at the end of this paper that the symmetrical pair is stable if ν is greater than 24·9599, but otherwise unstable.

Fig. 1 represents the critical curves of the family $2\Omega = C$, for the case $\nu = 10$. The points in the curves were determined, as explained above, by the solution of a number of cubic equations. I have only drawn the critical curves, because the addition of other members of the family would merely complicate the figure.

An important classification of orbits may be derived from this figure. When C is greater than 40·1821 the third body must be either a superior planet moving outside the large oval, or an inferior planet moving inside the larger internal oval, or a satellite moving inside the smaller internal oval; and it can never exchange one of these parts for either of the other two. The limiting case $C = 40·1821$ gives superior limits to the radii vectores of inferior planets and of satellites, which cannot sever their connections with their primaries.

When C is less than 40·1821 but greater than 38·8760, the third body may be a superior planet, or an inferior planet or satellite, or a body which moves in an orbit which partakes of the two latter characteristics; but it can never pass from the first condition to any of the latter ones.

When C is less than 38·8760 and greater than 34·9054, the body may move anywhere save inside a region shaped like a horse-shoe. The distinction between the two sorts of planetary motion and the motion as a satellite ceases to exist, and if the body is started in any one of these three ways it is possible for it to exchange the characteristics of its motion for either of the two other modes.

When C is less than 34·9054 and greater than 33, the forbidden region consists of two strangely shaped portions of space on each side of SJ.

Lastly when C is equal to 33, the forbidden regions have shrunk to a pair of infinitely small closed curves enclosing the third angles of a pair of equilateral triangles erected on SJ as a base.

For smaller values of C no portion of space is forbidden to the third body.

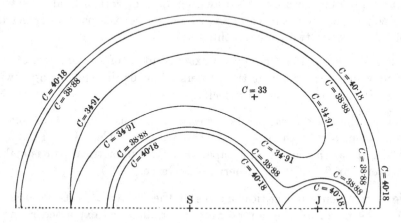

FIG. 1. Curves of zero velocity, $10\left(r^2+\dfrac{2}{r}\right)+\left(\rho^2+\dfrac{2}{\rho}\right)=C.$

§ 4. *A certain partition of space according to the nature of the curvature of the orbit.*

It appears from (4) of § 2 that the curvature of an orbit is given by

$$\frac{V^2}{R}=P-2nV,\quad\text{where }P=-\frac{\partial\Omega}{\partial x}\cos\phi-\frac{\partial\Omega}{\partial y}\sin\phi$$

Now if V_0 denotes any constant velocity, the equation $2\Omega=C+V_0^2$ defines a curve of constant velocity; it is one of the family of curves considered in § 3. We have seen that this family consists of a large oval enclosing two smaller ones, or of curves arising from the coalescence of ovals. In the mathematical sense of the term the "interior" of the curve of constant velocity consists of the space inside either of the smaller ovals or outside the large one, or the corresponding spaces when there is coalescence of ovals. It is a convenient and ordinary convention that when the circuit of a closed curve is described in a positive direction, the "interior" of the curve is on the left-hand side. According to this convention the meaning of the "inward" normal of one of these curves of constant velocity is clear, for it is directed towards the "interior." Similarly the inward normal of an orbit is towards the left-hand side, as the body moves along its path.

It is clear then that P is the component of effective force estimated along the inward normal of the orbit. Also if T be the resultant effective force $T^2 = (\partial\Omega/\partial x)^2 + (\partial\Omega/\partial y)^2$; and if χ be the angle between T and the inward normal to the orbit, $P = T\cos\chi$.

Hence
$$\frac{V^2}{R} = T\cos\chi - 2nV$$

If we consider curvature as a quantity which may range from infinite positive to infinite negative, it may be stated that of all the orbits passing through a given point the curvature is greatest for that orbit which is tangential to the curve of constant velocity, when the motion takes place in a positive direction along that curve.

If χ lies between $\pm\chi_0$, where $\cos\chi_0 = 2nV/T$, the orbit has positive curvature; if $\chi = \pm\chi_0$, there is a point of contrary flexure in the orbit; and if χ lies outside the limits $\pm\chi_0$, the curvature is negative.

If however T be less than $2nV$, there are no orbits, passing through the point under consideration, which have positive curvature. Hence the equation $T = 2nV$ defines a family of curves which separate the regions in which the curvature of orbits is necessarily negative, from those in which it may be positive.

Since
$$n^2 = \nu + 1, \quad V^2 = \nu\left(r^2 + \frac{2}{r}\right) + \left(\rho^2 + \frac{2}{\rho}\right) - C, \quad T^2 = \left(\frac{\partial\Omega}{\partial x}\right)^2 + \left(\frac{\partial\Omega}{\partial y}\right)^2$$

the equation $T = 2nV$ becomes,

$$\nu^2\left(\frac{1}{r^2} - r\right)^2 + \left(\frac{1}{\rho^2} - \rho\right)^2 + 2\nu\left(\frac{1}{r^2} - r\right)\left(\frac{1}{\rho^2} - \rho\right)\cos(\theta - \psi)$$

$$= 4(\nu + 1)\left[\nu\left(r^2 + \frac{2}{r}\right) + \left(\rho^2 + \frac{2}{\rho}\right) - C\right]$$

Since $2r\rho\cos(\theta - \psi) = r^2 + \rho^2 - 1$, it may be written

$$\nu^2\left(\frac{1}{r^4} - \frac{10}{r} - 3r^2\right) + \left(\frac{1}{\rho^4} - \frac{10}{\rho} - 3\rho^2\right)$$

$$+ \nu\left[\left(\frac{1}{r^3} - 1\right)\left(\frac{1}{\rho^3} - 1\right)(r^2 + \rho^2 - 1) - 4(r^2 + \rho^2) - 8\left(\frac{1}{r} + \frac{1}{\rho}\right)\right] + 4C(\nu + 1) = 0$$
$$\ldots\ldots(16)$$

This equation is reducible to the sextic equation,

$$\rho^6\left[3(\nu + 1)r^4 + \nu r\right]$$

$$+ \rho^4\left[3\nu(\nu + 1)r^6 - (4\nu C + 4C - \nu)r^4 + (10\nu^2 + 9\nu)r^3 - \nu r - \nu^2\right]$$

$$+ \rho^3\left[(9\nu + 10)r^4 - \nu r\right] + \rho\nu r(1 - r^2)(1 - r^3) - r^4 = 0 \ldots\ldots\ldots(16)$$

It may also be written as a sextic in r, by interchanging r and ρ and by writing $1/\nu$ for ν and C/ν for C.

It would require a great deal of computation to trace the curves represented by (16), and for the present I have not thought it worth while to undertake the task.

When however we adopt Mr Hill's approximate value for the potential Ω, the equation becomes so much simpler that it may be worth while to consider it further.

If m, a, n be the mass, distance from Sun and orbital angular velocity of Jove, the expression for Ω reduces to

$$\Omega = \frac{m}{\rho} + \tfrac{3}{2}n^2 (x-a)^2 + \tfrac{3}{2}n^2 a^2$$

The last term is constant, so that if C be replaced by C_0, where $C_0 = C - 3n^2 a^2$, we may omit the last term in Ω and use C_0 in place of C.

Now taking units of length and time such that $m = 1$, $n = 1$; also writing $\xi = (x-a)$, $\eta = y$; we have

$$\Omega = \frac{1}{\rho} + \tfrac{3}{2}\xi^2, \quad V^2 = 2\Omega - C_0, \quad \xi^2 + \eta^2 = \rho^2 \dots\dots\dots\dots(17)$$

Then
$$T^2 = \left(\frac{\partial\Omega}{\partial\xi}\right)^2 + \left(\frac{\partial\Omega}{\partial\eta}\right)^2 = 3\left(3 - \frac{2}{\rho^3}\right)\xi^2 + \frac{1}{\rho^4}$$

Hence the equation (16) becomes

$$3\left(3 - \frac{2}{\rho^3}\right)\xi^2 + \frac{1}{\rho^4} = 4\left(\frac{2}{\rho} + 3\xi^2 - C_0\right)$$

or
$$\xi^2\left(1 + \frac{2}{\rho^3}\right) = \tfrac{4}{3}\left(C_0 - \frac{2}{\rho} + \frac{1}{4\rho^4}\right) \dots\dots\dots\dots\dots(18)$$

Since $\xi = \rho\cos\psi$, the polar equation to the curve is

$$\cos^2\psi = \tfrac{4}{3}C_0\,\frac{\left(\rho - \frac{2}{C_0} + \frac{1}{4C_0\rho^3}\right)}{\rho^3 + 2} \dots\dots\dots\dots\dots(18)$$

Mr Hill's curve $2\Omega = C_0$ gives

$$\left.\begin{array}{l} \xi^2 = \tfrac{1}{3}C_0\left(1 - \frac{2}{C_0\rho}\right) \\[2mm] \cos^2\psi = \tfrac{1}{3}\frac{C_0}{\rho^3}\left(\rho - \frac{2}{C_0}\right) \end{array}\right\} \dots\dots\dots\dots\dots(19)$$

or

It is clear that the two curves present similar characteristics, but the former is the more complicated one.

The asymptotes of (18) are $\xi = \pm 2\sqrt{(\tfrac{1}{3}C_0)}$, whilst those of (19) are $\xi = \pm\sqrt{(\tfrac{1}{3}C_0)}$.

Again to find where the curves cut the positive half of the axis of η, we put $\xi = 0$, $\rho = \eta$ and find that (18) becomes

$$\eta^4 - \frac{2}{C_0}\eta^3 + \frac{1}{4C_0} = 0 \dots\dots\dots\dots\dots\dots(20)$$

whilst (19) becomes simply $\eta = 2/C_0$.

The condition that (20) shall have equal roots is $4\eta = 6/C_0$, or $1/\eta = \frac{2}{3}C_0$. But $C_0 = 2/\eta - 1/4\eta^4$, and therefore $C_0 = 3/2^{\frac{2}{3}}$.

The quartic for η has two real roots if C_0 is less than $3/2^{\frac{2}{3}}$ or 1.8899, but no real roots if it is greater than this value.

It is easy to show that when the roots are real, one is greater than and the other less than $3/2C_0$.

It follows that if C_0 is greater than 1.8899 the curve does not cut the axis of η, but if less it does so twice.

To find the critical values of C_0 in the case of Mr Hill's curve (19), we put (as in § 3) $\eta = 0$ and therefore $\rho = \xi$, and we then find the condition that the equation shall have equal roots.

Now with $\rho = \xi$, (19) becomes

$$\xi^2 = \tfrac{1}{3}C_0 - \frac{2}{3\xi}$$

This has equal roots when $\xi = 1/3^{\frac{1}{3}}$. Hence $C_0 = 3\xi^2 + 2/\xi = 3^{\frac{4}{3}} = 4.3267$. If C_0 be greater than 4.3267 the curve consists of an internal oval and of two asymptotic branches. With smaller values of C_0 the oval has coalesced with the two external branches.

Following the same procedure with our curve (18), we have to find when

$$\xi^2 \left(1 + \frac{2}{\xi^3}\right) = \tfrac{4}{3}\left(C_0 - \frac{2}{\xi} + \frac{1}{4\xi^4}\right)$$

has equal roots.

The condition is that $3\xi^6 - 7\xi^3 + 2 = 0$, and the solutions are $\xi^3 = 2$, $\xi^3 = \frac{1}{3}$.

Now
$$C_0 = \tfrac{3}{4}\left(\xi^2 + \frac{2}{\xi}\right) + \frac{2}{\xi} - \frac{1}{4\xi^4}$$

Hence when $\xi^3 = 2$, $C_0 = \dfrac{39}{8 \cdot 2^{\frac{1}{3}}} = 3.8693$

and when $\xi^3 = \frac{1}{3}$, $C_0 = 3^{\frac{4}{3}} = 4.3267$

Thus there are three critical values of C_0, viz.: $C_0 = 1.8899$, which separates the curves which do from those which do not intersect the axis of η; $C_0 = 3.8693$ when two branches coalesce; and $C_0 = 4.3267$ when two branches again coalesce. The last is also a critical value of C_0 in the case of Mr Hill's curve.

It would seem then that if these curves were traced for the values $C_0 = 1.5, 3, 4, 5$ a good idea might be obtained of their character, but I have not yet undertaken the task.

§ 5. *Formulæ of interpolation and quadrature.*

The object of this paper is to search for periodic orbits, but no general method has been as yet discovered by which they may be traced. I have therefore been compelled to employ a laborious method of tracing orbits by quadratures, and of finding the periodic orbits by trial. The formulæ of integration used in this process will now be exhibited.

According to the usual notation of the calculus of finite differences, u_x is to denote a function of x, and the operators E and Δ are defined by

$$Eu_x = u_{x+1}, \quad \Delta u_x = u_{x+1} - u_x = (E - 1) u_x$$

It is obvious that $E = e^{\frac{d}{dx}}$, where e is the base of Napierian logarithms, and that $E^x u_0 = u_x$.

In most of the work, as it presents itself in this investigation, the series of values ... u_{n-2}, u_{n-1}, u_n are known, but u_{n+1}, u_{n+2}, ... are as yet unknown.

Now $$E = 1 + \Delta = (1 - \Delta E^{-1})^{-1}$$

and $$u_x = E^x u_0 = (1 - \Delta E^{-1})^{-x} u_0$$

so that

$$u_x = \left(1 + x\Delta E^{-1} + \frac{x(x+1)}{2!} \Delta^2 E^{-2} + \frac{x(x+1)(x+2)}{3!} \Delta^3 E^{-3} + ...\right) u_0$$

$$......(21)$$

In the course of the work occasion will arise for finding $u_{-\frac{1}{2}}$ by interpolation ; putting then $x = -\frac{1}{2}$ in (21), we have

$$u_{-\frac{1}{2}} = (1 - \tfrac{1}{2}\Delta E^{-1} - \tfrac{1}{8}\Delta^2 E^{-2} - \tfrac{1}{16}\Delta^3 E^{-3} - \tfrac{5}{128}\Delta^4 E^{-4} - \tfrac{7}{256}\Delta^5 E^{-5} ...) u_0 ...(22)$$

In a subsequent section the two following well-known formulæ of interpolation will be of service,

$$u_x = \left\{1 + x \cdot \tfrac{1}{2}(\Delta + \Delta E^{-1}) + \frac{x^2}{2!}\Delta^2 E^{-1}\right.$$

$$\left. + \frac{x(x^2 - 1)}{3!} \cdot \tfrac{1}{2}(\Delta^3 E^{-1} + \Delta^3 E^{-2}) + \frac{x^2(x^2 - 1)}{4!}\Delta^4 E^{-2} ...\right\} u_0 \quad ...(23)$$

$$u_x = \left\{1 + x\Delta + \frac{x(x - 1)}{2!} \cdot \tfrac{1}{2}(\Delta^2 + \Delta^2 E^{-1})\right.$$

$$\left. + \frac{x(x - 1)(x - \frac{1}{2})}{3!}\Delta^3 E^{-1} + \frac{x(x^2 - 1)(x - 2)}{4!} \cdot \tfrac{1}{2}(\Delta^4 E^{-1} + \Delta^4 E^{-2}) ...\right\} u_0 \quad (23)$$

Of these formulæ the first is the better when the interpolated value of u_x lies between $x = -\frac{1}{4}$ and $x = +\frac{1}{4}$; and the second is the better when it lies between $x = +\frac{1}{4}$ and $x = +\frac{3}{4}$.

In order to obtain a formula of integration we require to prove that

$$-\frac{1}{\log(1-\alpha)} = \sum_{r=0}^{r=\infty} (-)^r \alpha^{r-1} \int_0^1 \frac{v^{(r)}}{r!} dv$$

where $v^{(r)}$ denotes the factorial $v(v-1)\ldots(v-r+1)$.

This is easily proved as follows :—

$$\int_0^1 (1-\alpha)^v dv = \left[\frac{e^{v\log(1-\alpha)}}{\log(1-\alpha)}\right]_0^1 = \frac{-\alpha}{\log(1-\alpha)}$$

But
$$\int_0^1 (1-\alpha)^v dv = \sum \int_0^1 (-)^r \alpha^r \frac{v^{(r)}}{r!} dv$$

If the last two forms of this integral be equated to one another, we obtain the required formula.

Now
$$e^{\frac{d}{dx}} = (1 - \Delta E^{-1})^{-1}$$

and therefore
$$\frac{d}{dx} = -\log(1 - \Delta E^{-1})$$

Hence
$$\int dx = \left(\frac{d}{dx}\right)^{-1} = -\frac{1}{\log(1-\Delta E^{-1})} = \sum (-)^r \Delta^{r-1} E^{-r+1} \int_0^1 \frac{v^{(r)}}{r!} dv$$

If the definite integrals on the right-hand side be evaluated, we find

$$\int_0^n u_x dx = (\Delta^{-1}E - \tfrac{1}{2} - \tfrac{1}{12}\Delta E^{-1} - \tfrac{1}{24}\Delta^2 E^{-2} - \tfrac{19}{720}\Delta^3 E^{-3}$$
$$- \tfrac{3}{160}\Delta^4 E^{-4} - \tfrac{863}{60480}\Delta^5 E^{-5} \ldots)(u_n - u_0)$$

Since Δ^{-1} contains an arbitrary constant we may choose

$$\Delta^{-1}u_1 = \tfrac{1}{2}u_0 + \tfrac{1}{12}\Delta u_{-1} + \tfrac{1}{24}\Delta^2 u_{-2} + \tfrac{19}{720}\Delta^3 u_{-3} + \ldots \ldots\ldots\ldots(24)$$

and we then have as our formula of integration,

$$\int_0^n u_x dx = \Delta^{-1}u_{n+1} - \tfrac{1}{2}u_n - \tfrac{1}{12}\Delta u_{n-1} - \tfrac{1}{24}\Delta^2 u_{n-2}$$
$$- \tfrac{19}{720}\Delta^3 u_{n-3} - \tfrac{3}{160}\Delta^4 u_{n-4} - \tfrac{863}{60480}\Delta^5 u_{n-5} \ldots(24)$$

This is the most convenient formula of integration when only the integral from n to 0 is wanted, and the integrals from $n-1$ to 0, $n-2$ to 0, etc. are not also wanted. But in the greater part of the work the intermediate integrals are also required. Now on applying the operator Δ to (24), we have

$$\int_n^{n+1} u_x dx = u_{n+1} - \tfrac{1}{2}\Delta u_n - \tfrac{1}{12}\Delta^2 u_{n-1} - \tfrac{1}{24}\Delta^3 u_{n-2} - \tfrac{19}{720}\Delta^4 u_{n-3}\ldots \ldots(25)$$

If this be added to the integral from n to 0 we have the integral from $n+1$ to 0.

I have found that a table of integration may be conveniently arranged as follows :—

Let us suppose that the integral from $n-1$ to 0 has been already found, and that the integral from n to 0 is required; write u_n and its differences Δu_{n-1}, $\Delta^2 u_{n-2}$, $\Delta^3 u_{n-3}$ in vertical column; below write $-\frac{1}{2}\Delta u_{n-1}$, $-\frac{1}{12}\Delta^2 u_{n-2}$, $-\frac{1}{24}\Delta^3 u_{n-3}$, and add them together; add u_n to the last; multiply the last sum by the common difference Δx, and the result is the integral from n to $n-1$; add to this the integral from $n-1$ to 0, and the result is the required integral from n to zero.

Thus each integration requires 13 lines of a vertical column, and the successive columns follow one another, headed by the value of the independent variable to which it applies.

A similar schedule would apply when the formula (24) is used; but when the initial value of Δ^{-1} has been so chosen as to insure the vanishing of the integral from 0 to 0, the final value of Δ^{-1} is to be found by adding to it the successive u's, so that the intermediate columns need not be written down.

When the successive values of u depend on their precursors, it is necessary at the first stage to take Δx small, because in the first integration it is only possible to take the first difference into account. At the second stage the second difference may be included and at the third the third difference.

But in almost every case I begin integration with such a value of the independent variable (say $x=0$), that we either have u_x an even function of x, or an odd function of x; in the first case $u_x=u_{-x}$, in the second $u_x=-u_{-x}$. Both these cases present special advantages for the commencement of integration, for in the first integration we may take second differences into account. Thus when u_x is an even function, the second difference involved in the table of integration from 1 to 0 is $2\Delta u_0$; and when u_x is an odd function it is zero. In both cases third differences may be included in the second integration.

It is of course desirable to use the largest value of the increment of the independent variable consistent with adequate accuracy. If at any stage of the work it appears by the smallness of the second and third differences involved in the integrals, that longer steps may safely be employed, it is easy to double the value of Δx, by forming a new difference table with omission of alternate entries amongst the values already computed. Thus if the change is to be made at the stage where $x=n$, the new difference table will be formed from u_{n-4}, u_{n-2}, u_n; and thereafter Δx will have double its previous value.

When on the other hand it appears by the growth of the second and third differences that Δx is becoming too large, Δx can be halved, and the new difference table must be formed by interpolation. The formula (22) enables us to find $u_{n-\frac{1}{2}}$ from u_n, u_{n-1}, u_{n-2},... with sufficient accuracy for the purpose of obtaining the differences of $u_{n-\frac{3}{2}}$, u_{n-1}, $u_{n-\frac{1}{2}}$, u_n. The process of halving the value of Δx is therefore similar to that of doubling it.

In some of the curves which I have to trace there are sharp bends or quasi-cusps, and in these cases the process is very tedious. It is sometimes necessary repeatedly to halve the increments of the independent variable, which is the arc s of the curve. Thus if (s) denotes the function of the arc to be integrated, and if s be the value of the arc at the point where the curvature begins to increase with great rapidity, and if δ be the previous increment of arc; then in integrating (s) from s to $s + \frac{1}{2}\delta$, the difference table is to be formed from $(s - \delta)$, $(s - \frac{1}{2}\delta)$, (s), the middle one of these three being an interpolated value. At the next step (s) has to be integrated from $s + \frac{1}{2}\delta$ to $s + \frac{3}{4}\delta$, and the difference table is formed from (s), $(s + \frac{1}{4}\delta)$, $(s + \frac{1}{2}\delta)$, the middle term being again an interpolation. This process may clearly be employed over and over again. In some of the curves traced the increment of arc has been 32 times less in one part than in another.

But the chief difficulty about these quasi-cusps arises when they are past, and when it is time to double the arc again. For the fact that the earlier values of the function to be used in the more open ranked difference tables are thrown back nearly to the cusp or even beyond it, makes the higher differences very large. Now the correctness of the formula of integration depends on the correctness of the hypothesis that an algebraic curve will give a good approximation to actuality. But in the neighbourhood of a quasi-cusp, and with increasing arcs this is far from correct. I have found then that in these cases of doubling the arc, a better result is obtained in the first and second integration by only including the second difference in the table of integration.

If we are tracing one member of a family of curves which are widely spaced throughout the greater part of their courses, but in one region are closely crowded into quasi-cusps, it is difficult to follow one member of the family through the crowded region, and on emerging from the region we shall probably find ourselves tracing a closely neighbouring member, and not the original one. I have applied the method to trace the curve drawn by a point attached to a circle at nine-tenths of its radius from the centre, as the circle rolls along a straight line. After the passage of the quasi-cusp I found that I was no longer exactly pursuing the correct line; nevertheless on a figure of the size of this page the difference between the two lines would be barely discernible. But the orbits which it is my object to trace do not quite resemble this case, since their cusps do not lie crowded together in one region of space. I believe therefore that these cases have been treated with substantial accuracy.

Another procedure has however been occasionally employed which I shall explain in § 7.

§ 6. *On the method of tracing a curve from its curvature.*

It will be supposed that the curve to be traced is symmetrical with respect to the x axis, and starts at right angles to it so that $x = x_0$, $y = 0$, $\phi = 0$, $s = 0$. This is not a necessary condition for the use of the method, but it appears from § 5 that the start is thus rendered somewhat easier than would be the case otherwise. The curvature at each point of the curve is supposed to be a known function of the coordinates x, y of the point, and of the direction of the normal defined by the angle ϕ.

The first step is to compute the initial curvature $1/R_0$; it is then necessary to choose such a value for the increment of arc δs as will give the requisite degree of accuracy.

I have found that it is well to take, as a rule, δs of such a size that $\delta s/R_0$ shall not be greater than about 8°; but later, when all the differences in the tables of integration have come into use, I allow the increments of ϕ to increase to about 12°.

It is obvious that the curvature is even, when considered as a function of s. When nothing further is known of the nature of the curve, it is necessary to assume that the curvature is constant throughout the first arc δs, but it is often possible to make a conjecture that the curvature at the end of the arc δs will be say $1/R_1$. By the formula of integration with first and second differences we then compute $\phi = \phi_1$ at the end of the arc, by the first of equations (5) in § 2.

With this value of ϕ we find $\sin \phi_1$, $\cos \phi_1$, and observing that $\sin \phi_0 = 0$, $\cos \phi_0 = 1$, we compute x_1, y_1 by means of the second and third of (5), using first and second differences.

We next compute $1/R_1$ with these values of x, y, and if it agrees with the conjecture the work is done; and if not so, the work is repeated until there is agreement between the initial and final values of the curvature.

After the first arc, a second is computed, and higher differences are introduced into the tables of integration. We thus proceed by steps along the curve.

The approximation to the final result is usually so rapid, that in the recalculation it commonly suffices to note the changes in the last significant figure of the numbers involved in the original computation, without rewriting the whole.

The correction of the tables of integration is also very simple; for suppose that the first assumed value of the function to be integrated is u, and that the second approximation shows that it should have been $u + \delta u$; then all the differences in the column of the table have to be augmented by δu, and therefore the integral has to be augmented by

$$(1 - \tfrac{1}{2} - \tfrac{1}{12} - \tfrac{1}{24} - \ldots)\, \delta u\, \delta s$$

If we stop with third differences, this gives the simple rule that the integral is to be augmented by $\tfrac{3}{8}\delta u\, \delta s$.

It has been shown in § 5 how the chosen arc δs is to be increased or diminished according to the requirements of the case.

This method is the numerical counterpart of the graphical process described by Lord Kelvin in his Popular Lectures*, but it is very much more accurate, and when the formula for the curvature is complex it is hardly if at all more laborious. In the present investigation it would have been far more troublesome to use the graphical method, with such care as to attain the requisite accuracy, than to follow the numerical method †.

In order to trace orbits I first computed auxiliary tables of $r^2 + 2/r$, and of $\log(1/r^2 - r)$ for $r < 1$, and of $\log(r - 1/r^2)$ for $r > 1$; the tables extend from $r = 0$ to $1\cdot5$ at intervals of $\cdot001$, but they will ultimately require further extension.

The following schedule shows the arrangement for the computation of the curvature at any point. The table has been arranged so as to be as compact as possible, and is not in strictly logical order; for the calculation of V^2 should follow that of r, ρ, but is entered at the foot of the first column. It will be observed that the calculation is in accordance with the formula (4) of § 2.

L denotes logarithm and C denotes cologarithm; ν the sun's mass is taken as 10, and L $2n = \cdot8217$, being L $2\sqrt{11}$, a constant. The brackets indicate that the numbers so marked are to be added together.

Popular Lectures, Vol. ɪ., 2nd ed., pp. 31—42; *Phil. Mag.*, Vol. xxxɪv., 1892, pp. 443—448.

† [Dr Carl Burrau has devised an entirely different method of effecting the quadratures. Although it seems to possess considerable advantages, especially in cases of abrupt curvature, I have always adhered to my own procedure. It is doubtless a matter of custom that each of us prefers his own method. See *Astronomische Nachrichten*, Nos. 3230, 3251, 3289 and *Vierteljahrschrift Astron. Gesellschaft*, 41st year, part 4, p. 261, 1906.]

Schedule for computation of curvature.

$$s$$

ϕ	$x-1$
x	y

$$\left.\begin{array}{c} Ly \\ Cx \end{array}\right\} \qquad \left.\begin{array}{c} Ly \\ C(x-1) \end{array}\right\}$$

$L\tan\theta$	$L\tan\psi$
θ	ψ
$\phi-\theta$	$\phi-\psi$

$$\left.\begin{array}{c} L\sec\theta \\ Lx \end{array}\right\} \qquad \left.\begin{array}{c} L\sec\psi \\ L(x-1) \end{array}\right\}$$

Lr	$L\rho$
r	ρ

$$\left.\begin{array}{c} L\left(\dfrac{1}{r^2}-r\right) \\ L\nu\cos(\phi-\theta) \\ CV^2 \end{array}\right\} \qquad \left.\begin{array}{c} L\left(\dfrac{1}{\rho^2}-\rho\right) \\ L\cos(\phi-\psi) \\ CV^2 \end{array}\right\}$$

La	Lb
a	
b	

$$\left.\begin{array}{c} CV \\ L\,2n \end{array}\right\}$$

$$\left.\begin{array}{c} \nu\left(r^2+\dfrac{2}{r}\right) \\[6pt] \rho^2+\dfrac{2}{\rho} \end{array}\right\} \qquad \begin{array}{c} L\,\dfrac{2n}{V} \end{array}$$

$$\left.\begin{array}{c} -\dfrac{2n}{V} \\ a+b \end{array}\right\}$$

V^2+C	
V^2	$\dfrac{1}{R}$

The formulæ $r = y\operatorname{cosec}\theta$, $\rho = y\operatorname{cosec}\psi$ are used, when the values of θ or ψ show that these are the better forms.

The tables of integration are kept on separate sheets in the forms indicated in § 5.

As the computation proceeds I keep tables of differences of x, y, ϕ, r, ρ, V^2, and this check has been of immense advantage in detecting errors.

The auxiliary tables of logarithms are computed to 5 figures, but the last figure is not always correct to unity, and the fifth figure is principally of use in order to make correct interpolation possible.

The conversion of ϕ from circular measure to degrees and the values of $\sin \phi$ and $\cos \phi$ are obtained from Bottomley's four-figured table.

Most of the work has been done with these tables, but as it appears that the principal source of error lies in the determination of r and ρ, five-figured logarithms have generally been used in this part of the work, and the values of θ and ψ are written down to $0' \cdot 1$.

In those parts of an orbit in which V^2 becomes small I have often ceased to use the auxiliary table for $\nu (r^2 + 2/r)$; for since the auxiliary table of this function only contains four decimal places and since ν is 10, it follows that only three places are obtainable from the table, and of course there may be an error of unity or even of 2 in the last significant figure of V^2.

In order to test the method, I computed an unperturbed elliptic orbit by means of the curvature. The formulæ were $V^2 = (2/r) - \frac{1}{10}, 1/R = P/V^2$, where $P = (1/r^2) \cos (\phi - \theta)$, and the initial values were $x_0 = 5, y_0 = 0, \phi_0 = 0, s_0 = 0$.

The curve described should be the ellipse of semiaxes 10 and $5 \sqrt{3}$, and x, y ought to satisfy the equation

$$\left(\frac{x + 5}{10} \right)^2 + \left(\frac{y}{5 \sqrt{3}} \right)^2 = 1$$

I take the square root of the left-hand side of this equation diminished by unity, with computed x, y, as one measure of the error of position in the ellipse.

Again if $\tan \chi = \frac{4}{3} y/(x + 5)$, χ ought to be identical with ϕ; hence $\chi - \phi$ measures the error in the direction of motion.

Lastly the area conserved h is $5 \sqrt{\frac{3}{10}}$ or $2 \cdot 7386$; but it is also $Vr \cos (\phi - \theta)$, if the computation gives perfect results. Hence $h - Vr \cos (\phi - \theta)$ measures the error in the equable description of areas. The semi-period should be $\pi \sqrt{1000}$ or $99 \cdot 346$.

The computations were made partly with five-figured and partly with four-figured logarithms, and the process followed the lines of my other work very closely.

The following table exhibits the results together with the errors. It will be observed that when $s = 24$ there is a sudden increase in the second column of errors, but I have not been able to detect the arithmetical mistake which is probably responsible for it. The accordance still remains so close, that it appeared to be a waste of time to work any longer at this example.

Computed positions in an ellipse described under the action of a central force.

s	x	y	ϕ	$\chi - \phi$	$\left[\left(\dfrac{x+5}{10}\right)^2 + \left(\dfrac{y}{5\sqrt{3}}\right)^2\right]^{\frac{1}{2}} - 1$	$h - Vr\cos(\phi - \theta)$
0	5·0000	·0000	0° 0′	0′·0	+·00000	·0000
1	4·9337	·9971	7° 37′	+0′·3	+·00002	·0000
2	4·7364	1·9768	15° 8′	+0′·8	+·00005	−·0001
3	4·4137	2·9227	22° 29′	+0′·3	+·00004	−·0001
4	3·9749	3·8205	29° 35′	−0′·3	+·00004	−·0002
5	3·4304	4·6586	36° 23′	0′·0	+·00004	−·0001
6	2·7925	5·4281	42° 53′	+0′·1	+·00004	−·0001
8	1·2843	6·7363	55° 1′	+0′·2	−·00002	+·0001
10	− ·4567	7·7147	66° 9′	+1′·0	−·00001	+·0002
12	− 2·3497	8·3507	76° 36′	+0′·6	·00000	+·0003
14	− 4·3259	8·6407	86° 39′	+0′·1	−·00001	·0000
16	− 6·3225	8·5845	96° 35′	+0′·4	+·00003	·0000
18	− 8·2787	8·1823	106° 43′	+0′·6	+·00010	+·0003
20	− 10·1305	7·4349	117° 21′	+0′·8	+·00012	+·0003
22	− 11·8051	6·3481	128° 47′	+1′·0	+·00001	+·0004
24	− 13·2181	4·9385	141° 17′	+0′·8	+·00028	+·0004
25	− 13·7968	4·1237	148° 0′	−0′·4	+·00027	+·0003
26	− 14·2740	3·2456	155° 0′	−0′·8	+·00027	+·0001
27	− 14·6385	2·3151	162° 15′	−0′·5	+·00023	+·0003
28	− 14·8808	1·3456	169° 43′	−0′·5	+·00021	+·0003
29	− 14·9938	·3526	177° 19′	−0′·6	+·00020	+·0002
30	− 14·9740	− ·6465	184° 57′	−0′·6	+·00019	+·0004
29·3546	− 15·0020	·0000	180° 1′	+1′·0		

The last line in the above table was found by interpolation.

The computed values of the semiaxes of the ellipse (both involving interpolations) were found to be 10·0010 and ·86604; their correct values are 10·0000 and ·866026. The computed semi-period (requiring another integration and interpolation) was found to be 99·346, agreeing with the correct value to the last place of decimals.

Considering that a considerable part of the computation was done with four-figured tables, the accuracy shown in this table is surprising.

This calculation is exactly comparable with the best of my calculations of orbits, but there has been from time to time a good deal of variety in my procedure. My object has been throughout to cover a wide field with adequate accuracy rather than a far smaller one with scrupulous exactness, for economy of labour is of the greatest importance in so heavy a piece of work. I shall in the appendix generally indicate which are the more exact and which the less exact computations. I do not think it would in any case have been possible in the figures to show the difference between an exactly computed and a roughly computed curve, because the lines would be almost or quite indistinguishable on the scale of the plates of figures.

This however might not be quite true of the orbits which have very sharp bends in them.

§ 7. *Development in powers of the time; the form of cusps**.

In a few cases the quasi-cusps of orbits have been computed by means of series; the mode of development will therefore now be considered.

If for brevity we write

$$2n = m, \qquad \frac{dx}{dt} = u, \qquad \frac{dy}{dt} = v,$$

the equations of motion (1) become

$$\frac{du}{dt} = mv + \frac{\partial\Omega}{\partial x}, \qquad \frac{dv}{dt} = -mu + \frac{\partial\Omega}{\partial y} \quad\ldots\ldots\ldots\ldots\ldots(26)$$

Now let $\qquad D_i = \dfrac{d^i u}{dt^i}\dfrac{\partial}{\partial x} + \dfrac{d^i v}{dt^i}\dfrac{\partial}{\partial y}$, where i is 0, 1, 2, 3 ...

Then total differentiation of a function of x, y, t or of x, y, u, v is expressed in terms of partial differentials as follows:

$$\frac{d}{dt} = \frac{\partial}{\partial t} + D_0.$$

It is obvious that $\dfrac{\partial}{\partial t} D_i = D_{i+1}$, and $\dfrac{d}{dt}$ performed on a function of x, y, but not of u, v, is simply D_0.

If we differentiate (26) repeatedly with respect to the time, we have

$$\frac{d^{i+1}u}{dt^{i+1}} = m\frac{d^i v}{dt^i} + \left(\frac{d}{dt}\right)^i\frac{\partial\Omega}{\partial x}, \qquad \frac{d^{i+1}v}{dt^{i+1}} = -m\frac{d^i u}{dt^i} + \left(\frac{d}{dt}\right)^i\frac{\partial\Omega}{\partial y} \quad\ldots(27)$$

Now $\partial\Omega/\partial x$ and $\partial\Omega/\partial y$ are functions of x, y only, and not also of u, v; therefore in the last terms of these equations,

$$\left.\begin{array}{ll}
\text{when } i = 1, & \dfrac{d}{dt} = D_0 \\[2mm]
\text{when } i = 2, & \left(\dfrac{d}{dt}\right)^2 = D_1 + D_0^2 \\[2mm]
\text{when } i = 3, & \left(\dfrac{d}{dt}\right)^3 = D_2 + 3D_0 D_1 + D_0^3 \\[2mm]
\text{when } i = 4, & \left(\dfrac{d}{dt}\right)^4 = D_3 + 4D_0 D_2 + 3D_1^2 + 6D_0 D_1 + D_0^4 \\[2mm]
& \qquad\qquad\qquad\qquad \text{and so forth}
\end{array}\right\}\ldots(27)$$

The function Ω consists of two parts, one being a function of r, the other of ρ; if in the latter part we write $\xi = (x-1)$, $\eta = y$,

$$\Omega = \tfrac{1}{2}\nu(x^2 + y^2) + \tfrac{1}{2}(\xi^2 + \eta^2) + \frac{\nu}{r} + \frac{1}{\rho}$$

---

* [Formulæ for cusps are given below in Paper 3.]

The partial differentials of Ω with respect to x, y may be regarded also as consisting of two parts, viz. of the partial differentials with respect to x, y of $\frac{1}{2}\nu\,(x^2 + y^2) + \nu/r$, and of the partial differentials with respect to ξ, η of $\frac{1}{2}(\xi^2 + \eta^2) + 1/\rho$. These two parts may be considered separately, since, except as regards the factor ν, the one is the exact counterpart of the other.

The partial differentials of $\frac{1}{2}\nu\,(x^2 + y^2)$ disappear after the first two orders, and those of ν/r are exactly those functions which occur in the theory of spherical harmonic analysis.

Thus
$$\frac{\partial}{\partial x}\frac{1}{r} = -\frac{1}{r^2}\cos\theta, \qquad\qquad \frac{\partial}{\partial y}\frac{1}{r} = -\frac{1}{r^2}\sin\theta$$

$$\frac{\partial^2}{\partial x^2}\frac{1}{r} = \frac{1}{r^3}(3\cos^2\theta - 1), \qquad \frac{\partial^2}{\partial x\partial y}\frac{1}{r} = \frac{3}{r^3}\sin\theta\cos\theta$$

$$\frac{\partial^2}{\partial y^2}\frac{1}{r} = \frac{1}{r^3}(3\sin^2\theta - 1)$$

$$\frac{\partial^3}{\partial x^3}\frac{1}{r} = \frac{3}{r^5}(3\cos\theta - 5\cos^3\theta), \qquad \frac{\partial^3}{\partial x^2\partial y}\frac{1}{r} = \frac{3}{r^5}(\sin\theta - 5\sin\theta\cos^2\theta)$$

$$\frac{\partial^3}{\partial x\partial y^2}\frac{1}{r} = \frac{3}{r^5}(\cos\theta - 5\cos\theta\sin^2\theta), \qquad \frac{\partial^3}{\partial y^3}\frac{1}{r} = \frac{3}{r^5}(3\sin\theta - 5\sin^3\theta)$$

and so forth.

It thus appears that the calculation of the successive differentials of u, v with regard to the time is easy, although laborious. These differentials, when appropriately divided by the factorials of 1, 2, 3, 4 etc., are the successive coefficients of the powers of the time in the developments of x, y. If the series for x, y be differentiated, we obtain those for u, v.

The Jacobian integral is useful as a control to the applicability of the series; for the square of the velocity corresponding to any position computed from the series for x and y should agree with the value of $u^2 + v^2$ as computed from the series for u and v.

The computation of an orbit by series is however so tedious, that I have made very little use of this method.

I have also obtained a less extended development for x, y in terms of powers of the arc of the orbit, but the formulæ are so cumbrous as to be of little service.

The development in powers of the time becomes much less laborious if we start from a point in the curve of zero velocity, and in this case the symbols D_i may be replaced by their full expressions in terms of the partial differentials of Ω. But it does not seem worth while to give these special forms, except as regards the first two terms.

If we have initially $x = x_0$, $y = y_0$, $u = 0$, $v = 0$, D_0 and all its powers vanish, and

$$\frac{du}{dt} = \frac{\partial\Omega}{\partial x}, \qquad \frac{dv}{dt} = \frac{\partial\Omega}{\partial y}$$

$$\frac{d^2 u}{dt^2} = m\frac{\partial\Omega}{\partial y}, \qquad \frac{d^2 v}{dt^2} = -m\frac{\partial\Omega}{\partial x}$$

Hence as far as the cube of the time,

$$x - x_0 = \tfrac{1}{2}t^2\frac{\partial\Omega}{\partial x} + \tfrac{1}{6}t^3 m\frac{\partial\Omega}{\partial y}$$

$$y - y_0 = \tfrac{1}{2}t^2\frac{\partial\Omega}{\partial y} - \tfrac{1}{6}t^3 m\frac{\partial\Omega}{\partial x}$$

These may be written

$$(x - x_0)\frac{\partial\Omega}{\partial y} - (y - y_0)\frac{\partial\Omega}{\partial x} = \tfrac{1}{6}t^3 . mT^2$$

$$(x - x_0)\frac{\partial\Omega}{\partial x} + (y - y_0)\frac{\partial\Omega}{\partial y} = \tfrac{1}{2}t^2 . T^2$$

where $T^2 = \left(\dfrac{\partial\Omega}{\partial x}\right)^2 + \left(\dfrac{\partial\Omega}{\partial y}\right)^2$.

By elimination of t, and substitution of $2n$ for m, we obtain the equation to the cusp,

$$8n^2\left[(x - x_0)\frac{\partial\Omega}{\partial x} + (y - y_0)\frac{\partial\Omega}{\partial y}\right]^3 = 9T^2\left[(x - x_0)\frac{\partial\Omega}{\partial y} - (y - y_0)\frac{\partial\Omega}{\partial x}\right]^2$$

The cusp is therefore a semicubical parabola, with the tangent at the cusp normal to the curve $2\Omega = C$.

§ 8.   Variation of orbit.

The object of this paper is not only to discover periodic orbits but also to consider their stability.

Now the stability of a periodic orbit is determinable by discovering whether the motion is oscillatory or not, when the path varies by infinitely little from that of the periodic orbit. The variation of an orbit may be of two kinds, for the constant of relative energy may be varied, or the planet may be displaced from the periodic orbit.

Suppose that the constant C undergoes a small variation and becomes $C + \delta C$; then there must be a periodic orbit, corresponding to $C + \delta C$, which differs by very little from that corresponding to C.

Now if a planet is moving in a periodic orbit, and if C suddenly becomes $C + \delta C$, we may henceforth refer the motion to the varied periodic orbit, and may consider the constant of relative energy as $C + \delta C$ and invariable. The periodic orbit of reference then varies *per saltum*, but the instantaneous position of the planet is unvaried, and therefore the planet is now displaced from its orbit of reference. Hence the result of a variation of C will virtually be determined by regarding C as constant, and by supposing the planet to be displaced from the periodic orbit. This subject is considered in the present section.

The whole of the following investigation is founded on the work of Mr Hill*, but it is presented in a different form.

If the Jacobian integral (2) be differentiated with respect to the time, and if the equations $dx/dt = - V \sin \phi$, $dy/dt = V \cos \phi$ be used in the result, we obtain

$$\frac{dV}{dt} = - \sin \phi \frac{\partial \Omega}{\partial x} + \cos \phi \frac{\partial \Omega}{\partial y} \quad \ldots\ldots\ldots\ldots\ldots(28)$$

Again if the first of the equations of motion (1) be multiplied by $- \cos \phi$, and the second by $- \sin \phi$, and if the two be added together, the result may be written

$$\cos \phi \frac{d}{dt} (V \sin \phi) - \sin \phi \frac{d}{dt} (V \cos \phi) + 2nV = - \cos \phi \frac{\partial \Omega}{\partial x} - \sin \phi \frac{\partial \Omega}{\partial y}$$

Completing the differentiations on the left-hand side, we have

$$V \left(\frac{d\phi}{dt} + 2n \right) = - \cos \phi \frac{\partial \Omega}{\partial x} - \sin \phi \frac{\partial \Omega}{\partial y} \quad \ldots\ldots\ldots\ldots(29)$$

Let s be the arc of the orbit, and p the arc of an orthogonal trajectory of the orbit, estimated in the direction of the outward normal of the orbit; then

$$\left. \begin{aligned} \frac{\partial}{\partial s} &= - \sin \phi \frac{\partial}{\partial x} + \cos \phi \frac{\partial}{\partial y} \\ \frac{\partial}{\partial p} &= \cos \phi \frac{\partial}{\partial x} + \sin \phi \frac{\partial}{\partial y} \end{aligned} \right\} \quad \ldots\ldots\ldots\ldots\ldots(30)$$

Accordingly (28), (29) and the Jacobian integral become

$$\left. \begin{aligned} \frac{dV}{dt} &= \frac{\partial \Omega}{\partial s} \\ V \left(\frac{d\phi}{dt} + 2n \right) &= - \frac{\partial \Omega}{\partial p} \\ V^2 &= 2\Omega - C \end{aligned} \right\} \quad \ldots\ldots\ldots\ldots\ldots(30)$$

The equations (30) are equivalent to (1) and (2).

Now suppose that x, y are the coordinates of a point on an orbit, and

* " On the part of the motion of the moon's perigee, etc.," *Acta Mathem.*, Vol. VIII., pp. 1—36.

that $x + \delta x$, $y + \delta y$ are the coordinates of a point on an adjacent orbit. Then if we put

$$\delta p = \quad \delta x \cos \phi + \delta y \sin \phi$$

$$\delta s = - \delta x \sin \phi + \delta y \cos \phi$$

δp, δs are the distances measured along the outward normal and along the arc of the unvaried orbit, from the original point x, y to the adjacent point $x + \delta x$, $y + \delta y$.

If, with x, y as origin, rectangular axes be drawn along the outward normal and along the arc of the unvaried orbit, we may regard δp, δs as the coordinates of the new point relatively to the old one.   The new axes rotate with angular velocity $n + d\phi/dt$, the second term representing the angular velocity of the normal and the first that of our original axes of x and y.

The well-known formulæ for the component accelerations of a point along two directions, which instantaneously coincide with a pair of rotating rectangular axes by reference to which the position of the point is determined, give the accelerations

$$\left.\begin{array}{l} \dfrac{d^2 \delta p}{dt^2} - \delta p \left(\dfrac{d\phi}{dt} + n\right)^2 - 2\dfrac{d\delta s}{dt}\left(\dfrac{d\phi}{dt} + n\right) - \delta s \dfrac{d^2\phi}{dt^2}, \text{ along the normal} \\[3mm] \dfrac{d^2 \delta s}{dt^2} - \delta s \left(\dfrac{d\phi}{dt} + n\right)^2 + 2\dfrac{d\delta p}{dt}\left(\dfrac{d\phi}{dt} + n\right) + \delta p \dfrac{d^2\phi}{dt^2}, \text{ along the tangent} \end{array}\right\} \dots(31)$$

These are the accelerations of the new point relatively to the old, estimated along lines fixed in space which coincide instantaneously with the normal and tangent of the unvaried orbit.

The function Ω includes the potential of the rotation n of the original axes of x and y.   Hence $\Omega - \tfrac{1}{2}n^2 r^2$ is the true potential of the forces under which the body moves in the unvaried orbit, and

$$\frac{\partial}{\partial p}(\Omega - \tfrac{1}{2}n^2 r^2), \qquad \frac{\partial}{\partial s}(\Omega - \tfrac{1}{2}n^2 r^2)$$

are the components of force in the unvaried orbit along the normal and along the arc.

Therefore the excesses of the forces in the varied orbit above those in the unvaried orbit are

$$\left(\delta p \frac{\partial^2}{\partial p^2} + \delta s \frac{\partial^2}{\partial p \partial s}\right)(\Omega - \tfrac{1}{2}n^2 r^2) \text{ and } \left(\delta p \frac{\partial^2}{\partial p \partial s} + \delta s \frac{\partial^2}{\partial s^2}\right)(\Omega - \tfrac{1}{2}n^2 r^2)$$

Now by considering the meaning (30) of the operations $\partial/\partial p$, $\partial/\partial s$, it is easy to prove that

$$\tfrac{1}{2}\frac{\partial^2 r^2}{\partial p^2} = \tfrac{1}{2}\frac{\partial^2 r^2}{\partial s^2} = 1, \qquad \tfrac{1}{2}\frac{\partial^2}{\partial p \partial s} r^2 = 0$$

Hence the excesses of the forces in the varied orbit above those in the unvaried orbit are

$$\delta p \frac{\partial^2 \Omega}{\partial p^2} + \delta s \frac{\partial^2 \Omega}{\partial p \partial s} - n^2 \delta p, \text{ and } \delta p \frac{\partial^2 \Omega}{\partial p \partial s} + \delta s \frac{\partial^2 \Omega}{\partial s^2} - n^2 \delta s$$

along the normal and along the arc of the unvaried orbit.

But these are necessarily equal to the accelerations (31) of which they are the cause. Then transferring $-n^2 \delta p$, $-n^2 \delta s$ to the left-hand sides of the equations, we have

$$\left.\begin{array}{l} \dfrac{d^2 \delta p}{dt^2} + \delta p \left[n^2 - \left(\dfrac{d\phi}{dt} + n \right)^2 \right] - 2 \dfrac{d\delta s}{dt} \left(\dfrac{d\phi}{dt} + n \right) - \delta s \dfrac{d^2\phi}{dt^2} \\[2mm] \qquad\qquad = \delta p \dfrac{\partial^2 \Omega}{\partial p^2} + \delta s \dfrac{\partial^2 \Omega}{\partial p \partial s} \\[4mm] \dfrac{d^2 \delta s}{dt^2} + \delta s \left[n^2 - \left(\dfrac{d\phi}{dt} + n \right)^2 \right] + 2 \dfrac{d\delta p}{dt} \left(\dfrac{d\phi}{dt} + n \right) + \delta p \dfrac{d^2\phi}{dt^2} \\[2mm] \qquad\qquad = \delta p \dfrac{\partial^2 \Omega}{\partial p \partial s} + \delta s \dfrac{\partial^2 \Omega}{\partial s^2} \end{array}\right\} \dots\dots(32)$$

These are the equations of motion in the varied orbit.

The variation of the last of (30), the Jacobian integral, gives

$$V\delta V = \delta p \frac{\partial \Omega}{\partial p} + \delta s \frac{\partial \Omega}{\partial s} \dots\dots\dots\dots\dots\dots(33)$$

Now δV is the tangential velocity of the point $x + \delta x$, $y + \delta y$ in the varied orbit, relatively to the original point x, y. But as we only want to consider a velocity relatively to the axes of x and y, which themselves rotate with angular velocity n, our p, s axes must be regarded as rotating with angular velocity $d\phi/dt$, instead of $(d\phi/dt) + n$.

Accordingly

$$\delta V = \frac{d\delta s}{dt} + \delta p \frac{d\phi}{dt} \dots\dots\dots\dots\dots\dots(34)$$

This may also be proved by putting $V\delta V = \dfrac{dx}{dt}\dfrac{d\delta x}{dt} + \dfrac{dy}{dt}\dfrac{d\delta y}{dt}$, and by substituting for the differentials in terms of δp, δs, V, ϕ.

The formula (34) enables us to get rid of δV in (33) but we may also get rid of $\partial\Omega/\partial p$ and $\partial\Omega/\partial s$ by means of the equations of motion (30). Thus the variation of the Jacobian integral leads to

$$V\left(\frac{d\delta s}{dt} + \delta p \frac{d\phi}{dt} \right) = - V\left(\frac{d\phi}{dt} + 2n \right) \delta p + \frac{dV}{dt} \delta s$$

Therefore

$$\left.\begin{array}{l} \dfrac{d\delta s}{dt} + 2\delta p \left(\dfrac{d\phi}{dt} + n \right) - \dfrac{1}{V}\dfrac{dV}{dt} \delta s = 0 \\[4mm] V\dfrac{d}{dt}\left(\dfrac{\delta s}{V} \right) + 2\delta p \left(\dfrac{d\phi}{dt} + n \right) = 0 \end{array}\right\} \dots\dots\dots\dots(35)$$

or

The equations (35) are two forms of the varied Jacobian integral.

A great simplification of the equations of motion (32) is possible by reference to the unvaried motion.

Let us suppose then that δp, δs are no longer displacements to a varied orbit, but are the actual displacements occurring in time δt in the unvaried orbit. Thus $\delta p = 0$, $\delta s = V \delta t$.

The equations (32) then give

$$\left. \begin{aligned}
-2 \frac{dV}{dt}\left(\frac{d\phi}{dt} + n\right) - V \frac{d^2\phi}{dt^2} &= V \frac{\partial^2 \Omega}{\partial p \partial s} \\
\frac{d^2 V}{dt^2} + V\left[n^2 - \left(\frac{d\phi}{dt} + n\right)^2\right] &= V \frac{\partial^2 \Omega}{\partial s^2}
\end{aligned} \right\} \quad \dots\dots\dots\dots(36)$$

The first of (36) may be written

$$\frac{d^2\phi}{dt^2} + \frac{\partial^2 \Omega}{\partial p \partial s} = -\frac{2}{V} \frac{dV}{dt}\left(\frac{d\phi}{dt} + n\right)$$

These two terms, multiplied by δs, occur in the first of (32), which may therefore be written

$$\frac{d^2 \delta p}{dt^2} + \delta p\left[n^2 - \left(\frac{d\phi}{dt} + n\right)^2\right] - 2 \frac{d\delta s}{dt}\left(\frac{d\phi}{dt} + n\right) + \frac{2\delta s}{V}\frac{dV}{dt}\left(\frac{d\phi}{dt} + n\right) - \delta p \frac{\partial^2 \Omega}{\partial p^2} = 0$$

The terms in this which involve δs may now be eliminated by the first of (35), and we have

$$\frac{d^2 \delta p}{dt^2} + \delta p\left[n^2 - \left(\frac{d\phi}{dt} + n\right)^2 + 4\left(\frac{d\phi}{dt} + n\right)^2 - \frac{\partial^2 \Omega}{\partial p^2}\right] = 0$$

If then we put

$$\Theta = n^2 + 3\left(\frac{d\phi}{dt} + n\right)^2 - \frac{\partial^2 \Omega}{\partial p^2} \quad \dots\dots\dots\dots\dots\dots(37)$$

we have

$$\left. \begin{aligned}
\frac{d^2 \delta p}{dt^2} + \Theta \delta p &= 0 \\
\frac{d}{dt}\left(\frac{\delta s}{V}\right) + 2 \frac{\delta p}{V}\left(\frac{d\phi}{dt} + n\right) &= 0
\end{aligned} \right\} \quad \dots\dots\dots\dots(37)$$

The differential equation for δp is Mr Hill's well-known result.

We have now to consider the form of the function Θ.

Let us write $\nabla^2 = \partial^2/\partial x^2 + \partial^2/\partial y^2 = \partial^2/\partial p^2 + \partial^2/\partial s^2$; then adding $V \partial^2 \Omega / \partial p^2$ to each side of the second of (36), we have

$$\frac{1}{V} \frac{d^2 V}{dt^2} + n^2 - \left(\frac{d\phi}{dt} + n\right)^2 + \frac{\partial^2 \Omega}{\partial p^2} = \nabla^2 \Omega$$

so that

$$n^2 - \frac{\partial^2 \Omega}{\partial p^2} = \frac{d}{dt}\left(\frac{dV}{V dt}\right) + \left(\frac{dV}{V dt}\right)^2 - \left(\frac{d\phi}{dt} + n\right)^2 + 2n^2 - \nabla^2 \Omega$$

Substituting in (37),

$$\Theta = 2n^2 - \nabla^2 \Omega + 2\left(\frac{d\phi}{dt} + n\right)^2 + \frac{d}{dt}\left(\frac{dV}{V dt}\right) + \left(\frac{dV}{V dt}\right)^2$$

If we put $u = x + y\iota$, $s = x - y\iota$, $d/dt = \iota D$, where $\iota = \sqrt{(-1)}$, it is easy to show that $Du = Ve^{\phi\iota}$, $Ds = -Ve^{-\phi\iota}$, and

$$2\frac{d\phi}{dt} = \frac{D^2u}{Du} - \frac{D^2s}{Ds}, \qquad 2\frac{dV}{Vdt} = \iota\left(\frac{D^2u}{Du} + \frac{D^2s}{Ds}\right)$$

Mr Hill's form for the function Θ follows at once from these transformations.

Another form for Θ, deducible directly from (37), is

$$\Theta = n^2 - \tfrac{1}{2}\nabla^2\Omega - \tfrac{1}{2}\left(\frac{\partial^2\Omega}{\partial x^2} - \frac{\partial^2\Omega}{\partial y^2}\right)\cos 2\phi - \tfrac{1}{2}\frac{\partial^2\Omega}{\partial x\partial y}\sin 2\phi + 3\left(\frac{d\phi}{dt} + n\right)^2$$

whence

$$\Theta = \frac{\nu}{r^3} + \frac{1}{\rho^3} - \frac{3\nu}{r^3}\cos^2(\phi-\theta) - \frac{3}{\rho^3}\cos^2(\phi-\psi) + 3V^2\left(\frac{1}{R} + \frac{n}{V}\right)^2$$

§ 9. *Change of independent variable from time to arc of orbit.*

For the purpose of future developments it is now necessary to change the independent variable from the time t to the arc s.

Let
$$\delta q = \delta p\, V^{\frac{1}{2}} \quad\dots\dots\dots\dots\dots\dots\dots\dots\dots\dots\dots(38)$$

Then
$$\frac{d^2\delta p}{dt^2} = V\frac{d}{ds}\left[V\frac{d}{ds}\left(\frac{\delta q}{V^{\frac{1}{2}}}\right)\right] = V\frac{d}{ds}\left(V^{\frac{1}{2}}\frac{d\delta q}{ds} - \frac{1}{2V^{\frac{1}{2}}}\delta q\frac{dV}{ds}\right)$$

$$= V^{\frac{3}{2}}\frac{d^2\delta q}{ds^2} - \tfrac{1}{2}\delta q V\frac{d}{ds}\left(\frac{1}{V^{\frac{1}{2}}}\frac{dV}{ds}\right)$$

But
$$V\frac{d}{ds}\left(\frac{1}{V^{\frac{1}{2}}}\frac{dV}{ds}\right) = \frac{d}{dt}\left(\frac{1}{V^{\frac{3}{2}}}\frac{dV}{dt}\right) = -\frac{3}{2V^{\frac{5}{2}}}\left(\frac{dV}{dt}\right)^2 + \frac{1}{V^{\frac{3}{2}}}\frac{d^2V}{dt^2}$$

$$= -\frac{3}{2V^{\frac{1}{2}}}\left(\frac{dV}{ds}\right)^2 + \frac{1}{V^{\frac{3}{2}}}\frac{d^2V}{dt^2}$$

Hence
$$\frac{d^2\delta p}{dt^2} = V^{\frac{3}{2}}\frac{d^2\delta q}{ds^2} + \frac{3}{4V^{\frac{1}{2}}}\left(\frac{dV}{ds}\right)^2\delta q - \frac{\delta q}{2V^{\frac{3}{2}}}\frac{d^2V}{dt^2}$$

Also
$$\Theta\delta p = \frac{\Theta\delta q}{V^{\frac{1}{2}}}$$

If these two be added together, and divided by $V^{\frac{3}{2}}$, we obtain

$$\left.\begin{aligned}\frac{d^2\delta q}{ds^2} + \Psi\,\delta q &= 0 \\[2mm] \Psi = \frac{\Theta}{V^2} + \tfrac{3}{4}\left(\frac{dV}{Vds}\right)^2 - \frac{1}{2V^3}\frac{d^2V}{dt^2}\end{aligned}\right\}\quad\dots\dots\dots\dots\dots\dots(39)$$

where

It remains to obtain the expression for the function Ψ.

Since $$\frac{d\phi}{ds} = \frac{1}{R}, \quad \text{and} \quad n^2 = \nu + 1$$

$$\Theta = \nu + 1 + 3\left(\frac{V}{R} + n\right)^2 - \frac{\partial^2 \Omega}{\partial p^2}$$

Now from the first of (30) and the second of (36),

$$V\frac{dV}{ds} = \frac{\partial \Omega}{\partial s}$$

$$\frac{1}{V}\frac{d^2 V}{dt^2} = \left(\frac{V}{R} + n\right)^2 + \frac{\partial^2 \Omega}{\partial s^2} - \nu - 1$$

Then by substitution in the second of (39),

$$\Psi V^2 = \tfrac{3}{2}(\nu + 1) + \tfrac{5}{2}\left(\frac{V}{R} + n\right)^2 + \tfrac{3}{4}\left(\frac{dV}{ds}\right)^2 - \frac{\partial^2 \Omega}{\partial p^2} - \tfrac{1}{2}\frac{\partial^2 \Omega}{\partial s^2}$$

Also $$\frac{\partial^2 \Omega}{\partial p^2} + \tfrac{1}{2}\frac{\partial^2 \Omega}{\partial s^2} = \tfrac{1}{2}\nabla^2 \Omega + \tfrac{1}{2}\frac{\partial^2 \Omega}{\partial p^2}$$

Now $2\Omega = \nu\left(r^2 + \frac{2}{r}\right) + \left(\rho^2 + \frac{2}{\rho}\right)$, and

$$\frac{\partial^2 \Omega}{\partial x^2} = \nu + 1 - \frac{\nu}{r^3} - \frac{1}{\rho^3} + \frac{3\nu}{r^3}\cos^2\theta + \frac{3}{\rho^3}\cos^2\psi$$

$$\frac{\partial^2 \Omega}{\partial x \partial y} = \frac{3\nu}{r^3}\sin\theta\cos\theta + \frac{3}{\rho^3}\sin\psi\cos\psi$$

$$\frac{\partial^2 \Omega}{\partial y^2} = \nu + 1 - \frac{\nu}{r^3} - \frac{1}{\rho^3} + \frac{3\nu}{r^3}\sin^2\theta + \frac{3}{\rho^3}\sin^2\psi$$

Hence $$\nabla^2 \Omega = 2(\nu + 1) + \frac{\nu}{r^3} + \frac{1}{\rho^3}$$

and $$\frac{\partial^2 \Omega}{\partial p^2} = \cos^2\phi\,\frac{\partial^2 \Omega}{\partial x^2} + 2\sin\phi\cos\phi\,\frac{\partial^2 \Omega}{\partial x \partial y} + \sin^2\phi\,\frac{\partial^2 \Omega}{\partial y^2}$$

$$= \nu + 1 - \frac{\nu}{r^3} - \frac{1}{\rho^3} + \frac{3\nu}{r^3}\cos^2(\phi - \theta) + \frac{3}{\rho^3}\cos^2(\phi - \psi)$$

Therefore

$$\Psi = \tfrac{5}{2}\left(\frac{1}{R} + \frac{n}{V}\right)^2 - \frac{3}{2V^2}\left[\frac{\nu}{r^3}\cos^2(\phi - \theta) + \frac{1}{\rho^3}\cos^2(\phi - \psi)\right] + \tfrac{3}{4}\left(\frac{dV}{Vds}\right)^2 \quad \text{...(40)}$$

Also since $$V\frac{dV}{ds} = \frac{\partial \Omega}{\partial s} = -\sin\phi\,\frac{\partial \Omega}{\partial x} + \cos\phi\,\frac{\partial \Omega}{\partial y}$$

$$\frac{dV}{Vds} = \frac{\nu}{V^2}\left(\frac{1}{r^2} - r\right)\sin(\phi - \theta) + \frac{1}{V^2}\left(\frac{1}{\rho^2} - \rho\right)\sin(\phi - \psi) \quad \text{......(40)}$$

This completes the formula for Ψ in terms of the coordinates, the velocity, the curvature and of ϕ.

It may be useful to obtain the expressions for δs and $\delta \phi$ in terms of the new independent variable s.

The second of (37) may be written down at once, namely

$$\frac{d}{ds}\left(\frac{\delta s}{V}\right) = -\frac{2\delta q}{V^{\frac{3}{2}}}\left(\frac{1}{R} + \frac{n}{V}\right) \dots\dots\dots\dots\dots(41)$$

Also it is clear from geometrical considerations that

$$\delta\phi = -\frac{d\delta p}{ds} + \frac{\delta s}{R}$$

whence

$$\delta\phi = -\frac{1}{V^{\frac{1}{2}}}\left[\frac{d\delta q}{ds} - \tfrac{1}{2}\delta q\left(\frac{dV}{Vds}\right)\right] + \frac{\delta s}{R}\dots\dots\dots\dots(42)$$

§ 10.   *The solution of the differential equation for δq*.

The function Ψ has a definite value at each point of a periodic orbit whose complete arc is S.   Therefore Ψ is a function of the arc s of the orbit, measured from any point therein, and when s has increased from zero to S, Ψ has returned to its initial value.   Also since a periodic orbit is symmetrical with respect to the x-axis, Ψ is an even function of the arc s, when s is measured from an orthogonal intersection of the orbit with the x-axis.   If the periodic orbit only goes once round S or J, or round both, all the intersections with the x-axis are necessarily orthogonal.   I call such an orbit simply periodic, but the term must have its meaning extended so as to embrace the possibility of loops.   But when there are loops all the intersections with the x-axis are not necessarily orthogonal, and if the orbit is only periodic after several revolutions some of the intersections cannot be orthogonal.

With the understanding that s is measured from an orthogonal intersection with the x-axis, Ψ is an even function of s and is expressible by the Fourier series

$$\Psi = \Psi_0 + 2\Psi_1\cos\frac{2\pi s}{S} + 2\Psi_2\cos\frac{4\pi s}{S} + \dots$$

Now multiply the differential equation (39) for δq by S^2/π^2, write σ for $\pi s/S$, and put $\Phi = (S^2/\pi^2)\Psi$, and we have

$$\frac{d^2\delta q}{d\sigma^2} + \Phi\delta q = 0 \dots\dots\dots\dots\dots(43)$$

Also if $\Phi_j = \frac{S^2}{\pi^2}\Psi_j$,

$$\Phi = \Phi_0 + 2\Phi_1\cos 2\sigma + 2\Phi_2\cos 4\sigma + \dots$$

If then we write $\zeta = e^{\sigma\sqrt{-1}}$,

$$\zeta\frac{d}{d\zeta} = \frac{1}{\sqrt{(-1)}}\frac{d}{d\sigma}$$

* [Another method of treating the problem will be found in § 1 of Paper 3 hereafter.]

and the equation (43) becomes

$$\left(\zeta \frac{d}{d\zeta}\right)^{2} \delta q = \Phi \delta q \dots\dots\dots\dots\dots\dots\dots(44)$$

where $\Phi = \Sigma_j \Phi_j \zeta^{2j}$, the summation being taken from $j = +\infty$ to $j = -\infty$, and Φ_{-j} being equal to Φ_j.

Let us assume as the solution of (44)

$$\delta q = \Sigma_j [(b_j + e_{-j}) \cos(c + 2j) \sigma + (b_j - e_{-j}) \sqrt{(-1)} \sin(c + 2j) \sigma]$$
$$= \Sigma_j [b_j \zeta^{c+2j} + e_j \zeta^{-c+2j}]$$

The equation (44) must be separately satisfied for the terms involving b and for those involving e; hence we need only regard one series of terms.

On substituting in (44) the assumed expression for δq, and equating to zero the coefficients of the several powers of ζ, we have

$$b_j (c + 2j)^2 = \Sigma_i b_{j-i} \Phi_i \text{ *}$$

written *in extenso* this is

$$\dots - b_{j-2} \Phi_2 - b_{j-1} \Phi_1 + b_j [(c + 2j)^2 - \Phi_0] - b_{j+1} \Phi_1 - b_{j+2} \Phi_2 - \dots = 0$$

There are an infinite number of equations like the above, but the infinity must be regarded as an odd number.

If from these equations the b's be eliminated, we have an infinite determinantal equation for determining c. If we write

$$(c + 2j)^2 - \Phi_0 = \{j\}$$

the equation is

$$\begin{vmatrix} \dots\dots\dots\dots\dots\dots\dots \\ \dots & \{-1\}, & -\Phi_1, & -\Phi_2 \dots \\ \dots & -\Phi_1, & \{0\}, & -\Phi_1 \dots \\ \dots & -\Phi_2, & -\Phi_1, & \{1\} \dots \\ \dots\dots\dots\dots\dots\dots\dots \end{vmatrix} = 0$$

This is the same in form as Mr Hill's determinantal equation.

As much has been written on the subject, it is unnecessary to reproduce the arguments by which it may be shown that if

$$[j] = \Phi_0 - 4j^2$$

---

* The equation of condition for the e's is easily shown to be

$$e_{-j}(c + 2j)^2 = \Sigma_i e_{i-j} \Phi_{-i}$$

and since $\Phi_i = \Phi_{-i}$, this is exactly the same as that for the b's save that e_{-j} corresponds with b_j.

$$
\text{and} \qquad \Delta = \begin{vmatrix}
\cdots \quad 1 \;, & \dfrac{\Phi_1}{[1]}, & \dfrac{\Phi_2}{[1]} \quad \cdots \\[1em]
\cdots \quad \dfrac{\Phi_1}{[0]}, & 1 \;, & \dfrac{\Phi_1}{[0]} \quad \cdots \\[1em]
\cdots \quad \dfrac{\Phi_2}{[1]}, & \dfrac{\Phi_1}{[1]}, & 1 \quad \cdots
\end{vmatrix} \qquad \cdots\cdots\cdots\cdots\cdots (45)
$$

the solution of the determinantal equation is given by

$$
\sin^2 \tfrac{1}{2}\pi c = \Delta \sin^2 \tfrac{1}{2}\pi \sqrt{\Phi_0} \quad \cdots\cdots\cdots\cdots\cdots (45)
$$

§ 11. *On the stability or instability of an orbit.*

When c is real, δq is expressible by a series of sines and cosines of multiples of the arc. Since V is an even function of the arc, it is expressible by a series of cosines of the same form as that for Φ; hence δp, which is equal to $V^{\frac{1}{2}} \delta q$, is expressible in a series, similar in form to that for δq.

But δp denotes normal displacement from the periodic orbit, and therefore the motion in the varied orbit is oscillatory with reference to the periodic orbit. In other words the periodic orbit is stable.

If c_0 be any one value of c, all its infinite values are comprised in the formula $\pm c_0 \pm 2i$, where i is an integer. It is however convenient to choose one value of c as fundamental. When the choice has been made we may refer to the terms in the series for δq of which the argument is c_0 as the principal terms, although it does not appear to be necessary that these terms should have the largest coefficients. In fact since two arbitrary constants are involved in the specification of a definite variation of orbit, it is probable that the terms, which are numerically the most important in one variation, will not be so in another.

If the body be considered as moving in an elliptic orbit, it will be at its pericentre or apocentre, when δp is a negative or positive maximum, respectively. The principal terms of δq, and therefore also of δp, have the argument $c\sigma$ or $c\pi s/S$; hence if we may assume that the principal term is also the most important, the body has passed through a complete anomalistic circuit when s has increased from zero to $2S/c$. Since S is the synodic arc in the relative orbit, $\tfrac{1}{2}c$ is the ratio of the anomalistic to the synodic arc, both arcs being measured on the orbit as drawn with reference to the moving axes.

Now I propose to adopt as a convention that the fundamental value of c shall be that value which lies nearest to $\sqrt{\Phi_0}$, where Φ_0 denotes the mean value of Φ. This convention certainly attributes to $\tfrac{1}{2}c$ a physical meaning, which is correct in all those cases which have any resemblance to the motion of an actual satellite in the solar system. I shall accordingly use the value of c which lies nearest to $\sqrt{\Phi_0}$ as fundamental.

We have just arrived at a physical meaning for c by considering the principal term in the series; now in so doing we were in effect considering only the mean motion of the body with reference to the moving axes; therefore $\tfrac{1}{2}c$ is also the ratio of the synodic to the anomalistic period*.

If T denotes the synodic period, the mean motion of the body referred to axes fixed in space is $(2\pi/T) + n$; and if $d\omega/dt$ denotes the mean angular velocity of the pericentre with reference to axes fixed in space, the mean motion of the body with reference to the pericentre is $(2\pi/T) + n - (d\omega/dt)$. Then, since angular velocities vary inversely as periods,

$$\tfrac{1}{2}c = \frac{\dfrac{2\pi}{T} + n - \dfrac{d\omega}{dt}}{\dfrac{2\pi}{T}}$$

where $n^2 = \nu + 1$.

Therefore†

$$\left. \begin{aligned} \frac{d\omega}{dt} &= n - \frac{2\pi}{T}(\tfrac{1}{2}c - 1) \\[2mm] T\left(n - \frac{d\omega}{dt}\right) &= 2\pi(\tfrac{1}{2}c - 1) \end{aligned} \right\} \quad \ldots\ldots\ldots\ldots\ldots\ldots(46)$$

or

Mr Hill's c is equal to one-half of my c, and accordingly the first of (46) is identical with the formula from which Mr Hill derives "a part of the motion of the lunar perigee‡."

The angular velocity of regression of the pericentre being $n - (d\omega/dt)$, it follows from (46) that $2\pi(\tfrac{1}{2}c - 1)$ is the amount of that regression with respect to the moving axes in the synodic period.

Whilst the pericentre regredes with reference to the moving axes, it advances with reference to fixed axes; the advance in the synodic period is $nT - 2\pi(\tfrac{1}{2}c - 1)$, and in the sidereal period the advance is

$$2\pi\left[1 - \frac{\tfrac{1}{2}c}{1 + nT/2\pi}\right]$$

In the numerical treatment of stable periodic orbits I tabulate the apparent regression $2\pi(\tfrac{1}{2}c - 1)$, and the actual advance $nT - 2\pi(\tfrac{1}{2}c - 1)$ in the synodic period; also $2\pi\left[1 - \dfrac{\tfrac{1}{2}c}{1 + nT/2\pi}\right]$ the advance in the sidereal period.

---

* It may be observed that when V is constant (as in the case when we only consider mean motion) $V^2\Psi = \Theta$, and Mr Hill's equation for δp becomes identical with the present one for δq. It is well to remark that what I denote by c is $2c$ of Mr Hill's notation.

† [These formulæ need some modification for retrograde orbits. See § 2 of Paper 3.]

‡ *Acta Mathem.*, Vol. VIII.

Let us now consider the case where c is imaginary, so that the motion is no longer oscillatory with respect to the periodic orbit, and the periodic orbit is unstable.

The form of (45) shows that c becomes imaginary either when $\Delta \sin^2 \frac{1}{2}\pi \sqrt{\Phi_0}$ is negative, or when it is greater than unity; this function will therefore be described below as the criterion of stability.

If Φ_0 were negative it would indicate that the mean force of restitution towards the periodic orbit was negative. Hence it seems obvious that the body would then depart from the periodic orbit, which would therefore be unstable. If however Δ were negative as well as Φ_0, it would seem as if it were possible to have a real value for c; but it is not easy to see how this condition could lead to a stable orbit.

I have not yet come on any case where Φ_0 is negative and accordingly that condition is left out of consideration for the present. We are left then with the two conditions, Δ negative or $\Delta \sin^2 \frac{1}{2}\pi \sqrt{\Phi_0}$ greater than unity; these lead to two kinds of instability.

In instability of the first kind Δ is negative; for reasons which will appear below, I shall call this "even instability."

In this case let us put
$$\Delta \sin^2 \tfrac{1}{2}\pi \sqrt{\Phi_0} = - D^2$$
so that (45) becomes $\sin \frac{1}{2}\pi c = \pm D \sqrt{-1}$.

The sine in this case is hyperbolic, and if we write $c = 2i + k\sqrt{-1}$, where i is an integer, the equation for k becomes $\sinh \frac{1}{2}\pi k = \pm D$.

Since the values of c occur in pairs, equal in magnitude and opposite in sign, it is only necessary to consider the upper sign and the result may be written

$$\left. \begin{aligned} e^{\frac{1}{2}\pi k} &= \sqrt{(D^2 + 1)} + D \\ k &= \frac{2}{\pi} \log_e [\sqrt{(D^2 + 1)} + D] \end{aligned} \right\} \quad \dots\dots\dots\dots\dots (47)$$

or

I shall return in § 12 to the form of solution adapted to the case of "even instability."

Turning to the instability of the second kind, which I shall call "uneven instability," we have
$$\sin^2 \tfrac{1}{2}\pi c = \Delta \sin^2 \tfrac{1}{2}\pi \sqrt{\Phi_0} = D^2$$
where D^2 is greater than unity, so that c is imaginary.

The sine in this case also becomes a hyperbolic function, and if we write $c = 2i + 1 + k\sqrt{-1}$, where i is an integer, we have
$$\sin \tfrac{1}{2}\pi c = (-)^i \cosh \tfrac{1}{2}\pi k$$
a hyperbolic cosine.

Hence $\cosh \frac{1}{2}\pi k = \pm D$

Taking only the upper sign as before, this may be written

$$e^{\frac{1}{2}\pi k} = \sqrt{(D^2 - 1)} + D$$

or $k = \dfrac{2}{\pi} \log_e [\sqrt{(D^2 - 1)} + D]$ $\quad\bigg\}\quad$ ....(48)

I shall return in § 12 to the form of solution adapted to the case of " uneven instability," but I wish now to consider the nature of the transitions from instability to stability.

Suppose that we are considering a family of periodic orbits, the members of which are determined by the continuous increase or decrease of the constant C of relative energy; and let us suppose that $\Delta \sin^2 \frac{1}{2}\pi \sqrt{\Phi_0}$, being at first negative, increases and reaches the value zero. At the moment of the transition of this function from negative to positive, there is transition from even instability to stability. If on the other hand this function were positive and less than unity, and were to increase up to and beyond unity there would be a transition from stability to uneven instability.

In all the cases of stability which I have investigated, except one*, the fundamental value of c lies between 2 and 3, and the apparent regression of pericentre in the synodic period, namely $2\pi (\frac{1}{2}c - 1)$, lies between 0 and 180°, these extreme values corresponding with transitional stages.

It will now conduce to brevity to regard c as lying between 2 and 3, instead of regarding it as a multiple-valued quantity.

If we refer back to the form of solution assumed for the equation (44), we see that when $c = 2$, the solution is

$$\delta q = (b_{-1} + e_1) + (b_0 + e_0 + b_{-2} + e_2) \cos \frac{2\pi s}{S} \ldots$$

$$+ (b_0 - e_0 - b_{-2} + e_2) \sqrt{(-1)} \sin \frac{2\pi s}{S} \ldots$$

and that when $c = 3$, it is

$$\delta q = \quad (b_1 + e_{-1} + b_{-2} + e_2) \cos \frac{\pi s}{S} + (b_0 + e_0 + b_{-3} + e_3) \cos \frac{3\pi s}{S} \ldots$$

$$+ (b_1 - e_{-1} - b_{-2} + e_2) \sqrt{(-1)} \sin \frac{\pi s}{S} + (b_0 - e_0 - b_{-3} + e_3) \sqrt{(-1)} \sin \frac{3\pi s}{S} \ldots$$

In the first case it is clear that when $s = S$, δq has gone through a complete period and has returned to its initial value; but in the second case whilst δq is equal in value, it is opposite in sign to what it was at first.

Consider then the first case where $c = 2$, and suppose that the body is displaced from the periodic orbit along the normal, at a conjunction. Then

---

* The orbit in question is $C = 40.0$, $x_0 = 1.0334$; see Appendix.

the body starts moving at right angles to the line of syzygies, and when $s = S$ it has again returned to the same point, and is again moving at right angles to the line of syzygies.

Hence it follows that we have found a new periodic orbit differing by infinitely little from the original one. Thus the original orbit is a double solution of the problem, and the interpretation to be put on the result $c = 2$ is, that we have found a periodic orbit which is a member of two distinct families.

The $\Delta \sin^2 \frac{1}{2}\pi \sqrt{\Phi_0}$ corresponding to our family of orbits has been supposed to be increasing from a negative to a positive value; at the instant of transition the same function for the other family must also be passing through the value zero.

If C be the value of the constant of relative energy for the critical orbit which gives $c = 2$, there must be *two* orbits, infinitely near to one another, for which the constant is $C - \delta C$.

If the orbits were classified according to values of the parameter $\Delta \sin^2 \frac{1}{2}\pi \sqrt{\Phi_0}$, instead of according to values of C, these two families would have to be regarded as a single family, and the critical stage would be that in which C reached a maximum or minimum value.

But when the classification is according to values of C, we say that there are two families which coalesce at the critical value of C; it is also clear that, as the orbit we were following was unstable up to this critical value, the other must have been stable.

An interesting example of this will be found below, where the families of orbits B and C spring from a single orbit.

Now reverting again to the question of the transition from instability to stability, let us suppose that as the constant C varies, $\Delta \sin^2 \frac{1}{2}\pi \sqrt{\Phi_0}$, being at first greater than unity, diminishes, passes through the value unity and continues diminishing. Then the orbit was at first unstable with uneven instability and c of the form $3 + k\sqrt{-1}$; it becomes stable at the critical stage with c less than 3. But there is now no real double solution at the moment of transition and no coalescence of families*. It is probable that there is coalescence with another family of imaginary orbits at this crisis, but I do not discuss this, since I am not looking at the subject from the point of view of the theory of differential equations. Accordingly in our figures of orbits there will be nothing to mark the transition from uneven instability to stability, and it will only be by the consideration of the function $\Delta \sin^2 \frac{1}{2}\pi \sqrt{\Phi_0}$ that we shall be aware of the change.

---

* When I explained the results at which I have arrived to M. Poincaré, he suggested that there may be coalescence between a doubly periodic orbit and a singly periodic one, when the two circuits of the former become identical with one another and with the latter.

The conclusions arrived at in this section seem to accord with those of M. Poincaré in his *Mécanique Céleste*, who remarks that periodic orbits will disappear in pairs.

It is clear from this discussion that uneven instability can never graduate directly into even instability, but the transition must take place through a range of stability*.

§ 12. *Modulus of instability, and form of solution.*

The cases of instability will now be considered.

When the instability is of the first or even kind, we have $c = 2i + k\sqrt{(-1)}$, and

$$\left. \begin{array}{l} e^{\frac{1}{2}\pi k} = \sqrt{(D^2 + 1)} + D \\ e^{-\frac{1}{2}\pi k} = \sqrt{(D^2 + 1)} - D \end{array} \right\} \quad \ldots\ldots\ldots\ldots\ldots\ldots(49)$$

where $D^2 = -\Delta \sin^2 \frac{1}{2}\pi \sqrt{\Phi_0}$.

The solution of (44) was

$$\delta q = \Sigma_j \left[(b_j + e_{-j}) \cos (c + 2j)\,\sigma + (b_j - e_{-j})\sqrt{(-1)} \sin (c + 2j)\,\sigma \right]$$

Now if we take the integer i involved in the expression for c as zero,

$$\cos (c + 2j)\,\sigma = \quad \cosh k\sigma \cos 2j\sigma - \sqrt{(-1)} \sinh k\sigma \sin 2j\sigma$$

$$\sqrt{(-1)} \sin (c + 2j)\,\sigma = - \sinh k\sigma \cos 2j\sigma + \sqrt{(-1)} \cosh k\sigma \sin 2j\sigma$$

Therefore when the sign of summation only runs from ∞ to 0, instead of to $-\infty$, and when b_0 and e_0 are supposed to be the halves of their values when the summation ran from $+\infty$ to $-\infty$, the solution may be written

$$\delta q = \sum_0^\infty \{\cosh k\sigma \left[(b_j + e_{-j} + b_{-j} + e_j) \cos 2j\sigma + (b_j - e_{-j} - b_{-j} + e_j)\sqrt{(-1)} \sin 2j\sigma \right]$$

$$+ \sinh k\sigma \left[-\sqrt{(-1)}(b_j + e_{-j} - b_{-j} - e_j) \sin 2j\sigma - (b_j - e_{-j} + b_{-j} - e_j) \cos 2j\sigma \right]\}$$

Putting

$$b_j + b_{-j} = B_j, \qquad\qquad e_{-j} + e_j = E_j$$

$$b_j - b_{-j} = \beta_j \sqrt{-1}, \qquad e_{-j} - e_j = \epsilon_j \sqrt{-1}$$

and writing the hyperbolic functions as exponentials, we have

$$\delta q = \sum_0^\infty \{ e^{k\sigma} (E_j \cos 2j\sigma + \epsilon_j \sin 2j\sigma) + e^{-k\sigma} (B_j \cos 2j\sigma - \beta_j \sin 2j\sigma) \} \ldots (50)$$

By means of (49) this may be written

$$\delta q = \sum_0^\infty \{ [\sqrt{(D^2 + 1)} + D]^{\frac{2\sigma}{\pi}} [E_j \cos 2j\sigma + \epsilon_j \sin 2j\sigma]$$

$$+ [\sqrt{(D^2 + 1)} - D]^{\frac{2\sigma}{\pi}} [B_j \cos 2j\sigma - \beta_j \sin 2j\sigma] \} \ldots\ldots(50)$$

* [An erroneous passage in the original paper is here omitted. It related to the subject considered in Part IV. of Paper 3, below.]

In (50) it is not safe to assume that the most important term is that for which $j = 0$; indeed this will usually not be the case. All that we know is that the series contains sines and cosines of even multiples of σ, that one set of terms increases without limit and that the other set diminishes.

In the numerical treatment of unstable periodic orbits it will be well to have a modulus of the degree of instability; and these considerations afford a convenient means of obtaining such a modulus.

This modulus may be taken to be the number of synodic revolutions in which the augmenting factor doubles its initial value; that is to say we are to put

$$e^{k\sigma} = [\sqrt{(D^2 + 1)} + D]^{\frac{2\sigma}{\pi}} = 2$$

Therefore
$$\frac{s}{S} = \frac{\sigma}{\pi} = \frac{\log \sqrt{2}}{\log [\sqrt{(D^2 + 1)} + D]} \quad \dots\dots\dots\dots(51)$$

This is the modulus of instability, when it is of the even kind.

A consideration of the form of the series for δq shows that it increases without limit, and that the planet or satellite crosses and recrosses the periodic orbit an even number of times in a single circuit; it is on this account that I have called this "even instability."

When the instability is of the second or uneven kind, we have $c = 2i + 1 + k\sqrt{-1}$, or if we take i as zero, $c = 1 + k\sqrt{-1}$; also

$$\left.\begin{array}{l} e^{\frac{1}{2}\pi k} = D + \sqrt{(D^2 - 1)} \\ e^{-\frac{1}{2}\pi k} = D - \sqrt{(D^2 - 1)} \end{array}\right\} \quad \dots\dots\dots\dots\dots(52)$$

where $D^2 = \Delta \sin^2 \frac{1}{2}\pi \sqrt{\Phi_0}$.

Then

$$\cos(c + 2j)\sigma = \cos(2j + 1)\sigma \cosh k\sigma - \sqrt{(-1)} \sin(2j + 1)\sigma \sinh k\sigma$$

$$\sqrt{(-1)}\sin(c + 2j)\sigma = -\cos(2j + 1)\sigma \sinh k\sigma + \sqrt{(-1)}\sin(2j + 1)\sigma \cosh k\sigma$$

And the solution, expressed with singly infinite summation and with the proper change in the meanings of b_0 and e_0, is

$$\delta q = \overset{\infty}{\underset{0}{\Sigma}} \{\cosh k\sigma \, [(b_j + b_{-j-1} + e_{-j} + e_{j+1}) \cos(2j + 1)\sigma$$

$$+ (b_j - b_{-j-1} - e_{-j} + e_{j+1}) \sqrt{(-1)} \sin(2j + 1)\sigma]$$

$$+ \sinh k\sigma \, [-\sqrt{(-1)}(b_j - b_{-j-1} + e_{-j} - e_{j+1}) \sin(2j + 1)\sigma$$

$$- (b_j + b_{-j-1} - e_{-j} - e_{j+1}) \cos(2j + 1)\sigma]\}$$

Putting
$$b_j + b_{-j-1} = B_j, \qquad\qquad e_{-j} + e_{j+1} = E_j$$
$$b_j - b_{-j-1} = \beta_j \sqrt{(-1)}, \qquad e_{-j} - e_{j+1} = \epsilon_j \sqrt{(-1)}$$

and writing the hyperbolic functions as exponentials, we have

$$\delta q = \overset{\infty}{\underset{0}{\Sigma}} \{e^{k\sigma} \, [E_j \cos(2j + 1)\sigma + \epsilon_j \sin(2j + 1)\sigma]$$

$$+ e^{-k\sigma} \, [B_j \cos(2j + 1)\sigma - \beta_j \sin(2j + 1)\sigma]\} \dots(52)$$

By means of (52) this may be written

$$\delta q = \sum_0^\infty \{[D + \sqrt{(D^2 - 1)}]^{\frac{2\sigma}{\pi}} [E_j \cos (2j + 1) \sigma + \epsilon_j \sin (2j + 1) \sigma]$$

$$+ [D - \sqrt{(D^2 - 1)}]^{\frac{2\sigma}{\pi}} [B_j \cos (2j + 1) \sigma - \beta_j \sin (2j + 1) \sigma]\}...(53)$$

In this case again the terms for which $j = 0$ are not usually the most important ones, but we see that the series contains sines and cosines of odd multiples of σ; and that one set of terms increases without limit and that the other diminishes.   As in the first sort of instability, a convenient modulus is the number of synodic revolutions in which the amplitude of the increasing oscillation doubles its initial value; that is to say we put

$$e^{k\sigma} = [D + \sqrt{(D^2 - 1)}]^{\frac{2\sigma}{\pi}} = 2$$

Therefore $$\frac{s}{S} = \frac{\sigma}{\pi} = \frac{\log \sqrt{2}}{\log [D + \sqrt{(D^2 - 1)}]} \qquad ....(54)$$

where $$D^2 = \Delta \sin^2 \tfrac{1}{2} \pi \sqrt{\Phi_0}$$

This is the modulus of instability, when it is of the uneven kind.   A consideration of the principal term has shown us that there is an oscillation, whose amplitude increases without limit.   The planet or satellite crosses and recrosses the periodic orbit an odd number of times in a single circuit, making ever increasing excursions on each side; it is on this account that I have called this "uneven instability."

It is interesting to consider the form which the equations of condition assume in the two sorts of instability.

In the case of even instability, we have $c = k \sqrt{(-1)}$, and the equations for the determination of the b's are given by

$$b_j (c + 2j)^2 = \Sigma_i b_{j-i} \Phi_i$$

$$= b_0 \Phi_j + \sum_1^\infty b_i \Phi_{j-i} + \sum_1^\infty b_{-i} \Phi_{j+i} ....(55)$$

We now have

$$2b_j = B_j + \beta_j \sqrt{(-1)}, \qquad 2b_{-j} = B_j - \beta_j \sqrt{(-1)}$$

$$2b_j (c + 2j)^2 = (4j^2 - k^2) B_j - 4jk\beta_j + \sqrt{(-1)} [4jkB_j + (4j^2 - k^2) \beta_j]$$

Then noting that β_0 is necessarily zero, and equating to zero the real and imaginary parts of the equation of condition (55), we have

$$\left.\begin{array}{l} (4j^2 - k^2) B_j - 4jk\beta_j = B_0 \Phi_j + \sum_1^\infty B_i (\Phi_{j-i} + \Phi_{j+i}) \\[4mm] 4jkB_j + (4j^2 - k^2) \beta_j = \qquad \sum_1^\infty B_i (\Phi_{j-i} - \Phi_{j+i}) \end{array}\right\} ....(56)$$

In the case of $j = 0$, the second equation is identically true, and the first becomes

$$- k^2 B_0 = B_0 \Phi_0 + 2 \sum_1^\infty B_i \Phi_i \quad \ldots\ldots\ldots\ldots\ldots(55)$$

It is easy to show that if we take j as negative, we are led to the same equations; thus it is only necessary to consider the case of j positive.

These equations suffice to determine all the B's and β's in terms of one of them, say B_0, which is an arbitrary constant of the solution.

We have already seen that the equations of condition for e_{-j} are exactly the same as those for b_j. Hence bearing in mind the definitions of E_j and ϵ_j, we see that the equations of condition for E_j, ϵ_j are the same as those for B_j, β_j. Then since $\epsilon_0 = 0$, E_j, ϵ_j are the same multiples of E_0 as B_j, β_j are of B_0. Thus E_0 is the second arbitrary constant of the solution.

Suppose that we put $B_0 = 1$, and solve the equations finding $B_j = \Lambda_j$, $\beta_j = \lambda_j$, then the general solution is

$$\delta q = \sum_0^\infty [E_0 e^{k\sigma} (\Lambda_j \cos 2j\sigma + \lambda_j \sin 2j\sigma) + B_0 e^{-k\sigma} (\Lambda_j \cos 2j\sigma - \lambda_j \sin 2j\sigma)]$$
$$\ldots\ldots(57)$$

Now turn to the case of uneven instability where $c = 1 + k\sqrt{(-1)}$; the equation of condition may be written

$$b_j (c + 2j)^2 = \sum_0^\infty b_i \Phi_{j-i} + \sum_0^\infty b_{-i-1} \Phi_{j+i+1} \ldots\ldots\ldots\ldots(58)$$

where $\qquad 2b_j = B_j + \beta_j \sqrt{(-1)}, \qquad 2b_{-j-1} = B_j - \beta_j \sqrt{(-1)}$

$$2b_j (c + 2j)^2 = [(2j+1)^2 - k^2] B_j - 2(2j+1) k\beta_j$$
$$+ \sqrt{(-1)} \{2(2j+1) kB_j + [(2j+1)^2 - k^2] \beta_j\}$$

Then equating to zero the real and imaginary parts of the equation of condition (58),

$$\left.\begin{aligned}
[(2j+1)^2 - k^2] B_j - 2(2j+1) k\beta_j &= \sum_0^\infty B_i (\Phi_{j-i} + \Phi_{j+i+1}) \\
2(2j+1) kB_j + [(2j+1)^2 - k^2] \beta_j &= \sum_0^\infty \beta_i (\Phi_{j-i} - \Phi_{j+i+1})
\end{aligned}\right\} \ldots\ldots\ldots(59)$$

It is easy to show that it is only necessary to consider the positive values of j.

These equations suffice to determine all the B's and β's in terms of B_0, which is one of the arbitrary constants of the solution.

From the definitions of E_j, ϵ_j it is easy to see that the equations of condition are the same as (59), and that E_j, ϵ_j are the same multiples of E_0 (the second arbitrary constant), that B_j, β_j are of B_0.

Suppose that (59) are solved with $B_0 = 1$, and that we find $B_j = \Lambda_j, \beta_j = \lambda_j$; then the general solution is

$$\delta q = \overset{\infty}{\underset{0}{\Sigma}}_j \{E_0 e^{k\sigma} [\Lambda_j \cos (2j+1)\,\sigma + \lambda_j \sin (2j+1)\,\sigma]$$
$$+ B_0 e^{-k\sigma} [\Lambda_j \cos (2j+1)\,\sigma - \lambda_j \sin (2j+1)\,\sigma]\}\ldots\ldots(60)$$

It follows therefore that when k has been found from the infinite determinant the solutions for the varied orbit are expressible by means of two arbitrary constants in both kinds of instability. Such solutions would of course only express the true motion for a short time.

I have actually applied this method to one of the unstable periodic orbits which was computed, but as the work leads to no useful conclusion I shall not give the details of it.

§ 13. *Numerical determination of stability.*

When a periodic orbit has been found by quadratures, it is not obvious by mere inspection whether it is stable or not, and we must consider the numerical processes requisite to obtain an answer to the question.

The points which are determined by quadratures in a periodic orbit do not divide the arc S into a number of equal parts. The distance along the arc from the first orthogonal crossing of the x-axis to the second orthogonal crossing is $\frac{1}{2}S$; this may be determined by interpolation, for we may find what value of s makes y vanish.

In general there are two orbits computed, which differ from exact periodicity in opposite directions by small amounts. The arc $\frac{1}{2}S$, measured from the first orthogonal crossing to the second, which is not exactly orthogonal, is determined in each of these cases. The subsequent proceedings are then carried out in duplicate, and the final step is an interpolation between the two results to obtain the result for the exactly periodic orbit. In many cases however the computed orbit differs from a truly periodic one by an amount which is so small, that it may be attributed to the errors inherent to the method of calculation. In such cases the duplicate computation is unnecessary, and since the operations on the approximately periodic orbits are exactly like those on the truly periodic ones, we may henceforth speak as if the true orbit had been found.

The next step is the computation of Φ corresponding to each computed point of the orbit. In order to take advantage of the work already carried out in the quadratures, I arrange the computation of Φ in the following form:

Computation of Φ.

$\phi - \theta$	$\phi - \psi$
Lr	$L\rho$
Lr^3	$L\rho^3$

$\phi - \theta$	$\phi - \psi$
$L\left(\dfrac{1}{r^2} - r\right)$ ⎱	$L\left(\dfrac{1}{\rho^2} - \rho\right)$ ⎱
$Lv \sin(\phi - \theta)$ ⎰	$L \sin(\phi - \psi)$ ⎰
CV^2	CV^2

La	Lb
a ⎱	$\dfrac{1}{R}$ ⎱
b ⎰	$\dfrac{n}{V}$ ⎰
$\dfrac{dV}{Vds}$	c
$\left(\dfrac{dV}{Vds}\right)^2$	c^2
	$\tfrac{10}{6}c^2$ ⎱
	$\tfrac{1}{2}\left(\dfrac{dV}{Vds}\right)^2$ ⎰
	A

$\phi - \theta$	$\phi - \psi$
$Lv \cos^2(\phi - \theta)$ ⎱	$L \cos^2(\phi - \psi)$ ⎱
Cr^3 ⎰	$C\rho^3$ ⎰
CV^2	CV^2

Ld	Le
d ⎱	A ⎱
e ⎰	$-B$ ⎰
B	$A - B$
$L\Psi$ ⎱	$\tfrac{1}{2}(A - B)$ ⎰
$L\dfrac{S^2}{\pi^2}$ ⎰	Ψ
$L\Phi$	Φ

As before L, C stand for logarithm and cologarithm, and the brackets indicate additions.

It would be tedious to find the Fourier's series for Φ from its computed values, and it is best to find interpolated values of Φ at exact sub-multiples of the arc S. I therefore interpolate Φ at points for which the arc is $\frac{1}{24}S$,

$\frac{2}{24}S \ldots \frac{12}{24}S$, 13 values in all. These interpolations are made by one of the formulæ (23).

The next step is the harmonic analysis of these 13 values of Φ, which is an even function of the arc.

The analysis may be conveniently arranged in a schedule of the following form.

Harmonic analysis of an even function of which 24 values
$$a_0, \; a_1 \ldots a_{11}, \; a_{12}, \; a_{11} \ldots a_1 \; \text{are given.}$$

i	ii	iii	M	iv	M	v	M	vi
		i – ii		M × iii		M × iii		M × iii
a_0	a_{12}	$a_0 - a_{12} \; (a)$	1	a	1	a	1	a
a_1	a_{11}	$a_1 - a_{11} \; (\beta)$	σ_5	$\sigma_5 \beta$	σ_1	$\sigma_1 \beta$	$-\sigma_1$	$-\sigma_1 \beta$
a_2	a_{10}	$a_2 - a_{10} \; (\gamma)$	σ_4	$\sigma_4 \gamma$	$-\sigma_4$	$-\sigma_4 \gamma$	$-\sigma_4$	$-\sigma_4 \gamma$
a_3	a_9	$a_3 - a_9 \; (\delta)$	σ_3	$\sigma_3 \delta$	$-\sigma_3$	$-\sigma_3 \delta$	σ_3	$\sigma_3 \delta$
a_4	a_8	$a_4 - a_8 \; (\epsilon)$	1	ϵ	1	ϵ	1	ϵ
a_5	a_7	$a_5 - a_7 \; (\zeta)$	σ_1	$\sigma_1 \zeta$	σ_5	$\sigma_5 \zeta$	$-\sigma_5$	$-\sigma_5 \zeta$
a_6	a_6	0	0	0	0	0	0	0

Sum 0 to 6	Sum 7 to 12		24	Sum	24	Sum	24	Sum
	Sum 0 to 6			Φ_1		Φ_5		Φ_7
	Sum 0 to 12							

$2 \times$ Sum 0 to 12

$$-(a_0 + a_{12})$$

24	Sum
	Φ_0

$$\Phi_3 = \tfrac{1}{24} \left[a - 2\epsilon + \sigma_3 (\beta - \delta - \zeta) \right]$$
(see iii)

$\sigma_1 = 2 \sin 15° = \cdot 5176$
$\sigma_3 = 2 \sin 45° = 1\cdot 4142$
$\sigma_4 = 2 \sin 60° = 1\cdot 7321$
$\sigma_5 = 2 \sin 75° = 1\cdot 9319$

vii	viii	ix	x
i + ii	Last 4 of vii reversed	vii – viii	vii + viii
$a_0 + a_{12}$	$a_6 + a_6$	$(a_0 + a_{12}) - (a_6 + a_6) \; (\eta)$	$(a_0 + a_{12}) + (a_6 + a_6) \; (\lambda)$
$a_1 + a_{11}$	$a_5 + a_7$	$(a_1 + a_{11}) - (a_5 + a_7) \; (\theta)$	$(a_1 + a_{11}) + (a_5 + a_7) \; (\mu)$
$a_2 + a_{10}$	$a_4 + a_8$	$(a_2 + a_{10}) - (a_4 + a_8) \; (\kappa)$	$(a_2 + a_{10}) + (a_4 + a_8) \; (\nu)$
$a_3 + a_9$	$a_3 + a_9$	0	$(a_3 + a_9) + (a_3 + a_9) \; (\rho)$
$a_4 + a_8$			
$a_5 + a_7$			
$a_6 + a_6$			

$$\Phi_2 = \tfrac{1}{24} \left[\eta + \kappa + \sigma_4 \theta \right], \qquad \Phi_4 = \tfrac{1}{24} \left[(\lambda + \mu) - (\nu + \rho) \right]$$
$$\Phi_6 = \tfrac{1}{24} \left[\eta - 2\kappa \right], \qquad \Phi_8 = \tfrac{1}{24} \left[(\lambda - \mu) - (\nu - \rho) \right]$$
$$\text{(see ix)} \qquad\qquad\qquad \text{(see x)}$$

If we write $\theta = 2\sigma = \dfrac{2\pi s}{S}$, the function Φ is equal to

$$\Phi_0 + 2\Phi_1 \cos \theta + 2\Phi_2 \cos 2\theta + \ldots + 2\Phi_8 \cos 8\theta$$

In order to test the accuracy of the work and the convergency of the series, it is well to compute the values of several of the a's directly from the harmonic expansion. For this purpose we have

$$\begin{cases} a_0 \\ a_{12} \end{cases} = \Phi_0 + 2\left(\Phi_2 + \Phi_4 + \Phi_6 + \Phi_8\right) \pm 2\left(\Phi_1 + \Phi_3 + \Phi_5 + \Phi_7\right)$$

$$\begin{cases} a_2 \\ a_{10} \end{cases} = \Phi_0 + \Phi_2 - \Phi_4 - 2\Phi_6 - \Phi_8 \pm \sigma_4\left(\Phi_1 - \Phi_5 - \Phi_7\right)$$

$$\begin{cases} a_3 \\ a_9 \end{cases} = \Phi_0 - 2\Phi_4 + 2\Phi_8 \pm \sigma_3 (\Phi_1 - \Phi_3 - \Phi_5 + \Phi_7)$$

$$\begin{cases} a_4 \\ a_8 \end{cases} = \Phi_0 - \Phi_2 - \Phi_4 + 2\Phi_6 - \Phi_8 \pm (\Phi_1 - 2\Phi_3 + \Phi_5 + \Phi_7)$$

$$a_6 = \Phi_0 + 2(\Phi_4 + \Phi_8) - 2(\Phi_2 + \Phi_6)$$

It may be remarked that if the harmonic expansion of Φ is convergent, the determinant from which the stability is determinable is also convergent.

But if the representation of Φ by the harmonic expansion up to the eighth harmonic is very imperfect, it is necessary to give up the attempt to determine the stability numerically. In such cases however it is nearly always possible to see that the orbit is unstable, although it may not some- times be so easy to perceive whether the instability is even or uneven.

We next have to calculate the several members of the determinant Δ by the formula

$$\frac{\Phi_i}{\Phi_0 - 4j^2}$$

This is the entry for the jth row above or below the centre of the determinant, and it is the ith member to the right and to the left of the leading diagonal, all the members on the diagonal being unity. The values of Φ_i computed by the preceding analysis suffice to enable us to write down 17 columns and rows of Δ. The method of computing Δ will be considered in the next section.

§ 14. *The calculation of a determinant of many columns and rows.*

The following transformation contains the principle by which the number of columns and rows of a determinant may be diminished by unity

$$\Delta = \begin{vmatrix} a_1, & a_2, & a_3, \ldots \\ b_1, & b_2, & b_3, \ldots \\ c_1, & c_2, & c_3, \ldots \\ \ldots \end{vmatrix} = a_1 \begin{vmatrix} 1, & \dfrac{a_2}{a_1}, & \dfrac{a_3}{a_1}, \ldots \\ 0, & b_2 - b_1 \dfrac{a_2}{a_1}, & b_3 - b_1 \dfrac{a_3}{a_1}, \ldots \\ 0, & c_2 - c_1 \dfrac{a_2}{a_1}, & c_3 - c_1 \dfrac{a_3}{a_1}, \ldots \\ \ldots \end{vmatrix}$$

$$= a_1 \begin{vmatrix} b_2 - b_1 \dfrac{a_2}{a_1}, & b_3 - b_1 \dfrac{a_3}{a_1}, \ldots \\ c_2 - c_1 \dfrac{a_2}{a_1}, & c_3 - c_1 \dfrac{a_3}{a_1}, \ldots \\ \ldots \end{vmatrix}$$

Now if we write $b_2' = b_2 - b_1 \dfrac{a_2}{a_1}$, and so on, and then extract the factor b_2', another column and row may be removed, and the process may be repeated until the determinant is reduced to a single member, say z_n; then

$$\Delta = a_1 b_2' c_3'' \ldots z_n$$

If the determinant is convergent and if the rows and columns be removed in proper succession, the factors tend to unity.

By interchanges of columns and rows any member of a determinant may be brought to stand at a corner, but if the number of interchanges is odd the sign of the determinant is changed.

It is not therefore necessary to work from a corner, as in the above example, but any column and any row may be chosen for elimination. The member which stands at the intersection of the chosen column and row may be called the centre of elimination. Then if the centre of elimination be at an odd or even number of moves from a corner, the sign of the whole is or is not changed.

In the determinants which arise in this investigation the centre of elimination is always taken on the diagonal, and thus no change of sign is introduced.

Let us suppose that the determinant to be evaluated is a symmetrical one, and that the columns and rows are numbered, as in the following example:

$$\begin{array}{c|ccccc} & -2 & -1 & 0 & 1 & 2 \\ \hline -2 & C, & c_1, & c_2, & c_3, & c_4 \\ -1 & b_1, & B, & b_1, & b_2, & b_3 \\ 0 & a_2, & a_1, & A, & a_1, & a_2 \\ 1 & b_3, & b_2, & b_1, & B, & b_1 \\ 2 & c_4, & c_3, & c_2, & c_1, & C \end{array}$$

Let $(-1, -1)$ be the first centre of elimination, and $(1, 1)$ the second; then if the double elimination be carried out and algebraic reductions effected, it will be found that the result is

$$B^2 \left(1 - \frac{b_2^2}{B^2}\right) \begin{array}{|ccc|c} & -2 & 0 & 2 \\ B', & b_1', & b_2' & -2 \\ a_1', & A', & a_1' & 0 \\ b_2', & b_1', & B' & 2 \end{array}$$

Where

$$B' = C - \frac{b_3 c_1 + b_1 c_3}{B + b_2} - \frac{(b_1 - b_3)(c_1 - c_3)}{B - b_2^2/B}, \qquad b_1' = c_2 - \frac{b_1(c_1 + c_3)}{B + b_2}$$

$$b_2' = c_4 - \frac{b_1 c_1 + b_3 c_3}{B + b_2} + \frac{(b_1 - b_3)(c_1 - c_3)}{B - b_2^2/B}, \qquad a_1' = a_2 - \frac{a_1(b_1 + b_3)}{B + b_2}$$

$$A' = A - \frac{2a_1 b_1}{B + b_2}$$

If the determinant is convergent, with an odd number of columns and rows, $(0, 0)$ is the heart of the determinant; if the elimination proceeds away from the heart, at any stage of the process the approximation consists of the product of all the factors extracted, multiplied by $(0, 0)$, the heart of the remaining determinant.

Thus in the above example after one double elimination the approximation is

$$B^2 \left(1 - \frac{b_2^2}{B^2}\right)\left(A - \frac{2a_1 b_1}{B + b_2}\right)$$

This is in fact the full expression for the determinant

$$\begin{vmatrix} B, & b_1, & b_2 \\ a_1, & A, & a_1 \\ b_2, & b_1, & B \end{vmatrix}$$

I have found it most convenient in practice first to extract a squared factor, such as B^2 (thus reducing $(-1, -1)$ and $(1, 1)$ to unity), and afterwards to extract a single factor, such as $1 - (b_2^2/B^2)$.

This process cannot of course be applied with advantage, when the work is algebraical, but some process of the kind seems to be practically necessary, when the approximate numerical value is to be found of a determinant of a large number of columns and rows.

It will be noticed that after each pair of eliminations the primitive symmetry is restored; but the work might equally well be arranged otherwise. For we might first eliminate from the centre $(0, 0)$, which would not affect the symmetry, and we might then take the pair $(-1, -1)$ and $(1, 1)$. This variation of procedure would afford a valuable check on the arithmetic.

Where the outer fringe of the determinant obviously has but little influence on the final result, and where we are in any case going to use all the members in the original determinant, I have found it best to begin from the outside. In such a case four or five columns and rows may, as it were, be shelled off the outside, with scarcely any alteration of the central entries.

The actual numerical work of evaluating a determinant may be arranged as follows:

The number of decimal places to be retained is first fixed on. A paper is then marked with a gridiron of columns and rows, numbered from zero at the centre upwards and downwards. Each square should be large enough to contain four or five rows of figures. The original determinant is then written in the squares, the numbers being put as near the top of each square as possible. I have found it convenient to omit decimal points, and to express the numbers in units of the last decimal place retained. In most of my work, where only a rough result was required, I have adopted three places of decimals; thus the unit in which the entries are expressed is ·001, and the diagonal members are all written as 1000.

The pair of symmetrical diagonal members, which is to form the first pair of centres, is then chosen. As stated above, I have in my later work usually worked from the outside. In the first pair of eliminations these diagonals are already unity, but this is not so subsequently, and we first reduce them to unity by dividing the rows on which they stand by their values, and by extracting a squared factor.

It will be found convenient to run a red line through the column and row to be removed. If the red lines be regarded as coordinate axes, the row being x and the column y, any member of the determinant may be specified by its x and y. If the member of the determinant whose coordinates are x, y be a; and if the member whose coordinates are $x, 0$ be b; and if the member whose coordinates are $0, y$ be c; then the number which has to be substituted for a is $a - bc$.

In other words each number on the horizontal red line has to be multiplied by each number on the vertical red line, and the products have to be subtracted from the numbers which stand at the remote corners of the rectangles.

In effecting this process I form a separate table of the subtrahends, and write down the differences immediately under the numbers which they displace.

After the first elimination, which has rendered the determinant unsymmetrical, a single factor corresponding to the other chosen diagonal member is extracted, its row is correspondingly altered, red lines are drawn to mark the column and row to be removed, and the similar process is repeated. The symmetry of the determinant should now be restored, but any pair of numbers which should agree are arrived at by different numerical processes.

The restoration of symmetry affords a very valuable check on arithmetical processes which I have found singularly difficult to work correctly.

As only a limited number of decimal places are employed there is often a discrepancy of unity in the last significant figure between two numbers which ought to agree. It is sometimes possible to determine by inspection which of the two numbers is arrived at by the less risky series of operations, and I then adopt that number to represent both entries. But where there is no obvious reason for choosing one result more than the other, I choose one or other at hazard, and restore the perfect symmetry.

The process of elimination is continued until the determinant is reduced to (0, 0), but in the last two or three stages it is well to increase the number of decimals retained.

If at any stage the factor to be extracted becomes small, the whole row to which it belongs becomes large, and the symmetry may perhaps be seriously affected. In this case it is well not to choose this pair of centres of elimination, but to take another pair, leaving this pair to a later stage in the calculation.

If the determinant is negative, a negative factor will be extracted at some stage. In all the cases which have been worked out it is easy to see that no other negative factor will ever arise, and thus the determinant will remain clearly negative. Most of the determinants have been written with 17 columns and rows; then beginning with $(-8, -8)$ and $(8, 8)$ I find that it is often possible to erase 8 columns and 8 rows on a single sheet of paper, with scarcely any modification of the central part of the determinant. Thus the determinant which at first had 289 spaces (although many only contain zeros) is reduced to 81 spaces, with but little labour.

The multiplications have been done with Crelle's table, but a specially computed auxiliary table of products, from $\cdot000 \times \cdot000$ up to $\cdot040 \times \cdot040$ to three places of decimals, has rendered the work much more rapid.

I believe that the values obtained by this process are correct to within about one per cent. For the same determinant when reduced with different order of elimination agrees with its previous determination within less than that amount of discrepancy.

Part II.

§ 15. *Periodic Orbits.*

An orbit in which the third body can continually revolve, so as always to present the same character relatively to the two other bodies, is said to be periodic. If the motion is referred to a plane which is carried round with Jove and revolves about the Sun as a centre, any re-entrant orbit of the third body is periodic. Periodic orbits may consist of any number of revolutions round either of the primaries, or round other points in space. Periodic orbits, which are only re-entrant after several circuits, are much more difficult to discover than those which only make a single one; as hardly anything is known up to the present time about this subject, I determined to confine my attention to " simple periodic orbits," which are re-entrant after a single circuit. This definition of a simply periodic orbit must not preclude the consideration of orbits with loops, for the inclusion of such loops is necessary to the comprehension of the subject.

It appears from the differential equations of motion that periodic orbits must in general be symmetrical with respect to the line of syzygy; or if any periodic orbit consists of a closed circuit round a point which does not lie on this line, there must be a similar closed circuit round a symmetrical point on the other side of it.

Periodic orbits are critical cases which separate the orbits of one class from those of another, and the chief difficulty in tracing them consists in the fact that it is necessary to trace the gradual change of an orbit, as its parameters change, and to discover its form at the instant of its transformation into an orbit of a different character.

The partition of space derived from the Jacobian integral (§ 3) shows that the constant of relative energy C is of primary importance in the classification of orbits. The work of this investigation being numerical, I was compelled to assume a definite ratio for the mass of the Sun in terms of that of Jove; this ratio is taken as 10. The mass of the actual Sun in terms of that of the actual planet Jupiter is about 1000, and accordingly all the phenomena of perturbation are greatly exaggerated in our figures as compared with the real solar system. This exaggeration appeared to me advantageous for the purpose of giving a clear view of the phenomena.

The mass of the Sun being 10, that of Jove being unity and the distance between them being unity, we found in (9) that when C is greater than 40·1821 the third body must be either a superior planet, or an inferior planet, or a satellite, but cannot change from one of these conditions to another.

These larger values of C then bring us to those cases which are treated in the Planetary and Lunar Theories; I therefore cease my consideration of the problem for all values of C which are greater than 40·5. On the other hand, so long as a portion of space is forbidden to the third body, C must be greater than 33. For such cases the whole field to be treated is covered by the values of C between 33 and 40·5, and the problem is to obtain a complete synopsis of simply periodic orbits and of their stabilities between these limits. It is of course also desirable to trace periodic orbits for values of C less than 33.

But even with this limitation the field of investigation is still so large that in the present paper I am compelled to make further restrictions. In the first place, the case of superior planets has not been touched at all; although, at the point at which I have now arrived, they must soon be taken into account.

Secondly all the orbits considered are direct; the retrograde orbits would afford an interesting field of research *.

Lastly the present paper only covers the field from C equal to 38 to 40·5; and even this has occupied me for three years.

* [A beginning of the investigation of superior planets and of retrograde orbits will be found in §§ 3, 4 of Paper 3.]

The slowness with which results are attained by arithmetical processes has been very tantalising, but the interest of the work has been sustained by the fact that the results have presented a succession of surprises. I have over and over again been deceived when I imagined I could foresee the shape which would be assumed by the next orbit to be treated, and thus the subject was continually presenting itself under a new light. Nevertheless a point has, I think, been now reached at which some forecasts are possible, and I shall venture to say something hereafter in § 19 on this head, with the full knowledge however that the conjectures may prove erroneous.

Being ignorant of the nature of the orbits of which I was in search, I determined to begin by a thorough examination of one case. It seemed likely that the most instructive results would be obtained from cases in which it should be possible for an inferior planet and satellite to interchange their parts. Now when C is greater than $38\cdot8760$ but less than $40\cdot1821$, the two interior ovals of the curve of zero velocity coalesce into the shape of an hour-glass, and thus interchange of parts is possible. I therefore began by the consideration of the case where C is 39, and traced a large number of orbits which start at right angles to SJ, and in some cases I also traced the orbit with reference to axes fixed in space.

The two curves, which represent the orbit in space and with reference to the moving plane, contain a complete solution of the problem.

For if the curve on the moving plane be drawn as a transparency, and if the Sun in the two figures be made to coincide, and if the transparent figure be made to revolve uniformly about the sun, the intersection of the two curves will give the position of the body both in time and place.

In order to exhibit this I show in fig. 2 a certain orbit with reference to axes fixed in space and also the same orbit referred to rotating axes. In the former figure the simultaneous positions of the planet and of Jove are joined by dotted lines. It is interesting to observe how the body hangs in the balance between the two centres, before the elliptic form of the orbit asserts itself, as the body approaches the Sun.

This figure, and others of the same sort, are instructive as illustrating the usual sequence of events in orbits of this class.

If a planet be started to move about the Sun in an orbit of a certain degree of eccentricity, it will at first move with more or less exactness in an ellipse with advancing perihelion. But as the aphelion approaches conjunction with Jove the perturbations will augment at each passage of the aphelion. At length the perturbation becomes so extreme that the elliptic form of the orbit is entirely lost for a time, and the body will either revert to the Sun, or it will be drawn off and begin a circuit round Jove. In either case after the approximate concurrence of aphelion with conjunction, the orbit will have lost all resemblance to its previous form.

The figure 2 exhibits the special case in which the body only makes a single circuit round Jove, and where the heliocentric elliptic orbit before and

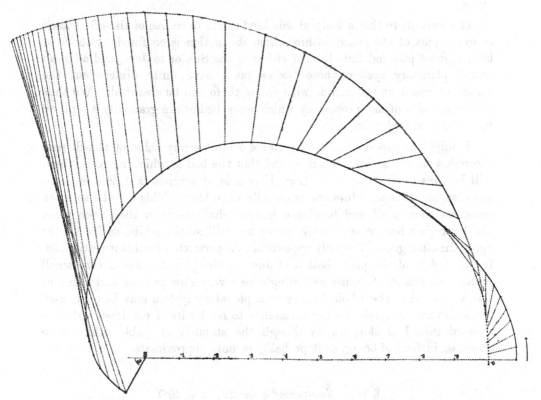

Orbit $x_0 = 1 \cdot 080$, $C = 39 \cdot 0$ referred to axes fixed in space, together with the simultaneous positions of Jove.

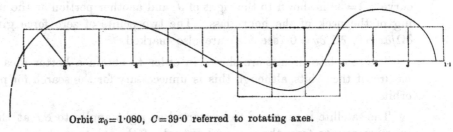

Orbit $x_0 = 1 \cdot 080$, $C = 39 \cdot 0$ referred to rotating axes.

FIG. 2.

after the crisis has the same form; the perihelion has however advanced through twice the angle marked ω on the figure. In general the body would, after parting from the Sun, move several times round Jove until a concurrence of apojove with conjunction produced a severance of the connection, but in the figure this concurrence happens after the first circuit. If the neck of the hour-glass defining the curve of zero velocity be narrow, the body may move hundreds of times round one of the centres before its removal to the other.

It seems likely that a body of this kind would in course of time find itself in every part of the space within which its motion is confined. Sooner or later it must pass indefinitely near either to the Sun or to Jove, and as in an actual planetary system those bodies must have finite dimensions, the wanderer would at last collide with one of them and be absorbed. We thus gain some idea of the process by which stray bodies are gradually swept up by the Sun and planets.

It might be supposed that all possible orbits for any value of C will pass through a similar series of changes and that the bodies which move in them will be thus finally absorbed. Lord Kelvin is of opinion that this must be the case, and that all orbits are essentially unstable[*]. This may be so when sufficient time is allowed to elapse, but we shall see later that, even when the hour-glass has an open neck, there are still stable orbits, as far as our approximation goes. The only approximation permitted in this investigation is the neglect of the perturbation of Jove by the planet. For a very small planet the instability must accordingly be a very slow process, and I cannot but believe that the whole history of a planetary system may be comprised in the interval required for the instability to render itself manifest. Henceforward then I shall speak as though the stability of stable orbits were absolute, instead of being, as it probably is, only approximate.

§ 16. *Non-periodic orbits; $C = 39.0$.*

(a) *Orbits round Jove.* Fig. 3.

The Sun S is outside the figure towards the left. A small portion of the curve $2\Omega = 39$ is shown to the right of J, and another portion at the narrowing of the neck of the hour-glass. The two points of zero force given by $\partial\Omega/\partial x = 0$, $\partial\Omega/\partial y = 0$ (see § 3) are also marked.

The complete circuits are shown in order to obtain a better idea of the nature of the orbits, although this is unnecessary for the search for periodic orbits.

The satellite is supposed to be started at right angles to SJ at the conjunction remote from the sun, and enough of the orbits is shown to obtain a synopsis of the class. Here and elsewhere I define the orbits by the initial value of x, which is denoted by x_0; in this case the final value of x after the complete circuit may be called x_1.

* Sir William Thomson, "On the Instability of Periodic Motion," *Philosophical Magazine*, Vol. xxxii., 1891, p. 555. M. Poincaré also considers that orbits may have a temporary, but not a secular stability. *Acta Mathem.*, T. xiii., 1890, p. 101.

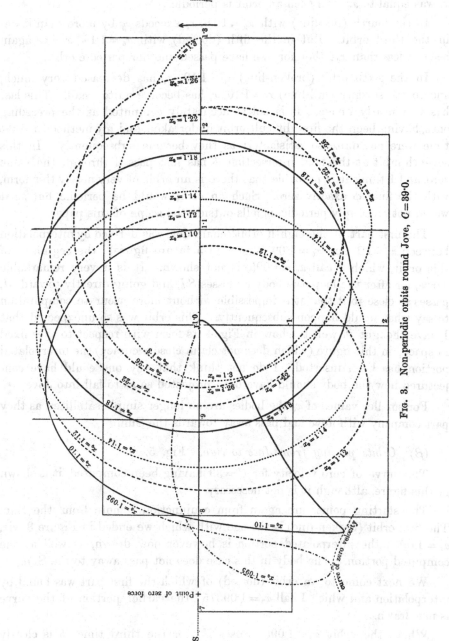

Fig. 3. Non-periodic orbits round Jove, $C = 39.0$.

The first on the right (dotted-line) starts with $x_0 = 1\cdot3$, and x_1 is much less than x_0. The second (chain-dotted) has $x_0 = 1\cdot26$, and x_1 has considerably increased so as to approach x_0. The third (broken-line) has $x_0 = 1\cdot22$, and x_1 has now become greater than x_0; therefore we have passed an orbit for which x_1 was equal to x_0, and such an orbit is periodic.

In the fourth (full-line) with $x_0 = 1\cdot18$, x_1 exceeds x_0 by more than it did in the third orbit. But in the fifth (dotted) with $x_0 = 1\cdot14$, x_1 has again become less than x_0; therefore we have passed another periodic orbit.

In the sixth orbit (broken-line) $x_0 = 1\cdot12$, x_1 has decreased very much, and in the seventh (full-line) $x_0 = 1\cdot10$, x_1 has become quite small. This last has very nearly a cusp. It is not so accurately computed as the preceding ones, having been the first difficult orbit undertaken, and my methods at that time were not quite so satisfactory as they became subsequently. In this seventh orbit at the final intersection ϕ has just passed through the value zero, and I think it is probable that there is an orbit of very nearly this form, with the final ϕ exactly zero. Such an orbit would be periodic, but as it would not be simply periodic, it falls outside the scope of this paper.

The first part of the eighth orbit (chain-dot) was derived by interpolation between $x_0 = 1\cdot1$ and $x_0 = 1\cdot09$ (shown in a future figure); the beginning of this orbit, which I call $x_0 = 1\cdot095$, is not shown. It is a very remarkable curve, for after the loop, the body recrosses SJ, and going directly towards J, passes so close to it that it is impossible without more accurate computation to say what would happen subsequently. This orbit was so unexpected that I have thought it well to show in Fig. 4 its form with respect to axes fixed in space; in this figure (which does not claim close accuracy) the interpolated portion has been inserted. I do not think that any one could have conjectured how the body should have been projected so as to fall into Jove.

For smaller values of x_0 the bodies are no longer simple satellites, as they part company with Jove and pass away towards the Sun.

(β) *Orbits passing from Jove to Sun.* Fig. 5.

The curve of zero velocity for $C = 39$ having been computed, it is shown in this figure, although it is not necessary.

The starting points are again from conjunction remote from the Sun. The first orbit (broken-line) is the one with which we ended in Figure 3, viz. $x_0 = 1\cdot095$; the interpolated portion is however now drawn, as well as the computed portion. The body in this case does not pass away to the Sun.

We next come to an orbit (dotted) of which the first part was found by interpolation and which I call $x_0 = 1\cdot09375$; the earlier portion of the curve is not drawn.

Where the orbit $x_0 = 1\cdot095$ crosses SJ for the third time, ϕ is clearly negative, but where the orbit $x_0 = 1\cdot09375$ crosses for the third time ϕ is

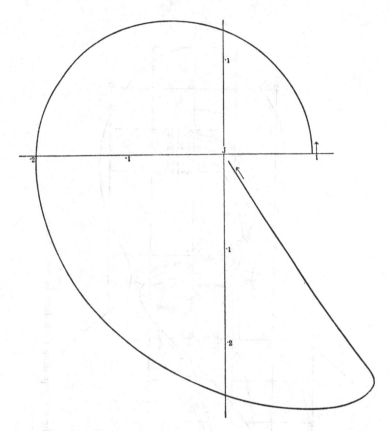

FIG. 4. Orbit round Jove referred to axes fixed in space ($x_0 = 1 \cdot 095$, $C = 39 \cdot 0$).

positive. There must therefore be an intermediate case for which ϕ vanishes, and this will give us a third periodic orbit round J. The orbit $x_0 = 1 \cdot 09375$ passes away to the sun; and we then come to four more orbits $x_0 = 1 \cdot 09$, $1 \cdot 08$, $1 \cdot 06$, $1 \cdot 04$ which follow a similar course, but with diminishing depression towards the negative side of SJ. The next orbit is $x_0 = 1 \cdot 02$, in which the depression has disappeared. This curve has a slight hump in the place of the depression; it is the sort of feature which would present itself in a computed curve, when there has been an arithmetical error in the calculation, but we shall soon see that this hump is not explicable in this way.

The next curve which is traced (although others have been computed) starts with $x_0 = 1 \cdot 001$ (chain-dot); in a figure of this scale, it apparently starts actually from J. It will be observed that we now have a remarkable cusp, and it becomes obvious that the hump referred to above was an incipient elevation towards the cusp*.

* [These curves are more interesting when studied in connection with Figs. 8, 9 of Paper 3.]

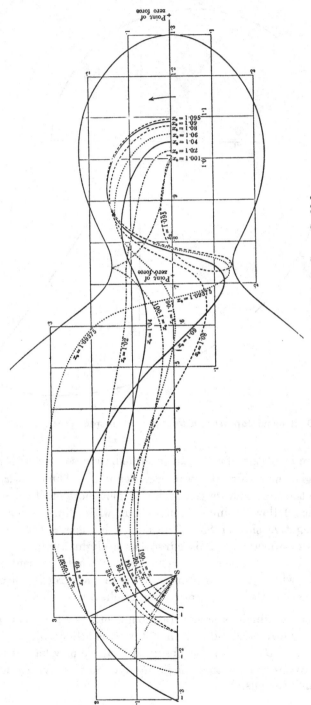

Fig. 5. Non-periodic orbits passing from Jove to Sun; $C = 39 \cdot 0$.

Passing now to the other end of the figure where the body passes round the Sun, we see from the incidence of the perihelia (which are indicated by radii from the Sun) that there can be no periodic orbit which is partly the path of a satellite and partly that of a planet; for such an orbit must have the longitude of the perihelion 180°.

The positions of the perihelia and the perihelion-distances seem to be almost chaotic in the figure, but I believe that the calculations are substantially correct, and a consideration of the numbers representing the positions of the perihelia shows that the chaos is rather apparent than real.

The following table gives the results:

Name of orbit	Longitude of Perihelion	Perihelion Distance
$x_0 = 1\cdot001$	$\pi - 32°\ 45'$	·058
$= 1\cdot02$	$\pi - 34°$	·125
$= 1\cdot04$	$\pi - 35°\ 45'$	·093
$= 1\cdot06$	$\pi - 39°\ 15'$	·078
$= 1\cdot08$	$\pi - 52°\ 15'$	·115
$= 1\cdot09$	$\pi - 64°\ 15'$	·240
$= 1\cdot09375$	$\pi - 30°\ 45'$	·222

Now if we were to plot out the defects of the longitudes from 180°, taking x_0 as abscissa, we should obtain a sweeping curve starting from a minimum of 33°, rising to a maximum of 64°, and falling abruptly to 31°. If the perihelion-distances be treated similarly, we find a somewhat less satisfactory curve, for there is a small maximum, then a minimum and then a large maximum, followed by a fall in value. As I have said above, I believe that these results are substantially correct; but as each one of these curves represents three or four weeks' hard work, I have not thought it good economy of labour to pursue the inquiry further in this respect.

(γ) *Orbits round the Sun*; $C = 39\cdot0$. Fig. 6.

These curves are drawn with less accuracy than the others, being computed with three-figured logarithms. I thought that sufficient accuracy would be attainable with this degree of approximation, but when I found that the saving of labour was not considerable, whilst the loss of accuracy was very great, I returned to the use of four-figured tables. It did not however seem necessary to recompute these curves.

The complete circuit is drawn for four of the curves, but the rest are only carried half-way round.

The orbits start to the left of the Sun at the conjunction remote from Jove. The first orbit is $x_0 = -\cdot6$ (full-line), and at the second crossing of the

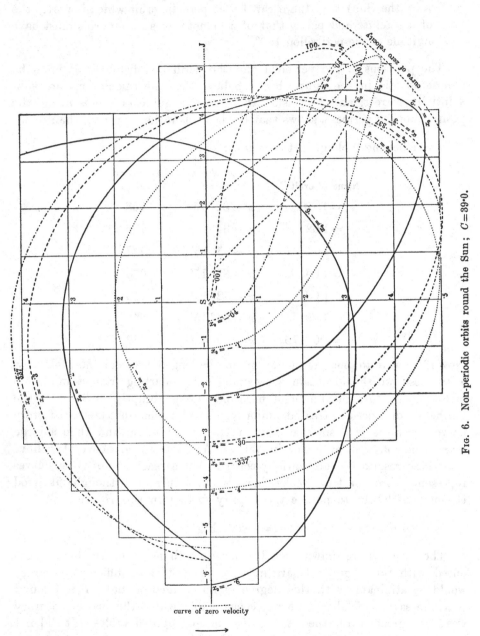

FIG. 6. Non-periodic orbits round the Sun; $C = 39\cdot0$.

line of conjunction the angle ϕ is negative. The second orbit $x_0 = -\cdot4$ (dotted) has ϕ positive, but small, at the second crossing; hence there is a periodic orbit for a value of x_0 a little less than $-\cdot4$.

All the succeeding orbits, viz. $x_0 = -\cdot337, -\cdot3, -\cdot2, -\cdot1, -\cdot04, -\cdot001$, have ϕ positive and successively increasing at the second crossing; and thus there is no other periodic orbit. The last two of these orbits have loops.

The orbit $x_0 = -\cdot337$ was found in part by interpolation. It has been inserted because the third crossing of the line SJ appears to be orthogonal, and therefore the orbit is periodic, but not *simply* periodic. No search was being made for this sort of orbit, and the discovery was accidental.

§ 17. *Periodic Orbits classified according to values of C.*

Plates I, II, III.

Plate I, fig. 1. $C = 40\cdot0$.

When C is greater than $40\cdot18$, the inner branches of the curve of zero velocity, $2\Omega = C$, consist of two ovals, as seen in fig. 1; the periodic orbits then consist of two approximately circular orbits round S and J respectively. These cases may be treated by the methods of the Planetary and Lunar Theories, and fall outside the scope of this paper.

When $C = 40\cdot18$ there is a third periodic orbit consisting of the point $x = \cdot7175$, $y = 0$. At this point a body is in unstable equilibrium, and this point is the beginning of a family of orbits; for, whilst in general periodic orbits begin in pairs, a single orbit may begin at a point.

In discussing these figures I shall denote the initial value of any function by the suffix 0; the suffix 1 will denote the value after the completion of a half circuit, and the suffix 2 the value on the completion of the whole circuit.

The planet A starts from $x_0 = -\cdot414$, $\phi_0 = \pi$, and ϕ increases.

When $x_0 \lessgtr -\cdot414$, $\phi_1 \lessgtr 0$, $x_2 \gtrless x_0$ ($x_0 < -\cdot414$ of course denotes a starting point more remote from S, with x_0 numerically greater than $\cdot414$).

This orbit is stable with $c = 2\cdot81$.

The satellite A starts from $x_0 = 1\cdot03341$, $\phi_0 = 0$, and ϕ increases.

This orbit changes its shape rapidly with changes of C, as will appear below in the classification by families. Great care was bestowed on this case, and it was very troublesome to compute, since a considerable variation of x_0 corresponded with a small variation of ϕ_1.

When $x_0 \gtrless 1\cdot03341$, $\phi_1 - \pi \lessgtr 0$, $x_2 \lessgtr x_0$.

The orbit is stable, but borders closely on instability, with $c = 3\cdot7$.

The third orbit is the oscillating satellite (a), moving slowly with a retrograde revolution round the point of zero force $x = \cdot7175$, $y = 0$, which was described above as the commencement of a series of orbits.

The orbit (a) starts from $x_0 = \cdot705$, $\phi_0 = 0$, and ϕ diminishes.

When $x_0 \gtrless \cdot705$, $\phi_1 - \pi \gtrless 0$. That is to say if the body starts too near to Jove the change of direction at the sharp turn is not quite sufficient for periodicity; and if it starts too near the Sun the converse is true. In the first case after one or more circuits the body passes away towards J, and in the second case towards S.

This orbit is very unstable, and the instability is almost certainly of the even type.

Plate I, fig. 2. $C = 39\cdot5$ and $39\cdot3$.

The planetary orbit A ($C = 39\cdot5$) differs little from the preceding case.

It starts from $x_0 = -\cdot424$, $\phi_0 = \pi$, and ϕ increases.

When $x_0 \lessgtr -\cdot424$, $\phi_1 \lessgtr 0$, $x_2 \gtrless x_0$.

The orbit is stable with $c = 2\cdot90$; but it is less stable than when $C = 40$.

The classification by families below shows that as C falls below $40\cdot0$, the orbit of the satellite A stretches out rapidly towards S, and at the same time the oval (a) expands. [It seemed to me when this paper was published that when C had fallen to a value a very little greater than $39\cdot5$, these two curves A and (a) would have expanded until they touched one another. If this were possible the body might at this stage of contact either move entirely on A or entirely on (a), or alternately on A or on (a), so as to describe a figure-of-8. But the path of a satellite is absolutely determinate, so that such an alternative is not permissible. Hence the scheme of classification for the family A of orbits, as originally suggested, must be erroneous. This oversight was pointed out by Mr Hough in a paper which is included as No. 2 in this volume. The true development of the family A of satellites is shown in detail in Paper 3 below, in accordance with Mr Hough's suggestion, and is also represented in Fig. 1, Plate IV of this paper. But having been led astray by the fact that when C had diminished to $39\cdot5$ there existed a periodic orbit of the form of the figure 8, I regarded it as the continuation of the family A. It is now clear that this kind of orbit is a new one and therefore it demands a new distinctive letter. In order to maintain as much continuity as possible with this paper in its original form I now describe it as belonging to a family A'. But since periodic orbits must arise in pairs, the existence of the family A' involves that of another family of figure-of-8 orbits to which the letter A" is now attached. Thus the complete classification of the orbits of the families A, A', A" may be gathered from the combination of the figures in this paper with Figs. 8 and 9, pp. 172—3, of Paper 3.]

The satellite [now called] A' starts from $x_0 = 1\cdot0650$, $\phi_0 = 0$, and ϕ begins increasing. When the body has passed half round J so that y vanishes, ϕ is equal to $\pi - 15° 37'$; shortly after this ϕ diminishes and continues doing so until when y again vanishes $\phi_1 = 0$.

We have $x_0 \gtrless 1\cdot0650$, $\phi_1 \lessgtr 0$. When the body starts too far from J, it will move in some orbit round J, and when it starts too near J it will pass away to S.

This orbit is very unstable with even instability.

The oscillating orbit a was not computed for $C = 39\cdot5$ *; during one part of its course it would be indistinguishable from part of A′, and the rest is shown conjecturally by a dotted line.

This orbit is very unstable, with even instability.

It has already been remarked that, after the first half circuit of satellite A′, ϕ was $\pi - 15°\ 37'$, or as we may now write it $\pi - \phi_1 = 15°\ 37'$. Now when x_0 is made to increase from $1\cdot0650$ until it reaches the curve $2\Omega = C$, $\phi_1 - \pi$ will always be negative, or $\pi - \phi_1$ positive. It appears however that $\pi - \phi_1$ has a minimum value, which very nearly reaches zero. In fact when $x_0 = 1\cdot140$, $\phi_1 = \pi - 0°\ 20'$.

Since $\pi - \phi_1$ is large when x_0 approaches $2\Omega = C$, and is $15°\ 37'$ when $x_0 = 1\cdot0650$, it follows that if it vanishes at all, it must vanish twice. That is to say there must be two new periodic orbits, if any.

As C diminishes the minimum value of $\pi - \phi_1$ falls, and I found that when $C = 39\cdot4$ the minimum was reached when x_0 is about $1\cdot15$; for this value of x_0, $\pi - \phi_1$ is $0°\ 9'$, and there is still no value of x_0 for which $\pi - \phi_1$ vanishes.

But when $C = 39\cdot3$ I computed the four orbits $x_0 = 1\cdot18$, $1\cdot17$, $1\cdot16$, $1\cdot15$ and found that for the two middle ones $\pi - \phi_1$ was negative. By interpolation the pair of periodic orbits B and C were found.

The orbit B is given by

$$x_0 = 1\cdot1575, \qquad \phi_0 = 0$$

and the orbit C by $\qquad x_0 = 1\cdot1751, \qquad \phi_0 = 0$

In both cases ϕ increases.

The relationship to the neighbouring orbits is given by the inequalities

$$x_0 > 1\cdot1751, \qquad \phi_1 - \pi < 0, \qquad x_2 < x_0$$

$$x_0 \begin{matrix} < 1\cdot1751 \\ > 1\cdot1575 \end{matrix}, \qquad \phi_1 - \pi > 0, \qquad x_2 > x_0$$

$$x_0 < 1\cdot1575, \qquad \phi_1 - \pi < 0, \qquad x_2 < x_0$$

The orbit B is slightly unstable, with even instability, and $c = \cdot156\ \sqrt{(-1)}$; the orbit C is stable, but approaches instability, and $c = 2\cdot163$.

Plate II, fig. 1. $C = 39\cdot0$.

These are the periodic orbits which belong to the families of non-periodic orbits shown in figs. 3, 4, 6 above.

* At least the computation was not completed, for it was found to be so troublesome, that it appeared that the work could be better bestowed elsewhere.

The planetary orbit A starts from $x_0 = -\cdot434$, $\phi_0 = \pi$, and ϕ increases. The incidence amongst the neighbouring orbits is shown by the inequalities

$$x_0 \lessgtr -\cdot434, \quad \phi_1 \lessgtr 0, \quad x_2 \gtrless x_0$$

This orbit is unstable with slight uneven instability and $c = 1 + \cdot10 \sqrt{(-1)}$. It thus appears that for some value of C between 39·5 and 39·0 we should find the passage of the planetary orbit A from stability to instability. It is certainly surprising to find that the instability of the planet sets in when the planet is a little less than half-way to Jove at conjunction.

[As already explained the satellite A was erroneously omitted. It is however very nearly represented by the orbit described as $C = 38\cdot85$ in fig. 1, Plate IV, but I have not thought it necessary to insert it in this figure.]

The satellite A′ starts from $x_0 = 1\cdot0941$, $\phi_0 = 0$, and ϕ increases until when y vanishes it is equal to about $\pi - 13° 30'$; it then diminishes to zero.

Its incidence among neighbouring orbits (figs. 3 and 4) is given by the inequalities

$$x_0 \gtrless 1\cdot0941, \quad \phi_1 \lessgtr 0$$

When it starts too far from Jove it will move in some orbit round J, and when it starts too near Jove it will pass away towards S.

This orbit is very unstable, with even instability and $c = \cdot46 \sqrt{(-1)}$. The orbit of the oscillating satellite (a) is indistinguishable from A′ throughout part of its course, but falls more remote from J on the side towards S. It starts from $x_0 = \cdot687$, $\phi_0 = 0$, and ϕ diminishes.

When $x_0 \gtrless \cdot687$, $\phi_1 - \pi \gtrless 0$; thus if the body starts too near Jove the total change of direction is insufficient for periodicity; and if it starts too near the Sun the converse is true. In the first case it passes away towards Jove, and in the second towards the Sun.

This orbit is very unstable with even instability, and c is about $2\sqrt{(-1)}$.

The satellite B starts with $x_0 = 1\cdot1500$, $\phi_0 = 0$, and ϕ increases.

When $x_0 \gtrless 1\cdot1500$, $\phi_1 - \pi \gtrless 0$, $x_2 \gtrless x_0$.

This orbit is unstable, with even instability, and $c = \cdot38 \sqrt{(-1)}$.

The satellite C starts with $x_0 = 1\cdot2338$, $\phi_0 = 0$, and ϕ increases.

When $x_0 \gtrless 1\cdot2338$, $\phi_1 - \pi \lessgtr 0$, $x_2 \lessgtr x_0$.

This orbit is stable, with $c = 2\cdot46$.

Plate II, fig. 2. $C = 38\cdot5$.

The planet A starts from $x_0 = -\cdot444$, $\phi_0 = \pi$, and ϕ increases.

When $x_0 \lessgtr -\cdot444$, $\phi_1 \lessgtr 0$, $x_2 \gtrless x_0$.

The orbit is unstable, with uneven instability, and $c = 1 + \cdot18 \sqrt{(-1)}$.

[The satellite A has become retrograde, and therefore does not appear amongst these direct orbits.]

The satellite A′ starts from $x_0 = 1\cdot1164$, $\phi_0 = 0$, and ϕ increases until when y vanishes it is equal to about $\pi - 12°$; it then diminishes to zero. It will be observed that at the first vanishing of y, the curve cuts the axis more nearly at right angles than was the case when $C = 39\cdot0$ and $39\cdot5$. When $x_0 \gtrless 1\cdot1164$, $\phi_1 \lessgtr 0$. When it starts too far from Jove it will move in some orbit round J, and when it starts too near Jove it will pass away to the Sun. The orbit is very unstable, with even instability.

The oscillating satellite a starts with $x_0 = \cdot6814$, $\phi_0 = 0$, and ϕ diminishes. When $x_0 \gtrless \cdot6814$, $\phi_1 - \pi \gtrless 0$. In the first case it passes away towards Jove, in the second towards the Sun. The orbit is very unstable with even instability.

The satellite B starts with $x_0 = 1\cdot1497$, $\phi_0 = 0$ and ϕ increases.

When $x_0 \gtrless 1\cdot1497$, $\phi_1 - \pi \gtrless 0$, $x_2 \gtrless x_0$.

The orbit is unstable with even instability, and $c = \cdot70 \sqrt{(-1)}$.

The satellite C starts with $x_0 = 1\cdot2760$, $\phi_0 = 0$, and ϕ increases.

When $x_0 \gtrless 1\cdot2760$, $\phi_1 - \pi \lessgtr 0$, $x_2 \lessgtr x_0$.

This orbit is very unstable, and as will appear below the instability is uneven. There has in fact been a passage from stability to uneven instability for some value of C between $39\cdot0$ and $38\cdot75$.

This orbit is interesting because it corresponds almost exactly to the cusped orbit described by Mr Hill as the moon of greatest lunation. It would seem however that this description is incorrect, for the satellite C moves with a still longer period when the cusp is replaced by a loop. Mr Hill's orbit was, on account of his approximation, necessarily a symmetrical one with reference to the line of quadratures, but it will be observed that when the solar parallax is taken into account the orbit is very unsymmetrical.

[As will be pointed out in Part V, Paper 3, p. 175, there must be two new families of figure-of-8 orbits called C′ and C″. They are not drawn in this figure.]

When $C = 38\cdot88$ a new periodic orbit arises in the point $x_0 = 1\cdot3470$, $y = 0$ marked in the figure. This is the beginning of a second family of oscillating satellites, referred to here as (b).

When $C = 38\cdot5$ this orbit begins with $x_0 = 1\cdot2919$, $\phi_0 = 0$, and ϕ diminishes.

When $x_0 \gtrless 1\cdot2919$, $\phi_1 - \pi \gtrless 0$. That is to say if the body starts too far from Jove for periodicity, it will pass away in an orbit as a superior planet;

if on the other hand it starts too near Jove for periodicity, it will pass to some orbit about Jove. This orbit is very unstable.

Plate III, fig. 1. $C = 38\cdot0$.

The planet A starts from $x_0 = -\cdot455$, $\phi_0 = \pi$, and ϕ increases.

When $x_0 \lessgtr -\cdot455$, $\phi_1 \lessgtr 0$, $x_2 \gtrless x_0$.

The orbit is unstable, with uneven instability, and $c = 1 + \cdot193 \sqrt{(-1)}$.

[The satellite A is retrograde, and does not appear.]

The satellite A′ starts from $x_0 = 1\cdot1305$, $\phi_0 = 0$, and ϕ increases.

When $x_0 \gtrless 1\cdot1305$, $\phi_1 \lessgtr 0$. The remarks concerning this orbit in previous cases apply again here.

At the point where the orbit crosses the axis of x for the second time $\pi - \phi$ is less than it was in the preceding case.

The oscillating satellite (a) starts from $x_0 = \cdot6760$, $\phi_0 = 0$ and ϕ decreases.

When $x_0 \gtrless \cdot6760$, $\phi_1 - \pi \gtrless 0$. It is very unstable, with even instability.

The satellite B starts from $x_0 = 1\cdot1470$, $\phi_0 = 0$, and ϕ increases.

When $x_0 \gtrless 1\cdot1470$, $\phi_1 - \pi \gtrless 0$, $x_2 \gtrless x_0$.

The orbit is very unstable with even instability, and $c = \cdot96 \sqrt{(-1)}$.

[The figure-of-8 orbits C′ and C″ are not drawn in this figure.]

The satellite C starts from $x_0 = 1\cdot2480$, $\phi_0 = 0$, and ϕ increases.

When $x_0 \gtrless 1\cdot2480$, $\phi_1 - \pi \lessgtr 0$, $x_2 \lessgtr x_0$.

This orbit was very troublesome, and is not computed with a high degree of accuracy. A very small variation of C would make a large change in the size of the loops in the curve.

The orbit is very unstable with uneven instability.

The oscillating satellite (b) starts with $x_0 = 1\cdot2595$, $\phi_0 = 0$, and ϕ decreases.

When $x_0 \gtrless 1\cdot2595$, $\phi_1 - \pi \gtrless 0$. The remarks made concerning this curve for $C = 38\cdot5$ apply again here.

This orbit is very unstable.

§ 18. *Classification of orbits by families.*

Several orbits are given in this classification which were not included in § 17. [In order to give the schedule as completely as is as yet possible I add some results which were obtained by means of the investigations to be explained in Paper 3. *These additional results are distinguished by being printed in italic type.*]

TABLE OF RESULTS.

Constant of Energy C	Coord. of starting point x_0	Synodic Period nT	Criterion of Stability $\Delta \sin^2 \tfrac{1}{2}\pi \sqrt{/\Phi_0}$	Apparent advance of pericentre in synodic period $2\pi(\tfrac{1}{2}c-1)$	Regression of pericentre in sid. period $2\pi\left(1-\dfrac{\tfrac{1}{2}c}{1+(nT/2\pi)}\right)$	Description of instability	Modulus of instability $\dfrac{\log\sqrt{2}}{\log[D+\sqrt{D^2\pm1}]}$	Remarks
					Satellite A, Plate IV, fig. 1.			
40·5	1·1135	61° 20′	+ ·112	39° 0′	22° 20′			minimum of criterion
40·25	1·1150	65° 40′	+ ·063	29° 0′	31° 0′			maximum of x_0
40·2	1·1090	66° 50′	+ ·064	29° 10′	31° 40′			minimum of x_0
40·0	1·0334	98° 0′	+ ·226	303°	−161°			The value of x_0 is inferred from results obtained for family A″. These orbits are probably unstable
39·8	1·0103	99°?		99°?				
39·5	1·0036							
38·85	1·0000							
					Satellite A′ (see fig. 8, Paper 3).			
39·9	1·025	?				even		C and x_0 interpolated
39·8	1·04183	236°				even		
39·5	1·0650	229°	− ?			even	?	Figure-of-8 begins
39·0	1·0941	240°	−1·06			even	0·5	
38·5	1·1164	258°	− ?			even	?	
38·0	1·1305	299°	− ?			even	?	
					Satellite A″ (see fig. 9, Paper 3).			
39·9	1·025	?	stable ?					stability inferred; C and x_0 interpolated; see Paper 3
39·8	1·0103	280°						
39·5	1·0036							
38·8₄5	1·0000							probably unstable
					Satellite B, Plate IV, fig. 2.			
39·3	1·1575	87° 40′	− ·061			even	1·42	
39·0	1·1500	97° 0′	− ·402			even	0·58	
38·5	1·1497	113° 20′	−1·82			even	0·31	
38·0	1·1470	131° 50′	−4·5			even	0·23	

TABLE OF RESULTS (continued).

Constant of Energy C	Coord. of starting point x_0	Synodic Period nT	Criterion of Stability $\Delta \sin^2 \frac{1}{2}\pi\sqrt{\Phi_0}$	Apparent advance of pericentre of synodic period $2\pi(\frac{1}{2}c-1)$	Regression of pericentre in sid. period $2\pi\left(1 - \dfrac{\frac{1}{2}c}{1+(nT/2\pi)}\right)$	Description of instability	Modulus of instability $\dfrac{\log\sqrt{2}}{\log[D+\sqrt{D^2\pm 1}]}$	Remarks
				Satellite C, Plate IV, fig. 3.				
39·3	1·1751	89° 20′	+ ·064	81° 0′	24° 30′	....	....	
39·0	1·2338	114° 0′	+ ·435	82° 40′	23° 30′	....	....	
38·75	1·2873	179° 30′	+ 1·95	....	....	uneven	0·4	maximum of x_0
38·5	1·2760	210° 50′	> + 1	....	....	uneven	?	
38·0	1·2480	235° 20′	> + 1	....	....	uneven	?	
37·5	1·2225	243°	?	....	....	?	?	nearly ejectional; not accurately computed
				Satellites C′ and C″ (see Part V, Paper 3, p. 175).				
				Oscillating Satellite a.				
40·18	·7175	....	−	....	....	even	....	a point on SJ
40·0	·705	138°	−	....	....	even	?	
39·5	·693	?	−	....	....	even	?	
39·0	·687	146°	− 148	....	....	even	0·1	
38·5	·681	150°	−	....	....	even	?	
38·0	·676	....	−	....	....	even	?	
				Oscillating Satellite b.				
38·88	1·3470	....	?	....	....	?	?	a point on SJ
38·5	1·2919	214° ?	?	....	....	?	?	
38·0	1·2595	208° ?	?	....	....	?	−	
				Planet A, Plate IV, fig. 4.				
40·0	− ·414	154°	+ ·91	145°	6° 30′	....	....	
39·5	− ·424	165°	+ ·98	162°	2°	....	....	
39·0	− ·434	177°	+ 1·03	....	....	uneven	2·1	
38·5	− ·444	191°	+ 1·08	....	....	uneven	1·25	
38·0	− ·455	207°	+ 1·09	....	....	uneven	1·14	
37·5	− ·465	243°	?	....	....	?	?	not inserted in figure

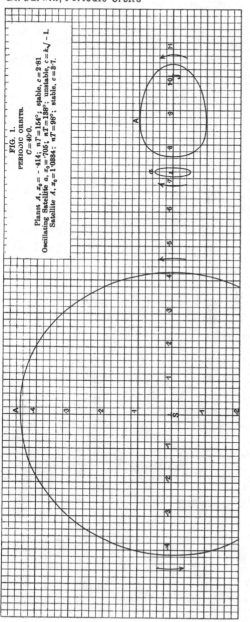

FIG. 1.
PERIODIC ORBITS.

$C = 40.0$.

Planet A, $z_0 = -\cdot414$; $nT = 154°$; stable, $c = 2\cdot81$.
Oscillating Satellite a, $x_0 = \cdot705$; $nT = 188°$; unstable, $c = k\sqrt{-1}$.
Satellite A, $z_0 = 1\cdot0894$; $nT = 98°$; stable, $c = 8\cdot7$.

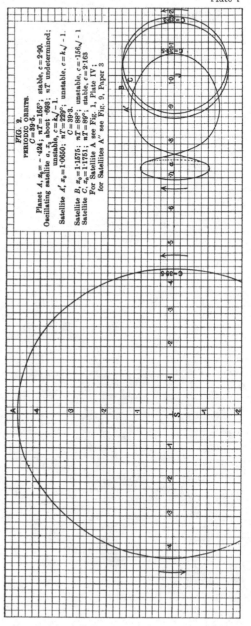

FIG. 2.
PERIODIC ORBITS.

$C = 39\cdot5$.

Planet A, $z_0 = -\cdot424$; $nT = 165°$; stable, $c = 2\cdot90$.
Oscillating satellite a, z_0 about $\cdot693$; nT undetermined;
unstable, $c = k\sqrt{-1}$.
Satellite A', $z_0 = 1\cdot0650$; $nT = 229°$; unstable, $c = k\sqrt{-1}$.

$C = 39\cdot3$.

Satellite B, $z_0 = 1\cdot1575$; $nT = 89°$; unstable, $c = \cdot156\sqrt{-1}$
Satellite C, $z_0 = 1\cdot1751$; $nT = 89°$; stable, $c = 2\cdot163$
For Satellite A see Fig. 1, Plate IV;
for Satellites A" see Fig. 9, Paper 3

Plate II.

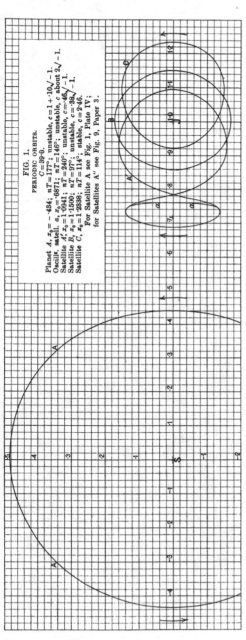

FIG. 1.
PERIODIC ORBITS.

$C = 39\cdot0$.

Planet A, $x_0 = -\cdot434$; $nT = 177^\circ$; unstable, $c = 1 + \cdot10\sqrt{-1}$.
Oscillg. satell. a, $x_0 = \cdot6871$; $nT = 146^\circ$; unstable, c about $2\sqrt{-1}$.
Satellite A', $x_0 = 1\cdot0941$; $nT = 240^\circ$; unstable, $c = -\cdot46\sqrt{-1}$.
Satellite B, $x_0 = 1\cdot1500$; $nT = 97^\circ$; unstable, $c = -\cdot38\sqrt{-1}$.
Satellite C, $x_0 = 1\cdot2838$; $nT = 114^\circ$; stable, $c = 2\cdot46$.
        For Satellite A see Fig. 1, Plate IV;
        for Satellites A'' see Fig. 9, Paper 3.

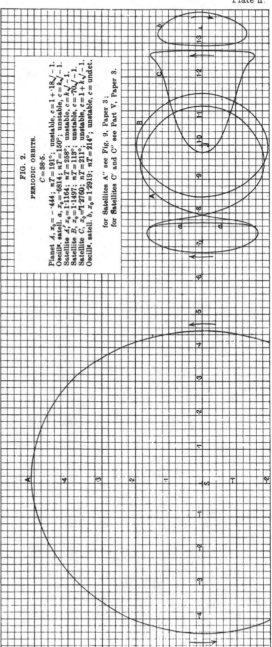

FIG. 2.
PERIODIC ORBITS.

$C = 38\cdot5$.

Planet A, $x_0 = -\cdot444$; $nT = 191^\circ$; unstable, $c = 1 + \cdot18\sqrt{-1}$.
Oscillg. satell. a, $x_0 = \cdot6814$; $nT = 150^\circ$; unstable, $c = k\sqrt{-1}$.
Satellite A', $x_0 = 1\cdot1164$; $nT = 268^\circ$; unstable, $c = k\sqrt{-1}$.
Satellite B, $x_0 = 1\cdot1497$; $nT = 113^\circ$; unstable, $c = -\cdot70\sqrt{-1}$.
Satellite C, $x_0 = 1\cdot2760$; $nT = 211^\circ$; unstable, $c = 1 + k\sqrt{-1}$.
Oscillg. satell. b, $x_0 = 1\cdot2919$; $nT = 214^\circ$; unstable, $c = $ undet.
        for Satellites A'' see Fig. 9, Paper 3;
        for Satellites C' and C'' see Part V, Paper 3.

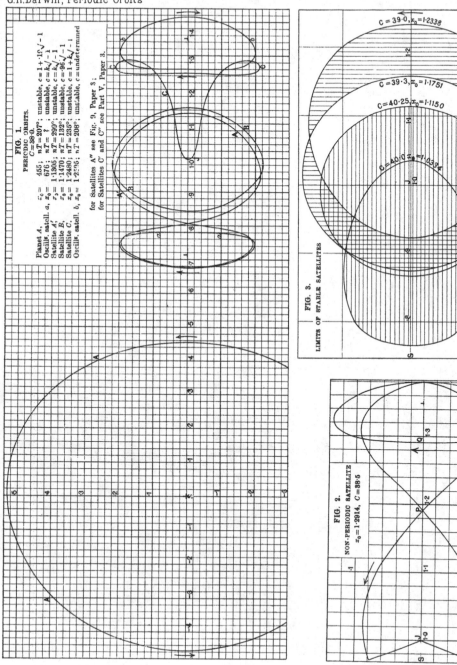

FIG. 1.
PERIODIC ORBITS.
$C = 38.0$.

Planet A.
Oscill. satell. a, $\bar{z}_0 = $ 455; $nT = 207°$; unstable, $c = 1 + 1\sqrt{-1}$
Satellite A', $\bar{z}_0 = $ 676; $nT = ?$;   unstable, $c = k_1\sqrt{-1}$
Satellite B, $x_0 = $ 1·1305; $nT = 290°$; unstable, $c = k_2\sqrt{-1}$
Satellite C, $x_0 = $ 1·1470; $nT = 132°$; unstable, $c = 96\sqrt{-1}$
Oscill. satell. b, $z_0 = $ 1·2480; $nT = 235°$; unstable, $c = 1 + k_4\sqrt{-1}$
                   $z_0 = $ 1·2505; $nT = 208°$; unstable, $c = $ undetermined

for Satellites A'' see Fig. 9, Paper 3;
for Satellites C' and C'' see Part V, Paper 3.

FIG. 3.
LIMITS OF STABLE SATELLITES

$C = 39·0$, $x_0 = 1·2338$
$C = 39·3$, $x_0 = 1·1751$
$C = 40·25$, $x_0 = 1·1150$
$C = 40·0$, $x_0 = 1·0334$

FIG. 2.
NON-PERIODIC SATELLITE
$x_0 = 1·2014$, $C = 38·5$

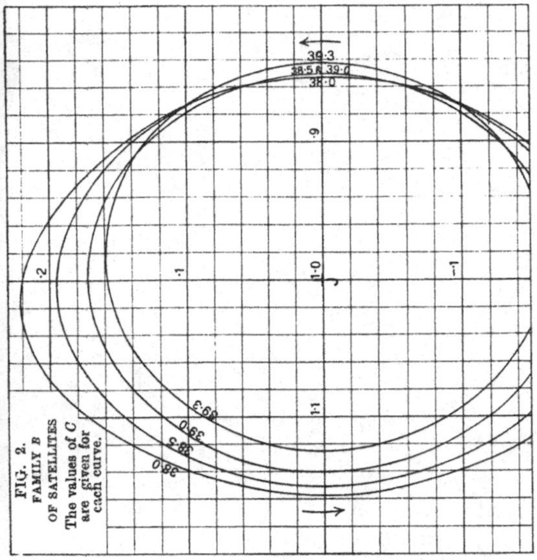

FIG. 2.
FAMILY B
OF SATELLITES
The values of C
are given for
each curve.

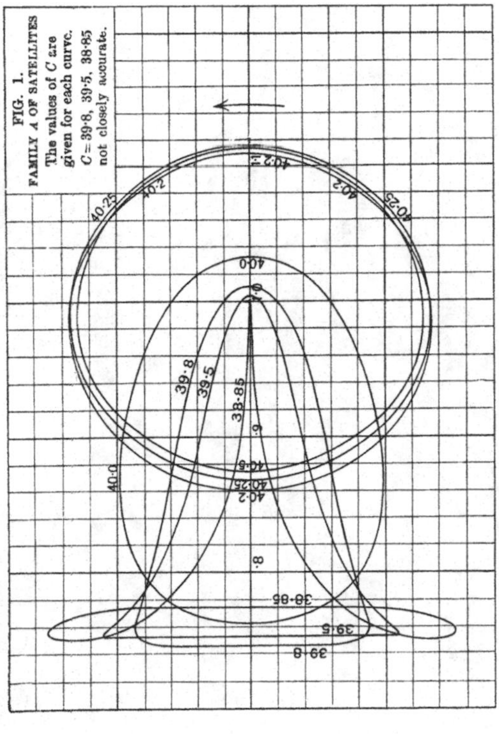

FIG. 1.
FAMILY A OF SATELLITES
The values of C are
given for each curve.
C = 39·8, 39·5, 38·85
not closely accurate.

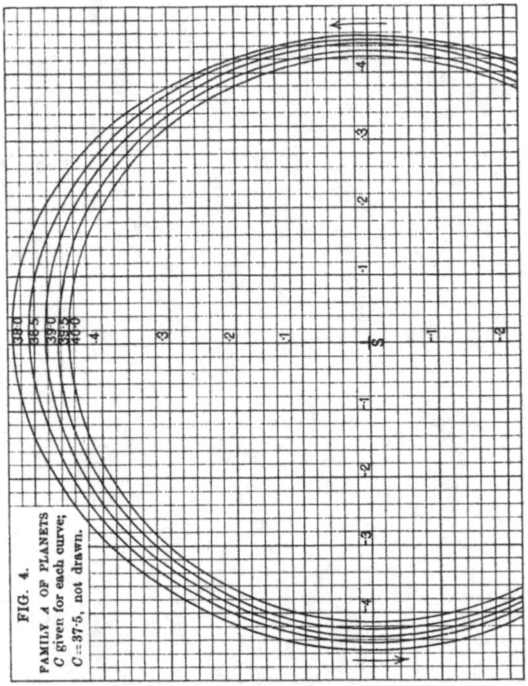

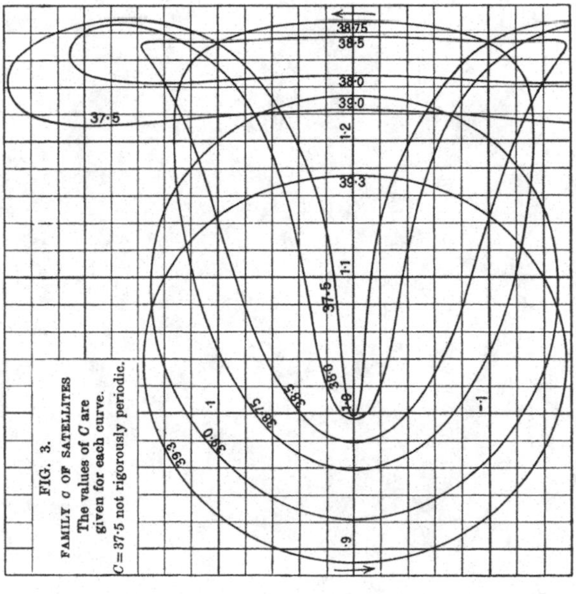

FIG. 4.
FAMILY A OF PLANETS
C given for each curve;
C = 37·5, not drawn.

FIG. 3.
FAMILY C OF SATELLITES
The values of C are
given for each curve.
C = 37·5 not rigorously periodic.

The above table gives most of the facts, [but a discussion of the families A, A', A'', C', C'' will be found in Part IV of Paper 3, p. 169.]

Mr Hill has drawn an interesting family of orbits of satellites, beginning with the orbit of the moon and ending with a cusped orbit. Now our moon undoubtedly belongs to the family A, whilst the cusped and looped orbits belong to the family C. He neglects the solar parallax, and this approximation has in fact led to the absorption of two families into one another. It appears now that it is not possible to comprehend the part played by this class of orbit without the inclusion of the solar parallax, for the asymmetry of the family C with regard to the line of quadratures is an essential feature in it.

Mr Hill draws attention to the minimum of distance at syzygies in the orbits of satellites, and this is observable in our family C, but we also find a maximum of distance in the family A at the superior syzygy.

§ 19. *On the probable forms of periodic orbits for values of C less than* 38.

[The subject of this paragraph is considered in Part V of Paper 3, and since several erroneous conjectures were made in this place the greater part of the paragraph is now omitted. I leave standing the curious non-periodic orbit shown in fig. 2, Plate III. In the original paper I drew some erroneous deductions from it.]

It appears from fig. 1 that when $C = 34\cdot91$, there is a new periodic orbit consisting of the point $x = -\cdot9469$, $y = 0$. This point is the origin of a new family of oscillating planets, say (c), which describe ovals with retrograde revolution round the point of zero force, for values of C less than $34\cdot91$.

§ 20. *Classification of stable orbits of satellites.*

We have seen that amongst satellites there are two classes of stable orbits, namely those of the A and C families. Plate III, fig. 3 exhibits the limits of the orbits which have been shown to be stable. The exact orbits which possess limiting stability would of course differ slightly from those drawn in this figure.

When C is large the stable orbits of the A family are approximate circles of small radius. As C decreases the orbits swell, but when C reaches $40\cdot25$ the radius vector at superior syzygy reaches a maximum. Hence the orbit $x_0 = 1\cdot1150$, $C = 40\cdot25$ gives one limit of the stable orbits of this family. The orbit $x_0 = 1\cdot0334$, $C = 40\cdot0$ gives approximately another limit as regards the inferior syzygy. The shaded space between these two orbits is filled with stable orbits.

The stable orbits of the C family begin when C is a little greater than 39·3, and the first one traced is that for which $x_0 = 1·1751$ and $C = 39·3$. The stability of these orbits still subsists when $C = 39·0$, but this orbit is already very unstable when C has fallen to 38·75. Accordingly I take for the other limit of orbits of this kind $x_0 = 1·2338$, $C = 39·0$. The shaded space between these two is filled with stable orbits.

It will be observed that there remains an unshaded tract within which no stable orbit can exist. I think moreover that it is probable that with a smaller mass for Jove we should have found a complete annulus within which stability is impossible.

This conclusion is interesting when viewed in connection with the distribution of the satellites and planets of our system, and it appears to me to be the first exact result, which throws any light on Bode's empirical law as to the mean distances of planets and satellites from their primaries.

It is as yet too soon to make a similar classification of stable planetary orbits, [and I have not been able to pursue the investigation to such a point as to throw any light on the subject.]

We have seen in an earlier section that unstable orbits are such as ultimately to lead to the absorption of bodies moving in them into one or other of the perturbing centres. If there were a large number of perturbing centres, as in our planetary system, the problem would become incomparably more difficult, but I think that the present investigation affords evidence that if we were to have a system consisting of a large planet moving round the sun, and of a cloud of infinitesimal bodies circling about them, a system would ultimately be evolved where there would be inferior and superior planets and satellites moving in certain zones indicated by our figures.

POSTSCRIPT.

I have to thank Mr S. S. Hough for pointing out to me that if the third body be placed at the vertex of the equilateral triangle drawn on SJ, it is not necessarily stable. It appears that if Jove is greater in mass than one twenty-fifth of the Sun, such a body is unstable.

This may be proved as follows:

The coordinates of the point for which $r = \rho = 1$ are $x = \frac{1}{2}$, $y = \frac{1}{2}\sqrt{3}$; also $\frac{\partial \Omega}{\partial x} = \frac{\partial \Omega}{\partial y} = 0$, but $\frac{\partial^2 \Omega}{\partial x^2} = \frac{3}{4}(\nu + 1)$, $\frac{\partial^2 \Omega}{\partial x \partial y} = \frac{3}{4}\sqrt{3}(\nu - 1)$, $\frac{\partial^2 \Omega}{\partial y^2} = \frac{9}{4}(\nu + 1)$. Hence at a point whose coordinates are $x = \frac{1}{2} + \xi$, $y = \frac{1}{2}\sqrt{3} + \eta$,

$$2\Omega = 3(\nu + 1) + \tfrac{3}{4}(\nu + 1)\xi^2 + \tfrac{3}{2}\sqrt{3}.(\nu - 1)\xi\eta + \tfrac{9}{4}(\nu + 1)\eta^2 + \ldots$$

and the equations of motion are

$$\frac{d^2\xi}{dt^2} - 2n\frac{d\eta}{dt} = \tfrac{3}{4}(\nu+1)\,\xi + \tfrac{3}{4}\sqrt{3}\,.\,(\nu-1)\,\eta$$

$$\frac{d^2\eta}{dt^2} + 2n\frac{d\xi}{dt} = \tfrac{3}{4}\sqrt{3}\,.\,(\nu-1)\,\xi + \tfrac{9}{4}(\nu+1)\,\eta$$

Noting that $n^2 = \nu+1$, and assuming $\xi = ae^{\lambda t}$, $\eta = be^{\lambda t}$, we easily find

$$\lambda^4 + (\nu+1)\,\lambda^2 + \tfrac{27}{4}\nu = 0$$

It is clear that if $(\nu+1)^2 > 27\nu$, λ^2 is negative, and the motion is oscillatory; but if $(\nu+1)^2 < 27\nu$, λ is complex and the solution will represent an oscillation with increasing amplitude.

The limiting value of ν consistent with stability is therefore given by $(\nu+1)^2 = 27\nu$, the solution of which is $\nu = 24{\cdot}9599$. The second solution is of course the reciprocal of the first.

In the numerical work in this paper I have taken $\nu = 10$, and there will accordingly be no stable orbits encircling the point $r = \rho = 1$.

[The approximate treatment of the equations in the foregoing investigation is susceptible of further extension, for it is not necessary that the point, in the neighbourhood of which we attribute approximate forms to the forces, should be one of the points of zero force.

Suppose then that we wish to determine the form of an orbit in the neighbourhood of a point defined by x_0, y_0 or by r_0, ρ_0.

For the sake of brevity let the differential coefficients of Ω at the point x_0, y_0 be written

$$K = \frac{\partial\Omega}{\partial x}, \quad L = \frac{\partial\Omega}{\partial y}, \quad M = \frac{\partial^2\Omega}{\partial x^2}, \quad N = \frac{\partial^2\Omega}{\partial y^2}, \quad Q = \frac{\partial^2\Omega}{\partial x\,\partial y}$$

From the formula for Ω we then easily find

$$K = \nu x_0\left(1 - \frac{1}{r_0^3}\right) + (x_0 - 1)\left(1 - \frac{1}{\rho_0^3}\right), \quad L = \nu y_0\left(1 - \frac{1}{r_0^3}\right) + y_0\left(1 - \frac{1}{\rho_0^3}\right)$$

$$M = (\nu+1) - \left(\frac{\nu}{r_0^3} + \frac{1}{\rho_0^3}\right) + 3\left[\frac{\nu x_0^2}{r_0^5} + \frac{(x_0-1)^2}{\rho_0^5}\right]$$

$$N = (\nu+1) - \left(\frac{\nu}{r_0^3} + \frac{1}{\rho_0^3}\right) + 3\left[\frac{\nu y_0^2}{r_0^5} + \frac{y_0^2}{\rho_0^5}\right]$$

$$Q = \qquad\qquad 3\left[\frac{\nu x_0 y_0}{r_0^5} + \frac{(x_0-1)\,y_0}{\rho_0^5}\right]$$

In the given neighbourhood let

$$x = x_0 - \lambda + \xi, \quad y = y_0 - \mu + \eta$$

so that

$$\Omega = \Omega_0 + K(\xi - \lambda) + L(\eta - \mu) + \tfrac{1}{2}M(\xi - \lambda)^2$$
$$+ Q(\xi - \lambda)(\eta - \mu) + \tfrac{1}{2}N(\eta - \mu)^2$$

and

$$\frac{\partial \Omega}{\partial x} = K + M(\xi - \lambda) + Q(\eta - \mu); \quad \frac{\partial \Omega}{\partial y} = L + Q(\xi - \lambda) + N(\eta - \mu)$$

The quantities λ, μ are as yet arbitrary; let them be defined by

$$\lambda = \frac{LQ - KN}{Q^2 - MN}, \quad \mu = \frac{KQ - LM}{Q^2 - MN}$$

and we have

$$\frac{\partial \Omega}{\partial x} = M\xi + Q\eta, \quad \frac{\partial \Omega}{\partial y} = Q\xi + N\eta$$

Hence the approximate equations of motion become

$$\frac{d^2\xi}{dt^2} - 2n\frac{d\eta}{dt} = M\xi + Q\eta$$

$$\frac{d^2\eta}{dt^2} + 2n\frac{d\xi}{dt} = Q\xi + N\eta$$

These equations remain true so long as ξ and η are nearly equal to λ and μ respectively.

By assuming as a solution

$$\xi = Ae^{\sigma t}, \quad \eta = Be^{\sigma t}$$

we easily find the frequency equation

$$\sigma^4 - (M + N - 4n^2)\sigma^2 + MN - Q^2 = 0$$

If α^2, γ^2 be the two roots of this equation, the solution of the differential equations is

$$\xi = C\cosh \alpha t + D\sinh \alpha t + E\cosh \gamma t + F\sinh \gamma t$$

$$\eta = C'\cosh \alpha t + D'\sinh \alpha t + E'\cosh \gamma t + F'\sinh \gamma t$$

where

$$C' = \frac{2n\alpha D - QC}{N - \alpha^2}, \quad D' = \frac{2n\alpha C - QD}{N - \alpha^2}$$

$$E' = \frac{2n\gamma F - QE}{N - \gamma^2}, \quad F' = \frac{2n\gamma E - QF}{N - \gamma^2}$$

If we substitute for M, N, Q their values in the frequency equation, we find that

$$\left.\begin{matrix}\alpha^2\\\gamma^2\end{matrix}\right\} = \tfrac{1}{2}\Gamma - n^2 \pm \tfrac{1}{2}\sqrt{\left[\Gamma(9\Gamma - 8n^2) - \frac{36\nu}{r_0^4\rho_0^4}\sin\theta_0\sin\Psi_0\right]}$$

where $\Gamma = \dfrac{\nu}{r_0^3} + \dfrac{1}{\rho_0^3}$, and of course $n^2 = \nu + 1$.

Suppose that when $t = 0$, $\xi = \xi_0$, $\eta = \eta_0$, $\dfrac{d\xi}{dt} = u_0$, $\dfrac{d\eta}{dt} = v_0$; then

$$C + E = \xi_0, \qquad D\alpha + F\gamma = u_0$$
$$C' + E' = \eta_0, \qquad D'\alpha + F'\gamma = v_0$$

whence we find

$$(\alpha^2 - \gamma^2)\, C = (M - \gamma^2)\, \xi_0 + Q\eta_0 + 2nv_0$$
$$(\alpha^2 - \gamma^2)\, C' = Q\xi_0 + (N - \gamma^2)\, \eta_0 - 2nu_0$$
$$\alpha\,(\alpha^2 - \gamma^2)\, D = 2n\,(Q\xi_0 + N\eta_0) - (N - \alpha^2)\, u_0 + Qv_0$$
$$\alpha\,(\alpha^2 - \gamma^2)\, D' = - 2n\,(M\xi_0 + Q\eta_0) + Qu_0 - (M - \alpha^2)\, v_0$$
$$E = \xi_0 - C, \quad E' = \eta_0 - C', \quad F\gamma = u_0 - \alpha D, \quad F'\gamma = v_0 - \alpha D'$$

Portion of one of the orbits of the family A″, referred to in the Appendix to this paper and in Part IV of Paper 3, was traced in this way, but the method is rather laborious.

If we apply this method to the tracing of orbits in the neighbourhood of one of the three points of zero force in the line of syzygies, so that x_0, y_0 refers to one of those points, we have K, L and Q equal to zero, and therefore λ and μ are also zero.

Suppose we wish to trace the orbit of one of the oscillating satellites as in Mr Hough's paper which follows this one. For example take the case of the oscillating satellite (a), and suppose that we start at right angles to the line of syzygy from a point between the point of zero force and the sun. In this case ξ_0 is negative, v_0 is positive, and η_0 and u_0 are zero.

In the frequency equation we now have

$$r_0 = \cdot71751, \quad \rho_0 = \cdot28249, \quad \theta_0 = \psi_0 = 0$$

Then it is easy to see that one of the two roots, say γ^2, is negative. I therefore write $\beta^2 = - \gamma^2$, and find

$$(\alpha^2 + \beta^2)\, C = (M + \beta^2)\, \xi_0 + 2nv_0$$
$$E = \xi_0 - C$$
$$\alpha\,(\alpha^2 + \beta^2)\, D' = - 2nM\xi_0 - (M - \alpha^2)\, v_0$$
$$F'\beta \sqrt{-1} = v_0 - \alpha D'$$
$$C' = E' = D = F = 0$$

The solution is therefore

$$\xi = C \cosh \alpha t + E \cos \beta t$$
$$\eta = D' \sinh \alpha t + F' \sqrt{-1} \sin \beta t$$

It is interesting to trace such curves, and see the effect of the hyperbolic terms in carrying the body away or towards the point of zero force, but I refer to Mr Hough's paper for the consideration of this point.

If the orbit belongs to the family (a) of oscillating satellites, it must be a re-entrant oval*. This necessitates the vanishing of C, which carries with it the vanishing of D'. One of the semiaxes of the oval is accordingly E and the other $F' \sqrt{-1}$.

Since C vanishes $\qquad \xi_0 = -\dfrac{2nv_0}{M + \beta^2} = -\dfrac{(M - \alpha^2)\, v_0}{2nM}$

If we attribute to M and α^2, β^2 their values we find

$$\xi_0 = \frac{4n^2 + 3\Gamma - \sqrt{(9\Gamma^2 - 8n^2\Gamma)}}{4n\,(n^2 + 2\Gamma)}\quad v_0 = \frac{4nv_0}{4n^2 + 3\Gamma + \sqrt{(9\Gamma^2 - 8n^2\Gamma)}}$$

If the point on the line of syzygy from which we start is distant r from the sun, we have

$$\xi_0 = r_0 - r = \cdot 71751 - r$$

Also if C now denotes as usual the constant of relative energy, we have

$$v_0^2 = \nu \left(r^2 + \frac{2}{r} \right) + (1 - r)^2 + \frac{2}{1 - r} - C$$

Hence on squaring the foregoing expression for ξ_0 we find

$$8n^2 \left[(n^2 - 1)\left(r^2 + \frac{2}{r} \right) + (1 - r)^2 + \frac{2}{1 - r} - C \right]$$
$$= [8n^2 \,(n^2 + \Gamma) + 9\Gamma^2 + (4n^2 + 3\Gamma)\sqrt{(9\Gamma^2 - 8n^2\Gamma)}]\,[r_0 - r]^2$$

We have therefore an equation connecting the point of projection with the constant of relative energy such as to ensure that the orbit shall be periodic.

This equation is of course merely approximate, but even in this simple case it is of the fourth order. It must have two imaginary roots and two real ones, which define the two crossing points on the line of syzygy.

If any general equation could be formulated for the connection between C and x_0 in periodic orbits, an enormous advance would be made in the subject. In Paper 3, p. 170, the curve in Fig. 7 is one in which the ordinates are C and the abscissae are x_0; it is the graphical analogue of the equation which has been found above.]

* The reader will find this subject treated fully in Charlier's *Die Mechanik des Himmels*, Vol. II., Section IX. (Veit and Co., Leipzig, 1902).

APPENDIX.

Computations of Periodic Orbits, and of their Stability.

Explanation.

The orbits are given in families, arranged according to descending values of C, the constant of relative energy. The families are distinguished by the initials A, A', A'', B, C, a, b. The initial A is attached to one of the families of satellites and also to the family of planets, because the satellite A appears to bear the same relationship to Jove and the Sun that the planet A bears to the Sun and Jove.

The data for the orbits are given as follows:—The first column is the arc of the relative orbit measured from conjunction; the second and third are the rectangular coordinates $x - 1$, y for satellites, or x, y for planets; the fourth gives ϕ the inclination of the outward normal to the line SJ; the fifth and sixth are the coordinates ρ, ψ for satellites, or r, θ for planets; the last column contains the function $2n/V$.

The last column is given so that the reader may be enabled to complete the solution, by drawing the orbit with reference to axes fixed in space. The integral $\frac{1}{2}\int \frac{2n}{V} ds$ would give nt, that is to say the angle turned through by the rotating axes since conjunction; then the polar coordinates with reference to Jove are ρ, $\psi + nt$, or with reference to the Sun are r, $\theta + nt$.

In the case of the oscillating bodies (families a and b) the polar coordinates are not given, but the rectangular coordinates with reference to axes fixed in space are clearly

$$x \cos nt - y \sin nt, \quad x \sin nt + y \cos nt$$

for heliocentric origin, and

$$(x - 1) \cos nt - y \sin nt, \quad (x - 1) \sin nt + y \cos nt$$

for jovicentric origin.

The last line of these tables gives the value of the arc and of ϕ when y vanishes. If the orbit were rigorously periodic and were computed with absolute accuracy, this angle would be 180° or 0°. It may be remarked that in some cases a small change in the initial value of x leads to a large change in the final value of ϕ, and in other cases the converse is true. Thus in some cases it is necessary to continue the search until the final value of ϕ only differs from 180° or 0° by a few minutes of arc, and in others even an error of a degree of arc is unimportant. The coordinates are certainly given with sufficient accuracy to draw the figures on a large scale.

Finally there is given the time-integral nT, being twice the angle turned through by the rotating axes between the first orthogonal crossing of SJ and the second (closely approximate) orthogonal crossing. Since the circuit is completed at the third crossing T is the period, and the ratio of nT to 360° is the ratio of the period of the body to that of Jove.

After the coordinates the discussion of the stability is given.

In order to test the sufficiency of the harmonic expansion of Φ to represent that function, a comparison is given between nine of the equi-distant values of Φ with the corresponding values derived from a synthesis of the harmonic series, which has been calculated as far as the eighth order inclusive. Following this comparison is Φ_0 the mean value of Φ.

In the cases where the orbit is stable the value of c is given, and certain functions of it. The function $\Delta \sin^2 \frac{1}{2}\pi\sqrt{\Phi_0}$ or $\sin^2 \frac{1}{2}\pi c$ is what is called in the table of § 18 the Criterion of Stability. The function $2\pi(\frac{1}{2}c-1)$ gives the retrogression of the pericentre, with respect to the rotating axes, in the synodic period. The function $nT - 2\pi(\frac{1}{2}c-1)$ gives the advance of pericentre, with respect to fixed axes, in the synodic period. And $2\pi\left(1 - \dfrac{\frac{1}{2}c}{1+(nT/2\pi)}\right)$ gives the advance of the pericentre, with respect to fixed axes, in the sidereal period. [In the case of retrograde orbits these formulæ need some modifica-tion, as pointed out in § 2 of Paper 3.]

Where the orbit is unstable, when the determinant Δ is negative the instability is of the even type, and when $\Delta \sin^2 \frac{1}{2}\pi\sqrt{\Phi_0}$ is greater than unity the instability is of the uneven type. The modulus of instability, or the number of synodic circuits, in which the amplitude of displacement increases to twice its primitive value, is given.

When the instability is of the even type c is of the form $2n + k\sqrt{-1}$, and when of the uneven type it is of the form $2n+1+k\sqrt{-1}$; in the tables c is given in one or other of these two forms.

[*Certain results computed for Paper 3 are included in the present appendix and are printed in italics.*]

FAMILY A OF SATELLITES.

C = 40·5　　　　　　　　　　$x_0 = 1\cdot1135.$

s	x − 1	y	φ	ρ	ψ	2n/V
·00	+·1135	+·0000	0° 0′	·1135	0° 0′	2·423
3	102	298	12° 56′	41	15° 7′	·441
6	002	580	25° 58′	58	30° 4′	·492
9	·0841	832	39° 15′	83	44° 5′	·574
·12	625	·1040	52° 56′	·1213	58° 59′	·679
5	366	189	67° 10′	44	72° 54′	·792
8	+ 078	269	82° 0′	71	86° 30′	·893
·21	− ·0222	271	π − 82° 42′	90	π − 80° 7′	·960
4	511	194	67° 20′	98	66° 51′	·975
7	769	044	52° 24′	96	53° 36′	·936
·30	981	·0833	38° 17′	87	40° 19′	·870
3	·1137	578	24° 58′	76	26° 55′	·803
6	233	+ 294	π − 12° 16′	66	π − 13° 24′	·757
·39	− ·1265	− ·0004	π + 0° 6′	·1265	π + 0° 11′	2·740
·3896		·0000	π − 0° 3′			

$$nT = 61° 23'$$

Stability of $x_0 = 1\cdot1135$, $C = 40\cdot5$.

Comparison

	Computed Φ	Synthesis		Computed Φ	Synthesis
a_0	3·19	3·18	a_8	7·28	7·22
a_2	3·84	3·84	a_9	5·80	5·91
a_3	4·67	4·66	a_{10}	5·01	4·86
a_4	5·81	5·83	a_{12}	3·19	3·00
a_6	8·04	8·04			

$$\Phi_0 = 5\cdot479$$

The harmonic series represents Φ well.

The determinant gives $\Delta \sin^2 \tfrac{1}{2}\pi \sqrt{\Phi_0} = \cdot1119$, $c = 2\cdot217$,

$$2\pi\,(\tfrac{1}{2}c - 1) = 39° 4', \quad nT - 2\pi\,(\tfrac{1}{2}c - 1) = 19° 42', \quad 2\pi\left(1 - \frac{\tfrac{1}{2}c}{1 + (nT/2\pi)}\right) = 22° 19'.$$

The orbit is stable.

C = 40·25　　　　　　　　　　$x_0 = 1\cdot1150.$

s	x − 1	y	φ	2n/V
·00	+·1150	+·0000	0° 0′	2·418
3	118	298	12° 24′	·437
6	022	583	24° 53′	·496
9	·0867	839	37° 34′	·587
·12	659	·1054	50° 36′	·708
5	407	216	64° 12′	·846
8	+ 124	312	78° 29′	·978
·21	− ·0175	333	π − 86° 39′	3·079
4	469	277	71° 31′	·120
7	739	146	56° 43′	·097
·30	966	·0952	42° 42′	·033
3	·1142	710	29° 37′	2·954
6	260	435	17° 20′	·886
9	320	+ 141	π − 5° 33′	·854
·42	− ·1319	− ·0158	π + 6° 4′	2·855
·4042		·0000	π − 0° 1′	

$$nT = 65° 40'$$

FAMILY A OF SATELLITES *continued.*

Stability of $x_0 = 1\cdot1150$, $C = 40\cdot25$.

Comparison

	Computed Φ	Synthesis		Computed Φ	Synthesis
a_0	2·928	2·936	a_8	7·839	7·865
a_2	3·652	3·650	a_9	6·050	6·036
a_3	4·574	4·580	a_{10}	4·383	4·384
a_4	5·885	5·881	a_{12}	2·947	2·932
a_6	8·718	8·730			

$\Phi_0 = 5\cdot574$

The harmonic series represents Φ well.

The determinant gives $\Delta \sin^2 \tfrac{1}{2}\pi \sqrt{\Phi_0} = \cdot0630$, $c = 2\cdot161$,

$$2\pi\left(\tfrac{1}{2}c - 1\right) = 29° \, 3', \; nT - 2\pi\left(\tfrac{1}{2}c - 1\right) = 36° \, 37', \; 2\pi\left(1 - \frac{\tfrac{1}{2}c}{1 + (nT/2\pi)}\right) = 30° \, 58'.$$

The orbit is stable.

[The stability of this orbit was also determined by a different method, as explained in Part I. of Paper 3.]

C = 40·2 $x_0 = 1\cdot1090$.

s	$x - 1$	y	ϕ	$2n/V$
·00	+ ·1090	+ ·0000	0° 0'	2·276
3	058	298	12° 30'	·298
6	·0961	581	25° 1'	·362
9	806	837	37° 38'	·467
·12	598	·1052	50° 28'	·609
5	346	215	63° 45'	·780
8	+ 064	314	77° 41'	·958
·21	− ·0234	340	$\pi -$ 87° 41'	3·119
4	529	289	72° 35'	·225
7	800	163	57° 36'	·255
·30	·1031	·0972	43° 22'	·219
3	210	732	30° 10'	·155
6	331	458	17° 55'	·092
9	394	+ 166	$\pi -$ 6° 18'	·055
·42	− ·1397	− ·0134	$\pi +$ 5° 6'	3·053
·4066		·0000	$\pi -$ 0° 0'	

$nT = 66° \, 52'$

Stability of $x_0 = 1\cdot1090$, $C = 40\cdot2$.

Comparison

	Computed Φ	Synthesis		Computed Φ	Synthesis
a_0	2·627	2·573	a_8	8·640	8·753
a_2	3·300	3·296	a_9	6·692	6·635
a_3	4·186	4·139	a_{10}	4·760	4·758
a_4	5·498	5·601	a_{12}	3·093	3·033
a_6	8·184	8·345			

$\Phi_0 = 5\cdot593$

The harmonic series represents Φ fairly well.

The determinant gives $\Delta \sin^2 \tfrac{1}{2}\pi \sqrt{\Phi_0} = \cdot0636$, $c = 2\cdot162$,

$$2\pi\left(\tfrac{1}{2}c - 1\right) = 29° \, 14', \; nT - 2\pi\left(\tfrac{1}{2}c - 1\right) = 37° \, 38', \; 2\pi\left(1 - \frac{\tfrac{1}{2}c}{1 + (nT/2\pi)}\right) = 31° \, 44'.$$

The orbit is stable.

FAMILY A OF SATELLITES *continued.*

C = 40·0 $x_0 = 1\cdot03341.$

s	$x-1$	y	ϕ	ρ	ψ	$2n/V$
·00	+ ·03441	+ ·00000	0° 0′	·03341	0° 0′	·939
1	3257	0995	9° 40′	3406	17° 0′	·950
2	3010	1963	18° 52′	3594	33° 7′	·981
3	2617	2882	27° 16′	3893	47° 46′	1·031
4	2101	3738	34° 44′	4288	60° 40′	·096
5	1484	4525	41° 17′	4762	71° 50′	·172
6	+ 0787	5241	47° 0′	5300	81° 28′	·259
8	− ·00785	6472	56° 20′	6519	$\pi -$ 83° 5′	·458
·10	2518	7467	63° 41′	7880	71° 22′	·690
2	·4355	8256	69° 39′	9334	62° 11′	·958
4	6259	8866	74° 42′	·10852	54° 47′	2·269
6	8207	9316	79° 12′	2416	48° 37′	·640
8	·10184	9617	83° 29′	4007	43° 22′	3·093
·20	2178	9769	87° 49′	5613	38° 44′	·664
2	4177	9762	$\pi -$ 87° 17′	7213	34° 33′	4·410
4	6166	9564	81° 3′	8783	30° 37′	5·406
6	8111	9111	72° 11′	·20274	26° 42′	6·745
7	9046	8758	66° 12′	0963	24° 42′	7·525
8	9934	8300	58° 59′	1594	22° 36′	8·330
9	·20752	7726	50° 39′	2143	20° 25′	9·030
·30	1474	7035	41° 46′	2596	18° 8′	·516
1	2081	6241	33° 11′	2946	15° 47′	·730
2	2570	5369	25° 34′	3200	13° 23′	·710
3	2949	4444	19° 9′	3375	10° 58′	·563
4	3231	3485	13° 50′	3491	8° 32′	·376
5	3431	2505	9° 21′	3564	6° 6′	·209
6	3559	1514	5° 26′	3608	3° 41′	·095
7	3622	+ 0516	$\pi -$ 1° 49′	3628	$\pi -$ 1° 15′	·032
·38	− ·23623	− ·00484	$\pi +$ 1° 41′	·23628	$\pi +$ 1° 10′	9·032
·37516		·00000	$\pi -$ 0° 1′			

$$nT = 97° 58′$$

Stability of $x_0 = 1\cdot03341,\ C = 40\cdot0.$

Comparison

	Computed Φ	Synthesis		Computed Φ	Synthesis
a_0	− 2·49	− 0·95	a_8	17·58	20·28
a_2	2·32	2·21	a_9	41·05	39·50
a_3	2·74	3·89	a_{10}	33·03	32·56
a_4	2·93	1·61	a_{12}	0·48	− 2·63
a_6	4·70	5·88			

$$\Phi_0 = 10\cdot124$$

The representation of Φ by the harmonic series is not very satisfactory, nevertheless it will serve to give the result with some approach to accuracy, for the following shows the gradual approximation to a definite value as the number of rows of the determinant is increased:—

No. of rows	Value of Δ
5	·000
9	·052
13	·233
15	·243
17	·246

FAMILY A OF SATELLITES *continued*.

The determinant gives $\Delta \sin^2 \tfrac{1}{2}\pi \sqrt{\Phi_0} = \cdot 2264$, and $c = 3\cdot 684$;

$$2\pi(\tfrac{1}{2}c - 1) = 303° 10', \quad nT - 2\pi(\tfrac{1}{2}c - 1) = -205° 12',$$

$$2\pi\left(1 - \frac{\tfrac{1}{2}c}{1 + (nT/2\pi)}\right) = -161° 18'.$$

The margin of stability is obviously small.

FAMILY A′ OF SATELLITES.

C = 39·8   *Figure-of-eight orbit, $x_0 = 1\cdot 04183$.*

s	x	y	ϕ
·00	1·04183	·00000	0° 0′
·01	4114	0997	7° 56′
·02	3909	1975	15° 40′
·03	3578	2918	22° 55′
·04	3135	3814	29° 39′
·06	1969	5435	41° 21′
·08	0525	6815	50° 54′
·10	·98889	7962	58° 45′
·12	7122	8896	65° 21′
·14	5265	9636	71° 4′
·16	3346	·10197	76° 14′
·18	1386	0589	81° 9′
·20	·89399	0812	86° 6′
·22	7400	0856	$\pi - 88° 31'$
·24	5408	0698	82° 12′
·26	3451	0296	74° 13′
·28	1583	·09590	63° 56′
·30	·79892	8530	51° 33′
·32	8475	7123	38° 59′
·34	7368	5458	28° 52′
·36	6520	3648	21° 55′
·38	5852	1764	17° 33′
·40	5293	− ·00156	15° 16′
·42	4778	2088	14° 55′
·44	·74232	− ·04012	$\pi - 17° 17'$

Continuation by an interpolated orbit accepted as being continuous with the foregoing.

s	x	y	ϕ
·44	·74426	− ·04059	$\pi - 14° 45'$
·45	4155	5022	16° 52′
·46	3834	5969	21° 1′
·465	3641	6430	24° 34′
·47	3413	6874	30° 17′
·475	3132	7282	40° 49′
·48	2749	7593	$\pi - 64° 3'$
·485	·72274	− ·07659	76° 4′

Continuation by an orbit computed backwards from $x_0 = \cdot 7$.

s	x	y	ϕ
·485	·72158	− ·07659	75° 55′
·4875	1930	7560	58° 30′
·49	1732	7408	47° 6′
·4925	1561	7226	39° 31′
·4975	1279	6814	30° 5′
·5075	0863	5907	20° 13′
·5275	0347	3979	10° 36′
·5475	0084	1998	4° 48′
·5675	·70000	·00000	0° 0′

FAMILY A' OF SATELLITES *continued.*

I have not thought it necessary to add the rather complicated data for determining the period; I find however

$$nT = 236°.$$

The discussion of the stability of this orbit will be found in Paper 3.

C = 39·5   Figure-of-eight orbit, $x_0 = 1·065$.

s	x − 1	y	φ	ρ	ψ	2n/V
·00	+ ·0650	+ ·0000	0° 0'	·0650	0° 0'	1·434
2	631	199	10° 52'	662	17° 29'	·452
4	576	390	21° 17'	696	34° 9'	·508
6	487	570	31° 0'	750	49° 27'	·598
8	371	732	39° 52'	821	63° 7'	·718
·10	233	876	47° 53'	907	75° 8'	·870
2	+ 076	·1000	55° 11'	·1004	85° 40'	2·053
4	− ·0095	104	61° 57'	109	π − 85° 6'	·268
6	276	188	68° 21'	220	76° 55'	·522
8	466	252	74° 35'	336	69° 36'	·820
·20	661	294	80° 56'	453	62° 57'	3·167
2	860	315	87° 38'	571	56° 48'	·573
4	·1060	310	π − 85° 0'	685	51° 2'	4·039
6	257	279	76° 41'	793	45° 29'	·543
8	447	217	67° 20'	891	40° 4'	5·054
·30	624	125	57° 15'	975	34° 42'	·463
2	783	002	47° 17'	·2045	29° 20'	·738
4	917	·0855	38° 17'	100	24° 2'	·856
6	·2030	690	30° 49'	145	18° 47'	·882
8	123	513	24° 58'	184	13° 36'	·876
·40	200	329	20° 30'	224	8° 30'	·922
2	265	+ 139	17° 14'	269	π − 3° 31'	6·053
4	320	− ·0053	15° 1'	321	π + 1° 18'	·287
6	369	247	13° 54'	382	5° 56'	·690
8	417	441	14° 4'	458	10° 20'	7·425
9	442	538	14° 47'	501	12° 25'	·950
·50	469	634	16° 8'	550	14° 24'	8·688
1	498	729	18° 27'	603	16° 16'	9·685
2	533	823	22° 26'	663	18° 0'	11·243
3	577	913	30° 4'	733	19° 31'	13·953
·535	604	955	36° 54'	773	20° 8'	16·633
4	637	992	48° 27'	818	20° 36'	19·562
425	657	·1007	56° 52'	841	20° 45'	21·083
45	679	019	67° 54'	866	20° 49'	23·755
475	703	025	π − 82° 2'	891	20° 46'	24·553
5	728	026	+ 83° 33'	915	20° 36'	·220
525	753	020	78° 35'	935	20° 20'	22·752
·555	775	009	60° 13'	954	19° 59'	21·190
·56	814	·0979	46° 34'	979	19° 11'	17·987
65	848	942	37° 23'	999	18° 19'	15·257
7	876	901	31° 33'	·3014	17° 23'	13·597
75	900	857	27° 13'	024	16° 28'	12·425
8	922	812	23° 55'	033	15° 31'	11·445
9	958	719	18° 51'	044	13° 40'	9·910
·60	987	624	15° 15'	052	11° 47'	·280
1	·3011	526	12° 22'	057	9° 55'	8·666
2	030	428	9° 56'	060	8° 3'	·220
4	058	230	5° 50'	067	4° 18'	7·727
6	072	− ·0031	+ 2° 22'	072	π + 0° 35'	·540
8	− ·3074	+ ·0169	− 0° 53'	·3079	π − 3° 9'	7·595
·66308	− ·3073	·0000	+ 1° 49'			

$$nT = 229° 19'$$

FAMILY A′ OF SATELLITES *continued*.

The above is not strictly periodic, since the final value of ϕ is $1°$ $49'$; but I find that when $x_0 = 1\cdot066$ the final value of ϕ is $-62°$ $24'$, hence the periodic orbit should be $x_0 = 1\cdot065028$. Since the above only differs from the true periodic in the fifth place of decimals of x_0, I accept it as periodic. It would seem however as if the final value of $x - 1$ in the periodic orbit is about $-\cdot305$ instead of $-\cdot3073$, as in the above.

Stability of $x_0 = 1\cdot065$, $C = 39\cdot5$.

The determinantal method fails, because Φ varies from about -20 in one part of the orbit to more than 3000 in another, and the harmonic series gives so insufficient a representation of Φ, when we stop with the term of the eighth order, that it does not seem worth while to form and evaluate the determinant.

The orbit is clearly very unstable, with instability of the even type, as appears below in the case when $C = 39\cdot0$.

C = 39·0   Figure-of-eight orbit, $x_0 = 1\cdot0941$.

It appeared from various computations that the periodic orbit should commence with $x_0 = 1\cdot0941$.

Accordingly after the latter part of the orbit had been computed the first part was calculated.

s	$x-1$	y	ϕ	ρ	ψ	$2n/V$
·00	$+\cdot0941$	$+\cdot0000$	$0°$ $0'$	·0941	$0°$ $0'$	1·875
2	927	200	$7°$ $56'$	948	$12°$ $9'$	·888
4	886	395	$15°$ $42'$	970	$24°$ $2'$	·928
6	819	583	$23°$ $20'$	·1005	$35°$ $27'$	·991
8	728	761	$30°$ $38'$	054	$46°$ $17'$	2·081
·12	485	·1077	$44°$ $27'$	181	$65°$ $49'$	·340
6	$+$ 174	329	$57°$ $29'$	340	$82°$ $32'$	·717
·20	$-\cdot0184$	504	$70°$ $41'$	515	$\pi-83°$ $0'$	3·227
4	574	589	$84°$ $51'$	690	$70°$ $8'$	·880
8	971	565	$\pi-76°$ $44'$	842	$58°$ $10'$	4·562
·32	·1337	407	$56°$ $48'$	942	$46°$ $26'$	·904
6	633	139	$39°$ $50'$	991	$34°$ $53'$	·807
·40	853	·0806	$27°$ $55'$	·2020	$23°$ $30'$	·606
4	·2013	440	$19°$ $35'$	061	$12°$ $20'$	·525
8	127	$+$ 057	$13°$ $59'$	128	$\pi-$ $1°$ $32'$	·639
·52	$-\cdot2211$	$-\cdot0334$	$\pi-10°$ $49'$	·2237	$\pi+$ $8°$ $36'$	5·048

$\int_0^{\cdot52} \dfrac{2n}{V}\, ds = 109°$ $10'$. Also the value of ϕ where the curve crosses the axis of x for the second time is $\pi - 13°$ $22'$.

The following results in square parentheses were found by interpolation, between $x_0 = 1\cdot09$ and $x_0 = 1\cdot10$. Starting from these values the remainder of the orbit was computed as follows:—

FAMILY A′ OF SATELLITES *continued*.

s	$x-1$	y	ϕ	ρ	ψ	$2n/V$
⌈ ·44	− ·2020	·0437	$\pi - 20° \ 5'$ ⌉			
6	084	244	17° 4′			
8	138	+ 055	14° 42′			
⌊ ·50	186	− ·0139	12° 50′ ⌋	·2190	$\pi + \ 3° \ 38'$	4·847
2	228	334	11° 46′	252	8° 32′	5·104
4	268	530	11° 30′	329	13° 9′	·504
6	309	726	12° 19′	420	17° 27′	6·117
8	356	920	14° 50′	530	21° 20′	7·092
9	383	·1017	17° 10′	591	23° 6′	·839
·60	416	111	20° 46′	660	24° 42′	8·888
·605	435	158	23° 18′	695	25° 25′	9·565
1	456	203	26° 37′	734	26° 6′	10·476
15	480	247	31° 4′	776	26° 41′	11·582
2	508	288	37° 16′	820	27° 11′	13·008
25	541	326	46° 17′	866	27° 33′	14·945
3	580	356	59° 46′	914	27° 43′	16·959
35	627	374	$\pi - 78° \ 46'$	965	27° 37′	19·068
4	677	374	+ 79° 10′	·3009	27° 11′	18·399
45	724	357	62° 30′	043	26° 29′	16·379
5	765	330	51° 3′	068	25° 41′	14·408
55	801	295	42° 57′	087	24° 49′	12·815
6	833	257	37° 7′	099	23° 55′	11·582
65	862	216	32° 41′	109	23° 1′	10·638
7	888	173	29° 13′	117	22° 6′	9·932
75	911	129	26° 22′	121	21° 12′	·189
8	932	084	23° 58′	126	20° 17′	8·732
85	952	038	21° 52′	129	19° 22′	·305
9	969	·0991	20° 3′	131	18° 27′	7·986
·70	·3001	896	16° 56′	133	16° 38′	·379
1	028	800	14° 22′	132	14° 48′	6·953
2	050	702	12° 8′	130	12° 58′	·647
3	070	604	10° 8′	129	11° 8′	·380
5	099	406	6° 41′	126	7° 28′	·053
7	117	207	3° 37′	124	3° 48′	5·862
9	124	− 007	+ 0° 43′	124	$\pi + \ 0° \ 8'$	·789
·81	− ·3122	+ ·0193	− 2° 8′	·3128	$\pi - \ 3° \ 32'$	5·847

Integrating $2n/V$ from the completion of the half circuit to $s = ·52$, I find
$$\int_{·52}^{\frac{1}{2}S} \frac{2n}{V}\, ds = 130° \ 33',$$ and combining this with the previous integral, we have
$nT = 239° \ 43'$.

Stability of $x_0 = 1·0941$, $C = 39·0$.

Comparison

	Computed Φ	Synthesis		Computed Φ	Synthesis
a_0	2·59	1·76	a_6	5·51	8·34
a_1	4·27	5·24	a_7	− 8·43	−11·01
a_2	8·89	7·68	a_8	− 13·95	− 13·86
a_3	18·68	19·65	a_9	− 0·87	+ 3·55
a_4	44·10	44·18	a_{10}	+ 31·93	+ 39·87
a_5	41·49	39·87	a_{11}	− 18·92	− 4·86
			a_{12}	− 18·28	− 33·96

$$\Phi_0 = 8·74$$

The computed and synthetic values of Φ present some concordance, but the representation of Φ by the harmonic series is unsatisfactory.

FAMILY A' OF SATELLITES *continued*.

The harmonic constituents being however used in the determinant give $\Delta \sin^2 \frac{1}{2}\pi \sqrt{\Phi_0} = -1\cdot063$, $c = \cdot46 \sqrt{(-1)}$, modulus $= \cdot48$.

The orbit is very unstable with even instability.

C = 38·5   Figure-of-eight orbit, $x_0 = 1\cdot1164$.

This orbit was exceedingly troublesome, and the coordinates were found by several interpolations. After the calculations were completed an error was discovered which may be substantially corrected by increasing all the arcs by ·0001. The following figures to three places of decimals suffice for drawing the curve with fairly close accuracy. I have not thought it worth while to recompute the whole, and only give the interpolated coordinates and function $2n/V$.

s	$x-1$	y	$2n/V$
·00	$+\cdot1164$	$+\cdot000$	2·20
4	12	40	·25
8	·099	78	·39
·12	79	·112	·63
6	52	41	·99
·20	$+$ 19	65	3·49
4	$-\cdot017$	80	4·13
8	57	85	·81
·32	96	77	5·12
6	·129	55	4·85
·40	56	25	·39
4	75	·090	·07
8	90	53	3·90
·52	·201	$+$ 15	·92
6	09	$-\cdot024$	4·13
·60	16	64	·63
4	22	·103	5·65
8	32	42	8·30
·70	42	59	11·83
2	60	67	15·43
4	76	58	10·58
6	90	43	8·20
8	98	24	6·88
·80	·304	06	·08
2	09	·086	5·59
4	13	66	·26
6	15	47	·05
8	17	27	4·88
·90	$-\cdot318$	$-\cdot007$	4·86

When y vanishes between $s = \cdot52$ and $\cdot56$, $\phi = \pi - 12°\ 6'$.

$$nT = 258°.$$

The stability was not worked out, but the orbit is obviously evenly unstable.

FAMILY A′ OF SATELLITES *continued.*

C = 38·0 Figure-of-eight orbit, $x_0 = 1·1305$.

The calculation of this orbit proved excessively troublesome, and the results given below are only obtained with sufficient accuracy to draw a good figure.

Two sets of curves were traced; in the first set I travelled in a positive direction, starting from points on the line SJ for which x_0 is greater than unity; in the second set I travelled in a negative direction, starting from points on the line SJ for which x_0 is less than unity. One member of each of these two families was finally selected, such that they might be approximately parts of a single orbit.

The first of these two orbits is found by interpolation between the two, namely $x_0 = 1·126$ and $x_0 = 1·134$.

(arc increasing)			(arc diminishing)		
s	$x-1$	y	s	$x-1$	y
·00	+ ·1305	+ ·000	·00	− ·3225	− ·000
4	27	40	− ·04	21	40
8	16	78	8	16	80
·12	·098	·114	·12	07	·119
6	75	47	6	·294	56
·20	47	75	8	83	73
4	+ 14	97	·20	70	88
8	− ·023	·211	1	61	93
·32	63	12	2	52	94
6	99	·196	3	42	92
·40	·128	68	4	34	85
4	50	34	5	29	77
8	67	·098	6	24	68
·52	81	61	7	21	59
6	90	+ 22	8	18	49
·60	97	− ·017	− ·30	− ·214	− ·129
4	·201	57			
8	05	96			
·72	− ·210	− ·135			

The period of the whole periodic orbit is given in round numbers by $nT = 299°$.

The orbit is obviously very unstable, and the instability is doubtless of the even type.

FAMILY A″ OF SATELLITES (see Paper 3).

C = 39·8 *Figure-of-eight orbit, $x_0 = 1·0103$.*

This orbit was computed as an unperturbed orbit with $x_0 = 1·01$, and was then completed backwards by quadratures from the point where the original quadratures begin. Hence the arcs are not measured from the line of syzygy but from the beginning of the original quadratures.

FAMILY A″ OF SATELLITES *continued.*

s	x	y	ϕ	
− ·05	1·01030	− ·00098	1° 42′	
[490	1029	·00000	1° 11′]	interpolated
45	0981	0398	12° 36′	
4	0817	0869	25° 27′	
35	0559	1296	36° 9′	
3	0238	1679	43° 26′	
− ·02	·99485	2335	53° 33′	
·00	7757	3330	65° 29′	
·00	·97757	·03330	65° 29′	
2	5891	4051	71° 55′	
4	3969	4603	75° 25′	
6	2024	5068	77° 37′	
8	0064	5474	78° 48′	
·10	·88102	5854	79° 12′	
2	6138	6232	78° 54′	
4	4178	6632	77° 56′	
6	2228	7076	76° 19′	
8	0294	7584	74° 12′	
·20	·78379	8162	72° 24′	
1	7425	8463	72° 47′	
15	6946	8606	74° 11′	
2	6461	8729	77° 48′	
225	6215	8775	81° 19′	
25	5967	8801	87° 3′	
275	5717	8795	$\pi - 83°$ 27′	
3	5475	8738	68° 44′	
325	5259	8613	50° 38′	
35	5090	8428	34° 56′	
375	4968	8210	24° 49′	
4	4877	7978	18° 39′	
425	4806	7738	14° 39′	
45	4749	7495	11° 58′	
5	4661	7003	8° 49′	
55	4593	6508	7° 2′	
6	4536	6011	6° 1′	
7	4443	5015	5° 8′	
·28	4357	4019	4° 53′	
·30	4178	2027	6° 36′	
2	3957	+ ·00040	7° 24′	
·34	·73648	− ·01936	$\pi - 10°$ 39′	

This orbit is nearly continuous with one which begins with $x_0 = ·6983$, $y = 0$, $\phi = \pi$, of which the values are given below. Continuity between the two is established with sufficient accuracy by

$$x = ·7354 \qquad y = - ·03000$$
$$= ·7320 \qquad\quad = - ·05000$$

These values are derived from the approximate analytical solution for a body moving in the neighbourhood of the point of zero force with $x_0 = ·6983$, as follows:—

s	x	y	ϕ
·00	·69830	·00000	π
2	9902	− ·01998	$\pi + 4°$ 9′
4	·70127	3985	8° 56′
·06	0550	5950	$\pi +16°$ 4′

FAMILY A″ OF SATELLITES *continued.*

s	x	y	ϕ
·07	·70876	− ·06893	$\pi + 22°\ 36'$
75	1089	7345	28° 24'
775	1216	7560	32° 3'
·08	1365	7761	40° 37'
·0825	1549	7929	56° 20'
·085	·71775	− ·07992	$\pi + 90°$ nearly

The periodic time may be computed with fair approximation by adding to the period of a satellite oscillating about the point of zero force twice the time occupied by the body in passing from the point $x_0 = 1·01029$, $y_0 = 0$ until y vanishes again for the first time. From this it appears that

$$nT = 280°.$$

C = 39·5　　　　　A″, *figure-of-eight orbit, $x_0 = 1·0036$.*

This orbit was extremely troublesome and perhaps is not very exact. After many trials it was computed entirely from the side towards S, and it appeared that x_0 should be ·6921.

For the earlier portion this is indistinguishable from another of which the following are the data:—

x	y
·6918	·0000
924	− ·0200
949	399
970	597
·7015	792
·7045	− ·0887

A number of approximate solutions were determined analytically by regarding the third body as moving in the neighbourhood of the point of zero force. Of these the orbits beginning with $x_0 = ·6920$ and ·6922 afford the following values:—

($x_0 = ·6920$)		($x_0 = ·6922$)	
x	y	x	y
·7020	− ·0801		
070	926		
126	998		
184	− ·1019		
244	− ·0987		
294	904		
351	774		
393	604	·7419	− ·0593
427	400	·7461	− ·0387
·7454	− ·0176	·7495	− ·0158

On pursuing these orbits further it appeared that ·6920 was too small and ·6922 too great. Accordingly an interpolation was effected for $x_0 = ·6921$, and this orbit was continued by the method sketched in the postscript to the

FAMILY A″ OF SATELLITES *continued.*

*foregoing paper (see p. 74).   I give the coordinates found in this way, as
follows :*—

x	y
·7511	+ ·0400
517	599
5201	788
5203	938
5196	·1045
522	104
535	114
567	078
·7631	·1002

Quadratures were resumed from here with the following results :—

x	y
·7665	·0966
701	931
738	897
777	865
857	805
940	751
·8116	655
300	577
489	512
681	457
875	409
·9071	366
266	324
462	281
655	232
·9846	·0167

From this point the orbit was computed as unperturbed, with the result
that perijove was found where $\rho = ·0036$, $\psi = 0°\ 30'$.   Hence within the limits
of error of computation the orbit cuts the line of syzygies at right angles.   The
period was not computed.

C = 38·85   *Orbit of Ejection from J towards S.*

To find the ejectional member of this family several orbits of ejection, viz.
for $C = 38·7,\ 39,\ 38·9$, were traced by the formulæ of *Part III of Paper 3*,
p. 164.

$C = 38·7$

The detailed values begin where the analytical formulæ of *Part III of
Paper 3* cease to be applicable.

s	x	y	φ
·10	·90017	·00578	81° 8′
2	·88027	0946	77° 39′
4	6090	1442	73° 32′
6	4198	2087	68° 41
8	2373	2903	63° 2′
·20	·80644	·03908	56° 36′

FAMILY A″ OF SATELLITES *continued.*

s	x	y	φ
·22	·79045	·05107	49° 33′
4	7612	6500	42° 1′
6	6378	8072	34° 16′
8	5367	9796	26° 31′
·30	4597	·11640	18° 48′
2	4085	3571	10° 49′
3	3934	4559	6° 25′
4	3867	5556	1° 4′
45	3873	6056	− 2° 41′
475	3891	6305	5° 31′
·35	·73939	·06549	−19° 43′

The curvature then becomes very great and the body moves off towards S.

C = 39

s	x	y	φ
·10	·90017	·00582	81° 2′
2	·88038	0955	77° 28′
4	6103	1459	73° 11′
6	4216	2120	68° 5′
8	2402	2960	62° 4′
·20	0695	4000	55° 6′
2	·79137	5251	47° 11′
4	7778	6715	38° 24′
6	6670	8378	28° 48′
7	6228	9275	23° 36′
8	5872	·10209	18° 0′
9	5615	1175	11° 42′
·30	5476	2164	+ 3° 59′
05	5462	2664	− 0° 54′
1	5496	3163	7° 12′
125	5536	3410	11° 14′
1375	5563	3532	13° 36′
15	5595	3653	16° 15′
·31625	·75633	·13772	−19° 17′

The continuation shows that the curvature is not sufficiently abrupt and that the body will return towards the neighbourhood of J.

Between the curves for C = 38·7 and 39 an interpolated orbit was computed beginning with s = ·22; as follows:—

C = 38·9

s	x	y	φ
·22	·79116	·05188	48° 10′
4	7725	6623	39° 54′
6	6564	8249	31° 3′
8	5677	·10039	21° 34′
·30	5117	1956	10° 41′
1	4989	2947	+ 3° 46′
15	4975	3447	− 0° 37′
2	5004	3946	6° 20′
25	5094	4437	15° 8′
275	5173	4674	22° 9′
3	5290	4895	33° 37′
3125	5367	4994	42° 37′
·3325	5461	5076	55° 34′

FAMILY A″ OF SATELLITES *continued*.

s	x	y	ϕ
·33375	·75573	·15130	$-73°\ 26'$
35	5697	5144	$\pi+86°\ 25'$
3625	5821	5116	$68°\ 42'$
375	5931	5057	$56°\ 11'$
3875	6029	4980	$47°\ 57'$
4	6117	4892	$42°\ 4'$
4125	6197	4796	$37°\ 40'$
425	6270	4695	$34°\ 15'$
45	6401	4482	$29°\ 7'$
475	6516	4260	$25°\ 39'$
5	6619	4032	$23°\ 2'$
55	6799	3566	$19°\ 22'$
6	6954	3091	$16°\ 49'$
7	7214	2125	$13°\ 42'$
8	7432	1149	$11°\ 43'$
·40	7797	·09182	$9°\ 51'$
2	8126	7209	$9°\ 20'$
4	8456	5236	$9°\ 50'$
6	8817	3269	$11°\ 5'$
·48	·79229	·01312	$\pi+12°\ 52'$

The curvature is not quite sharp enough and C should have been taken a little smaller.

The rest of the orbit A″ was constructed by the considerations adduced in Part IV of Paper 3, pp. 169, 171.

FAMILIES B AND C OF SATELLITES.

$C = 39·3$

These are two orbits which nearly coalesce. It would have been more interesting to find the orbits for that critical value of C for which they exactly coalesce, but on account of the difficulty of the search I have only found two orbits nearly coalescent.

Four orbits were computed, viz. $x_0 = 1·15$, $1·16$, $1·17$, $1·18$; the values of $\phi - \pi$ after a semi-circuit were found to be $- 6'·5$, $+ 1'·5$, $+ 2'·8$, $- 5'·4$.

If u_0, u_1, u_2, u_3 denote any functions connected respectively with the four orbits $x_0 = 1·15$, $1·16$, $1·17$, $1·18$ it appears that the two orbits for which the value of $\phi - \pi$ is exactly zero are given by

$$u_1 + ·1188\,(u_0 - u_1) + ·2127\,(u_1 - u_2) + ·0394\,(u_3 - u_1)$$

and

$$u_2 + ·0628\,(u_0 - u_2) + ·3133\,(u_2 - u_1) + ·3193\,(u_3 - u_2)$$

Putting the u's equal to $1·15$, $1·16$, $1·17$, $1·18$ we find $x_0 = 1·15747$, $x_0 = 1·17506$ for the two periodics.

The four computed orbits gave nT equal to $87°\ 15'$, $87°\ 52'$, $88°\ 46'$, $89°\ 51'$ respectively.

FAMILIES B AND C OF SATELLITES *continued*.

On applying the formulæ of interpolation to the values of $x-1$, y and nT I find the two periodics as follows :—

	orbit B		orbit C	
s	$x-1$	y	$x-1$	y
·00	+ ·15747	+ ·00000	+ ·17506	+ ·00000
3	5499	2986	7257	2986
6	4756	5889	6512	5888
9	3526	8620	5270	8614
·12	1825	·11085	3539	·11058
5	·09675	3172	1348	3098
8	7136	4756	·08761	4604
·21	4299	5717	5893	5462
4	+ 1317	5962	+ 2902	5616
7	− 1638	5475	− ·00043	5082
·30	4398	4315	2807	3923
3	6845	2588	5279	2234
6	8902	0412	7384	0102
9	·10519	·07889	9053	·07615
·42	1658	5119	·10232	4860
5	2296	+ 2191	0877	+ 1936
·48	− ·12418	− ·00802	− ·10961	− ·01058
	$nT=87°\ 41'$		$nT=89°\ 18'$	

The semi-arc of the periodic orbit B is ·47197, and that of C is ·46941.

The fifth place of decimals in the coordinates has been given, although it is perhaps frequently inaccurate.

Stability of orbit B, $x_0 = 1·15747$, $C = 39·3$.

Comparison

	Computed Φ	Synthesis		Computed Φ	Synthesis
a_0	2·887	2·879	a_8	7·427	7·418
a_2	4·240	4·243	a_9	4·594	4·602
a_3	6·165	6·152	a_{10}	2·676	2·677
a_4	9·024	9·042	a_{12}	1·209	1·215
a_6	12·925	12·931			

$$\Phi_0 = 6·393$$

The harmonic expansion represents Φ well.

The determinant Δ is negative, and $\Delta \sin^2 \frac{1}{2}\pi \sqrt{\Phi_0} = - ·0612$.

The modulus is 1·415, and the instability is not great; $c = ·156 \sqrt{(-1)}$.

The orbit is unstable.

Stability of orbit C, $x_0 = 1·17506$, $C = 39·3$.

Comparison

	Computed Φ	Synthesis		Computed Φ	Synthesis
a_0	3·736	3·725	a_8	6·123	6·119
a_2	5·507	5·517	a_9	3·948	3·956
a_3	7·862	7·834	a_{10}	2·430	2·431
a_4	10·715	10·749	a_{12}	1·199	1·185
a_6	11·641	10·663			

$$\Phi_0 = 6·489$$

FAMILIES B AND C OF SATELLITES *continued.*

The harmonic expansion represents Φ well.

The determinant gives, $\Delta \sin^2 \tfrac{1}{2}\pi \sqrt{\Phi_0} = \cdot0644$, $c = 2\cdot163$,

$$2\pi\left(\tfrac{1}{2}c - 1\right) = 80° \ 57', \ nT - 2\pi\left(\tfrac{1}{2}c-1\right) = 30° \ 31', \ 2\pi\left(1 - \frac{\tfrac{1}{2}c}{1 + (nT/2\pi)}\right) = 24° \ 27'.$$

The orbit is stable.

FAMILY B OF SATELLITES.

C = 39·0 $x_0 = 1\cdot1500.$

s	$x-1$	y	ϕ	ρ	ψ	$2n/V$
·00	+ ·1500	+ ·0000	0° 0'	·1500	0° 0'	2·975
4	459	397	11° 50'	512	15° 13'	3·016
8	337	777	23° 54'	546	30° 10'	·135
·12	136	·1122	36° 34'	597	44° 39'	·340
6	·0862	412	50° 29'	654	58° 36'	·611
·20	523	622	66° 27'	704	72° 8'	·876
4	+ 137	723	84° 44'	728	85° 27'	4·093
8	− ·0260	691	$\pi -$ 75° 30'	711	$\pi -$ 81° 16'	·021
·32	624	529	57° 27'	651	67° 48'	3·696
6	928	271	42° 13'	574	53° 52'	·335
·40	·1159	·0946	29° 3'	496	39° 13'	·174
4	316	579	17° 12'	438	23° 45'	2·832
8	395	+ 188	$\pi -$ 5° 28'	408	$\pi -$ 7° 41'	·738
·52	− ·1392	− ·0212	$\pi +$ 5° 56'	·1408	$\pi +$ 8° 40'	2·738
·4991		·0000	$\pi +$ 0° 1'			

$$nT = 96° \ 56'$$

Stability of $x_0 = 1\cdot1500$, $C = 39\cdot0$.

Comparison

	Computed Φ	Synthesis		Computed Φ	Synthesis
a_0	1·861	2·012	a_8	9·599	9·602
a_2	3·087	3·078	a_9	4·994	4·926
a_3	5·045	5·202	a_{10}	2·206	2·274
a_4	8·405	8·166	a_{12}	0·538	0·588
a_6	17·315	17·124			

$$\Phi_0 = 6\cdot924$$

The harmonic expansion represents Φ with fair accuracy.

The determinant Δ is negative, and $\Delta \sin^2 \tfrac{1}{2}\pi \sqrt{\Phi_0} = - \cdot4019$.

The instability is of the even type, the modulus is 0·58 and c is $0\cdot38 \sqrt{(-1)}$. The orbit is therefore very unstable.

C = 38·5 $x_0 = 1\cdot1497.$

The comparison of the orbits $x_0 = 1\cdot1500$ with a neighbouring orbit showed that the exactly periodic orbit would correspond with $x_0 = 1\cdot1497$, but the results for $x_0 = 1\cdot1500$ will be sufficiently exact.

FAMILY B OF SATELLITES *continued*.

s	$x-1$	y	ϕ	ρ	ψ	$2n/V$
·00	+·1500	+·0000	0° 0'	·1500	0° 0'	2·835
4	464	398	10° 24'	517	15° 12'	·880
8	356	782	20° 50'	566	29° 59'	3·020
·12	181	·1141	31° 27'	643	44° 1'	·264
6	·0941	460	42° 46'	737	57° 12'	·626
·20	639	721	55° 52'	837	69° 38'	4·119
4	+282	897	72° 23'	919	81° 33'	·668
8	−·0113	950	$\pi-$86° 53'	953	$\pi-$86° 41'	·972
·32	500	854	65° 19'	920	74° 54'	·708
6	831	631	48° 18'	830	63° 0'	4·106
·40	·1095	333	35° 24'	725	50° 36'	3·574
4	294	·0987	24° 28'	628	37° 20'	·191
8	427	610	14° 36'	552	23° 9'	2·946
·52	495	+217	$\pi-$ 5° 2'	511	$\pi-$ 8° 16'	·826
·56	−·1498	−·0183	$\pi+$ 4° 25'	·1509	$\pi+$ 6° 58'	2·821
·5418		·0000	$\pi+$ 0° 4'			

$$nT = 113° 20'$$

Stability of $x_0 = 1·1497$, $C = 38·5$.

The values of Φ were computed for $x_0 = 1·1500$, and were corrected by interpolation with values computed for $x_0 = 1·1475$, but the corrections were so small that they might have been omitted.

Comparison

	Computed Φ	Synthesis		Computed Φ	Synthesis
a_0	0·68	1·15	a_8	11·79	11·97
a_2	1·83	1·81	a_9	4·53	4·29
a_3	3·77	4·18	a_{10}	1·21	1·34
a_4	8·03	7·30	a_{12}	−0·82	−0·90
a_6	29·34	28·97			

$$\Phi_0 = 8·60$$

The representation of Φ by the harmonic series is fairly good.

The determinant is negative, and $\Delta \sin^2 \tfrac{1}{2}\pi \sqrt{\Phi_0} = -1·815$.

The orbit is very unstable with even instability; the modulus is ·313 and $c = ·70 \sqrt{(-1)}$.

C = 38·0 $x_0 = 1·1470$.

s	$x-1$	y	ϕ	ρ	ψ	$2n/V$
·00	+·1470	+·0000	0° 0'	·1470	0° 0'	2·660
4	437	398	9° 26'	491	15° 29'	·706
8	340	786	18° 40'	553	30° 23'	·850
·12	183	·1153	27° 39'	652	44° 17'	3·106
6	·0970	492	36° 36'	779	56° 58'	·497
·20	706	791	46° 19'	926	68° 29'	4·089
2	556	922	51° 58'	·2001	73° 54'	·482
4	391	·2036	58° 41'	073	79° 8'	·957
6	213	128	67° 4'	139	84° 16'	5·504
8	+023	189	77° 46'	189	89° 23'	6·042
·30	−·0175	209	$\pi-$89° 7'	216	$\pi-$85° 28'	·397
2	373	182	74° 56'	213	80° 18'	·353
·34	−·0558	·2108	$\pi-$62° 3'	·2181	$\pi-$75° 10'	5·932

FAMILY B OF SATELLITES continued.

s	x − 1	y	φ	ρ	ψ	$2n/V$
·36	− ·0735	+ ·1998	π − 51° 53′	·2126	π − 70° 3′	5·352
8	873	864	44° 6′	059	64° 53′	4·805
·40	·1004	713	37° 55′	·1986	59° 37′	·336
4	221	378	28° 22′	841	48° 28′	3·653
8	385	014	20° 7′	717	36° 13′	·214
·52	496	·0630	12° 19′	624	22° 51′	2·944
·56	− ·1555	+ ·0235	π − 4° 30′	·1573	π − 8° 36′	2·814
·5836	− ·1564	·0000	π − 0° 8′			

$$nT = 131° 45'$$

The final value of ϕ changes rapidly with the initial value of x, and therefore this is a very close approximation to the periodic orbit.

Stability of $x_0 = 1\cdot1470$, $C = 38\cdot0$.

Comparison

	Computed Φ	Synthesis		Computed Φ	Synthesis
a_0	− 0·402	2·265	a_8	12·358	13·083
a_2	0·670	− 0·363	a_9	3·160	1·931
a_3	2·899	5·403	a_{10}	− 0·241	1·439
a_4	6·413	2·487	a_{12}	− 2·174	2·271
a_6	59·339	56·777			

$$\Phi_0 = 12\cdot237$$

The representation of Φ by the harmonic series is poor, but it will suffice to give some idea of the degree of instability.

The determinant is negative, and $- \Delta \sin^2 \tfrac{1}{2}\pi \sqrt{\Phi_0} = 4\cdot55$.

The orbit is very unstable, with even instability; the modulus is about ·23 and $c = \cdot96 \sqrt{(-1)}$.

FAMILY C OF SATELLITES.

C = 39·0 $x_0 = 1\cdot2338$.

The periodic orbit was found by interpolation between $x_0 = 1\cdot230$ and $x_0 = 1\cdot235$, by the formula $\cdot24\,[x_0 = 1\cdot230] + \cdot76\,[x_0 = 1\cdot235]$. The following are the two computations:—

s	x − 1	y	φ	ρ	ψ	$2n/V$
·00	+ ·2300	+ ·0000	0° 0′	·2300	0° 0′	6·219
4	258	397	12° 10′	293	9° 59′	·259
8	128	774	25° 50′	265	19° 59′	·302
·12	·1905	·1105	42° 5′	202	30° 6′	·180
6	594	354	60° 24′	092	40° 21′	5·679
·20	221	494	77° 56′	·1929	50° 45′	4·833
4	·0824	526	π − 87° 10′	735	61° 38′	3·961
8	430	461	74° 5′	523	73° 36′	·223
·32	+ 060	311	61° 52′	312	87° 23′	2·652
6	− ·0269	085	49° 14′	118	π − 76° 4′	·221
·40	538	·0790	35° 13′	·0956	55° 46′	1·914
4	721	436	19° 19′	843	31° 11′	·719
8	795	+ 045	π − 1° 44′	795	π − 3° 14′	·643
·52	− ·0745	− ·0351	π + 16° 6′	·0823	π + 25° 12′	1·687
·4846		·0000	π + 0° 19′			

$$nT = 112° 26'$$

FAMILY C OF SATELLITES *continued*.

s	$x-1$	y	ϕ	ρ	ψ	$2n/V$
·00	+ ·2350	+ ·0000	0° 0′	·2350	0° 0′	6·594
4	306	397	12° 31′	340	9° 46′	·616
8	173	773	26° 41′	307	19° 35′	·640
·12	·1944	·1099	43° 32′	233	29° 30′	·434
6	627	340	62° 8′	108	39° 29′	5·780
·20	249	470	79° 9′	·1928	49° 39′	4·805
4	·0851	495	$\pi-$86° 43′	720	60° 21′	3·888
8	458	428	74° 15′	500	72° 13′	·147
·32	+ 087	281	62° 21′	284	86° 7′	2·581
6	− ·0244	059	49° 46′	087	$\pi-$77° 2′	·161
·40	516	·0766	35° 34′	·0945	56° 3′	1·859
4	700	413	19° 10′	813	30° 32′	·670
8	771	+ 021	$\pi-$ 1° 3′	771	$\pi-$ 1° 31′	·603
·52	− ·0715	− ·0374	$\pi+$17° 10′	·0807	$\pi+$27° 37′	1·661
·4821		·0000	$\pi-$ 0° 6′			

$$nT = 114° 4'$$

The interpolated coordinates for the periodic orbit are

$x-1$	y
·2338	+ ·0000
294	397
162	773
·1935	·1100
619	343
242	476
·0845	502
451	436
+ 081	288
− ·0250	065
521	·0772
705	418
777	+ 026
− ·0722	− ·0369

$$nT = 113° 41'$$

The arcs with which these orbits are computed are rather longer than is desirable, nor were quite sufficient pains taken to make the second approximations satisfactory. Thus the order of accuracy attained is not very high. It seemed however to be sufficient for the purpose.

Stability of $x_0 = 1·2338$, $C = 39·0$.

The values of Φ and of the determinant were computed for the two orbits between which the periodic orbit lies; the following are the results:—

$$x_0 = 1·230.$$

	Computed Φ	Comparison Synthesis		Computed Φ	Synthesis
a_0	5·40	5·57	a_8	4·47	4·58
a_2	10·65	10·71	a_9	3·06	3·04
a_3	16·30	16·40	a_{10}	1·93	1·99
a_4	18·44	18·38	a_{12}	0·47	0·47
a_6	9·69	9·70			

$$\Phi_0 = 8·065$$

FAMILY C OF SATELLITES *continued.*

The determinant gives $\Delta \sin^2 \frac{1}{2}\pi \sqrt{\Phi_0} = {\cdot}421$, $c = 2{\cdot}450$,

$$2\pi(\tfrac{1}{2}c - 1) = 80° 57', \quad nT - 2\pi(\tfrac{1}{2}c - 1) = 31° 29', \quad 2\pi\left(1 - \frac{\frac{1}{2}c}{1 + (nT/2\pi)}\right) = 23° 59'.$$

$$x_0 = 1{\cdot}235.$$

Comparison

	Computed Φ	Synthesis		Computed Φ	Synthesis
a_0	6·04	6·20	a_8	4·26	4·27
a_2	11·94	12·00	a_9	2·98	2·95
a_3	17·77	17·81	a_{10}	1·88	1·89
a_4	18·65	18·55	a_{12}	0·43	0·42
a_6	9·13	9·04			

$$\Phi_0 = 8{\cdot}176$$

The determinant gives $\Delta \sin^2 \frac{1}{2}\pi \sqrt{\Phi_0} = {\cdot}439$, $c = 2{\cdot}462$,

$$2\pi(\tfrac{1}{2}c - 1) = 83° 10', \quad nT - 2\pi(\tfrac{1}{2}c - 1) = 30° 54', \quad 2\pi\left(1 - \frac{\frac{1}{2}c}{1 + (nT/2\pi)}\right) = 23° 28'.$$

By interpolation between these two for $x_0 = 1{\cdot}2338$,

$$\Delta \sin^2 \tfrac{1}{2}\pi \sqrt{\Phi_0} = {\cdot}435, \quad c = 2{\cdot}459, \quad 2\pi(\tfrac{1}{2}c - 1) = 82° 38',$$

$$nT - 2\pi(\tfrac{1}{2}c - 1) = 31° 2', \quad 2\pi\left(1 - \frac{\frac{1}{2}c}{1 + (nT/2\pi)}\right) = 23° 35'.$$

The orbit is stable.

C = 38·75 $x_0 = 1{\cdot}28733.$

s	$x - 1$	y	ϕ	ρ	ψ	$2n/V$
·00	+ ·28733	+ ·00000	0° 0′	·28733	0° 0′	10·472
2	8693	1999	2° 18′	8763	3° 59′	·610
4	8568	3995	4° 55′	8846	7° 58′	11·044
6	8340	5982	8° 24′	8964	11° 55′	·862
7	8174	6968	10° 46′	9023	13° 53′	12·471
8	7962	7945	13° 54′	9069	15° 52′	13·239
9	7688	8906	18° 8′	9085	17° 50′	14·168
·10	7330	9839	24° 14′	9047	19° 48′	15·216
1	6856	·10719	32° 51′	8916	21° 46′	16·241
2	6237	1502	44° 17′	8647	23° 41′	·623
3	5465	2136	57° 2′	8210	25° 29′	15·899
4	4576	2590	68° 19′	7615	27° 8′	14·171
5	3621	2887	76° 31′	6907	28° 37′	12·217
6	2638	3068	82° 9′	6140	30° 0′	10·533
7	1644	3168	86° 8′	5335	31° 19′	9·156
8	0645	3210	89° 0′	4509	32° 37′	8·048
·20	·18648	3172	$\pi - 87°$ 1′	2830	35° 14′	6·421
2	6655	3018	84° 11′	1138	38° 1′	5·289
4	4670	2774	81° 47′	·19453	41° 3′	4·457
6	2697	2448	79° 27′	7781	44° 26′	3·815
8	0739	2040	76° 59′	6133	48° 16′	·302
·30	·08802	1544	74° 14′	4517	52° 41′	2·881
·32	+ ·06893	+ ·10949	$\pi - 71°$ 4′	·12938	57° 49′	2·527

FAMILY C OF SATELLITES *continued.*

s	$x-1$	y	ϕ	ρ	ψ	$2n/V$
·34	+·05023	+·10241	$\pi-67°$ 22'	·11406	63° 53'	2·225
6	3208	·09403	62° 56'	·09935	71° 10'	1·962
8	+1471	8413	57° 34'	8541	80° 5'	1·731
·40	−·00154	7248	50° 56'	7250	$\pi-88°$ 47'	·528
2	1614	5884	42° 39'	6101	74° 40'	·353
4	2833	4303	32° 13'	5152	56° 38'	·209
5	3321	3431	26° 5'	4774	45° 56'	·152
6	3707	2509	19° 20'	4477	34° 5'	·106
7	3978	1547	12° 3'	4269	21° 15'	·074
8	4122	0558	4° 24'	4159	$\pi-7°$ 43'	·057
·485	−·04143	+·00059	$\pi-0°$ 30'	·04143		1·054
·48559		·00000	$\pi-0°$ 2'			

$$nT = 179°\ 31'$$

Stability of $x_0 = 1·28733$, $C = 38·75$.

Comparison

	Computed Φ	Synthesis		Computed Φ	Synthesis
a_0	− 4·38	8·08	a_8	3·42	− 9·75
a_2	18·34	43·45	a_9	3·08	11·03
a_3	185·33	155·74	a_{10}	2·57	0·49
a_4	46·39	79·81	a_{12}	−3·08	8·88
a_6	6·22	15·65			

$$\Phi_0 = 23·02$$

The representation of Φ by the harmonic series is bad, but it may serve to give some idea of the degree of instability.

The determinant gives $\Delta \sin^2 \tfrac{1}{2}\pi \sqrt{\Phi_0} = 1·946$.

The instability is uneven; $c = 1 + ·55 \sqrt{(-1)}$; modulus = ·40.

$C = 38·5$ $x_0 = 1·2760.$

s	$x-1$	y	ϕ	ρ	ψ	$2n/V$
·00	+·2760	+·0000	0° 0'	·2760	0° 0'	7·516
2	759	200	0° 34'	766	4° 9'	·590
4	756	400	1° 3'	785	8° 15'	·829
6	752	600	1° 25'	816	12° 18'	8·258
8	746	800	1° 34'	861	16° 14'	·984
·10	741	·1000	1° 27'	918	20° 2'	10·212
2	737	200	1° 2'	988	23° 40'	12·467
3	735	300	0° 49'	·3028	25° 25'	14·561
4	734	400	1° 2'	071	27° 7'	18·411
45	732	450	1° 47'	093	27° 57'	22·00
5	730	500	4° 32'	115	28° 47'	29·20
525	727	524	8° 27'	124	29° 12'	36·46
55	721	549	21° 17'	131	29° 39'	53·80
5625	715	560	38° 47'	131	29° 52'	67·34
5750	705	567	72° 47'	126	30° 5'	81·66
5875	693	567	$\pi-71°$ 23'	115	30° 11'	62·13
6000	681	561	63° 45'	103	30° 12'	46·22
·16125	+·2671	+·1555	$\pi-58°$ 52'	·3090	30° 13'	37·74

FAMILY C OF SATELLITES continued.

s	$x-1$	y	ϕ	ρ	ψ	$2n/V$
·16250	+·2660	+·1549	$\pi-56°$ 23'	·3078	30° 12'	32·84
650	640	534	54° 39'	053	30° 10'	26·41
675	619	520	54° 4'	028	30° 7'	22·50
70	599	505	54° 2'	005	30° 7'	20·315
75	558	476	54° 53'	·2954	29° 59'	16·355
80	517	448	56° 8'	904	29° 54'	14·083
9	433	394	58° 59'	804	29° 49'	11·217
·20	346	344	61° 29'	704	29° 49'	9·406
2	167	256	65° 50'	505	30° 6'	7·150
4	·1982	179	69° 0'	306	30° 46'	5·748
8	603	050	72° 42'	·1916	33° 14'	4·027
·32	220	·0936	73° 32'	537	37° 29'	2·974
6	·0838	818	71° 41'	171	44° 18'	·234
·40	464	677	66° 35'	·0821	55° 36'	1·611
2	283	591	62° 9'	655	64° 24'	·412
4	+ 112	488	55° 11'	500	77° 4'	·182
6	− ·0041	360	44° 30'	362	$\pi-83°$ 27'	0·971
7	107	285	36° 53'	304	69° 25'	·876
8	160	200	27° 7'	256	51° 20'	·795
9	196	107	14° 56'	223	28° 36'	·737
·50	210	+ 008	$\pi-$ 0° 51'	210	$\pi-$ 2° 14'	·713
·51	−·0199	−·0091	$\pi+13°$ 29'	·0219	$\pi+24°$ 33'	0·729
·50084	−·02102	·0000	$\pi+$ 0° 21'			

$$nT = 210° 52'$$

A small change in x_0 makes a large change in the final value of ϕ, and it is therefore unnecessary to seek a more exact representation of the periodic orbit.

The stability was not computed, since the method would fail, but the orbit is obviously very unstable with uneven instability.

C = 38·0 $x_0 = 1·2480.$

s	$x-1$	y	ϕ	ρ	ψ	$2n/V$
·00	+·2480	+·0000	0° 0'	·2480	0° 0'	5·047
4	475	400	1° 32'	507	9° 11'	·176
8	460	800	2° 27'	586	18° 1'	·591
·12	444	·1199	+ 1° 50'	723	26° 9'	6·479
6	444	599	− 2° 30'	921	33° 12'	8·470
8	461	798	8° 1'	·3048	36° 10'	10·593
·20	510	991	22° 22'	204	38° 25'	15·63
1	561	·2076	41° 49'	297	39° 2'	22·07
15	599	108	60° 44'	345	39° 1'	25·62
2	646	122	−87° 11'	389	38° 44'	27·81
25	695	113	$\pi+65°$ 34'	424	38° 6'	26·60
3	736	084	46° 34'	440	37° 18'	22·64
35	768	046	34° 36'	442	36° 28'	19·60
4	793	003	25° 52'	437	35° 38'	17·03
5	827	·1908	$\pi+14°$ 12'	410	34° 2'	14·16
7	847	708	$\pi-$ 0° 26'	320	30° 58'	11·02
9	824	512	13° 15'	204	28° 9'	9·286
·31	759	323	24° 36'	060	25° 37'	8·070
3	660	150	34° 42'	·2898	23° 23'	7·072
·37	+·2384	+·0862	$\pi-51°$ 48'	·2535	19° 53'	5·462

FAMILY C OF SATELLITES continued.

s	x - 1	y	φ	ρ	ψ	2n/V
·41	+ ·2043	+ ·0655	π − 64° 26'	·2145	17° 46'	4·197
5	·1670	512	73° 2'	·1747	16° 50'	3·218
9	282	416	78° 16'	348	17° 58'	2·452
·53	·0889	343	80° 14'	·0953	21° 6'	1·824
5	692	309	79° 56'	758	24° 4'	·541
7	495	271	78° 21'	565	28° 45'	·266
8	398	250	76° 52'	470	32° 9'	·127
9	301	226	74° 39'	376	36° 56'	0·986
·60	205	196	71° 11'	284	43° 45'	·839
1	112	160	65° 24'	195	55° 0'	·682
2	+ 026	110	54° 0'	113	+76° 57'	·510
25	− ·0012	077	42° 46'	0783	π − 81° 6'	·420
30	039	036	23° 4'	0535	42° 22'	·347
325	− ·0047	+ ·0012	π − 9° 0'	·00481	π − 14° 32'	0·328
·63371	− ·0048	·0000	π − 1° 37'	·00478	π − 0° 0'	0·327

$$nT = 235° 17'$$

This orbit was not computed with high accuracy. As far as can be judged from other computations, the exactly periodic orbit would correspond to $x_0 = 1·2465$, but the calculations from which this is inferred were not conducted with the closest accuracy.

A very small difference in the initial value of x makes a considerable change in the size of the loop described. It would be very laborious to obtain the exact periodic orbit for this value of C, and the above appears to suffice.

The orbit is obviously very unstable, with uneven instability.

C = 37·5 $x_0 = 1·2225.$

s	x - 1	y	φ	2n/V
·00	·2225	·0000	0° 0'	3·934
·04	216	400	2° 33'	4·012
·08	191	799	4° 32'	4·258
·12	156	·1198	5° 20'	4·733
·16	122	596	+ 3° 56'	5·586
·20	111	995	− 1° 42'	7·280
·24	185	·2383	−23° 49'	11·82
·26	305	536	−56° 9'	16·78
·28	492	571	π + 71° 20'	16·51
·30	659	447	40° 1'	12·09
·32	762	377	27° 13'	9·92
·34	836	094	15° 14'	8·636
·36	872	·1897	π + 5° 53'	7·743
·38	884	697	π − 2° 51'	7·120
·42	838	307	19° 36'	6·193
·46	646	·0954	35° 14'	5·346
·50	378	660	49° 56'	4·508
·54	043	442	63° 34'	3·692
·58	·1675	287	72° 18'	2·948
·66	·0891	135	83° 29'	1·716
·70	493	097	85° 8'	1·165
·72	294	079	83° 59'	0·869
·74	096	049	77° 27'	0·498
·745	·0049	·0036	π − 72° 21'	0·368

FAMILY C OF SATELLITES *continued.*

This orbit was computed by Mr Craig before convenient formulæ were found for orbits of ejection. It is obviously nearly such an orbit and may be taken as a fairly close representation of the last member of the family C before those orbits pass into the retrograde class. It is shown in Fig. 3, Plate IV.

If we accept it as being periodic, its period in round numbers is

$$nT = 243°.$$

FAMILY A OF PLANETS.

C = 40·0 $x_0 = - \cdot 414.$

s	x	y	ϕ	r	θ	$2n/V$
·0	− ·4140	− ·0000	$\pi +$ 0° 0′	·4140	$\pi +$ 0° 0′	1·809
·1	032	992	12° 22′	152	13° 49′	·820
·2	·3715	·1938	24° 49′	191	27° 34′	·851
·3	199	·2793	37° 28′	246	41° 7′	·899
·4	·2507	·3512	50° 24′	314	54° 29′	·960
·5	·1670	·4055	63° 47′	385	67° 38′	2·030
·6	− ·0728	385	$\pi +$ 77° 42′	445	$\pi +$ 80° 34′	·093
·7	+ ·0265	474	− 87° 53′	482	− 86° 37′	·135
·8	·1249	309	73° 9′	486	73° 50′	·141
·9	·2159	·3901	58° 34′	459	61° 3′	·109
1·0	939	280	44° 27′	405	48° 8′	·045
·1	·3549	·2490	31° 9′	336	35° 4′	1·967
·2	969	·1585	18° 45′	274	21° 46′	·897
·3	·4191	·0612	7° 5′	235	8° 19′	·856
1·35	+ ·4228	− ·0114	− 1° 24′	·4229	− 1° 32′	1·848
1·3614	+ ·423	·000	− 0° 6′			

$$nT = 154° \ 13'$$

Although this is not strictly periodic, since the final value of ϕ is − 0° 6′, it is sufficiently near so to be accepted as such.

Stability of $x_0 = - \cdot 414$, $C = 40\cdot0$.

Comparison

	Computed Φ	Synthesis		Computed Φ	Synthesis
a_0	5·476	5·490	a_8	11·027	11·021
a_2	6·184	6·180	a_9	9·104	9·106
a_3	7·069	7·088	a_{10}	6·700	6·696
a_4	8·356	8·327	a_{12}	3·801	3·793
a_6	11·463	11·438			

$$\Phi_0 = 8\cdot051$$

The harmonic series represents Φ well. The determinant gives

$$\Delta \sin^2 \tfrac{1}{2}\pi \ \sqrt{\Phi_0} = \cdot9096, \ c = 2\cdot806, \ 2\pi \left(\tfrac{1}{2}c - 1\right) = 145° \ 0',$$

$$nT - 2\pi \left(\tfrac{1}{2}c - 1\right) = 9° \ 13', \ 2\pi \left(1 - \frac{\tfrac{1}{2}c}{1 + (nT/2\pi)}\right) = 6° \ 27'.$$

The orbit is stable.

FAMILY A OF PLANETS *continued*.

C = 39·5 $x_0 = -\cdot4240.$

The periodic orbit is found by interpolation between $x_0 = -\cdot426$ and $x_0 = -\cdot4$, by the formula $\cdot92228\,[x_0 = -\cdot426] + \cdot07772\,[x_0 = -\cdot4]$.

The following are the two computations:—

s	x	y	ϕ	r	θ	$2n/V$
0·0	− ·4260	− ·0000	$\pi +$ 0° 0′	·4260	$\pi +$ 0° 0′	1·861
·1	157	993	11° 52′	275	13° 26′	·874
·2	·3851	·1943	23° 49′	314	26° 46′	·905
·3	354	·2809	35° 51′	374	39° 57′	·959
·4	·2686	·3550	48° 18′	451	52° 53′	2·031
·5	·1871	·4127	61° 14′	531	65° 37′	·111
·6	− ·0947	501	74° 45′	600	$\pi +$78° 8′	·189
·7	+ ·0041	644	$\pi +$88° 52′	644	− 89° 30′	·242
·8	·1032	538	− 76° 35′	654	77° 11′	·256
·9	965	185	62° 5′	624	64° 50′	·221
1·0	·2783	·3614	48° 6′	560	52° 24′	·144
·1	·3443	·2866	34° 59′	481	39° 47′	·052
·2	924	·1991	22° 48′	399	26° 54′	1·962
·3	·4218	037	11° 30′	343	+ 13° 49′	·900
1·4	+ ·4324	− ·1044	− 0° 43′	·4324	− 0° 35′	1·878
1·4044	+ ·4324	·0000	− 0° 15′			

$$nT = 165°\ 0'$$

s	x	y	ϕ	r	θ	$2n/V$
·0	− ·4000	− ·0000	$\pi +$ 0° 0′	·4000	$\pi +$ 0° 0′	1·686
·1	·3899	993	11° 39′	024	14° 17′	·701
·2	599	·1945	23° 20′	091	28° 23′	·748
·3	111	·2817	35° 7′	197	42° 9′	·825
·4	·2455	·3570	47° 8′	333	55° 29′	·934
·5	·1654	·4165	59° 39′	481	68° 20′	2·067
·6	− ·0742	568	72° 57′	627	$\pi +$80° 47′	·218
·7	+ ·0241	740	$\pi +$87° 19′	746	− 87° 6′	·354
·8	·1234	655	− 77° 20′	817	75° 9′	·448
·9	·2167	304	61° 22′	819	63° 16′	·444
1·0	970	·3712	46° 7′	754	51° 20′	·348
·1	·3598	·2937	32° 15′	644	39° 13′	·204
·2	·4035	040	19° 51′	522	26° 49′	·063
·3	278	·1072	− 8° 33′	410	14° 4′	1·950
1·4	+ ·4334	− ·0075	+ 2° 10′	·4335	− 1° 0′	1·887
1·4075	+ ·4331	·0000	+ 2° 58′			

$$nT = 167°\ 31'$$

The interpolated coordinates for the periodic orbit are:—

x	y
− ·4240	− ·0000
137	993
·3831	·1943
335	·2810
·2668	·3552
·1854	·4130
− ·0931	506
+ ·0057	651
+ ·1048	− ·3547

FAMILY A OF PLANETS *continued*.

x	y
+ ·1981	− ·4194
·2798	·3622
·3455	·2872
933	·1995
·4223	040
+ ·4325	− ·0046

$$nT = 165° \; 12'$$

Stability of $x_0 = -·426,\; C = 39·5.$

The orbit $x_0 = -·426$ was treated for stability in place of the interpolated orbit $x_0 = -·424$.

Comparison

	Computed Φ	Synthesis		Computed Φ	Synthesis
a_0	5·73	5·73	a_8	11·94	11·94
a_2	6·54	6·54	a_9	9·42	9·42
a_3	7·59	7·59	a_{10}	6·57	6·57
a_4	9·14	9·14	a_{12}	3·25	3·25
a_6	12·71	12·71			

$$\Phi_0 = 8·565$$

The harmonic series represents Φ perfectly. The determinant gives

$$\Delta \sin^2 \tfrac{1}{2}\pi \sqrt{\Phi_0} = ·976, \quad c = 2·901, \quad 2\pi (\tfrac{1}{2}c - 1) = 162° \; 15',$$

$$nT - 2\pi (\tfrac{1}{2}c - 1) = 2° \; 47', \quad 2\pi \left(1 - \frac{\tfrac{1}{2}c}{1 + (nT/2\pi)}\right) = 1° \; 52'.$$

The orbit is stable, but approaches very near to instability.

The results would have been somewhat modified if we had operated on the true periodic orbit $x_0 = -·424$.

C = 39·0 $x_0 = -·434.$

(Computed with 3-figured logarithms and to tenths of degree.)

s	x	y	ϕ	r	$\theta + nt$
·0	− ·434	− ·000	$\pi + 0° \; 0'$	·434	$\pi + 0° \; 0'$
·1	24	99	11° 18′	36	18° 36′
·2	·395	·195	22° 36′	42	37° 12′
·3	48	·282	34° 12′	49	55° 36′
·4	·284	·359	46° 0′	57	$\pi + 74° \; 6'$
·5	04	·420	58° 24′	67	− 87° 24′
·6	·114	63	71° 30′	78	68° 54′
·7	− ·016	83	$\pi + 85° \; 24'$	84	52° 24′
·8	+ ·083	78	− 80° 0′	85	31° 30′
·9	·179	49	65° 24′	83	− 12° 42′
1·0	·264	·396	51° 12′	76	+ 5° 54′
·1	·334	25	38° 12′	67	24° 24′
·2	87	·241	26° 6′	56	42° 42′
·3	·422	·148	15° 6′	47	61° 6′
·4	·440	− ·048	− 4° 48′	43	+ 79° 36′
1·45	+ ·442	+ ·001	+ 0° 12′	·442	
1·446		·000	+ 0° 6′		

$$nT = 177° \; 0'$$

FAMILY A OF PLANETS *continued*.

Stability of $x_0 = - \cdot 434$, $C = 39 \cdot 0$.

Comparison

	Computed Φ	Synthesis		Computed Φ	Synthesis
a_0	5·489	5·434	a_8	13·595	13·609
a_2	6·507	6·527	a_9	10·271	10·247
a_3	7·442	7·529	a_{10}	6·507	6·527
a_4	9·721	9·637	a_{12}	2·627	2·638
a_6	14·870	14·828			

$$\Phi_0 = 9 \cdot 156$$

The harmonic expansion represents Φ well.

The determinant Δ is positive and $\Delta \sin^2 \tfrac{1}{2}\pi \sqrt{\Phi_0}$ is 1·027, and $c = 3 + \cdot 10 \sqrt{(-1)}$.

The modulus of instability is 2·1.

The orbit is unstable, with uneven instability, but the instability is slight.

C = 38·5 $x_0 = - \cdot 4440$.

s	x	y	ϕ	r	θ	2n/V
·00	− ·4440	− ·0000	$\pi +$ 0° 0′	·4440	$\pi +$ 0° 0′	1·916
·08	380	797	8° 33′	452	10° 19′	925
·16	203	·1576	17° 8′	489	20° 33′	955
·24	·3911	·2320	25° 47′	547	30° 41′	2·004
·32	509	·3011	34° 34′	624	40° 38′	071
·40	006	632	43° 35′	714	50° 23′	157
8	·2410	·4164	52° 57′	811	59° 57′	258
·56	·1733	589	62° 48′	906	69° 19′	368
·64	·0993	889	73° 15′	989	78° 31′	474
·72	− ·0210	·5045	$\pi +84°$ 22′	·5049	$\pi +87°$ 37′	560
·80	589	043	− 83° 58′	077	− 83° 20′	605
8	·1370	·4877	72° 5′	066	74° 18′	592
·96	·2101	555	60° 26′	016	65° 14′	523
1·04	755	095	49° 25′	·4935	56° 4′	413
·12	·3312	·3523	39° 14′	835	46° 46′	286
·20	765	·2864	29° 54′	730	37° 16′	163
8	·4109	143	21° 15′	634	27° 33′	058
·36	345	·1379	13° 13′	559	17° 37′	1·980
·44	475	− ·0591	− 5° 36′	514	− 7° 31′	935
1·52	− ·4502	+ ·0208	+ 1° 49′	·4507	+ 2° 39′	1·927
1·4992		·0000	− 0° 6′			

$$nT = 191° \ 21'$$

Stability of $x_0 = - \cdot 4440$, $C = 38 \cdot 5$.

After the computation of the stability had been completed a small mistake in the calculation of the orbit was detected in consequence of which the semi-arc of the periodic orbit was taken to be 1·4987 (instead of 1·4992 as above); it was not however thought to be worth while to recompute the stability.

FAMILY A OF PLANETS *continued*.

Comparison

	Computed Φ	Synthesis		Computed Φ	Synthesis
a_0	5·084	5·084	a_8	15·319	15·346
a_2	6·174	6·155	a_9	10·517	10·516
a_3	7·695	7·724	a_{10}	6·157	6·121
a_4	10·183	10·160	a_{12}	2·029	1·952
a_6	17·402	17·418			

$$\Phi_0 = 9 \cdot 786$$

The harmonic series represents Φ well.

The determinant Δ is positive and $\Delta \sin^2 \tfrac{1}{2}\pi \sqrt{\Phi_0} = 1 \cdot 078$, and $c = 3 + \cdot 176 \sqrt{(-1)}$.

The modulus is 1·25. The orbit is unstable, with uneven instability, but the instability is not great.

C = 38·0 $x_0 = - \cdot 455$.

s	x	y	ϕ	r	θ	$2n/V$
·00	− ·4550	− ·0000	$\pi +$ 0° 0′	·4550	$\pi +$ 0° 0′	1·954
·08	494	·0797	8° 4′	563	10° 4′	·964
·16	326	·1579	16° 10′	606	20° 3′	2·000
·24	050	·2329	24° 19′	672	29° 54′	·056
·32	·3669	·3032	32° 35′	760	39° 34′	·133
·40	190	672	41° 4′	864	49° 1′	·234
8	·2621	·4233	49° 50′	978	58° 14′	·354
·56	·1970	697	59° 14′	·5092	67° 14′	·496
·64	251	·5044	69° 18′	193	76° 4′	·631
·72	− ·0480	255	$\pi +$ 80° 15′	282	$\pi +$ 84° 47′	·770
·80	+ ·0316	310	− 87° 58′	310	− 86° 35′	·825
·88	·1107	197	75° 48′	316	77° 59′	·841
·96	856	·4921	63° 47′	259	69° 20′	·753
1·04	·2535	498	52° 34′	164	60° 36′	·611
·12	·3122	·3957	42° 15′	042	51° 44′	·445
·20	611	324	33° 7′	·4908	42° 38′	·281
·28	996	·2624	24° 43′	730	33° 17′	·140
·36	·4280	·1877	16° 58′	673	23° 41′	·031
·44	464	099	9° 39′	597	13° 50′	1·958
·52	550	− ·0304	− 2° 45′	561	− 4° 0′	·921
1·60	+ ·4540	+ ·0495	+ 4° 6′	·4567	+ 6° 14′	1·929
1·5505		·0000	− 0° 8′			

$$nT = 207° 9′$$

Stability of $x_0 = - \cdot 455$, $C = 38 \cdot 0$.

Comparison

	Computed Φ	Synthesis		Computed Φ	Synthesis
a_0	4·722	4·886	a_8	17·052	17·170
a_2	5·941	5·927	a_9	10·602	10·491
a_3	7·767	7·821	a_{10}	5·618	5·649
a_4	10·991	10·898	a_{12}	0·952	0·990
a_6	21·495	21·508			

$$\Phi_0 = 10 \cdot 666$$

FAMILY A OF PLANETS *continued*.

The representation of Φ by the harmonic series is good.

The determinant Δ is positive, and $\Delta \sin^2 \frac{1}{2}\pi \sqrt{\Phi_0}$ is 1·095.

The orbit is unstable and the instability is of the uneven type.

The modulus of instability is 1·14, and $c = 1 + ·193 \sqrt{(-1)}$.

C = 37·5

s	x	y	ϕ
·00	− ·4650	·0000	$\pi +$ 0° 0′
·08	·4597	− ·0798	7° 34′
·16	·4440	·1582	15° 8′
·24	·4181	·2338	22° 43′
·32	·3823	·3053	30° 23′
·40	·3373	·3713	38° 14′
·48	·2835	·4305	46° 26′
·56	·2216	·4810	55° 12′
·64	·1525	·5211	64° 51′
·72	− ·0774	·5482	75° 41′
·80	+ ·0016	·5598	$\pi + 87°$ 45′
·88	·0813	·5540	− 79° 24′
·96	·1577	·5308	66° 42′
1·04	·2273	·4917	55° 6′
1·12	·2884	·4402	44° 53′
1·20	·3401	·3792	35° 53′
1·28	·3821	·3112	27° 44′
1·36	·4146	·2382	20° 19′
1·44	·4377	·1616	13° 22′
1·52	·7517	·0829	6° 46′
1·60	·4566	− ·0031	− 0° 20′

$$nT = 225°$$

This orbit has not been added to Fig. 2, Plate IV. The stability was not computed, but the orbit is certainly unstable.

FAMILY (a) OF OSCILLATING SATELLITES.

C = 40·0 $x_0 = ·705$.

s	x	y	ϕ	$2n/V$
·00	+ ·7050	+ ·0000	− 0° 0′	14·622
1	053	100	3° 12′	·867
2	061	200	6° 43′	15·674
3	077	298	11° 3′	17·354
4	101	395	17° 30′	20·872
5	118	442	22° 44′	24·319
45	141	487	31° 8′	31·098
525	155	507	40° 55′	37·31
550	174	524	57° 32′	47·14
5625	185	529	− 71° 59′	54·71
·05750	+ ·7197	+ ·0531	$\pi + 88°$ 27′	55·66

FAMILY (a) OF OSCILLATING SATELLITES *continued.*

s	x	y	ϕ	$2n/V$
·05875	+ ·7210	+ ·0529	π + 69° 30′	54·34
6000	220	523	54° 9′	47·14
6125	230	514	44° 29′	41·87
6250	238	505	37° 8′	37·50
6375	245	495	31° 59′	33·98
6500	251	484	28° 7′	30·28
675	262	461	22° 43′	26·92
700	271	438	18° 55′	24·14
75	285	390	13° 17′	20·59
80	295	341	10° 9′	18·47
85	303	292	7° 55′	16·99
90	309	242	6° 10′	·027
95	313	192	4° 43′	15·335
·100	317	142	3° 27′	14·810
05	319	093	2° 20′	·516
10	321	+ 043	1° 18′	·329
·115	+ ·7322	− ·0007	π + 0° 18′	14·276
·11427		·0000	π + 0° 27′	

$$nT = 138° \; 20'$$

Stability of $x_0 = ·705$, $C = 40·0$.

The thirteen equidistant values of Φ show great irregularity. The values numbered 0, 1, 2, 3, 4 and 8, 9, 10, 11, 12 are all negative and lie between $-2·6$ and $-3·0$; the values numbered 5 and 7 are about $+8$, and the value numbered 6 is about $+800$.

The harmonic analysis led to results which showed that the representation of Φ by the series would be so bad that it would not be worth while to continue the calculation.

The orbit is obviously very unstable.

C = 39·0 $x_0 = ·6871$.

The coordinates for the periodic orbit were derived from the following by interpolation, as explained below.

s	x	y	ϕ	$2n/V$
·00	+ ·6870	+ ·0000	− 0° 0′	5·773
4	890	399	5° 44′	6·008
8	954	794	12° 58′	·893
·10	·7007	987	18° 4′	7·834
1	040	·1081	21° 29′	8·570
2	080	172	25° 58′	9·634
3	129	260	32° 31′	11·293
35	157	301	37° 14′	12·511
40	190	339	43° 46′	14·174
45	227	372	53° 38′	16·688
475	248	386	60° 40′	18·12
500	271	396	69° 41′	19·72
·1525	+ ·7295	+ ·1403	− 81° 8′	21·26

FAMILY (a) OF OSCILLATING SATELLITES *continued*.

s	x	y	ϕ	$2n/V$
·1550	+·7320	+·1404	π +85° 19′	21·96
575	344	399	71° 19′	·62
600	367	388	58° 48′	20·36
625	387	373	48° 46′	18·66
650	404	355	41° 4′	17·04
675	420	336	35° 6′	15·64
70	433	315	30° 25′	14·45
75	456	270	23° 37′	12·58
80	474	224	18° 59′	11·243
85	488	176	15° 26′	10·235
90	501	127	12° 47′	9·467
·20	519	029	9° 1′	8·350
1	533	·0930	6° 22′	7·584
2	542	830	4° 25′	·027
4	553	631	1° 54′	6·294
6	556	431	π + 0° 11′	5·875
8	555	231	π − 1° 7′	·653
·30	549	+ 031	2° 18′	·585
·32	+·7538	− ·0169	π − 3° 37′	5·656
·3031		·0000	π − 2° 23′	

$$nT = 146° \ 36'$$

The following are coordinates interpolated between the preceding and the loop of the figure-of-8 $x_0 = 1·0941$, in such a way as to give a periodic orbit:—

s	x	y
·00	+·6871	+·0000
4	892	400
8	956	795
·10	·7010	987
1	045	·1081
2	085	172
3	135	259
35	164	300
4	196	337
475	252	381
55	320	399
6	368	386
65	407	355
7	437	316
75	461	273
8	481	227
85	497	180
9	510	142
·20	534	027
1	549	·0929
4	573	645
6	583	446
8	587	246
·30	+·7588	+·0047

$$nT = 145° \ 40'$$

Stability of $x_0 = ·6870$, $C = 39·0$.

In order to try the determinantal process on one orbit which is obviously very unstable, I treated the first of the above as though it were periodic with the following results:—

FAMILY (a) OF OSCILLATING SATELLITES *continued*.

Comparison

	Computed Φ	Synthesis			Computed Φ	Synthesis
a_0	$-2\cdot7$	$+38\cdot6$		a_7	$+18\cdot2$	......
a_1	$-2\cdot7$	......		a_8	$-2\cdot2$	$+87\cdot0$
a_2	$-2\cdot9$	$-3\cdot2$		a_9	$-3\cdot3$	$+34\cdot7$
a_3	$-2\cdot9$	$+38\cdot3$		a_{10}	$-3\cdot3$	$+2\cdot6$
a_4	$-2\cdot4$	$+82\cdot9$		a_{11}	$-3\cdot3$	......
a_5	$+3\cdot7$	......		a_{12}	$-3\cdot3$	$+35\cdot8$
a_6	$+498\cdot9$	$+379\cdot5$				

$$\Phi_0 = 41\cdot2$$

The function Φ is obviously one which would require a very large number of terms of an harmonic series for adequate representation, and the above is very bad.

However with 17 rows I find $\Delta \sin^2 \tfrac{1}{2}\pi \sqrt{\Phi_0} = -148\cdot4$; $c = 2\cdot0 \sqrt{(-1)}$, modulus $= \cdot11$.

I think it is certain that the instability is of the even type, and is very great.

C = 38·5　　　　　　　　　　　$x_0 = \cdot6814$.

Two orbits were computed, namely, $x_0 = \cdot6817$, giving the final value of ϕ equal to $\pi + 5° 11'$ and $nT = 147° 46'$, and $\cdot6810$, giving final $\phi = \pi - 6° 26'$ and $nT = 151° 53'$. The arcs in the latter orbit were shorter than in the former throughout a portion of the curve. Interpolation between these two by the formula $\cdot446 (x_0 = \cdot6810) + \cdot554 (x_0 = \cdot6817)$ gives the following results :—

s	x	y	$2n/V$
$\cdot00$	$+\cdot6814$	$+\cdot0000$	$4\cdot85$
4	831	400	$\cdot98$
8	884	796	$5\cdot44$
$\cdot12$	982	$\cdot1183$	$6\cdot53$
4	$\cdot7055$	369	$7\cdot62$
6	153	543	$9\cdot70$
7	217	620	$11\cdot63$
8	295	675	$14\cdot46$
9	390	699	$17\cdot44$
$\cdot20$	482	662	$15\cdot27$
1	543	581	$11\cdot69$
2	584	491	$9\cdot61$
3	615	396	$8\cdot28$
4	637	299	$7\cdot36$
6	666	102	$6\cdot22$
8	682	$\cdot0903$	$\cdot50$
$\cdot30$	691	703	$5\cdot07$
2	695	504	$4\cdot79$
4	698	304	$\cdot61$
6	698	$+ 105$	$\cdot52$
$\cdot38$	$+\cdot7698$	$-\cdot0094$	$4\cdot52$
$\cdot37054$	$\cdot7698$	$\cdot0000$	

$$nT = 149° 36'$$

The orbit is obviously unstable, and the instability is of the even type.

FAMILY (a) OF OSCILLATING SATELLITES *continued*.

C = 38·0 $x_0 = ·676$.

This orbit was exceedingly troublesome, and the coordinates were found by several interpolations amongst the same orbits as those used in finding the figure-of-8 orbit $x_0 = 1·1305$. Two sets of curves were traced; in the first set I started from one side of the oval, and in the second from the other side. The two curves were so selected that they might join one another as nearly as may be. The period of this orbit was not determined.

(arc increasing)		(arc diminishing)	
x	y	x	y
+ ·676	+ ·000	+ ·778	+ ·009
77	40	78	+ ·011
82	80	79	31
90	·119	79	51
·704	56	79	71
13	74	78	·111
19	82	77	31
26	90	+ ·774	+ ·151
34	95		
43	98		
53	96		
60	89		
65	80		
68	71		
71	61		
73	51		
+ ·774	+ ·141		

nT undetermined

FAMILY (b) OF OSCILLATING SATELLITES.

C = 38·5 $x_0 = 1·2919$.

The following was computed :—

s	$x-1$	y	ϕ	$2n/V$
·00	+ ·29215	+ ·0000	− 0° 0′	8·52
4	932	400	2° 54′	9·00
8	971	797	9° 14′	10·84
·10	·3014	993	16° 10′	13·02
1	046	·1087	21° 56′	14·70
2	091	177	31° 49′	17·19
25	120	217	39° 5′	19·54
30	155	254	48° 21′	20·60
35	195	283	59° 56′	22·21
40	241	303	73° 40′	23·21
45	290	311	− 87° 38′	·00
50	340	307	π + 79° 36′	21·83
55	388	293	69° 0′	20·27
60	433	272	60° 51′	18·70
·165	+ ·3475	+ ·1245	π + 54° 15′	17·32

FAMILY (b) OF OSCILLATING SATELLITES *continued*.

s	$x-1$	y	ϕ	$2n/V$
·17	+ ·3514	+ ·1214	$\pi+48°$ 59′	16·21
8	584	143	40° 54′	14·40
9	645	064	34° 51′	13·09
·20	699	·0980	30° 5′	12·15
2	787	801	22° 20′	10·89
4	853	612	16° 22′	·12
6	900	418	11° 13′	9·63
8	931	220	6° 28′	·36
·30	+ ·3945	+ ·0021	$\pi+$ 1° 55′	9·25
·30209		·0000	$\pi+$ 1° 27′	

$$nT = 213° 52'$$

The above, not being exactly periodic, was corrected by extrapolation from the orbit $x_0 = 1·295$, which gave $\pi + 7° 58'$ as the final value of ϕ. The corrected coordinates are :—

s	$x-1$	y
·00	+ ·2919	+ ·0000
4	929	400
8	968	797
·10	·3009	993
1	041	·1088
2	085	178
25	113	219
3	147	256
35	187	286
4	233	306
45	282	314
5	332	311
55	380	297
6	425	275
7	505	216
8	575	145
9	635	065
·20	687	·0979
2	772	799
4	835	609
6	879	413
8	905	214
·30	+ ·3915	+ ·0014

C = 38·0 $x_0 = 1·25945$.

The following orbit was computed :—

s	$x-1$	y	ϕ	$2n/V$
·00	+ ·2600	+ ·0000	− 0° 0′	5·399
8	607	800	1° 4′	6·030
·12	625	·1199	4° 51′	7·152
6	693	592	16° 33′	9·480
8	772	776	29° 45′	11·822
9	829	858	40° 37′	13·133
·20	903	925	55° 9′	14·339
·21	+ ·2992	+ ·1970	− 72° 20′	14·822

FAMILY (b) OF OSCILLATING SATELLITES *continued.*

s	$x-1$	y	ϕ	$2n/V$
·22	+ ·3090	·1986	− 89° 9′	14·306
3	190	974	π +77° 10′	13·153
4	284	943	66° 53′	11·932
5	373	897	59° 2′	10·935
7	532	778	48° 3′	9·423
9	671	634	40° 12′	8·410
·33	892	309	28° 46′	7·231
7	·4056	·0945	20° 15′	6·567
·41	171	563	13° 15′	·202
·45	+ ·4241	+ ·0169	π + 6° 59′	6·005
·4670	+ ·4258	·0000	π + 4° 27′	

$$nT = 214° \ 40'$$

Interpolation between the above and a neighbouring orbit gave the following coordinates for the periodic orbit:—

s	$x-1$	y
·00	·2595	·0000
8	600	800
·12	616	·1199
6	681	593
8	757	778
9	812	861
·20	884	929
1	973	975
2	·3071	992
3	170	980
4	264	948
5	352	900
7	508	777
9	642	630
·33	852	299
7	·4001	·0931
·41	095	546
·45	·4139	·0154
·4656	·4149	·0000

$$nT = 208°$$

2.

ON CERTAIN DISCONTINUITIES CONNECTED WITH
PERIODIC ORBITS.

By S. S. HOUGH, His Majesty's Astronomer at the Cape of Good Hope.

[*Acta Mathematica*, Vol. XXIV. (1901), pp. 257—288.]

IN the final part of his work on *Celestial Mechanics* which has lately appeared M. Poincaré devotes some space to the consideration of the orbits discussed by Professor Darwin in his recent memoir on *Periodic Orbits* [Paper 1]. From considerations of analytical continuity M. Poincaré has been driven to the conclusion that Professor Darwin is in error in classifying together certain orbits of the form of a figure-of-8 and others which he designates as satellites of the class A. "Je conclus" says Poincaré "que les satellites A instables ne sont pas la continuation analytique des satellites A stables. Mais alors que sont devenus les satellites A stables?"

Besides the question here raised by Poincaré a second immediately presents itself. After explaining the disappearance of the stable orbits A it is necessary also to give a satisfactory account of the origin of the unstable orbits A. These questions had occupied my mind prior to the publication of M. Poincaré's work, and the present paper contains in substance the conclusions at which I had arrived in connection with them *.

It will be seen that the difficulties which have occurred in following up the changes in form of Darwin's orbits arise in some measure from the omission to take into account the orbits described in the present paper as "retrograde," and the failure to recognize the analytical continuity between these orbits and the direct orbits. It had been my intention to defer publication of my conclusions until I had made an exhaustive examination of the retrograde orbits with something approaching the completeness devoted by Darwin to the direct orbits, but as I see little prospect of

---

* [The error is corrected in the present republication, and the unstable orbits called A originally are now called A'. G. H. D.]

obtaining the necessary leisure for so vast an undertaking in the immediate future I have thought it desirable to announce the results at which I have arrived, with some confidence that a closer investigation will prove them to be correct in their essential features though possibly subject to modification as regards details largely of a speculative character.

A summary of the contents of the paper and of the conclusions derived will be found in the last section.

§ 1.   On the form of an orbit in the neighbourhood of a point of zero force.

We shall throughout adopt the notation of Professor Darwin. Thus S will denote the Sun, J a planet Jove, ν the ratio of the mass of the Sun to that of Jove whose mass is unity, n the angular velocity of J about S.

Then the equations of motion of a satellite of infinitesimal mass referred to rectangular axes rotating with uniform angular velocity n about the centre of gravity of S and J, the origin being at the point S and the axis of x coinciding with the line SJ, are

$$\left. \begin{aligned} \frac{d^2x}{dt^2} - 2n\frac{dy}{dt} &= \frac{\partial\Omega}{\partial x} \\ \frac{d^2y}{dt^2} + 2n\frac{dx}{dt} &= \frac{\partial\Omega}{\partial y} \end{aligned} \right\} \dots\dots\dots\dots\dots\dots\dots(1)$$

where
$$2\Omega = \nu\left(r^2 + \frac{2}{r}\right) + \rho^2 + \frac{2}{\rho}\dots\dots\dots\dots\dots(2)$$

the length SJ being taken as unity, and r, ρ denoting the distances of the satellite from S, J respectively.

These equations admit of Jacobi's integral

$$V^2 = \left(\frac{dx}{dt}\right)^2 + \left(\frac{dy}{dt}\right)^2 = 2\Omega - C\dots\dots\dots\dots\dots(3)$$

where V denotes the velocity of the satellite relatively to the moving axes.

The points of zero force at which a satellite might remain in a position of relative equilibrium are determined by the equations

$$\frac{\partial\Omega}{\partial x} = 0, \quad \frac{\partial\Omega}{\partial y} = 0 \dots\dots\dots\dots\dots\dots(4)$$

The positions of these points have been examined by Darwin who finds that there are three of them situated on the line SJ and two more at the vertices of the equilateral triangles described on the line SJ.

Now suppose that x_0, y_0 are the coordinates of one of these points and that ξ, η are the coordinates of the satellite referred to it as origin, so that

$$x = x_0 + \xi, \quad y = y_0 + \eta$$

Then if ξ, η be sufficiently small we may expand $\dfrac{\partial \Omega}{\partial x}$, $\dfrac{\partial \Omega}{\partial y}$ in ascending powers of ξ, η and by Taylor's theorem we shall obtain

$$\frac{\partial \Omega}{\partial x} = \xi \frac{\partial^2 \Omega}{\partial x_0^2} + \eta \frac{\partial^2 \Omega}{\partial x_0 \partial y_0} + \frac{1}{2}\left[\xi^2 \frac{\partial^3 \Omega}{\partial x_0^3} + 2\xi\eta \frac{\partial^3 \Omega}{\partial x_0^2 \partial y_0} + \eta^2 \frac{\partial^3 \Omega}{\partial x_0 \partial y_0^2} \right] + \dots$$

$$\frac{\partial \Omega}{\partial y} = \xi \frac{\partial^2 \Omega}{\partial x_0 \partial y_0} + \eta \frac{\partial^2 \Omega}{\partial y_0^2} + \frac{1}{2}\left[\xi^2 \frac{\partial^3 \Omega}{\partial x_0^2 \partial y_0} + 2\xi\eta \frac{\partial^3 \Omega}{\partial x_0 \partial y_0^2} + \eta^2 \frac{\partial^3 \Omega}{\partial y_0^3} \right] + \dots$$

Thus in the immediate neighbourhood of a point of zero force x_0, y_0 the motion of the satellite will be approximately determined by the equations

$$\left.\begin{aligned}\frac{d^2 \xi}{dt^2} - 2n\frac{d\eta}{dt} &= \xi\, \frac{\partial^2 \Omega}{\partial x_0^2} + \eta\, \frac{\partial^2 \Omega}{\partial x_0 \partial y_0} \\[2mm] \frac{d^2 \eta}{dt^2} + 2n\frac{d\xi}{dt} &= \xi\, \frac{\partial^2 \Omega}{\partial x_0 \partial y_0} + \eta\, \frac{\partial^2 \Omega}{\partial y_0^2}\end{aligned}\right\} \dots\dots\dots\dots\dots(5)$$

where terms involving squares and products of ξ, η have been omitted. These equations being linear, it will be possible to obtain particular solutions of them by assuming for ξ, η the forms $ae^{\lambda t}$, $be^{\lambda t}$. On substituting these forms in the differential equations (5) we find

$$\left.\begin{aligned}a\left(\lambda^2 - \frac{\partial^2 \Omega}{\partial x_0^2} \right) - b\left(2n\lambda + \frac{\partial^2 \Omega}{\partial x_0 \partial y_0} \right) &= 0 \\[2mm] b\left(\lambda^2 - \frac{\partial^2 \Omega}{\partial y_0^2} \right) + a\left(2n\lambda - \frac{\partial^2 \Omega}{\partial x_0 \partial y_0} \right) &= 0\end{aligned}\right\} \dots\dots\dots\dots(6)$$

whence, on eliminating a, b, we obtain for the determination of λ the biquadratic equation

$$\left(\lambda^2 - \frac{\partial^2 \Omega}{\partial x_0^2} \right)\left(\lambda^2 - \frac{\partial^2 \Omega}{\partial y_0^2} \right) + 4n^2\lambda^2 - \left(\frac{\partial^2 \Omega}{\partial x_0 \partial y_0} \right)^2 = 0$$

or

$$\lambda^4 + \lambda^2 \left\{ 4n^2 - \frac{\partial^2 \Omega}{\partial x_0^2} - \frac{\partial^2 \Omega}{\partial y_0^2} \right\} + \left\{ \frac{\partial^2 \Omega}{\partial x_0^2}\frac{\partial^2 \Omega}{\partial y_0^2} - \left(\frac{\partial^2 \Omega}{\partial x_0 \partial y_0} \right)^2 \right\} = 0 \dots\dots\dots(7)$$

The character of the motion indicated by the equations (5) will turn on the nature of the roots of this equation. Now the conditions implied by the equations (4) involve that the point x_0, y_0 is a singular point of the curve belonging to the family $\Omega = \text{const.}$, which contains it, while the nature of the singularity for the different points of zero force has been examined by Darwin. For those points which lie on the axis of x he finds that there will be two real intersecting branches, and thus at these points

$$\frac{\partial^2 \Omega}{\partial x_0^2}\frac{\partial^2 \Omega}{\partial y_0^2} - \left(\frac{\partial^2 \Omega}{\partial x_0 \partial y_0} \right)^2$$

will be negative.

Hence the values of λ^2 derivable from the equation (7) will be both real for these points, but they will be of opposite signs. Also at points on the axis of x

$$\frac{\partial^2 \Omega}{\partial x_0 \partial y_0} = 0$$

which introduces some simplification into the equations (6). We propose only to concern ourselves with the points of zero force which lie on the axis of x. Suppose that for one of these points the two values of λ^2 derivable from equation (7) are

$$\lambda^2 = \alpha^2, \quad \lambda^2 = -\beta^2$$

where α, β are real quantities. We then obtain the following particular solutions of (5)

$$\xi = a_1 e^{\alpha t}, \qquad \eta = a_1 \frac{\alpha^2 - \partial^2 \Omega / \partial x_0^2}{2n\alpha} e^{\alpha t} \quad \dots\dots\dots\dots(i)$$

$$\xi = a_2 e^{-\alpha t}, \qquad \eta = -a_2 \frac{\alpha^2 - \partial^2 \Omega / \partial x_0^2}{2n\alpha} e^{-\alpha t} \quad \dots\dots\dots\dots(ii)$$

$$\xi = a_3 e^{\beta i t}, \qquad \eta = -a_3 \frac{\beta^2 + \partial^2 \Omega / \partial x_0^2}{2n\beta i} e^{\beta i t} \quad \dots\dots\dots\dots(iii)$$

$$\xi = a_4 e^{-\beta i t}, \qquad \eta = a_4 \frac{\beta^2 + \partial^2 \Omega / \partial x_0^2}{2n\beta i} e^{-\beta i t} \dots\dots\dots\dots(iv)$$

and the general solution, involving the four arbitrary constants a_1, a_2, a_3, a_4 will be obtained by adding together these particular solutions.

If we put

$$a_1 + a_2 = h, \quad a_1 - a_2 = k, \quad a_3 + a_4 = H, \quad a_3 - a_4 = -Ki$$

and write for brevity γ^2 in place of $\dfrac{\partial^2 \Omega}{\partial x_0^2}$, we obtain as the general solution of (5) involving four arbitrary constants h, k, H, K

$$\left. \begin{aligned} \xi &= h \cosh \alpha t + k \sinh \alpha t + H \cos \beta t + K \sin \beta t \\ \eta &= \frac{\alpha^2 - \gamma^2}{2n\alpha} (h \sinh \alpha t + k \cosh \alpha t) - \frac{\beta^2 + \gamma^2}{2n\beta} (H \sin \beta t - K \cos \beta t) \end{aligned} \right\} \dots(8)$$

This solution being free from imaginary quantities is capable of a real physical interpretation.

To avoid circumlocution we shall speak of the region surrounding a point of zero force L within which the equations (5) may be regarded as giving an approximation to the motion as the "domain" of L. If then a satellite be initially in the domain of L it will be possible to determine four constants h, k, H, K so that the equations (8) will represent its motion at least for a finite time, but the terms involving the hyperbolic functions will rapidly increase with t so that the satellite will depart from the domain of L, and only a short length of its path will be sensibly represented by these equations.

In like manner whatever be the initial circumstances, if the satellite should at any instant enter the domain of L, it will be possible to determine four arbitrary constants h, k, H, K so that the equations (8) will represent its motion so long as it remains within this domain. Let us then examine the character of the motion represented by the equations (8).

First suppose that a satellite is describing a path within the domain of L such that $h = 0$, $k = 0$. Its motion will then be given by

$$\left. \begin{aligned} \xi &= H \cos \beta t + K \sin \beta t \\ \eta &= -\frac{\beta^2 + \gamma^2}{2n\beta} (H \sin \beta t - K \cos \beta t) \end{aligned} \right\} \quad \dots\dots\dots\dots(9)$$

Without loss of generality we may put $K = 0$, since this evidently only involves a change in the epoch from which t is measured. We thus have

$$\xi = H \cos \beta t, \qquad \eta = -\frac{\beta^2 + \gamma^2}{2n\beta} H \sin \beta t$$

The path of the satellite referred to the moving axes is therefore elliptic and the satellite will not tend to leave the domain of L.

Next suppose that $H = 0$, $K = 0$ so that the motion is given by

$$\left. \begin{aligned} \xi &= h \cosh \alpha t + k \sinh \alpha t \\ \eta &= \frac{\alpha^2 - \gamma^2}{2n\alpha} (h \sinh \alpha t + k \cosh \alpha t) \end{aligned} \right\} \quad \dots\dots\dots\dots(10)$$

The path of the satellite is now a hyperbola whose centre is at L. The satellite will enter the domain of L along one branch of the hyperbola and after traversing the part of the curve which lies within the domain it will recede along another branch. Of course the path before entering and after leaving the domain of L may depart rapidly from the infinite branches of the hyperbola, but we are at present only concerned with the form of the orbit within the domain of L.

A special case of importance occurs when the path within the domain of L is such that $h^2 = k^2$. The hyperbola represented by (10) then degenerates into a pair of straight lines, coincident with the asymptotes. If h, k have like signs the satellite will then be continually receding from L along a straight line, whereas if h, k have unlike signs it will be continually approaching L, but it will not reach L until $t = +\infty$. In fact the nearer it approaches to L the smaller does its velocity become, so that it will not be able to reach this point within a finite time.

If now the initial circumstances of projection undergo continuous change in such a manner that the satellite always enters the domain of L, and that its motion within the domain may be represented by the equations (10), the quantities h, k and therefore also $h^2 - k^2$ will vary continuously. It may happen that in the course of the change $h^2 - k^2$ will pass through the value zero and change sign. This will imply that the infinite branch of the hyperbola along which the satellite enters the domain will cross the asymptote, and that consequently immediately after the change in sign of $h^2 - k^2$ the satellite will recede along the second asymptote in the opposite direction to that in which it receded before the change. It follows that, if two satellites be projected simultaneously under initial circumstances which differ infinitesimally, but so that, when they enter the domain of L, the values of $h^2 - k^2$ for their two paths have infinitesimal values with opposite signs, the paths of the satellite though differing infinitesimally prior to entering the domain of L will have lost all similarity of character before they depart from this region. The nearer however the hyperbolic paths approach to the asymptotes the longer will the satellites take in passing round the vertices, and consequently the smaller the difference in the initial circumstances the longer will be the interval before the separation commences.

We have so far for simplicity supposed that the satellite moves within the domain of L either in an elliptic path (9) or in a hyperbolic path (10). In general however its path will be represented by (8) which indicates an elliptic path superposed upon a hyperbolic path. We may form a conception of this motion by supposing the satellite to move in an ellipse whose form is represented by (9) while the centre of this ellipse moves along the hyperbola represented by (10).

The character of the motion of the centre of the ellipse will then be to some extent shared by that of the satellite, but, when the major axis of the hyperbolic path becomes small, the time which the centre of the ellipse takes to move round the vertex of the hyperbola will increase and the elliptic element of the motion will commence to shew its independent existence by the formation of loops in the orbit of the satellite. The nearer the hyperbola approaches to its asymptote the greater will be the number of loops described by the satellite prior to leaving the domain of L, until when the hyperbola actually coincides with an asymptote the number of loops will become infinite. The orbit of the satellite will then approach closer and closer to the simple elliptic orbit represented by (9), and will in fact be asymptotic to this orbit in the sense in which the term is used by Poincaré.

Except in the case just considered the satellite will finally recede from the domain of L in one of two essentially different ways according as the centre of the ellipse recedes along a branch of the hyperbola which approximates to one or other of the infinite arms of the second asymptote.

The "asymptotic" orbit just dealt with is the limiting orbit which separates those which leave in one way from those which leave in the other.

§ 2.   *Application to the Orbits of Professor Darwin.*

The results proved in the last section rigorously apply only to the very limited region surrounding a point of zero force within which the motion can be sensibly represented by the approximate equations (5), but there can be no reasonable doubt that the general characteristics will be maintained over a far more extensive field.   A good illustration is furnished by the orbits traced by Darwin.

For example the orbits of the "oscillating satellites" figured in Darwin's plates are closely analogous to the elliptic orbits represented by our equations (8).   They are in fact the orbits at which we should arrive by the continuous deformation of the elliptic orbits which would result from diminution of the constant of relative energy C.   By the time C has attained the values for which the figures have been drawn, these orbits have lost their symmetrical form but are still roughly elliptic in character.

Next consider the non-periodic orbits traced on pp. [57 and 60] of Darwin's memoir, viz. those started at right angles to the line SJ on the side of J remote from S with C equal to 39·0.   As x_0 (the abscissa of the point of projection) decreases and reaches a value in the neighbourhood of 1·095 the orbit begins to approach the region of the point of zero force L.   It however recedes from this region towards the planet J after describing a loop.   But when x_0 has the value 1·09375, or a smaller value, the orbit after passing near the orbit of the oscillating satellite no longer recedes towards the planet J, but towards the Sun S.   As in the last section we may regard the motion of the satellite when in the neighbourhood of the point of zero force as consisting of two independent motions, (1) a motion in a closed periodic orbit similar to that of the oscillating satellite, (2) a bodily transference of this closed orbit in virtue of which each point of the orbit is carried along a curve analogous to the hyperbola of the last section.   We may refer to these two parts of the motion briefly as the "elliptic" and the "hyperbolic" elements. Evidently the fate of a satellite after passing near L will turn on the character of the hyperbolic element of its motion, and the satellite will recede towards J or towards S according as the branch of the hyperbolic curve along which the elliptic orbit is carried recedes towards J or towards S.

The critical case which separates orbits receding towards J from those receding towards S, occurs when the hyperbolic element of the motion takes place in a curve which plays the part of the asymptotes in the last section. Each point of the periodic "elliptic" orbit then tends towards a fixed limiting position, but it takes an infinite time for it to reach this position. Consequently the satellite will describe an infinite series of loops each of which approximates closer than the preceding to the orbit of the oscillating satellite (a).   If the hyperbolic element of the motion differs only very slightly from its asymptotic form a large but finite number of loops approximating to the orbit (a) will be described, but the satellite will ultimately recede either towards J or towards S according as the hyperbolic path lies on one side or the other of its critical form.

We are thus able to describe the manner in which the interval between the orbits $x_0 = 1\cdot095$, $x_0 = 1\cdot09375$ is to be filled up. As x_0 decreases from $1\cdot095$, loops, the first of which has already shewn its existence in the figure traced, will be formed in gradually increasing numbers and these will tend to approximate in figure to the orbit of the oscillating satellite; the path however will ultimately fall away in the direction of the planet J.

At length a stage will be reached when an infinite number of loops will be described before the satellite recedes. The orbit will then approach the orbit of the oscillating satellite asymptotically after the manner of the "asymptotic orbits" treated of by Poincaré*.

As x_0 still further decreases the satellite will after describing at first a large number of loops recede towards S. The number of loops described will however rapidly diminish with x_0, until when $x_0 = 1\cdot09375$ all trace of them will have disappeared.

It will save circumlocution if we make use of the terms "lunar" and "planetary" to distinguish those of our orbits which, after passing near the orbit of the oscillating satellite, recede towards J from those which recede towards S. These terms are however at present only to be used to describe the character of the motion in the course of a single revolution round the primary. Thus an orbit which is "lunar" so far as its first approach to L is concerned might become "planetary" after two or more revolutions round the primary.

The lunar and planetary orbits will be separated from one another by orbits which are asymptotic to the orbit of the oscillating satellite. These orbits we shall speak of briefly as "asymptotic" orbits.

With large values of C, Darwin has shewn that any infinitesimal body moving in the plane of the orbit of J about S may be regarded either as a satellite, as an inferior planet, or as a superior planet, but that with smaller values of C it may be transferred from one category to another. The circumstances described in the present section are those which occur when an orbit is undergoing the change from that of a satellite to that of an inferior planet. It is clear that a similar sequence of events will occur when the orbit changes from that of a satellite or an inferior planet to that of a superior planet.

---

* In consideration of the fact that the orbit is symmetrical with respect to the line SJ it will be asymptotic to the orbit of the oscillating satellite for $t = -\infty$ as well as for $t = +\infty$, and will thus furnish an interesting illustration of one of Poincaré's " doubly asymptotic orbits."

§ 3. *On the relative orbit of a satellite which approaches indefinitely close to its primary.*

Imagine a satellite P to be moving subject solely to the attraction of its primary J. The orbit will then be a conic section.

Let us suppose that the initial circumstances are such that the orbit is an ellipse of large eccentricity. At perijove the satellite will then pass very near to its primary. Further suppose that the initial circumstances are varied continuously in such a manner that the distance of the satellite at perijove diminishes without limit, while the length and position of the major axis remain invariable. The elliptic orbit will become more and more flattened until it becomes sensibly a straight line except through very small portions of its length at perijove and apojove.

The same will be true in whichever direction the satellite is moving in its orbit and the two orbits which correspond to the two different directions of projection will approach the same limiting rectilinear form.

It is clear that if we suppose all the circumstances remote from the primary to undergo continuous variation the two forms of orbit may be regarded as continuations of one another, the rectilinear orbit forming the connecting link between the direct and the retrograde orbits, though physically only the part of this path between two successive perijove passages can be regarded as having a real existence, owing to the collision which would ensue between the satellite and primary at the instant which corresponds to the time of perijove passage.

There will be a similar connection between the direct and the retrograde orbits if we suppose that the initial circumstances are more general in character, admitting of change in the length and position of the major axis. If the initial conditions vary continuously the length and position of the major axis will likewise vary continuously, and, provided only that the length retains a finite value at the critical rectilinear stage, the direct and the retrograde orbits will merge into one another.

Let us next consider the figures of the relative orbit, when the motion is referred to axes which rotate uniformly. The motion may then be regarded as taking place in a moving ellipse, the line of apses of which revolves uniformly in a direction opposite to that of the rotation of the axes. We wish to consider the form of the relative orbit when this ellipse is very much flattened and approximates to the rectilinear form.

When the satellite is at perijove it is evident that the motion in the ellipse takes place very rapidly and therefore that the form of the path will resemble closely that which occurs when the axes are at rest. This results from the fact that the axes can only be very slightly displaced during the passage of the satellite round Jove. The more eccentric the ellipse the more closely will the relative orbit correspond with the actual orbit. Thus the motion in the relative orbit will be direct or retrograde according as that in the actual orbit is direct or retrograde *.

On the other hand when the satellite is in apojove the motion in the actual orbit will be very slow and the apparent motion in the relative orbit will be chiefly that due to the rotation of the axes themselves. Thus the apparent motion will be retrograde whatever be the direction of motion in the true orbit. It is then evident that the path from apojove will be of the form indicated in the annexed diagrams (figs. 1 and 2) according as the motion in the true orbit is direct or retrograde. In these figures the axes are supposed to be rotating in a counter-clockwise direction, and the direction of motion of the satellite is indicated by arrowheads.

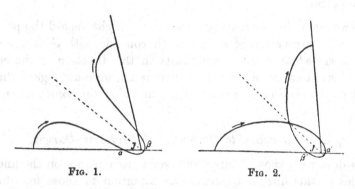

FIG. 1.   FIG. 2.

The critical form of orbit which corresponds with our previously rectilinear orbit, and which separates the orbits of the type represented in fig. 1 from those of the type represented in fig. 2, evidently possesses a cusp at J. Of course if a satellite moving in this orbit arrived at J a collision would occur and the physical continuity would be interrupted, but if we suppose that in the event of a collision the satellite rebounds without loss of energy in the direction opposite to that in which it fell into the primary the physical as well as the analytical continuity between the direct and retrograde orbits will be maintained.

So far we have regarded the satellite as moving subject to the attraction of J alone, but if it be moving under the disturbing influence of the Sun, its path when in the neighbourhood of J will still be governed by the laws of elliptic motion. If then all the circumstances remote from J undergo continuous variation in such a manner that the orbit approaches J and in the course of the change actually falls into J, it is evident that so far as its form in the immediate neighbourhood of J is concerned a series of changes will occur similar to that already described. The orbit will pass through the

* A " direct " motion implies motion in the same sense as that of the rotation of the axes.

cusped form and after the critical stage the direction of the motion round J will be reversed. If at first the satellite described an open loop round J as in fig. 1, it would afterwards describe a closed loop as in fig. 2 and *vice versâ*.

As the orbit approaches the critical form from either direction its form in the neighbourhood of J will approximate closer and closer to that of a parabola. Suppose α, α' are the points in which the orbits (figures 1 and 2) cut any line through J as the satellite approaches J and β, β' the points in which they cut this line as the satellite recedes. Then the points α, β, α', β' will all ultimately coincide with J, but the tangents to the orbits at α, β, and at α', β' being ultimately tangents at the extremities of a focal chord of a parabola will in their limiting position be at right angles. These limiting positions will be the two bisectors of the angles between the line $\alpha\beta$ and the limiting position of the axis of the parabola, i.e. the tangent at the cusp of the limiting orbit.

It follows from the figures that though we might regard the point α' as the analytical continuation of α since both coincide with J at the critical stage, if we wished to maintain continuity in the direction of the curve at the point where it crosses a given line, such as $\alpha\beta$, we must regard the point β' in fig. 2 as the continuation of α in fig. 1 and α' as the continuation of β.

§ 4. *Application to the orbits of Professor Darwin.*

Let us deal with those orbits which start from points on the line SJ at right angles to this line, and confine our attention to those for which the starting point lies outside SJ on the side of J remote from S. With a given value of the constant of relative energy two such orbits may be regarded as originating from each point, distinguishable by the initial direction of projection. We may without ambiguity designate these orbits as direct or retrograde according as the initial direction of motion is direct or retrograde.

It is now clear that if the starting point moves up towards J the direct and the retrograde orbits will approach the same limiting (cusped) form, and that so far as the circumstances in the remote parts of the orbits are concerned each form of orbit can be regarded as the analytical continuation of the other.

Next suppose that the region to which the starting point is confined is limited by a branch of the curve of zero velocity as in the figure (5) on p. [60] of Darwin's paper, and let us further suppose that the starting point moves up to its extreme limit in the opposite direction, viz. the point M where the curve of zero velocity cuts SJ. Now the form of an orbit in the neighbourhood of the curve of zero velocity has been dealt with by Darwin, who has shewn that at such a point the orbit will possess very large curvature. The limiting form will be cusped while the figures before and after the passage through the cusped form will be similar in character to those presented in our figs. 1 and 2 above.

Hence again we see that as the starting point approaches M the direct and retrograde orbits will approach the same limiting cusped form, the cusp being at M on the curve of zero velocity. Likewise also the direct and retrograde orbits may be regarded as continuations of one another.

As the starting point P moves along the line of syzygies the direct and retrograde orbits may then be regarded as forming a continuous cycle as P moves backwards and forwards between J and M. In this cycle when P arrives at J or M the orbit will assume the cusped form and an interchange from the direct to the retrograde will occur, or *vice versâ*. The tendency of the direct orbits to assume the cusped form is well indicated by the orbit $C = 39 \cdot 0$, $x_0 = 1 \cdot 001$ shewn by Professor Darwin (fig. 5).

§ 5. *Conjectural Forms of Retrograde Orbits**.

The cusped orbit $C = 40 \cdot 0$, $x_0 = 1$ has been computed in part by myself and independently by Professor Darwin. Its form is found to resemble that shewn in fig. 3 below. Again when x_0 reaches its extreme limit in the opposite direction the form of the cusped orbit is that shewn in fig. 4. We

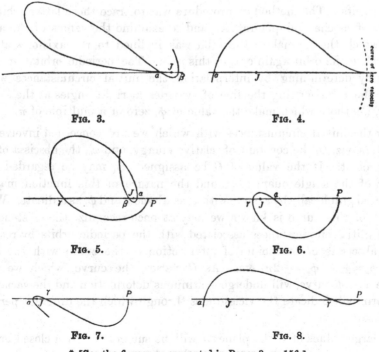

FIG. 3.　　　　　　　　　FIG. 4.

FIG. 5.　　　　　　　　　FIG. 6.

FIG. 7.　　　　　　　　　FIG. 8.

* [See the figures as computed in Paper 3, p. 156.]

may pass from one of these forms to the other either by following the direct orbits or by following the retrograde orbits.

The sequence of changes in the forms of the orbits which occur as we follow the direct orbits may be inferred from the results already given by Darwin, but as regards the retrograde orbits we have as yet no positive data available. It seems to me however that the probable sequence of changes is that indicated by figs. 5—8. In fig. 5 the starting point P has receded slightly from J and the direction of rotation round J at the next approach has been reversed by a passage through the cusped form.

In fig. 6 the large loop has diminished in size and the small loop (no longer represented in full) has increased, indicating another near approach of the satellite to the primary.

Fig. 7 represents the form of the orbit after this loop has passed through the cusped stage, while fig. 8 at once fills up the interval which remains between this orbit and that shewn in fig. 4.

§ 6. ϕ-curves.

The main object of Darwin's research was to investigate the forms of the different varieties of *periodic* orbits, but in order to discover these periodic orbits he has incidentally determined the forms of a large number of non-periodic orbits. The method of procedure was to trace these latter orbits by a process of mechanical quadratures, and to examine the values of the angle (ϕ_1) at which the normal to the orbit was inclined to the axis-of-x at the instant when the orbit again crosses this axis. The periodic orbits are then selected by determining by interpolation the initial circumstances which allow of the orbit cutting the line of syzygies at right angles at the second crossing, i.e. those which make the value of ϕ_1 zero or a multiple of π.

Now the initial circumstances with which we are concerned involve two parameters, viz. C, the constant of relative energy, and x_0, the abscissa of the starting point. If the value of C be assigned, ϕ_1 may be regarded as a function of the single quantity x_0, and the nature of this function may be represented graphically by a curve with x_0 as abscissa and ϕ_1 as ordinate. When the form of this curve is known we may at once recognize the existence of, and the initial circumstances associated with the periodic orbits by reading off the abscissae of the points of intersection of the curve with the lines $\phi_1 = 0$, $\phi_1 = \pm \pi$, $\phi_1 = \pm 2\pi$, &c. As C varies, the curve, which we shall describe as a ϕ-curve, will undergo continuous deformation and the variations in its form will indicate the vicissitudes through which the different periodic orbits pass.

For large values of C the planet J will be surrounded by a closed branch of the curve of zero velocity, and it therefore appears that two forms of discontinuity in the figures of the non-periodic orbits under discussion may present themselves; (1) where the satellite is instantaneously reduced to rest by attaining a point on the curve of zero velocity, and (2) where the satellite falls into the primary.

The former case has been frequently met with by Darwin and it is clear from his figures that the crisis concerned involves no abrupt change at points on the orbit other than that where it occurs. Thus no discontinuity in the value of ϕ_1 will result.

No instance of the occurrence of the second event has been found by Darwin prior to the satellite first crossing the line of syzygies. A case has however been found ($C = 39{\cdot}0$, $x_0 = 1{\cdot}095$) where the satellite passes very close to its primary after twice crossing the axis-of-x, while similar instances appear to occur among the retrograde orbits in the critical forms which separate orbits of the characters represented in figs. 5, 6, 7.

Now the angle ϕ made by the normal with the axis-of-x at the points where the orbit cuts this axis must be regarded analytically as a multiple-valued function of x_0 having an infinite number of determinations, since the orbit will evidently cut the axis-of-x an infinite number of times. The complete ϕ-curve will then consist of an infinite number of branches each one of which corresponds with a particular crossing. In dealing with the direct orbits we define ϕ_1 as that particular determination which corresponds to the first crossing of the axis after a semi-revolution round the primary, the values which correspond to subsequent crossings being denoted by different suffixes (ϕ_2, ϕ_3, &c.).

In so far as we can pass from the direct orbits to the retrograde orbits, by continuous deformation through the cusped form, the particular crossing, to which the angle ϕ_1 belongs in the case of the retrograde orbits, may be defined as the geometrical continuation of that previously defined so long as the point under consideration does not fall into the planet J. This definition will however lead to ambiguity when the orbit falls into J. To remove this ambiguity we will suppose that before and after the passage through the cusped form, the crossings corresponding to the approach of the satellite to the primary are continuations of one another, as also are those which correspond to the recession of the satellite from the primary. The angle ϕ_1 will then be defined without ambiguity, and it is evident that if the sequence of changes indicated in figs. 3—8 be the correct sequence through which the orbits pass, the points marked with similar letters on these figures will correspond with one another, the determination ϕ_1 being that which belongs to the point α throughout.

From the result proved at the end of § 3 it follows that the angle ϕ_1 will no longer be a continuous function of x_0, but when the crossing to which ϕ_1 belongs falls into J the ordinate of the ϕ-curve will change abruptly by $\frac{1}{2}\pi$.

The points selected as the continuations of one another are in fact not the true analytical continuations of one another so that when the orbit passes through the critical form we transfer our attention from one branch to another of the ϕ-curve.

The advantage gained by defining ϕ_1 in this manner, rather than by following the continuous changes in the angle ϕ, is that as we follow the cycle of orbits discussed in § 5, ϕ_1 will go through a series of cyclical changes, whereas if we maintained strict analytical continuity, on each passage through a cycle we should have to fix our attention on a different determination of the angle ϕ.

We are now in a position to examine that branch of the ϕ-curve which corresponds to the determination ϕ_1. Since the direct and retrograde orbits merge into one another when x_0 has its extreme values (1 and m) the curve will touch the lines $x_0 = 1$, $x_0 = m$.

For large values of C, Darwin finds only a single periodic orbit for which $\phi_1 = \pi$ among the direct orbits in question, while the angle ϕ_1 decreases with increasing values of x_0. Assuming that the forms of the retrograde orbits are similar to those represented in § 5 the ϕ_1-curve will then be as below, where two passages through the cusped form are indicated by abrupt diminutions by $\frac{1}{2}\pi$ as x_0 increases.

In this figure the point A corresponds with Darwin's periodic orbit, the points J and M with orbits which start from a cusp, while the points P, P', Q, Q' correspond with orbits which have a cusp at J, at some time after starting.

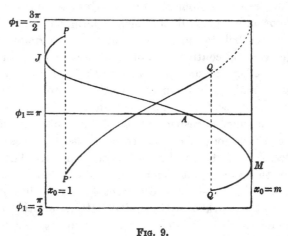

Fig. 9.

If we follow the retrograde orbits up to their limiting form when cusped at M it is evident that the normal at the crossing denoted by γ in figs. 7, 8, which may be regarded as the true analytical continuation of α in figs. 5 and 6, will approach nearer and nearer to a direction at right angles to the line of syzygies. At the final stage the crossing will disappear by coalescence with a second which occurs before the passage through the starting point. We thus conclude that the continuation of the branch $P'Q$ will touch the line $x_0 = m$ at the point where $\phi_1 = \frac{3}{2}\pi$, as indicated by the dotted part of the curve in the above diagram.

Further, since the points P, Q' necessarily lie between the lines $\phi_1 = \frac{3}{2}\pi$ and $\phi_1 = \frac{1}{2}\pi$, it may be readily seen that the branch of our curve which corresponds with the retrograde orbits must necessarily cut the line $\phi_1 = \pi$, indicating the existence of a retrograde periodic*.

If as in the above figure P, Q' lie on opposite sides of the line $\phi_1 = \pi$, P', Q will also lie on opposite sides of this line and the retrograde periodic will then be of the character indicated by figs. 5 and 6 in which the crossing α becomes rectangular, i.e. it will be a "doubly" periodic orbit.

On the other hand if P, Q' lie on the same side of $\phi_1 = \pi$ one of the branches JP, MQ' must have bent round so as to cross this line, and the form of the retrograde orbit will be modified in a manner which it is easy to trace.

§ 7.　First deformation of the ϕ-curve accounting for the disappearance of the orbit A.

As C varies the figure of the ϕ-curve given in the last section will undergo continuous deformation, and we may follow the fate of the orbit A by fixing our attention on the point A of this figure. Now from the figures given by Darwin we see that as C decreases the point A will approach the line $x_0 = 1$. Meanwhile the point J will move along the line $x_0 = 1$, and it is evident that the points A and J may at some stage coincide. When this occurs the cusped orbit J itself cuts the axis at right angles at the next crossing and may be regarded as periodic. Professor Darwin who is at present examining the forms of some of these cusped orbits informs me that the orbits in question appear to become periodic for a value of C about 39·5, but the actual numerical value has not yet been determined†.

After the critical stage the point J will cross the line $\phi_1 = \pi$ and the point A will no longer be found in that part of the curve which corresponds with the direct orbits, but in the part which corresponds with the retrograde

---

* [This retrograde periodic orbit was computed in the course of the investigations for Paper 3, and was found to be stable. G. H. D.]

† [This conjectural value of C is a good way from that ultimately found. It appears to be really 38·85. G. H. D.]

orbits. Subsequently the curve will bend up so as to again cut the line $\phi_1 = \pi$, indicating the growth of two new periodic orbits, the orbits B and C of Professor Darwin's paper. The form of the ϕ-curve, so far at least as regards that part of it with which we are concerned, will now be as below (fig. 10).

It appears then that the starting point of the orbit A will move up to the planet J, that the orbit will at first become cusped and that afterwards it will have a loop round J which is described by the satellite in a retrograde direction. The critical stage occurs when C is in the neighbourhood of 39·5 [really 38·85]. This explains why Professor Darwin, who confined his attention to the direct orbits alone, failed to find any trace of this orbit when $C = 39\cdot0$*.

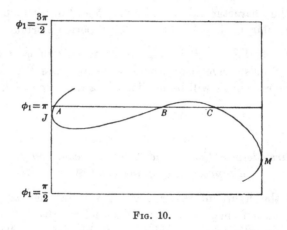

FIG. 10.

§ 8. *Form of the ϕ-curve in the neighbourhood of an*
asymptotic orbit.

So long as the curve of zero velocity possesses a closed branch round J, the only forms of discontinuity which the ϕ-curve can present are those dealt with in the preceding sections, but when the curve of zero velocity has assumed an hour-glass form more complex figures, resulting from the existence of asymptotic orbits, may occur. The nature of the singularity which appears in the neighbourhood of an asymptotic orbit may however be inferred from the curves given by Darwin for the case $C = 39\cdot0$.

Hitherto by defining ϕ_1 as a particular determination of a multiple-valued function we have been able to regard ϕ_1 as a single-valued function of the quantity x_0, but we must now attach a slightly extended meaning

---

* [The orbit really exists for $C = 39\cdot0$, but I had failed to detect it. If this figure be taken to correspond with $C = 39$, the point of contact J should be just above the line $\phi_1 = \pi$ instead of just below it. G. H. D.]

to the symbol ϕ_1, which will no longer permit us to regard it as single-valued.

Suppose that the crossing on which our attention is fixed and to which the determination ϕ_1 belongs approaches the point of zero force. We have seen (§ 1) that the orbit may then acquire loops, and may therefore cut the axis of x in several points before it recedes again either towards the planet or the Sun. When such loops exist, we define ϕ_1 as the angle made by the normal with the axis at any crossing prior to its again receding from the point of zero force. The number of real determinations of ϕ_1 will then depend on the number of loops which cut the line of syzygies.

Now on reference to Darwin's figures it will be seen that as x_0 decreases from the value $x_0 = m$, $\phi_1 - \pi$ is initially negative and twice vanishes and changes sign. The vanishing points determine two periodic orbits B and C. The corresponding portion of the ϕ-curve will then resemble that of the curves previously given.

If we confine our attention to the value of ϕ_1 at the first crossing it is evident that after passing the point B the value of ϕ_1 diminishes, until when x_0 is rather less than 1·06 it attains the value $\frac{1}{2}\pi$. For values of x_0 beyond this one, the corresponding crossing becomes imaginary by coalescence with a second. Hence the branch of the ϕ_1 curve, to which we limit ourselves in directing our attention only to the first crossing, will intersect the line $\phi_1 = \frac{1}{2}\pi$ at right angles and will at this stage identify itself with a second branch. This second branch applies to the values of ϕ_1 at the second crossing, and from the figures it is clear that it first appears when x_0 is rather larger than 1·095. The second value of ϕ_1 comes into existence simultaneously with a third and its initial value is $-\frac{1}{2}\pi$. The value of ϕ_1 as we pass along the second branch ranges between $-\frac{1}{2}\pi$ and $+\frac{1}{2}\pi$, and consequently vanishes at some stage. The vanishing stage is that which corresponds to the figure-of-8 orbit of Darwin's paper. It is clear that the second branch may be continued into a third, the third into a fourth and so on, that these various branches are in reality different parts of one and the same continuous curve, which possesses an infinite branch consisting of a series of waves of the form shewn in fig. 11.

This figure gives a concise summary of the results implied in Darwin's figs. 3 and 5, and we may interpret all the features of it in connection with the orbits represented in those figures.

Thus consider the intersections with the curve of an ordinate which moves from right to left.

The first critical stage which will occur after its foot has passed B will be when the ordinate just touches the crest of the first wave. The ordinate will then intersect the curve in two new coincident points indicating that the orbit will intersect the axis of x in two new coincident points. This results from the fact that the orbit has acquired a single loop which just comes into contact with the axis of x at the critical stage. Evidently the initial values of ϕ_1 will be each equal to $-\frac{1}{2}\pi$ and consequently the crest of the first wave lies on the line $\phi_1 = -\frac{1}{2}\pi$.

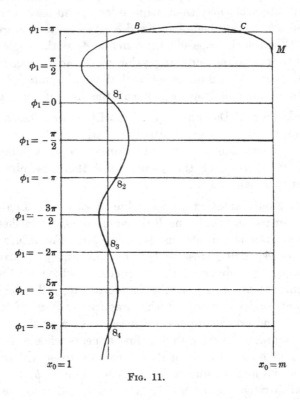

FIG. 11.

The new values of ϕ_1 will then separate and the next critical stage will occur when the larger of them attains the value zero, i.e. when the ordinate passes through the point marked 8_1 in the diagram. The orbit will then be periodic and of the form of a figure-of-8. It is the figure-of-8 orbit which has been found by Darwin.

For further decrease in x_0 the ordinate will touch the crest of a second wave which indicates that a second loop will have been formed in the orbit and that this second loop bends upwards so as to touch and afterwards cut the axis-of-x. Subsequently when the ordinate passes through the point 8_2 the orbit with two loops will have become periodic. It will then be one of the more complex figure-of-8 orbits whose existence has been foreshadowed by Professor Darwin*.

---

* [The passage in Paper 1 referred to is omitted in the present reprint, because the subject is considered more satisfactorily in Paper 3. G. H. D.]

Evidently the number of waves intersected by the ordinate will go on rapidly increasing in number, and with them the number of loops of the orbit intersecting the axis of x, until a critical stage will be reached when the ordinate attains a position shewn in the figure about which all the waves oscillate. When the ordinate attains this position the orbit will be the asymptotic orbit described in § 2 which possesses an infinite number of loops.

For further decrease in x_0 we see from the results of § 2 that the orbit will no longer be of the "lunar" type but of the "planetary" type. The number of real intersections of the ordinate with the ϕ-curve will rapidly diminish which implies that the orbit will shed its loops. Finally the ordinate will cease to intersect the curve in real points and the orbit will have attained the form shewn by Darwin for the cases $x_0 = 1\cdot04$, $1\cdot02$ and $1\cdot001$, where there are no real intersections with the axis of x prior to the recession towards the Sun.

The intersections of the infinite branch with each of the lines $\phi_1 = 0$, $\phi_1 = -\pi$, $\phi_1 = -2\pi$, &c. will indicate the existence of periodic orbits having 1, 2, 3, &c. loops. Since these orbits are all necessarily of the lunar type the points on the figure, marked 8_1, 8_2, 8_3, &c., which correspond to these periodic orbits, all lie to the right of the critical ordinate which separates the lunar orbits from the planetary.

We see then that the existence of an asymptotic orbit implies also the existence of an infinite number of complex figure-of-8 orbits, the first of which is that which has been found by Darwin.

§ 9.   *Completion of the ϕ-curve.*

The part of the curve shewn in our last figure is that which corresponds to the direct orbits. Darwin's investigations enable us to figure this part of the curve with certainty, but no attempt has been made to draw it to scale in order that the essential features may be exhibited in a somewhat exaggerated form. Thus the amplitudes of the successive waves will be very much smaller than they are represented, all the critical features being included between $x_0 = 1\cdot095$ and $x_0 = 1\cdot06$ (approximately). As regards the remaining part of the curve we however have no such data available, and we have to fall back entirely on considerations of continuity to supply them.

Now Darwin finds that when $x_0 = 1\cdot02$ and even when $x_0 = 1\cdot001$ the orbits are planetary and that they do not intersect the axis-of-x prior to their recession towards the Sun. It seems probable that they will retain this same character until x_0 reaches its limiting value (unity) and the orbit has a cusp at J^*. On the other hand when $x_0 = 1 \cdot 3$ (see fig. 3) the orbits are lunar, and it is probable that they will retain their lunar character until the starting point reaches the curve of zero velocity ($x_0 = m$).

If this be the case, as we pass from the orbit $x_0 = m$ to the orbit $x_0 = 1$ by continuous deformation, following the retrograde orbits instead of the direct orbits, we must pass through a stage where the orbits change from the lunar to the planetary form. Thus there must be a second asymptotic orbit among the retrograde orbits.

Whether or not the above assumptions as to the form of the cusped orbits be correct, it is clear that in such a cycle of orbits as that under consideration asymptotic orbits must occur in even numbers. Thus the existence of a single asymptotic orbit in the cycle necessarily involves the existence of a second.

Now it might appear at first sight that an asymptotic orbit could pass out of our cycle when the orbit falls into the planet and ϕ_1 undergoes an abrupt change by $\frac{1}{2}\pi$. Such however cannot be the case. To prove this let us suppose that the crossing on which our attention is fixed moves up to the planet J. A second crossing will reach J simultaneously with it. Now if our crossing corresponds with the approach to J, since a second crossing will occur after it on the opposite side of J, the orbit (so far as this crossing is concerned) will necessarily be of the lunar type both before and after the change. On the other hand if the crossing with which we are concerned belongs to the branch of the curve along which the satellite recedes from J, since the discontinuity in the geometrical form of the orbit is confined to the critical point and does not extend to remote parts of the curve, the orbit will be of the same character before and after its passage through the critical form. Thus such a discontinuity, as that represented by the passage from the point P to P' or from Q to Q' in fig. 9, can never involve a transition from the lunar to the planetary form or *vice versâ*.

It appears then that asymptotic orbits can only disappear from our cycle by coalescence and that the development of them will always occur in pairs.

§ 10. *Second deformation of the ϕ-curve.*

Having recognized the existence of the asymptotic orbits and the form which the ϕ-curve assumes in their neighbourhood we next proceed to examine the transitional forms through which the curve will pass when two such orbits coalesce. A reversal of the order of the events considered will then indicate the state of affairs prior and subsequent to the development of a pair of asymptotic orbits.

---

* [No, this is not so. It was the apparent certainty of the argument here adduced by Mr Hough which misled me. G. H. D.]

First let us consider the forms of the orbits which possess a cusp at J as C increases. When $C = 39 \cdot 0$ these belong to the planetary class, but when $C = 40 \cdot 0$ they are found to be no longer planetary but lunar. The cusped orbit must therefore, for intermediate values of C, acquire loops, pass through the asymptotic form and finally shed these loops again. This will occur when one of the two ordinates which correspond to the asymptotic orbits in our cycle moves up to the line $x_0 = 1$. The corresponding asymptotic orbit will then undergo a change from the direct to the retrograde form or *vice versâ*, and subsequently both asymptotic orbits will be direct or both retrograde. It seems probable that it is the retrograde orbit which passes through the critical form and becomes direct after the crisis. This assumption is however only made for the purpose of giving greater definiteness to our statements and is not essential to the arguments. After the passage of the asymptotic orbit through the cusped form the figure-of-8 orbits which accompany it will each in turn undergo a like change. For simplicity we will suppose that all these changes occur before the next critical stage is reached and that consequently the form of the ϕ-curve, so far as it applies to the direct orbits, is now as below, having two infinite branches.

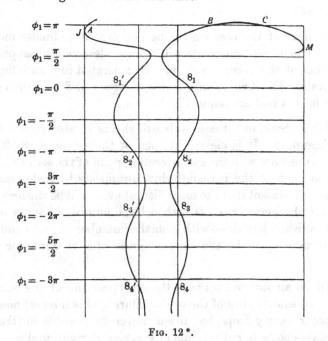

FIG. 12 *.

There will now be a second figure-of-8 orbit of a character similar to Professor Darwin's, which corresponds to the point $8_1'$ of the figure, while

* [The figure for the case of $C = 39$ must be modified by moving the point of contact J above the line $\phi_1 = \pi$. Mr Hough was not however aware of this. The curve denoted 8_1 is my orbit A', that marked $8_1'$ is my A''. G. H. D.]

there will be two infinite series of more complex figures-of-8 corresponding to the two series of points $8_2, 8_3, \ldots, 8_2', 8_3', \ldots.$ It should be noticed that none of these points can lie between the two critical ordinates which correspond with the asymptotic curves, since the periodic orbits all belong essentially to the lunar category. As C decreases further the next event of importance will be the coalescence of the crests of the two first waves in the above figure. After this coalescence the infinite branches of the curve will be severed from the remaining part, which will form a curve of the type shewn in figs. 9 and 10.

The significance of this change is that for larger values of C orbits of the type represented by $C = 39 \cdot 0$, $x_0 = 1 \cdot 04$, which do not intersect the axis of x before their recession, can no longer exist.

Subsequently as the two critical orbits approach one another the successive wave crests of the two branches will coalesce in turn and a series of isolated ovals will be formed. The curve will then consist of a branch similar to those of §§ 6 and 7, a finite number of isolated ovals and two infinite branches as below, where the curve is drawn on a reduced vertical scale with two ovals (fig. 13).

We arrive next at the case where the two critical ordinates move up to coincidence. The asymptotic orbits will then coalesce and disappear. The infinite branches of the ϕ-curve will have degenerated into an infinite series of isolated ovals. The two infinite series of figure-of-8 periodic orbits will however still have a real existence.

For further increase in C these ovals will shrink in size, reduce to points and finally disappear. It is clear that each of the points $8_1, 8_2, 8_3, \ldots$ will disappear by coalescence with the corresponding point of the series $8_1', 8_2', 8_3', \ldots$ indicating that each of the periodic orbits disappears by coalescence with a second. Also it is evident that the more distant ovals will be the first to vanish indicating that the periodic orbits with a large number of turns round the orbit (a) will vanish before those with a smaller number of turns and that the last orbits to vanish will be the simple figures-of-8 of Professor Darwin's paper*.

There will be an interval between the disappearance of the points $8_1, 8_1'$ &c. and the final evanescence of the ovals. During this interval non-periodic orbits may occur having loops, but it will never be possible for the loops to adjust themselves so as to cut the line of syzygies at right angles.

---

* [In Paper 3 I have supposed C to be decreasing instead of increasing, and my statement is in agreement with this. G. H. D.]

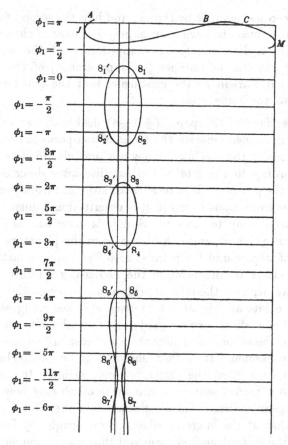

FIG. 13.

§ 11. *Summary and Conclusion.*

The problems dealt with arise as the result of the study of an apparently remarkable change in the forms of certain periodic orbits which have been examined by Professor Darwin, whereby an orbit originally in the form of a simple closed oval seems to have developed into the form of a figure-of-8. This change is of so surprising a character that M. Poincaré and others have concluded that Professor Darwin was misled in regarding the simple oval orbit and the figure-of-8 orbit as members of the same family. But if we regard them as members of two different families questions at once arise as to the earlier history of the second family and the later history of the former. On these points Darwin's results throw very little light, but by supplementing the numerical results obtained by him by arguments based on the consideration of geometrical continuity I have shewn that the history of either family may be traced in part, and have verified M. Poincaré's conclusion as to the independence of these two families. The results explain fully how the first family has been lost sight of by Darwin and how the second family comes into existence. It appears to have been largely a matter of chance that with the actual numerical data adopted the appearance of the second family coincided exactly with the appearance [? disappearance] of the first, a fact which naturally led Darwin to the conclusion that the two forms of orbit really belonged to the same group.

The earlier sections of the paper (§§ 1—4) deal with two critical cases which may occur in connection with the forms of non-periodic orbits and lead to two classifications of these orbits. They are first classified as "lunar" or "planetary" according to the fate of the satellite after describing a semi-revolution round the primary. If it recedes towards the primary and proceeds to describe further revolutions round it, it is described as "lunar." If on the other hand it passes away towards the Sun it is described as "planetary." The critical orbits which separate the lunar from the planetary describe an infinite number of loops round the point of zero force, approximating at each turn closer and closer to the orbit of the oscillating satellite. Adopting Poincaré's term we describe the critical orbit as an "asymptotic" orbit. We next classify the orbits as "direct" or "retrograde" according as the initial direction of motion is direct or retrograde, the critical orbits which separate the one class from the other being described as "cusped" on account of the forms which they assume. It is then shewn that by the inclusion of the retrograde orbits as well as the direct, we may arrange the orbits under consideration into a perfect self-contained cycle which may however involve certain abrupt discontinuities. The disappearance of the orbit A is accounted for by the fact that at the stage at which it was sought by Darwin it had passed through the critical cusped form and thus was no longer to be found in the part of the cycle examined by him, viz. the part which includes only the direct orbits.

We proceed to shew how all the results indicated in the figures of the non-periodic orbits traced by Darwin may be represented on a single curve, and how a study in the variations in the form of this curve will enable us to trace the history of the different families of periodic orbits. The method employed by Darwin to discover the periodic orbits in fact is equivalent to the examination of the form of this curve in regions where there were à priori grounds for suspecting the existence of a periodic orbit.

The chief point of interest in connection with these curves is the form which they assume in the neighbourhood of an asymptotic orbit, i.e. one of the critical orbits implied in our first classification. The form indicates that such an asymptotic orbit is accompanied not only by a simple figure-of-8 orbit of the kind found by Darwin, but also by an infinite series of complex figures-of-8 whose exact forms have not yet been examined but whose existence has been predicted by Darwin.

We next shew that in a complete cycle of orbits, such as that which we have found to exist, asymptotic orbits must occur in even numbers and consequently must appear or disappear in pairs. As Darwin's results indicate only one of such orbits among the direct orbits for the case $C = 39\cdot0$ we conclude that a second one must exist among the retrograde orbits. The two asymptotic orbits must have had a common origin and at the instant of their first appearance must have been both direct or both retrograde. For purposes of illustration it has been assumed that both are initially direct though this is not essential to the arguments employed.

The existence of two asymptotic orbits implies also the existence of two simple figure-of-8 orbits of the form found by Darwin, which must likewise have had a common origin. The converse is however not necessarily true. For if we trace back the changes described in the last section we see that the first indication of the development of a pair of asymptotic orbits will be the growth of loops, as in the orbit $x = 1\cdot095$ of Darwin's fig. 5. When these loops first appear it will not be possible for them to arrange themselves so as to cut the line of syzygies at right angles and render the orbit periodic, but with smaller values of C pairs of periodic looped orbits will appear which gradually separate from one another. The first to appear will be those with the smaller number of loops, the final ones being the asymptotic orbits which may be regarded as the limiting form of periodic looped orbits when the number of loops becomes infinitely great.

The passage of the second asymptotic orbit from the direct to the retrograde form will be preceded by a similar passage of the whole series of periodic looped orbits which accompany it, including the simple figure-of-8 orbit which at its origin coincided with that of Professor Darwin. Here again the failure to find any trace of this orbit is to be explained by the fact that the search was confined to the direct orbits alone, whereas at the stage under investigation ($C = 39\cdot0$) this orbit had already passed into the retrograde form*.

In conclusion I have to thank Professor Darwin not only for the unfailing courtesy with which he has placed at my disposal all the details of the prodigious amount of numerical work which formed the basis of his published memoir on this subject, but also for the readiness with which he has communicated to me his further results still under investigation and for valuable criticism which has saved me from numerous errors. Even should the present results be found to require modification on further investigation, and it is admitted that the detail is to be regarded as conjectural rather than proven, I shall feel that the paper will have served a useful purpose if it succeeds in enticing other investigators into the vast and hitherto unexplored field which seems to be opened up with each new development of this highly interesting subject.

* [The stage at which this occurs is really when $C = 38\cdot85$. G. H. D.]

3.

ON CERTAIN FAMILIES OF PERIODIC ORBITS.

[*Monthly Notices of the Royal Astronomical Society*, Vol. 70, 1909, pp. 108—143; with which is incorporated a note, *ibid.* Vol. 70, 1910, p. 604.]

INTRODUCTION.

IN my paper on " Periodic Orbits " in the restricted problem of the three bodies (*Acta Mathematica*, Vol. XXI., 1897, p. 101, or Paper 1) I attempted to make a classification of those orbits which are simply periodic, that is to say, re-entrant after a single circuit either round the Sun, or round the perturbing planet called Jove, or round one of the points of zero force. In that paper I only proposed to consider orbits in which the revolution is in the positive or " direct " sense, and the retrograde orbits were left for future consideration.

Amongst the orbits discussed there was a certain family of satellites which I called A. There were also certain other orbits resembling the figure 8, in which one loop was described round Jove and the second round that point of zero force which lies between the Sun and Jove. The results seemed to indicate that the orbits of these satellites, conjointly with the orbit of a satellite (a) oscillating about the point of zero force, and the figure-of-8 orbit belonged to the same family. But shortly after the publication of the paper it was pointed out both by Mr S. S. Hough and by M. Poincaré that this kind of continuity was not possible, so that the A orbits and the figure-of-8 orbits must in reality belong to different families; and indeed I had independently come to the same conclusion myself.

In consequence of this oversight my investigation left several points unexplained, for there was no indication as to the course of development of the A orbits after I had supposed them to cease to exist, and it remained to discover how they came to disappear from amongst the directly moving satellites. The origin of the figure-of-8 orbits also remained obscure.

To meet these deficiencies Mr Hough wrote a paper entitled "On certain discontinuities connected with Periodic Orbits" (*Acta Mathematica*, Vol. XXIV., 1901, p. 257, or Paper 2). He there suggested, with great ability, the manner in which my classification should be completed, and indicated the existence of another family of figure-of-8 orbits which I had not detected.

My own experience has shown how hard it is, without actual computation, to attain full confidence as to the changes which orbits will undergo when the constant of relative energy changes. It was therefore desirable to test the accuracy of even well-reasoned conjectures, such as those of Mr Hough.

In the present paper I shall show the results of my computations made with this object, and shall thereby confirm his conclusions in almost every detail. Indeed, he only went wrong, and that not seriously, where he relied implicitly on the completeness of my own work. I had not then considered what M. Burrau calls "orbits of ejection," which furnish the transitional forms between direct and retrograde orbits*. I had computed an orbit for $C = 39$ which started normally to the line of syzygy at a distance of only ·001 of the adopted unit of length, and not unnaturally it did not occur to me to suspect that if the initial pericentral distance were diminished yet further, the orbit would assume a very different form. Yet the computation of the orbit of ejection, in which the pericentral distance is reduced to zero, shows that such is the case. This oversight led to the erroneous conclusion (on which Mr Hough relied) that the A orbit does not exist when $C = 39$. Mr Hough supposed that it had already become retrograde when C had fallen to 39, whereas it will appear below that it only passes through the stage of ejection, on the way to becoming retrograde, when C has fallen to about 38·85.

The complete discussion of a family of orbits, many of whose members are direct, demands the consideration of retrograde orbits, because it would seem that sooner or later every family will pass through the ejectional stage. Moreover superior planets, which are really revolving in a positive direction in space, will appear as retrograde when the motion is referred to rotating axes.

Those who have read my previous paper on this subject will be aware that the determination of the stability of a periodic orbit presents an arduous task. Mr Hough has suggested to me a mode of procedure which differs completely from that which I had adopted; it is explained below, but in my own words. Whatever plan is followed, the computation must necessarily remain laborious, and for orbits with sharp flexures both methods become almost impracticable. In such cases Mr Hough's method appears to be somewhat less intractable than that of the infinite determinant, although hardly less laborious. On the whole, I think his plan is somewhat preferable to mine, although for orbits of fairly continuous changes of curvature it does

---

* See *Astronomische Nachrichten*, No. 3230, Vol. 135, and No. 3251, Vol. 136 (1894).

not seem to possess any conspicuous advantage.   I apply his method to the discussion of several of the orbits determined in this paper.

The various subjects considered here are so far connected together that they all aim at the completion of our knowledge of periodic orbits, but they do not form the matter of a single continuous argument, and therefore the paper is divided into several parts.

I begin in Part I. with the investigation of Mr Hough's method of determining the stability of an orbit, and reduce the formulæ to convenient forms for computation.

There follows next in Part II. the discussion of certain retrograde orbits. It had originally been my intention to obtain a complete knowledge of these orbits, but as I see no immediate prospect of continuing these laborious investigations, I set forth all the results obtained up to the present time. As regards the satellites, with which Mr Hough was concerned, the results are entirely confirmatory of his suggestions.   Several members of the family of superior planets (only apparently retrograde) are also shown, and incidentally we obtain some insight into other families of orbits which must exist.

It has been already remarked that orbits of ejection are transitional forms between direct and retrograde orbits, and they therefore possess, as M. Burrau perceived, a special degree of importance.   Accordingly, in Part III. analytical formulæ are found which enable us to trace these orbits for a considerable distance from the point of ejection.   A large number of such orbits, completed by quadratures, are presented.   I also add some results as to the form of the cusp when the body reaches the curve of zero velocity.

Part IV. is devoted to the subject which affords the principal reason for this paper, namely, the completion of the deficiencies in my previous work.

Since the figure-of-8 orbits were erroneously classified along with the satellites A, I now denote them by A'.   The new family of figures-of-8, foreseen by Mr Hough, is also determined and denoted A".   Figures of these two families are drawn, and I also show the fate of the satellites A, and how they become retrograde in passing through the ejectional form.   An attempt is made to discriminate between the orbits A' and A" near the point of their coalescence, and it is shown that it is probably A" which is stable.

Finally, Part V. contains a discussion of the light which is thrown on the development of several of the families of orbits by means of these investigations, and I indicate various new families of orbits which must spring into existence.

I give in an appendix the numerical values from which many of the figures were drawn, and others are given in the appendix to Paper 1, being distinguished from the results as given originally by being printed in italic type.

I. The Stability of Orbits.

§ 1. *Determination of the Stability of a Periodic Orbit.*

In my paper on "Periodic Orbits" (Paper 1) the criterion of stability of the orbits was found by means of Hill's infinite determinant, but the method virtually fails when the orbit has sharp flexures. A different method of treatment will therefore now be explained.

In § 10, p. 34, of that paper I put $\sigma = \pi s/S$, where s is the arc of orbit measured from the initial syzygy, and S is the complete arc of the periodic orbit. The equation from which the stability was determined was

$$\frac{d^2 \delta q}{d\sigma^2} + \Phi \delta q = 0$$

where Φ is an even function of σ completing its period when σ increases from 0 to π.

The following investigation, as far as regards the analytical principles involved, was given to me by Mr S. S. Hough, although I am responsible for the manner in which it is presented.

[I have also to thank Mr H. C. Plummer for drawing my attention to discussions of this equation by Korteweg (*Wiener Sitzungsb.* 93, or Whittaker's *Dynamics*) and by Liapounoff (*Comptes Rendus*, 1896 and 1899, or Forsyth's *Theory of Differential Equations*, Vol. IV.). Other references will be found in Chapter I. Vol. III. of Tisserand's *Mécanique Céleste*. Mr Plummer has also pointed out that the treatment of the equation in this paper as originally printed was not satisfactory, although the conclusions were correct. He kindly made suggestions which form the basis of the present treatment.

The differential equation remains unchanged when σ becomes $\sigma + \pi$. Hence if a certain function of σ is a solution, the same function of $\sigma + \pi$ is also one. Accordingly if $\phi(\sigma)$ and $\psi(\sigma)$ are two particular solutions, $\phi(\sigma + \pi)$ and $\psi(\sigma + \pi)$ are solutions also.

The general solution must be expressible in the form of the sum of two particular solutions each multiplied by an arbitrary constant. Hence it follows that

$$\left. \begin{array}{l} \phi(\sigma + \pi) = \alpha\phi(\sigma) + \beta\psi(\sigma) \\ \psi(\sigma + \pi) = \gamma\phi(\sigma) + \delta\psi(\sigma) \end{array} \right\} \quad \dots\dots\dots\dots\dots\dots(1)$$

where $\alpha, \beta, \gamma, \delta$ are constants.

It will now be proved that these equations imply that certain conditions are satisfied by $\alpha, \beta, \gamma, \delta$.

Since ϕ and ψ are solutions of the differential equation, we have

$$\frac{d^2\phi}{d\sigma^2} + \Phi\phi = 0, \qquad \frac{d^2\psi}{d\sigma^2} + \Phi\psi = 0$$

On multiplying the first of these by ψ, subtracting it from the second when multiplied by ϕ, and integrating, we find

$$\phi\psi' - \phi'\psi = \text{constant}$$

where accented symbols denote differential coefficients as usual.

By substitution from (1) we find

$$\phi(\sigma + \pi)\,\psi'(\sigma + \pi) - \phi'(\sigma + \pi)\,\psi(\sigma + \pi) = (\alpha\delta - \beta\gamma)\,[\phi\psi' - \phi'\psi]$$

It follows that the four constants must satisfy the condition

$$\alpha\delta - \beta\gamma = 1$$

Again suppose that $\chi(\sigma)$, $\omega(\sigma)$ are two other particular solutions, and that they are connected with the pair in (1) by the conditions

$$\phi(\sigma) = \epsilon\chi(\sigma) + \eta\omega(\sigma)$$

$$\psi(\sigma) = \zeta\chi(\sigma) + \theta\omega(\sigma)$$

where ϵ, η, ζ, θ are constants.

Then on substitution in (1) we find of course a relation similar to (1), and that if α', β', γ', δ' are the new constants which play the same parts as α, β, γ, δ in (1) $\alpha' + \delta'$ is equal to $\alpha + \delta$.

Hence it follows that $\alpha + \delta$ is constant and is independent of the pair of particular solutions used.

Now suppose that $F(\sigma)$ is a solution of the differential equation, and if possible let

$$F(\sigma + \pi) = \nu F(\sigma)$$

where ν is constant.

Since $F(\sigma)$ is a solution it must be expressible by means of the particular solutions ϕ and ψ, and two constants A, B, so that

$$F(\sigma) = A\phi(\sigma) + B\psi(\sigma)$$

The hypothesis then gives us

$$F(\sigma + \pi) = A\phi(\sigma + \pi) + B\psi(\sigma + \pi) = \nu A\phi(\sigma) + \nu B\psi(\sigma)$$

On substitution from (1) we have

$$[A(\alpha - \nu) + B\gamma]\,\phi(\sigma) + [A\beta + B(\delta - \nu)]\,\psi(\sigma) = 0$$

Since this must be satisfied for all values of σ, we obtain

$$A(\alpha - \nu) + B\gamma = 0$$

$$A\beta + B(\delta - \nu) = 0$$

On eliminating A and B, and noting that $\alpha\delta - \beta\gamma = 1$, we have

$$\nu^2 - \nu(\alpha + \delta) + 1 = 0$$

or

$$\nu + \frac{1}{\nu} = \alpha + \delta$$

Since $\alpha + \delta$ is independent of the choice of the solutions ϕ and ψ, ν is also independent thereof. The supposition that

$$F(\sigma + \pi) = \nu F(\sigma)$$

is therefore possible, and we see that if ν is one value of the constant, it has also another value which is the reciprocal of the first.

It follows that if ν is one of the solutions of the equation for determining it, and if one solution of the equation is $\nu^{\sigma/\pi}\Psi(\sigma)$, where Ψ is an even function of σ completing its period between 0 and π; then there is another solution $\nu^{-\sigma/\pi}X(\sigma)$, where X is a function of the same character as Ψ.

Hence the general solution of the differential equation is of the form

$$C\nu^{\sigma/\pi}\Psi(\sigma) + D\nu^{-\sigma/\pi}X(\sigma)$$

where C, D are arbitrary constants.

This is expressible more conveniently for further development if we replace ν by another constant c such that

$$\nu = e^{\pi c\sqrt{-1}}$$

It follows that the general solution may be written in the form

$$\Psi_1(\sigma)\cos c\sigma + X_1(\sigma)\sin c\sigma$$

where Ψ_1, X_1 are clearly even functions of σ periodic between 0 and π.

It is obvious that Ψ_1, X_1 are expressible by a Fourier expansion in cosines and sines of even multiples of σ. Hence the general solution of the equation is of the form

$$\delta q = \Sigma_j\left[(b_j + e_j)\cos(c + 2j)\sigma + (b_j - e_j)\sqrt{(-1)}\sin(c + 2j)\sigma\right]$$

where Σ_j denotes summation for all integral values of j between $\pm\infty$. Also the b's and e's satisfy conditions such that only two of them are independent.

This solution is written so that it shall be identical in form with the solution in § 10, p. 35 of Paper 1.

In that paper we eliminated the b's and e's by means of a determinant, and found c from Hill's infinite determinantal equation. But we have now found another way of determining c, for

$$\tfrac{1}{2}\left(\nu + \frac{1}{\nu}\right) = \cos \pi c = \tfrac{1}{2}(\alpha + \delta)$$

It only remains therefore to show how the values of α and δ are to be found from a convenient choice of solutions ϕ and ψ.

Let the particular solution ϕ be such that when $\sigma = 0$, $\delta q = 1$, $d\delta q/d\sigma = 0$; that is to say $\phi(0) = 1$, $\phi'(0) = 0$. And let the solution ψ be such that when $\sigma = 0$, $\delta q = 0$, $d\delta q/d\sigma = 1$; that is to say $\psi(0) = 0$, $\psi'(0) = 1$.

These suppositions give

$$\phi(0)\psi'(0) - \phi'(0)\psi(0) = 1$$

If we put $\sigma = 0$ in (1) we find

$$\phi(\pi) = \alpha, \quad \phi'(\pi) = \beta, \quad \psi(\pi) = \gamma, \quad \psi'(\pi) = \delta$$

Hence　　　　$\phi(\pi)\psi'(\pi) - \phi'(\pi)\psi(\pi) = \alpha\delta - \beta\gamma = 1]$

Since the chosen ϕ and ψ are respectively odd and even functions of σ,

$$\phi(-\sigma) = \phi(\sigma), \qquad \psi(-\sigma) = -\psi(\sigma)$$
$$\phi'(-\sigma) = -\phi'(\sigma), \qquad \psi'(-\sigma) = \psi'(\sigma)$$

On putting $\sigma = -\tfrac{1}{2}\pi$ in (1) and in the pair of equations derived therefrom by differentiation, we obtain

$$\phi(\tfrac{1}{2}\pi) = \alpha\phi(\tfrac{1}{2}\pi) - \beta\psi(\tfrac{1}{2}\pi), \quad \phi'(\tfrac{1}{2}\pi) = -\alpha\phi'(\tfrac{1}{2}\pi) + \beta\psi'(\tfrac{1}{2}\pi)$$
$$\psi(\tfrac{1}{2}\pi) = \gamma\phi(\tfrac{1}{2}\pi) - \delta\psi(\tfrac{1}{2}\pi), \quad \psi'(\tfrac{1}{2}\pi) = -\gamma\phi'(\tfrac{1}{2}\pi) + \delta\psi'(\tfrac{1}{2}\pi)$$

Whence　　　$\dfrac{\phi(\tfrac{1}{2}\pi)}{\psi(\tfrac{1}{2}\pi)} = \dfrac{\beta}{\alpha - 1} = \dfrac{\delta + 1}{\gamma}, \quad \dfrac{\psi'(\tfrac{1}{2}\pi)}{\phi'(\tfrac{1}{2}\pi)} = \dfrac{\alpha + 1}{\beta} = \dfrac{\gamma}{\delta - 1}$

Therefore　　　$\dfrac{\alpha + 1}{\alpha - 1} = \dfrac{\delta + 1}{\delta - 1} = \dfrac{\phi(\tfrac{1}{2}\pi)\psi'(\tfrac{1}{2}\pi)}{\phi'(\tfrac{1}{2}\pi)\psi(\tfrac{1}{2}\pi)}$

Since $\phi(\tfrac{1}{2}\pi)\psi'(\tfrac{1}{2}\pi) - \phi'(\tfrac{1}{2}\pi)\psi(\tfrac{1}{2}\pi) = 1$, we have

$$\alpha = \delta = \tfrac{1}{2}(\alpha + \delta) = \phi(\tfrac{1}{2}\pi)\psi'(\tfrac{1}{2}\pi) + \phi'(\tfrac{1}{2}\pi)\psi(\tfrac{1}{2}\pi)$$

We may therefore write the equation for c in any one of the five following forms:—

$$\cos\pi c = \phi(\pi) = \psi'(\pi) = \phi(\tfrac{1}{2}\pi)\psi'(\tfrac{1}{2}\pi) + \phi'(\tfrac{1}{2}\pi)\psi(\tfrac{1}{2}\pi)$$

$$= 1 + 2\phi'(\tfrac{1}{2}\pi)\psi(\tfrac{1}{2}\pi) = 2\phi(\tfrac{1}{2}\pi)\psi'(\tfrac{1}{2}\pi) - 1$$

The last two of these give us

$$\cos^2\tfrac{1}{2}\pi c = \phi(\tfrac{1}{2}\pi)\psi'(\tfrac{1}{2}\pi)$$

$$\sin^2\tfrac{1}{2}\pi c = -\phi'(\tfrac{1}{2}\pi)\psi(\tfrac{1}{2}\pi)$$

The last of these has the form used in "Periodic Orbits," where the result was

$$\sin^2\tfrac{1}{2}\pi c = \Delta \sin^2\tfrac{1}{2}\pi \sqrt{\Phi_0}$$

Δ being a certain infinite determinant.

If c is any solution of the equation, all others are included in $\pm c \pm 2j$, and I choose the fundamental value of c to be that which lies nearest to $\sqrt{\Phi_0}$, or the positive square root of the mean value of Φ.

The evaluation of c, and therefore the determination of stability, accordingly depends on the values of two special solutions, after half a circuit of the orbit has been performed when σ has increased from 0 to $\frac{1}{2}\pi$.

The periodic orbit having been determined by quadrature, the value of Φ may be found at a succession of points distributed along the orbit, and therefore it is a matter of quadrature to find the two required solutions.

We may start from the line of syzygies with two quadratures, denoted by suffixes 1 and 2; the first begins with $(\delta q_1)_0 = 1$, $\left(\dfrac{d\delta q_1}{d\sigma}\right)_0 = 0$, and the second with $(\delta q_2)_0 = 0$, $\left(\dfrac{d\delta q_2}{d\sigma}\right)_0 = 1$, where the suffix 0 indicates the initial value. If we denote the values of any quantity after passing half round the periodic orbit by the suffix s, we have

$$\sin^2 \tfrac{1}{2}\pi c = -(\delta q_2)_s \left(\frac{d\delta q_1}{d\sigma}\right)_s$$

For practical work the form of this solution may be modified with advantage. The arc of orbit s is the independent variable which arises naturally, and the differential equation (as shown in (39), p. 32 of " Periodic Orbits ") is then

$$\frac{d^2\delta q}{ds^2} + \Psi \delta q = 0$$

I subsequently, in § 10, changed the independent variable to $\sigma = \pi s/S$, where S is the complete arc of the periodic orbit. To effect this change I put $\Phi = S^2\Psi/\pi^2$. Hence with s as independent variable the second of the two solutions starts with $\left(\dfrac{d\delta q_2}{ds}\right)_0 = \dfrac{\pi}{S}$, and at every point the solutions satisfy the condition

$$\delta q_1 \frac{d\delta q_2}{ds} - \delta q_2 \frac{d\delta q_1}{ds} = \frac{\pi}{S}$$

The equation for c then becomes

$$\sin^2 \tfrac{1}{2}\pi c = -\frac{S}{\pi}(\delta q_2)_s \left(\frac{d\delta q_1}{ds}\right)_s$$

But for practical numerical work some further transformations will be found convenient.

In computing a periodic orbit I proceed step by step, the common difference of the arcs being Δs, the value of which may, however, change from time to time. In using the formulæ of finite differences it is convenient to make the common difference of the independent variable equal to unity. This will be secured by adopting z as independent variable where

$$dz = d\left(\frac{s - s_n}{\Delta s}\right)$$

s_n being the arc of the orbit measured from the initial syzygy up to the nth point in the orbit.   If we make

$$\Omega = \Psi (\Delta s)^2$$

the differential equation may be written

$$\frac{d^2 \delta q}{dz^2} + \Omega \delta q = 0$$

Since Δs is usually a simple fraction such as $\cdot 01$ or $\cdot 02$, the multiplication of the Ψ's to form the Ω's is very simple.

Suppose that $\Omega^{(n-1)}$, $\Omega^{(n)}$, $\Omega^{(n+1)}$, $\Omega^{(n+2)}$ are four successive values of Ω ; that ΔE^{-1}, Δ, ΔE are the three first differences; $\Delta^2 E^{-1}$, Δ^2 the two second differences; and $\Delta^3 E^{-1}$ the third difference.   The first difference falling between the nth and $(n+1)$th entry in the table of Ω is Δ, the mean second difference is $\frac{1}{2}(\Delta + \Delta E^{-1})$, the third difference is $\Delta^3 E^{-1}$.   If we call these three differences δ, δ^2, δ^3, Bessel's formula of interpolation is

$$\Omega^{(z)} = \Omega^{(n)} + z\delta + \tfrac{1}{2}z (z-1)\, \delta^2 + \tfrac{1}{6}z (z-1)(z-\tfrac{1}{2})\, \delta^3$$

If we write

$$\Omega_0 = \Omega^{(n)}, \quad \Omega_1 = \delta - \tfrac{1}{2}\delta^2 + \tfrac{1}{12}\delta^3, \quad \Omega_2 = \tfrac{1}{2}\delta^2 - \tfrac{1}{4}\delta^3, \quad \Omega_3 = \tfrac{1}{6}\delta^3$$

we have

$$\Omega^{(z)} = \Omega_0 + z\Omega_1 + z^2 \Omega_2 + z^3 \Omega_3 + \dots$$

Dropping the (z) on the left as being unnecessary, we may therefore write

$$\Omega = \sum_0^\infty \Omega_i z^i$$

If we assume

$$\delta q_1 = \sum_0^\infty p_i z^i$$

we have

$$\Omega \delta q_1 = \sum_{i=0}^{i=\infty} \sum_{j=0}^{j=i} \Omega_j p_{i-j} z^i$$

by integration of which we have from the differential equation,

$$\frac{d\delta q_1}{dz} = p_1 - \sum_{i=0}^{i=\infty} \sum_{j=0}^{j=i} \frac{1}{i+1} \Omega_j p_{i-j} z^{i+1}$$

$$\delta q_1 = p_0 + p_1 z - \sum_{i=0}^{i=\infty} \sum_{j=0}^{j=i} \frac{1}{(i+1)(i+2)} \Omega_j p_{i-j} z^{i+2}$$

By equating the coefficients of the assumed form of δq_1 with this result, we find

$$p_i = -\frac{1}{i(i+1)}[\Omega_0 p_{i-1} + \Omega_1 p_{i-2} + \dots + \Omega_{i-2}p_1 + \Omega_{i-1}p_0]$$

Let there be a second solution, and let it be expressed in the same form, save that q's replace p's.

The first solution starts with

$$(\delta q_1)_0 = 1, \quad \left(\frac{d\delta q_1}{dz}\right)_0 = 0$$

and the second with $(\delta q_2)_0 = 0, \quad \left(\dfrac{d\delta q_2}{dz}\right)_0 = \dfrac{\pi \Delta s}{S}$

At every point of the orbit we have

$$\delta q_1 \frac{d\delta q_2}{dz} - \delta q_2 \frac{d\delta q_1}{dz} = \frac{\pi \Delta s}{S}$$

As this is true at the point $z = 0$, which is the nth point in the periodic orbit,

$$p_0 q_1 - p_1 q_0 = \frac{\pi \Delta s}{S}$$

If we desire to record in the notation the meanings of p_0, p_1, q_0, q_1, this may be written

$$\left(\delta q_1 \frac{d\delta q_2}{dz} - \delta q_2 \frac{d\delta q_1}{dz}\right)_n = \frac{\pi \Delta s}{S}$$

This is the condition that at the nth point the solutions shall correspond to the supposed initial conditions. It will appear that if it holds good at the nth point, it does so also at the $(n+1)$th.

The value $z = 1$ corresponds to the passage from the nth to the $(n+1)$th point. Therefore

$$(\delta q_1)_{n+1} = \overset{\infty}{\underset{0}{\Sigma}} p_i, \qquad (\delta q_2)_{n+1} = \overset{\infty}{\underset{0}{\Sigma}} q_i$$

$$\left(\frac{d\delta q_1}{dz}\right)_{n+1} = \overset{\infty}{\underset{0}{\Sigma}} i p_i, \quad \left(\frac{d\delta q_2}{dz}\right)_{n+1} = \overset{\infty}{\underset{0}{\Sigma}} i q_i$$

It has been supposed that four computed values of Ω at the points $n-1$, n, $n+1$, $n+2$ give a sufficient representation of Ω throughout the range n to $n+1$. These four values afford the means of computing Ω_0, Ω_1, Ω_2, Ω_3 as given above. It will be supposed that the higher Ω's are negligible, and that it is unnecessary to consider their squares and products, excepting Ω_0^2 and $\Omega_0 \Omega_1$.

The solution of the successive equations for the p's gives

$$p_2 = -\frac{1}{1\,.\,2}\,\Omega_0 p_0, \quad p_3 = -\frac{1}{2\,.\,3}\,(\Omega_0 p_1 + \Omega_1 p_0)$$

$$p_4 = -\frac{1}{3\,.\,4}\,[(\Omega_2 - \tfrac{1}{2}\Omega_0^2)\,p_0 + \Omega_1 p_1]$$

$$p_5 = -\frac{1}{4\,.\,5}\,[(\Omega_3 - \tfrac{2}{3}\Omega_0\Omega_1)\,p_0 + (\Omega_2 - \tfrac{1}{6}\Omega_0^2)\,p_1]$$

$$p_6 = -\frac{1}{5\,.\,6}\,(\Omega_3 - \tfrac{1}{4}\Omega_0\Omega_1)\,p_1$$

To the degree of approximation adopted all higher p's are to be neglected, although it would be easy to give them to any desired degree of accuracy.

Substituting these values of the p's in the foregoing expression, and noting that $(\delta q_1)_n = p_0$, $\left(\dfrac{d\delta q_1}{dz}\right)_n = p_1$, we find

$$(\delta q_1)_{n+1} = A_0 (\delta q_1)_n + A_1 \left(\frac{d\delta q_1}{dz}\right)_n$$

$$\left(\frac{d\delta q_1}{dz}\right)_{n+1} = - B_0 (\delta q_1)_n + B_1 \left(\frac{d\delta q_1}{dz}\right)_n$$

Similar equations hold good with suffix 2, and the same A_0, A_1, B_0, B_1.

The values of these coefficients are found to be

$$A_0 = 1 - \frac{1}{1\cdot 2} \Omega_0 - \frac{1}{2\cdot 3} \Omega_1 - \frac{1}{3\cdot 4} \Omega_2 - \frac{1}{4\cdot 5} \Omega_3 + \frac{1}{24}\Omega_0{}^2 + \frac{1}{30} \Omega_0\Omega_1$$

$$A_1 = 1 - \frac{1}{2\cdot 3} \Omega_0 - \frac{1}{3\cdot 4} \Omega_1 - \frac{1}{4\cdot 5} \Omega_2 - \frac{1}{5\cdot 6} \Omega_3 + \frac{1}{120} \Omega_0{}^2 + \frac{1}{120} \Omega_0\Omega_1$$

$$B_0 = \Omega_0 + \tfrac{1}{2}\Omega_1 + \tfrac{1}{3}\Omega_2 + \tfrac{1}{4}\Omega_3 - \tfrac{1}{6}\Omega_0{}^2 - \tfrac{1}{6}\Omega_0\Omega_1$$

$$B_1 = 1 - \tfrac{1}{2}\Omega_0 - \tfrac{1}{3}\Omega_1 - \tfrac{1}{4}\Omega_2 - \tfrac{1}{5}\Omega_3 + \tfrac{1}{24}\Omega_0{}^2 + \tfrac{1}{20}\Omega_0\Omega_1$$

The form of the solution shows that

$$\left(\delta q_1 \frac{d\delta q_2}{dz} - \delta q_2 \frac{d\delta q_1}{dz}\right)_{n+1} = (A_0 B_1 + A_1 B_0) \left(\delta q_1 \frac{d\delta q_2}{dz} - \delta q_2 \frac{d\delta q_1}{dz}\right)_n$$

It is easy to show that, to the degree of approximation adopted,

$$A_0 B_1 + A_1 B_0 = 1$$

Hence, if the condition is satisfied at the nth point, it is so also at the $(n+1)$th.

Accordingly, the integration can be effected by computing the Ω's at each point, combining them to form A_0, A_1, B_0, B_1, and thence computing δq_1, δq_2, $d\delta q_1/dz$, $d\delta q_2/dz$.

If a suffix s attached to any symbol indicates its value after carrying the integration half round the periodic orbit, we have

$$\sin^2 \tfrac{1}{2}\pi c = - \frac{S}{\pi \Delta s} (\delta q_2)_s \left(\frac{d\delta q_1}{dz}\right)_s$$

If this quantity is positive and less than unity, the orbit is stable; if negative or greater than unity, it is unstable.

The computation is somewhat abridged if we do not actually compute Ω_2, Ω_3, for the results may be expressed in terms of δ's, which are easily computed from a difference table of the Ω_0's.

I find that

$$A_0 = 1 - \tfrac{1}{2}\left[(\Omega_0 - \tfrac{1}{12}\delta^2) + \tfrac{1}{3}\delta + \tfrac{1}{360}\delta^3\right] + \tfrac{1}{24}\Omega_0{}^2 + \tfrac{1}{30}\Omega_0\Omega_1$$

$$B_0 = (\Omega_0 - \tfrac{1}{12}\delta^2) + \tfrac{1}{2}\delta - \tfrac{1}{6}\Omega_0{}^2 - \tfrac{1}{6}\Omega_0\Omega_1$$

$$B_1 = A_0 - \tfrac{1}{6}\delta + \tfrac{1}{360}\delta^3 + \tfrac{1}{60}\Omega_0\Omega_1$$

$$A_1 = 1 - \tfrac{1}{6}(B_0 - \tfrac{1}{60}\delta^2) - \tfrac{7}{360}\Omega_0{}^2 - \tfrac{7}{360}\Omega_0\Omega_1$$

It is well to verify the relationship $A_0B_1 + A_1B_0 = 1$, and those between δq_1, δq_2 and their differentials at each stage of the computation.

In order to test this method of determining stability, I applied it to a case in which the stability seemed to be well determined by means of the infinite determinant. The orbit chosen was the second of those for which tabular values are given in my previous paper, viz. $C = 40\cdot25$, $x_0 = 1\cdot1150$. The determinantal method gave $\sin^2 \frac{1}{2}\pi c = \cdot0630$, and $c = 2\cdot161$. This new method gives $\cdot0625$ and $2\cdot157$ respectively; thus the agreement is excellent[*].

I have also applied the method to other orbits and to one of the figure-of-8 orbits, as will be explained hereafter.

It will now be shown that the values of $(\delta q_2)_s$ and $\left(\dfrac{d\delta q_1}{dz}\right)_s$ may be obtained otherwise than by the method of quadrature already explained. It will be proved that the computation of two orbits, having certain relationships to the periodic orbit, suffices for this end.

If our two solutions had corresponded with $(\delta q_1)_0 = \eta$, $\left(\dfrac{d\delta q_1}{dz}\right)_0 = 0$, and $(\delta q_2)_0 = 0$, $\left(\dfrac{d\delta q_2}{dz}\right)_0 = \eta'$, the former work, in which we took $\eta = 1$ and $\eta' = \dfrac{\pi\Delta s}{S}$, would have applied again with very slight changes. It is easy indeed to see that the solution becomes

$$\sin^2 \tfrac{1}{2}\pi c = \frac{1}{\eta\eta'} \cdot \frac{S}{\pi\Delta s}\left(\delta q_2 \frac{d\delta q_1}{dz}\right)_s$$

When the suffix s is attached to any symbol, or group of symbols, it is meant to denote the value after half the circuit of the periodic orbit has been performed, starting from the initial syzygy. The suffix 0 is to denote the corresponding value at that initial syzygy.

In the paper on "Periodic Orbits" (§ 8, p. 27), δp, δs denoted the displacements whereby we pass from an orbit of reference to the corresponding point in a varied orbit. The displacement along the outward normal is δp, that along the tangent is δs. The symbol $\delta\phi$ means the corresponding variation of ϕ.

In (41) and (42) of that paper it is shown that

$$\frac{d}{ds}\left(\frac{\delta s}{V}\right) = -\frac{2\delta q}{V^{\frac{3}{2}}}\left(\frac{1}{R} + \frac{n}{V}\right)$$

$$\delta\phi = -\frac{1}{V^{\frac{1}{2}}}\left[\frac{d\delta q}{ds} - \tfrac{1}{2}\delta q\left(\frac{dV}{Vds}\right)\right] + \frac{\delta s}{R}$$

also

$$\delta p = \frac{\delta q}{V^{\frac{1}{2}}}$$

---

[*] A table showing the march of δq_1, δq_2 and of their differential coefficients, and exhibiting the further steps in the computation, will be found in the Appendix to this paper.

The first of these three will not be needed, but it is given for the sake of completeness.

In the first solution we are supposed to start with

$$(\delta q_1)_0 = \eta, \quad \left(\frac{d\delta q_1}{dz}\right)_0 = \left(\frac{d\delta q_1}{ds}\right)_0 = 0$$

Suppose that at the initial syzygy $(\delta s_1)_0 = 0$, $(\delta p_1)_0 = \epsilon$ (say $\cdot 001$ as a convenient variation). Then we are proposing to trace an orbit starting from syzygy at distance from the sun S greater by ϵ than the initial starting-point x_0 of the periodic orbit. Thus we may say that $(x_1)_0 = x_0 + \epsilon$. It remains to consider the angle at which we shall start this new orbit. At syzygies V is a maximum or minimum, so that $dV/ds = 0$; also by hypothesis $(\delta s_1)_0 = 0$. Hence our second equation becomes

$$(\delta\phi_1)_0 = -\left(\frac{1}{V^{\frac{1}{2}}} \frac{d\delta q}{ds}\right)_0$$

If therefore $\left(\dfrac{d\delta q}{ds}\right)_0 = 0$, we shall have $(\delta\phi_1)_0 = 0$.

In our proposed first orbit this condition is fulfilled, and therefore $(\delta\phi_1)_0 = 0$. That is to say, we are to start at right angles to the line of syzygy as in the periodic orbit, but at a slightly greater distance from S.

The definition of δq shows us that

$$(\delta q_1)_0 = (V^{\frac{1}{2}} \delta p_1)_0$$

Hence $\qquad\qquad\qquad\qquad \eta = V_0^{\frac{1}{2}} \epsilon$

It follows that in tracing the proposed new orbit we are evaluating the first solution.

When we have traversed half the orbit $(\delta s_1)_s$ will not be zero, so that the corresponding point in the varied orbit will not be on the line of syzygies. But supposing the varied orbit to differ but slightly from the periodic orbit, we may denote the arc in the varied orbit from the starting-point to the second syzygy by $\frac{1}{2}S_1$. Then it is clear that $(\delta s_1)_s$ must be very nearly equal to $\frac{1}{2}S_1 - \frac{1}{2}S$, and we have merely to deduct the half arc of the periodic orbit from the half arc of the varied orbit in order to find it.

At this second syzygy V is again stationary, and the equation for $\delta\phi$ becomes

$$(\delta\phi_1)_s = -\left(\frac{1}{V^{\frac{3}{2}}} \frac{d\delta q_1}{ds}\right)_s + \left(\frac{\delta s_1}{R}\right)_s$$

$$= -\frac{1}{\Delta s}\left(\frac{1}{V^{\frac{3}{2}}} \frac{d\delta q_1}{dz}\right)_s + \left(\frac{\delta s_1}{R}\right)_s$$

But $(\delta\phi_1)_s$ is the excess of the final ϕ in the varied orbit above its value in the periodic orbit, and therefore can be determined from the orbit as traced.

Hence we have $\left(\dfrac{d\delta q_1}{dz}\right)_s = V^{\frac{3}{2}} \Delta s \left[\dfrac{\delta s_1}{R} - \delta\phi_1\right]_s$

If x_s is the value of the abscissa x at the second crossing of the line of syzygy in the periodic orbit, $x_s + (\delta x_1)_s$ may be taken to represent the abscissa at this second crossing in the varied orbit. It is clear that if the varied orbit differs but little from the periodic, and therefore cuts the syzygies nearly at right angles, and if x_0 (or for satellites $x_0 - 1$) is positive, $(\delta p_1)_s$ will be equal to $-(\delta x_1)_s$. In such a case as the figure-of-8 orbits A′ considered hereafter, however, $(\delta p_1)_s$ will be equal to $+(\delta x_1)_s$, but we may consider that case by a mere change of sign.

Hence we have $(\delta q_1)_s = -(V^{\frac{1}{2}} \delta x_1)_s$

It follows that the tracing of this varied orbit affords the means of computing δq_1 and its differential coefficient after half a circuit has been performed.

Now, consider the second solution denoted by suffix 2. Suppose we start with $(\delta p_2)_0 = 0$, $(\delta s_2)_0 = 0$, but with $(\delta\phi_2)_0$ equal to $-\epsilon'$ (say for example $-1°$). Thus we start from the same point as for the periodic orbit, but not quite at right angles to the line of syzygy.

We have initially $(\delta\phi_2)_0 = -\dfrac{1}{\Delta s}\left(\dfrac{1}{V^{\frac{1}{2}}}\dfrac{d\delta q_2}{dz}\right)_0$

This may be written $\epsilon' = \dfrac{1}{\Delta s}\dfrac{1}{V_0^{\frac{1}{2}}}\eta'$

So that $\eta' = \Delta s\, V_0^{\frac{1}{2}}\epsilon'$

Having found η' in terms of ϵ' (say 1°), the rest of the solution follows the same lines as the first, and therefore

$$(\delta q_2)_s = -(V^{\frac{1}{2}} \delta x_2)_s$$

$$\left(\dfrac{d\delta q_2}{dz}\right)_s = V^{\frac{3}{2}} \Delta s \left(\dfrac{\delta s_2}{R} - \delta\phi_2\right)_s$$

Thus the computation of these two additional orbits affords the means of determining the stability by means of the formula

$$\sin^2 \tfrac{1}{2}\pi c = -\dfrac{1}{\eta\eta'}\dfrac{S}{\pi\Delta s}\left(\delta q_2 \dfrac{d\delta q_1}{dz}\right)_s$$

I think that this would be the most accurate method of determining stability in cases where there are sharp flexures, but I have not carried out the laborious computations in any case.

A variation of this procedure would result from the consideration of two orbits neither of which is exactly periodic, but I leave the reader to discuss this.

§ 2. *Interpretation of the Formulæ for the Motion of the Pericentre.*

In determining the stability of direct orbits, c was evaluated, and if it involved an imaginary quantity the orbit was unstable. If it was real, certain functions gave the motion of the pericentre. In determining these functions the notation used was as follows: T synodic period, n mean motion of Jove, $d\omega/dt$ mean motion of pericentre in space. The functions were:—

(1) The regression of pericentre in the synodic period referred to moving axes, or

$$T\left(n - \frac{d\omega}{dt}\right) = 2\pi\left(\tfrac{1}{2}c - 1\right)$$

(2) The advance of pericentre in space in the synodic period, or

$$T\frac{d\omega}{dt} = nT - 2\pi\left(\tfrac{1}{2}c - 1\right)$$

(3) The advance of pericentre in space in the sidereal period, or

$$\frac{2\pi T}{2\pi + nT}\frac{d\omega}{dt} = 2\pi\left(1 - \frac{\tfrac{1}{2}c}{1 + nT/2\pi}\right)$$

When we consider orbits either really or apparently retrograde, some modification is needed in these functions.

It is well in all cases to adopt as the fundamental value of c that which lies nearest to $+\sqrt{\Phi_0}$ or $+\frac{S}{\pi}\sqrt{\Psi_0}$. As before, $\tfrac{1}{2}c$, regarded as positive, is the ratio of the synodic to the anomalistic period, although the corresponding angular velocities may have different signs.

Consider first the case of apparently retrograde orbits in which $2\pi/T$ is less than n. Here the orbit is direct in space. We must then change the sign of T in the above formulæ because the motion is apparently retrograde, and that of c because the two angular velocities are in opposite senses. Hence we get

(1) $$T\left(n - \frac{d\omega}{dt}\right) = 2\pi\left(\tfrac{1}{2}c + 1\right)$$

(2) $$T\frac{d\omega}{dt} = nT - 2\pi\left(\tfrac{1}{2}c + 1\right)$$

(3) $$\frac{2\pi T}{2\pi - nT}\frac{d\omega}{dt} = 2\pi\left(1 - \frac{\tfrac{1}{2}c}{nT/2\pi - 1}\right)$$

If the orbit is really retrograde $2\pi/T$ is greater than n, and the two angular velocities are in the same sense; hence we only change the sign of T.

We thus obtain

$$(1) \quad T\left(n - \frac{d\omega}{dt}\right) = -2\pi\left(\tfrac{1}{2}c - 1\right)$$

$$(2) \quad T\frac{d\omega}{dt} \qquad = nT + 2\pi\left(\tfrac{1}{2}c - 1\right)$$

$$(3) \quad \frac{2\pi T}{2\pi - nT}\frac{d\omega}{dt} = -2\pi\left(1 - \frac{\tfrac{1}{2}c}{1 - nT/2\pi}\right)$$

A detailed consideration of these cases shows that these interpretations are correct. My attention was called to the point by the difficulty which arose when, on using the old formulæ, I seemed to find a very large movement of pericentre in orbits which were clearly very stable.

II. RETROGRADE ORBITS.

§ 3. *Retrograde Satellites.*

In § 5 of Paper 2 Mr Hough drew conjecturally a number of retrograde satellites, and indicated the existence of a retrograde periodic orbit. My object was to test his conjectures, and with that view I chose the case of $C = 40$, and here as elsewhere took the Sun's mass ν as ten times as great as that of Jove, which was unity. The results of the computations are exhibited in Fig. 1.

If the point of departure were at the curve of zero velocity on the side of J remote from S, which in Fig. 1 and in all the other figures is supposed to be to the left of J at a distance unity, the orbit would be cusped and the satellite would revolve about J in the positive direction. I have not drawn this cusped curve, but the first orbit shown is one in which the projection is retrograde at a distance $x_0 = 1\cdot23$; it is shown in chain-dot. It will be seen that as this satellite falls towards J it loses its retrograde character, so that it revolves round Jove in the positive direction. The next one, $x_0 = 1\cdot22$, has the same character, but the pericentral distance would be very small, and a further diminution of x_0 would give a satellite which would strike Jove. When $x_0 = 1\cdot20$ we have a satellite which retains its retrograde character, with a very small pericentral distance. This portion of the orbit is not drawn, but the pericentre is reached before the satellite reaches the line of syzygy, and ϕ_1, the angle of crossing that line, is negative.

As x_0 continues to diminish ϕ_1 increases, passes through zero, and when $x_0 = 1\cdot050$ rises to $9°\ 8'$.

When $x_0 = 1\cdot08596$, $\phi_1 = -23'$, with a period given by $nT = 27°\ 13'$; and when $x_0 = 1\cdot10364$, $\phi_1 = -3°\ 15'$.

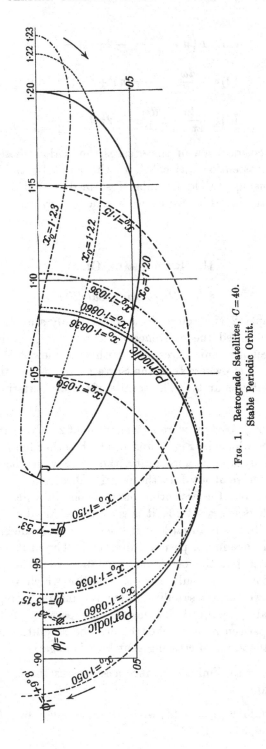

Fig. 1. Retrograde Satellites, $C = 40$.
Stable Periodic Orbit.

By extrapolation from these I find that $x_0 = 1.0836$ will make $\phi_1 = 0$, and the orbit periodic, as shown in the figure by the firm line.

The orbit $x_0 = 1.08596$ is so nearly periodic that it suffices for the computation of stability, and I find the orbit to be very stable. The numerical details are given in the appendix below.

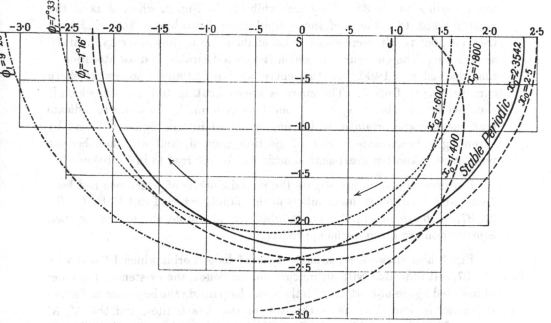

FIG. 2. Superior Planets, Direct in Space, Retrograde as to Moving Axes; when $C = 39$.

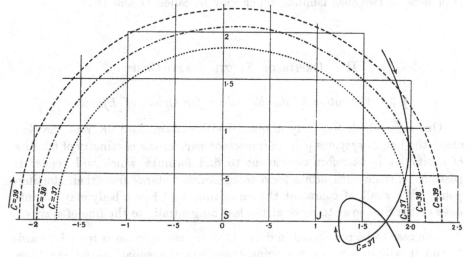

FIG. 3. Family D of Superior Planets, Direct in Space, Retrograde as to Moving Axes; All Stable. Non-periodic $x_0 = 1.25$, $C = 37$.

§ 4. *Superior Planets.*

(See p. 157 for figures.)

I began, as in my previous work, by computing a number of orbits corresponding to $C = 39$. They are exhibited in Fig. 2, where J is at the point marked 1·0. Some of the computations for orbits of this kind have been lost, but results were saved which enabled me to draw the curve marked as periodic. The only curve to which it seems desirable to draw attention is that marked $x_0 = 1·400$. In this curve we see initially a somewhat sharp curvature away from J. The curve is approximating to the shape which is assumed by an orbit of ejection from J away from S. It seems to indicate that the periodic orbits, of which one is shown in this figure, will end by becoming a heart-shaped orbit of ejection from J, and will then become retrograde. Another heart-shaped orbit will be referred to in § 6 below.

I next continued to investigate the periodic orbits of which one had been found, and determined the members of this family, which I call D, for $C = 39$, 38, 37, as shown in Fig. 3. These orbits are all stable, and details for their construction are given in the appendix.

Fig. 3 also shows a remarkable form of looped orbit which I found with $C = 37$. It indicates, without much room for doubt, the existence of families of inverted figure-of-8 orbits, with the small loop inside the large one, as further explained in Part V. As in the case of the A satellites, and the A', A" figure-of-8 orbits (considered in Part IV.), there must, in connection with the D planets, be two such families, which may be called D' and D".

III. ORBITS OF EJECTION AND CUSPS.

§ 5. *Formation of the Equations for Orbits of Ejection.*

Orbits in which the body is ejected either from Jove or from the Sun along the line of syzygies play an important part in the continuity of families of orbits; it is therefore convenient to find formulæ which will render it possible to trace such orbits for a considerable distance from their starting-point. We shall, of course, at the same time find how a body may fall into Jove or the Sun so as to move at the last tangentially to the line of syzygy.

The case to be considered in detail is where the ejection is from J towards S, and it will be easy to determine therefrom the equations for the three other cases.

The formulæ from which the solution will be found are those in § 1 of "Periodic Orbits," namely :—

$$\frac{1}{R} = \frac{P}{V^2} - \frac{2n}{V}$$

$$P = \nu \left(\frac{1}{r^2} - r \right) \cos(\phi - \theta) + \left(\frac{1}{\rho^2} - \rho \right) \cos(\phi - \psi)$$

$$V^2 = \nu \left(r^2 + \frac{2}{r} \right) + \left(\rho^2 + \frac{2}{\rho} \right) - C$$

$$\phi = \phi_0 + \int \frac{ds}{R}, \quad x = x_0 - \int \sin\phi\, ds, \quad y = y_0 + \int \cos\phi\, ds$$

In the case to be considered $\phi_0 = \frac{1}{2}\pi$, $x_0 = 1$, $y_0 = 0$.

Since the body starts straight away from J, the arc s must be at first nearly equal to the radius vector ρ. This suggests development in powers of ρ.

The change of independent variable from s to ρ may be made by means of the formula $ds = - \operatorname{cosec}(\phi - \psi)\, d\rho$, which is obvious from geometrical considerations.

If at a finite distance from J the body were still moving exactly along SJ, P would be zero and $1/R$ would have the form $2n/V$. Since near J the radius vector r is nearly equal to unity, ρ is small, and V^2 is approximately equal to $2/\rho - (C - 3\nu)$. Hence when $1/V$ is developed in powers of ρ, it must begin with $\rho^{\frac{1}{2}}$, and $1/R$ must be expressible by a series of the same form. Although P is not zero, this argument is found to have suggested the proper form for the series.

I assume, then,

$$\frac{1}{R} = - 5a\rho^{\frac{1}{2}} \left[1 + A_1\rho + A_2\rho^2 + A_3\rho^3 + \dots \right]$$

The 5 is introduced for convenience, and at present I do not go beyond ρ^3.

It is first necessary to find ds in powers of ρ.

The perpendicular on to the tangent to the orbit is $\rho \cos(\phi - \psi)$ and a known formula gives

$$\frac{1}{R} = \frac{1}{\rho} \frac{d}{d\rho} [\rho \cos(\phi - \psi)]$$

Hence $\cos(\phi - \psi) = \int \dfrac{\rho}{R}\, d\rho$

$$= - 2a\rho^{\frac{3}{2}} \left[1 + \tfrac{5}{7} A_1\rho + \tfrac{5}{9} A_2\rho^2 + \tfrac{5}{11} A_3\rho^3 + \dots \right]$$

From this we find

$$\sin^2(\phi - \psi) = 1 - 4\alpha^2\rho^3 + 4\frac{2.5}{7}\alpha^2 A_1\rho^4 - 4\alpha^2\left(\frac{2.5}{9}A_2 + \frac{5^2}{7^2}A_1^2\right)\rho^5$$

$$- 4\alpha^2\left(\frac{2.5}{11}A_3 + 2\cdot\frac{5}{7}\cdot\frac{5}{9}A_1A_2\right)\rho^6 - \dots$$

Whence, taking the negative sign of the reciprocal of the square root,

$$\frac{ds}{d\rho} = -\operatorname{cosec}(\phi - \psi)$$

$$= 1 + 2\alpha^2\rho^3 + 2\frac{2.5}{7}\alpha^2 A_1\rho^4 + 2\alpha^2\left(\frac{2.5}{9}A_2 + \frac{5^2}{7^2}A_1^2\right)\rho^5$$

$$+ 2\alpha^2\left(\frac{2.5}{11}A_3 + 2\cdot\frac{5}{7}\cdot\frac{5}{9}A_1A_2 + 3\alpha^2\right)\rho^6 + \dots$$

The determination of $1/R$ as far as ρ^3 gives $ds/d\rho$ as far as ρ^6, but at present it is only required as far as ρ^3.

Thus $ds = (1 + 2\alpha^2\rho^3)\,d\rho$

and $\dfrac{ds}{R} = -5\alpha\rho^{\frac{1}{2}}[1 + A_1\rho + A_2\rho^2 + (A_3 + 2\alpha^2)\rho^3]\,d\rho$

By integration we obtain

$$\phi = \tfrac{1}{2}\pi - \tfrac{10}{3}\alpha\rho^{\frac{3}{2}}[1 + \tfrac{3}{5}A_1\rho + \tfrac{3}{7}A_2\rho^2 + \tfrac{1}{3}(A_3 + 2\alpha^2)\rho^3]$$

The development of $\cos\phi$ and $\sin\phi$ may be effected in powers of $\tfrac{1}{2}\pi - \phi$; whence

$$\cos\phi = \frac{10}{3}\alpha\rho^{\frac{3}{2}}\left[1 + \frac{3}{5}A_1\rho + \frac{3}{7}A_2\rho^2 + \frac{1}{3}\left(A_3 - 2\cdot\frac{4^2}{3^2}\alpha^2\right)\rho^3\right]$$

$$\sin\phi = 1 - 2\frac{5^2}{3^2}\alpha^2\rho^3\left[1 + \frac{2.3}{5}A_1\rho + \left(\frac{2.3}{7}A_2 + \frac{3^2}{5^2}A_1^2\right)\rho^2\right.$$

$$\left. + \left(\frac{2}{3}A_3 + \frac{2.3^2}{5.7}A_1A_2 + \frac{11}{3^3}\alpha^2\right)\rho^3\right]$$

As yet, however, we only need $\sin\phi$ as far as ρ^3.

Thus

$$\cos\phi\,ds = \frac{10}{3}\alpha\rho^{\frac{3}{2}}\left[1 + \frac{3}{5}A_1\rho + \frac{3}{7}A_2\rho^2 + \frac{1}{3}\left(A_3 + 2\cdot\frac{11}{3^2}\alpha^2\right)\rho^3\right]d\rho$$

$$\sin\phi\,ds = \left[1 - 2\cdot\frac{4^2}{3^2}\alpha^2\rho^3\right]d\rho$$

Hence by integration

$$y = \frac{4}{3}\alpha\rho^{\frac{5}{2}}\left[1 + \frac{3}{7}A_1\rho + \frac{5}{3.7}A_2\rho^2 + \frac{5}{3.11}\left(A_3 + 2\cdot\frac{11}{3^2}\alpha^2\right)\rho^3\right]$$

$$x = 1 - \rho + \frac{8}{3^2}\alpha^2\rho^4$$

If we square these and add them together we obtain r^2. Since y^2 begins with ρ^5, r does not differ from x as far as the fourth power of ρ.

Accordingly
$$r = 1 - \rho + \frac{8}{3^2}\alpha^2\rho^4$$

We are now in a position to determine V^2, but I shall replace the constant C by another, E such that
$$C = 4E + 3\nu$$

The formula for V^2 thus becomes
$$V^2 = \frac{2}{\rho}\left[1 + \tfrac{1}{2}\rho^3 + \tfrac{1}{2}\nu\rho\left(r^2 + \frac{2}{r}\right) - 2E\rho - \tfrac{3}{2}\nu\rho\right]$$

On substituting for r its value in terms of ρ we obtain
$$V^2 = \frac{2}{\rho}\left[1 - 2E\rho + \tfrac{1}{2}(3\nu + 1)\rho^3 + \nu\rho^4 + \nu\rho^5\right]$$

Since $n^2 = \nu + 1$, $\tfrac{1}{2}(3\nu + 1) = \tfrac{3}{2}n^2 - 1$. I have no present use for the terms in ρ^4 and ρ^5; hence, as far as necessary,
$$V^2 = \frac{2}{\rho}\left[1 - 2E\rho - (\tfrac{3}{2}n^2 - 1)\rho^3\right]$$

Whence
$$\frac{1}{V^2} = \tfrac{1}{2}\rho\left[1 + 2E\rho + 4E^2\rho^2 + (8E^3 - \tfrac{3}{2}n^2 + 1)\rho^3\right]$$

$$\frac{1}{V} = \frac{1}{\sqrt{2}}\rho^{\frac{1}{2}}\left[1 + E\rho + \tfrac{3}{2}E^2\rho^2 + (\tfrac{5}{2}E^3 - \tfrac{3}{4}n^2 + \tfrac{1}{2})\rho^3\right]$$

In order to find $1/R$ it will be necessary to form $\nu\left(\frac{1}{r^2} - r\right)\Big/V^2$ and $\left(\frac{1}{\rho^2} - \rho\right)\Big/V^2$. As far as ρ^2,
$$\frac{1}{r^2} - r = 3\rho(1 + \rho)$$

and
$$\frac{\nu}{V^2}\left(\frac{1}{r^2} - r\right) = \tfrac{3}{2}\nu\rho^2\left[1 + (2E + 1)\rho\right], \text{ as far as } \rho^3 \quad \ldots\ldots\ldots(2)$$

This expression has to be multiplied by $\cos(\phi - \theta)$, and as it will only be required to develop $1/R$ as far as $\rho^{\frac{7}{2}}$, it is clear that it will only be necessary to use the development of $\cos(\phi - \theta)$ as far as $\rho^{\frac{3}{2}}$.

Now, $\sin\theta = y/r$, and the value of y found above shows that the first term in θ is $\tfrac{4}{3}\alpha\rho^{\frac{5}{2}}$. Hence to the order required $\theta = 0$, and
$$\cos(\phi - \theta) = \cos\phi = \tfrac{10}{3}\alpha\rho^{\frac{3}{2}}$$

Writing $\nu = n^2 - 1$, we have to the required order,
$$\frac{\nu}{V^2}\left(\frac{1}{r^2} - r\right)\cos(\phi - \theta) = -5\alpha\rho^{\frac{1}{2}}\left[-(n^2 - 1)\rho^3\right] \quad \ldots\ldots\ldots(3)$$

Since $\dfrac{1}{\rho^2} - \rho = \dfrac{1}{\rho^2}(1 - \rho^3)$,

$$\frac{1}{V^2}\left(\frac{1}{\rho^2} - \rho\right) = \frac{1}{2\rho}\left[1 + 2E\rho + 4E^2\rho^2 + (8E^3 - \tfrac{3}{2}n^2)\rho^3\right]$$

and $\qquad \cos(\phi - \psi) = -2\alpha\rho^{\frac{3}{2}}\left[1 + \tfrac{5}{7}A_1\rho + \tfrac{5}{9}A_2\rho^2 + \tfrac{5}{11}A_3\rho^3\right]$

Whence

$$\frac{1}{V^2}\left(\frac{1}{\rho^2} - \rho\right)\cos(\phi - \psi)$$

$$= -5\alpha\rho^{\frac{1}{2}}\left[\tfrac{1}{5} + (\tfrac{2}{5}E + \tfrac{1}{7}A_1)\rho + (\tfrac{4}{5}E^2 + \tfrac{2}{7}EA_1 + \tfrac{1}{9}A_2)\rho^2\right.$$

$$\left. + (\tfrac{8}{5}E^3 - \tfrac{3}{10}n^2 + \tfrac{4}{7}E^2A_1 + \tfrac{2}{9}EA_2 + \tfrac{1}{11}A_3)\rho^3\right]\ \ldots\ldots(4)$$

Also $\qquad -\dfrac{2n}{V} = -n\sqrt{2}\,.\,\rho^{\frac{1}{2}}\left[1 + E\rho + \tfrac{3}{2}E^2\rho^2 + (\tfrac{5}{2}E^3 - \tfrac{3}{4}n^2 + \tfrac{1}{2})\rho^3\right]\ \ldots\ldots(5)$

The sum of (3), (4), and (5) is equal to $1/R$.

If we consider only the term in $\rho^{\frac{1}{2}}$, we have

$$\frac{1}{R} = -\alpha\rho^{\frac{1}{2}} - n\rho^{\frac{1}{2}}\sqrt{2}$$

But our assumed expression for $1/R$ was $-5\alpha\rho^{\frac{1}{2}}$. Hence $\alpha = \tfrac{1}{4}n\sqrt{2}$.

On substituting this value for α, the sum of (3), (4), and (5) gives us

$$\frac{1}{R} = -\tfrac{5}{4}n\sqrt{2}\,.\,\rho^{\frac{1}{2}}\left[1 + (\tfrac{6}{5}E + \tfrac{1}{7}A_1)\rho + (2E^2 + \tfrac{2}{7}EA_1 + \tfrac{1}{9}A_2)\rho^2\right.$$

$$\left. + (\tfrac{18}{5}E^3 - \tfrac{19}{10}n^2 + \tfrac{7}{5} + \tfrac{4}{7}E^2A_1 + \tfrac{2}{9}EA_2 + \tfrac{1}{11}A_3)\rho^3\right]$$

By equating the coefficients inside the bracket to A_1, A_2, A_3 respectively, we find

$$A_1 = \tfrac{7}{5}E, \quad A_2 = \tfrac{27}{10}E^2, \quad A_3 = \tfrac{11}{2}E^3 - \tfrac{209}{100}n^2 + \tfrac{77}{50}$$

It is now easy to find the solution. Suppose that we write

$$\begin{array}{l}
\phi = \tfrac{1}{2}\pi - \tfrac{5}{8}n\sqrt{2}\,.\,\rho^{\frac{3}{2}}\left[1 + \Sigma B_i\rho^i\right] \\[2mm]
y = \tfrac{1}{3}n\sqrt{2}\,.\,\rho^{\frac{5}{2}}\left[1 + \Sigma C_i\rho^i\right] \\[2mm]
x = 1 - \rho + \Sigma D_i\rho^i \\[2mm]
\dfrac{ds}{d\rho} = \left[1 + \Sigma\alpha_i\rho^i\right] \\[2mm]
\dfrac{d\rho}{ds} = \left[1 + \Sigma\beta_i\rho^i\right]
\end{array}\right\}\ \ldots\ldots\ldots\ldots(6)$$

then I find

$$B_1 = \tfrac{21}{25}E, \quad B_2 = \tfrac{81}{70}E^2, \quad B_3 = \tfrac{11}{6}E^3 - \tfrac{46}{75}n^2 + \tfrac{77}{150}$$

$$C_1 = \tfrac{3}{5}E, \quad C_2 = \tfrac{9}{14}E^2, \quad C_3 = \tfrac{5}{6}E^3 - \tfrac{73}{270}n^2 + \tfrac{7}{30}$$

$$D_2 = 0, \quad D_3 = 0, \quad D_4 = \tfrac{1}{9}n^2$$

$$\alpha_1 = 0, \quad \alpha_2 = 0, \quad \alpha_3 = \tfrac{1}{4}n^2, \quad \alpha_4 = \tfrac{1}{2}En^2, \quad \alpha_5 = E^2n^2$$

$$\alpha_6 = 2E^3n^2 - \tfrac{61}{160}n^4 + \tfrac{7}{20}n^2$$

$$\beta_1 = 0, \quad \beta_2 = 0, \quad \beta_3 = -\tfrac{1}{4}n^2, \quad \beta_4 = -\tfrac{1}{2}En^2, \quad \beta_5 = -E^2n^2$$

$$\beta_6 = -(2E^3n^2 - \tfrac{157}{320}n^4 + \tfrac{7}{20}n^2)$$

In order to find the time we may proceed from the formula

$$nt = n\int \frac{1}{V}\frac{ds}{d\rho}\,d\rho$$

This involves the multiplication together of two series and subsequent integration. The series for $1/V$ and $ds/d\rho$ are given above, and if we write

$$nt = \tfrac{1}{3}\sqrt{2}\,.\,n\rho^{\frac{3}{2}}[1 + \Sigma G_i\rho^i]$$

we find $$G_1 = \tfrac{3}{5}E, \quad G_2 = \tfrac{9}{14}E^2, \quad G_3 = \tfrac{5}{6}E^3 - \tfrac{1}{6}\nu$$

Since we now have $ds/d\rho$ as far as ρ^6, we have the means of determining A_4, A_5, A_6. Thus we multiply $1/R$ by ds, and find ϕ as far as $\rho^{\frac{3}{2}}\rho^6$; $\cos(\phi - \psi)$ may be found by integrating ρ/R. In the further development of $\cos(\phi - \theta)$, it is useful to use the formula

$$r\cos(\phi - \theta) = \cos\phi + \rho\cos(\phi - \psi)$$

Finally, by following the same procedure as before we find equations from which A_4, A_5, A_6 may be found, and thence the successive B's, C's, D's. The analysis is tedious, and the results will merely be given in a tabular form. I shall moreover omit A_6, B_6, C_6, since the expressions become very long.

Thus far we have considered ejection from J towards S, and we will now consider ejection from J away from S.

The same forms of series serve again, but the constant of integration in finding ϕ becomes $-\tfrac{1}{2}\pi$. Hence $\sin\phi$ and $\cos\phi$ simply have their signs changed. When these are multiplied by ds and integrated, y is found to have the same form as before, but with opposite sign. However, x becomes $1 + \rho - \tfrac{8}{9}\alpha^2\rho^4$. Hence it follows that

$$V^2 = \frac{2}{\rho}[1 - 2E\rho + (\tfrac{3}{2}n^2 - 1)\rho^3 - (n^2 - 1)\rho^4 + (n^2 - 1)\rho^5]$$

In determining A_1, A_2, A_3 the last two terms are not needed, so that V^2 has the same form as before.

The expression $\dfrac{1}{r^2} - r$ becomes -3ρ, so that its sign is changed. But

11—2

$\cos(\phi - \theta)$ also has a change of sign, so that $\dfrac{1}{V^2}\left(\dfrac{1}{r^2} - r\right)\cos(\phi - \theta)$ remains as before.

The form of $\cos(\phi - \psi)$ clearly remains the same.

Thus as far as ρ^3 the result is obtained by simply changing certain signs. Further on, however, other changes are introduced.

Ejection from S may be treated by symmetry. Ejection towards J is symmetrical with ejection towards S. It is clear that r replaces ρ and ρ replaces r; $-y$ replaces y; $1 - x$ replaces x; π must be subtracted from ϕ; the Sun replaces Jove, and *vice versâ*, so that $1/\nu$ replaces ν, and hence n^2/ν replaces n^2. The n which occurs in the several coefficients at the beginning of each series represents the angular velocity of the axes and remains unchanged. The only difficulty is in the constant E or $\tfrac{1}{4}(C - 3\nu)$, and it will be found that this must be replaced by $F = \tfrac{1}{4}(C - 3)/\nu$.

The following table gives the results:—

ORBITS OF EJECTION FROM J.

The upper signs refer to ejection towards S, the lower away from S.

$$\frac{1}{R} = -\tfrac{5}{4}n(2\rho)^{\frac{3}{2}}[1 + \Sigma A_i\rho^i]$$

$$E = \tfrac{1}{4}(C - 3\nu)$$

$$A_1 = \tfrac{7}{5}E; \quad A_2 = \tfrac{27}{10}E^2; \quad A_3 = \tfrac{11}{2}E^3 - \tfrac{209}{100}\nu - \tfrac{11}{20}$$

$$A_4 = \tfrac{91}{8}E^4 - \tfrac{2119}{300}E\nu - \tfrac{39}{20}E \mp \tfrac{13}{6}\nu$$

$$A_5 = \tfrac{189}{8}E^5 - \tfrac{38711}{1960}E^2\nu - \tfrac{45}{8}E^2 \mp \tfrac{481}{70}E\nu - \tfrac{41}{14}\nu$$

N.B.—The coefficient of the highest power of E is $\tfrac{1}{6}(2i + 5)\left(\dfrac{2i\,!}{2^i\,(i\,!)^2}\right)$.

$$\phi = \pm\tfrac{1}{2}\pi - \tfrac{5}{8}n\sqrt{2}\,.\,\rho^{\frac{3}{2}}[1 + \Sigma B_i\rho^i]$$

$$B_1 = \tfrac{21}{25}E; \quad B_2 = \tfrac{81}{70}E^2; \quad B_3 = \tfrac{11}{6}E^3 - \tfrac{46}{75}\nu - \tfrac{1}{10}$$

$$B_4 = \tfrac{273}{88}E^4 - \tfrac{466}{275}E\nu - \tfrac{3}{10}E \mp \tfrac{13}{22}\nu$$

$$B_5 = \tfrac{567}{104}E^5 - \tfrac{12771}{3185}E^2\nu - \tfrac{3}{4}E^2 \mp \tfrac{111}{70}E\nu - \tfrac{123}{82}\nu$$

$$y = \pm\tfrac{1}{3}n\sqrt{2}\,.\,\rho^{\frac{5}{2}}[1 + \Sigma C_i\rho^i]$$

$$C_1 = \tfrac{3}{5}E; \quad C_2 = \tfrac{9}{14}E^2; \quad C_3 = \tfrac{5}{6}E^3 - \tfrac{73}{270}\nu - \tfrac{1}{27}$$

$$C_4 = \tfrac{105}{88}E^4 - \tfrac{199}{330}E\nu - \tfrac{1}{15}E \mp \tfrac{5}{22}\nu$$

$$C_5 = \tfrac{189}{104}E^5 - \tfrac{76313}{63700}E^2\nu - \tfrac{39}{350}E^2 \mp \tfrac{37}{70}E\nu - \tfrac{41}{182}\nu$$

$$x = 1 \mp \rho \pm \Sigma D_i \rho^i$$

$$D_2 = 0; \quad D_3 = 0; \quad D_4 = \tfrac{1}{9}n^2; \quad D_5 = \tfrac{2}{15}En^2; \quad D_6 = \tfrac{32}{175}E^2n^2$$

$$D_7 = n^2\left[\tfrac{256}{945}E^3 - \tfrac{131}{2430}\nu - \tfrac{1}{486}\right]$$

$$D_8 = n^2\left[\tfrac{2048}{4851}E^4 - \tfrac{3458}{22275}E\nu - \tfrac{2}{405}E - \tfrac{5}{99}\nu\right]$$

$$D_9 = n^2\left[\tfrac{2048}{3003}E^5 - \tfrac{1683998}{4729725}E^2\nu - \tfrac{46}{4725}E^2 \mp \tfrac{512}{3465}E\nu - \tfrac{41}{819}\nu\right]$$

$$ds = d\rho\left[1 + \Sigma \alpha_i \rho^i\right]$$

$$\alpha_1 = 0; \quad \alpha_2 = 0; \quad \alpha_3 = \tfrac{1}{4}n^2; \quad \alpha_4 = \tfrac{1}{2}En^2; \quad \alpha_5 = E^2n^2$$

$$\alpha_6 = n^2\left[2E^3 - \tfrac{61}{160}\nu - \tfrac{1}{32}\right]$$

$$d\rho = ds\left[1 + \Sigma \beta_i \rho^i\right]$$

$$\beta_1 = 0; \quad \beta_2 = 0; \quad \beta_3 = -\tfrac{1}{4}n^2; \quad \beta_4 = -\tfrac{1}{2}En^2$$

$$\beta_5 = -E^2n^2; \quad \beta_6 = -n^2\left[2E^3 - \tfrac{157}{320}\nu - \tfrac{9}{64}\right]$$

$$nt = \tfrac{1}{3}n\sqrt{2} \cdot \rho^{\frac{3}{2}}\left[1 + \Sigma G_i \rho^i\right]$$

$$G_1 = \tfrac{3}{5}E; \quad G_2 = \tfrac{9}{14}E^2; \quad G_3 = \tfrac{5}{6}E^3 - \tfrac{1}{6}\nu; \quad G_4 = \tfrac{105}{88}E^4 - \tfrac{3}{22}\nu(3E \pm 1)$$

$$G_5 = \tfrac{189}{104}E^5 - \nu\left(\tfrac{45}{32}E^2 \pm \tfrac{9}{26}E + \tfrac{3}{26}\right)$$

The same formulæ may be used for ejection from S, with the appropriate changes indicated above.

§ 6. *On certain Orbits of Ejection.*

This class of orbits was considered principally with the object of completing the classification of the family A of satellites, as indicated in the Introduction. It will therefore be more convenient to consider the orbits of ejection from J towards S in Part IV. below. We will here, therefore, only consider a group of orbits arising from ejection from S towards J. Several of these orbits are shown in Fig. 4, and we may see the gradual transformation which occurs as the constant of relative energy falls.

Being curious to understand the later development for smaller values of C, I computed at hazard the ejectional orbit for $C = 20$, and happened by chance on a very curious form of periodic orbit. This is exhibited in Fig. 5, and it will be observed that the body is not only ejected along the line of syzygy, but returns to S tangentially to the same line with extraordinary exactness. There is of course another branch, on the positive side of the line of syzygy but not shown in the figure, in which the ejection takes place away from J and where the return is on the side towards J. It seemed worth while to show how the body moves in space whilst performing this

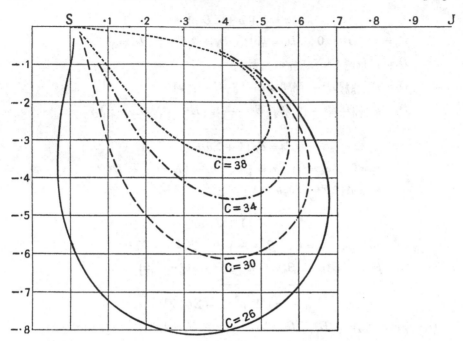

Fig. 4.   Orbits of Ejection from S towards J.

curious orbit with respect to moving axes, and the result is exhibited in an inset of the same figure. The body starts in a straight line towards J, and is gradually slightly deflected by J. By the time it has reached and passed the place where J was, J has passed away, and it falls back on S nearly in a

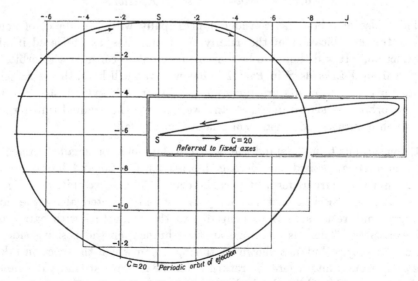

Fig. 5.   A Periodic Orbit of Ejection from S towards J.

straight line, so timing its fall that the bodies are again in a straight line as it reaches the Sun again. The time from ejection to return is obviously given by a value of nT a little greater than 180°, being in fact 190°.

It is easy to perceive that this orbit is the limiting form of four others which coalesce. A body may move in any one of the four following ways, all coalescent in the orbit in Fig. 5 :—

(1)   Initially direct, finally direct.

(2)   Initially direct, finally retrograde.

(3)   Initially retrograde, finally direct.

(4)   Initially retrograde, finally retrograde.

It is probable that the possibility of the simultaneous coalescence of these four kinds of orbits depends on the choice of a particular value of the Sun's mass, in this case ten times that of the large planet Jove. I give the numerical values for Fig. 5 in the appendix.

In the course of the further investigation of this class of orbit I computed one more, which proves the existence of a heart-shaped periodic orbit of ejection from S towards J.

When $C = 3$ the orbit was found to cut the line of syzygies where $x = -3.57$, $y = 0$, and with $\phi = 11\frac{1}{2}°$. Now when C was 20 the final ϕ was $-90°$, and therefore for a value of C a very little greater than 3 the final ϕ must vanish, and the orbit must be periodic. A figure exhibiting the results of the calculation shows that in this periodic orbit the crossing of the line of syzygy will occur when x is about equal to -3. A little consideration will show the reader that the orbit is heart-shaped, and that the body will fall back on S tangentially to the line of syzygy in the direction opposite to that along which it was ejected. It would be possible to give a fairly accurate drawing of this orbit, but it hardly seems worth while to do so, because Dr Carl Burrau has already given illustrations of this kind of orbit, and has shown that they are probably limiting forms of the families of oscillating satellites[*].

§ 7.   On the Form of a Cusp in an Orbit.

If the third body reaches the curve of zero velocity the orbit has a cusp, and the common tangent to the two branches of the orbit is at right angles to the curve of zero velocity.

If the point x_0, y_0 (or in polar co-ordinates r_0, θ_0 or ρ_0, ψ_0) lies on the curve $2\Omega = C$, and if ϕ_0 defines the direction of the normal to that curve, we have

$$\nu \left(\frac{1}{r_0^2} - r_0 \right) \cos (\phi_0 - \theta_0) + \left(\frac{1}{\rho_0^2} - \rho_0 \right) \cos (\phi_0 - \psi_0) = 0$$

* *Ast. Nach.*, No. 3251, Vol. 136 (1894).

I find that the form of the orbit at this point is given by equations of which I shall not give the proof.

In order to express the result the following notation is adopted :—

$$f = \nu \left(\frac{1}{r_0^2} - r_0 \right) \sin (\phi_0 - \theta_0) + \left(\frac{1}{\rho_0^2} - \rho_0 \right) \sin (\phi_0 - \psi_0)$$

$$g = \frac{\nu}{r_0^3} + \frac{1}{\rho_0^3}$$

$$h = \frac{\nu}{r_0^3} \sin^2 (\phi_0 - \theta_0) + \frac{1}{\rho_0^3} \sin^2 (\phi_0 - \psi_0)$$

$$k = \frac{\nu}{r_0^3} \sin (\phi_0 - \theta_0) \cos (\phi_0 - \theta_0) + \frac{1}{\rho_0^3} \sin (\phi_0 - \psi_0) \cos (\phi_0 - \psi_0)$$

also s denotes the arc of orbit measured from the cusp, and ϕ_0 is the inclination of the tangent of the curve $2\Omega = C$ to the x axis, and therefore ϕ_0 gives the initial value of ϕ in the orbit which is at right angles to the curve of zero velocity.

Then I find

$$\frac{1}{R} = \frac{k}{f} - \tfrac{1}{2} n \left(\frac{2}{fs} \right)^{\frac{1}{2}} \left(1 - \frac{15h - 7g}{4f} s \ldots \right)$$

$$\phi = \phi_0 - n \left(\frac{2s}{f} \right)^{\frac{1}{2}} + \frac{k}{f} s + \tfrac{1}{3} n \left(\frac{2}{f} \right)^{\frac{1}{2}} \left(\frac{15h - 7g}{4f} s^{\frac{3}{2}} \ldots \right)$$

$$x = r_0 \cos \theta_0 + \cos \phi_0 \left[\tfrac{2}{3} n \left(\frac{2}{f} \right)^{\frac{1}{2}} s^{\frac{3}{2}} - \frac{k}{2f} s^2 - \tfrac{2}{15} n \left(\frac{2}{f} \right)^{\frac{1}{2}} \frac{15h - 7g + 4n^2}{4f} s^{\frac{5}{2}} \ldots \right]$$

$$- \sin \phi_0 \left[s - \frac{n^2}{2f} s^2 + \tfrac{2}{5} \frac{nk}{f} \left(\frac{2}{f} \right)^{\frac{1}{2}} s^{\frac{5}{2}} \ldots \right]$$

$$y = r_0 \sin \theta_0 + \sin \phi_0 \left[\tfrac{2}{3} n \left(\frac{2}{f} \right)^{\frac{1}{2}} s^{\frac{3}{2}} - \frac{k}{2f} s^2 - \tfrac{2}{15} n \left(\frac{2}{f} \right)^{\frac{1}{2}} \frac{15h - 7g + 4n^2}{4f} s^{\frac{5}{2}} \ldots \right]$$

$$+ \cos \phi_0 \left[s - \frac{n^2}{2f} s^2 + \tfrac{2}{5} \frac{nk}{f} \left(\frac{2}{f} \right)^{\frac{1}{2}} s^{\frac{5}{2}} \ldots \right]$$

For small values of s the curve is a semicubical parabola.

These formulæ were of some use in the investigations of Part IV.

IV. The Families of Orbits of Satellites denoted A, A′, A″.

§ 8. *Completion of the Classification.*

A comparison of the orbit of the family C of satellites when $C = 38\cdot75$, as shown in Fig. 3, Plate IV. of my Paper on "Periodic Orbits," with the orbit of the family A when $C = 40$, as shown in Fig. 1, Plate IV., ought to have afforded a sufficient indication of the fate of the A satellites; it ought also to have indicated the existence of the A″ family of figure-of-8 orbits. Now that I know the result, the indication is so clear that I find it hard to understand how it came to be overlooked.

The A satellites are clearly tending to the ejectional form, and this was pointed out by Mr Hough. Accordingly, in order to investigate the fate of the A family it was necessary to find an orbit of ejection which should cut the line of syzygy at right angles. This would form the end of that part of the family A in which the revolution round J is direct.

Secondly, it was required to find an orbit of ejection which should cut the line of syzygy at an angle a little less than π, and from this intersection the satellite should pass on, describe half a loop round the point of zero force, and end by cutting the line of syzygy a second time at right angles. This would form the end of the A″ orbits considered as direct orbits.

With these objects in view a great many orbits of ejection were computed. Unfortunately, I had become convinced that the critical orbit would be found to occur for a value of C very nearly equal to 40, and as this idea proved to be incorrect a great deal of my work became almost useless. However, in Fig. 6 I exhibit several of these orbits from which the results may be inferred with close accuracy.

A comparison of the orbit when $C = 39$ with the small portion of the orbit for $C = 39\cdot2$ shows that C had to be diminished further. The orbit for $C = 38\cdot9$ shows that C is still a very little too large, but that for $C = 38\cdot7$ it is too small. I conjecture that when C is about $38\cdot85$ the downward branch of the curve would cut the line of syzygy at right angles, and this would give us the ejectional form of the A family of satellites.

Again, if we were to take C a very little less than $38\cdot85$, say perhaps $38\cdot845$, the curve would cut the line of syzygies at an angle a very little less than π. Now, all the upper part of the loop would hardly be sensibly changed by these small changes in C. Hence we know with pretty close accuracy all the first part of the A″ orbit of ejection. In order to construct the loop of this orbit, it is best to begin from the left-hand side. The table of values for the oscillating satellite (a) when $C = 39$ is given on pp. [108—9] of my previous paper. We know that for $C = 38\cdot845$ the loop must fall just outside the path of this oscillating satellite. Hence by drawing a curve just outside this

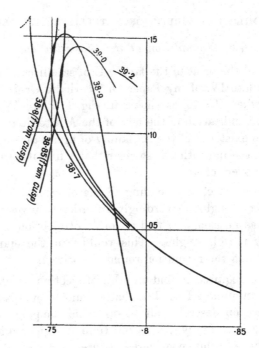

Fig. 6. Some Orbits of Ejection from J towards S.

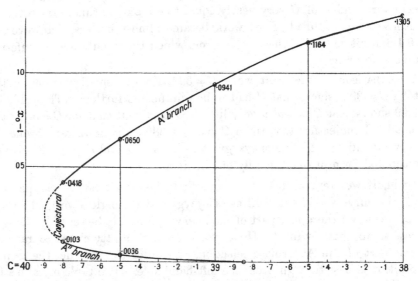

Fig. 7. Curve showing Coalescence of the A′ and A″ Families of Figure-of-8 Orbits.
Coalescence occurs for $x_0 = 1·025$, $C = 39·9$.

orbit (a), and joining it with a nearly straight line to the upper part of the loop of $C = 38\cdot9$, we shall obtain a very close representation of the ejectional form of the A″ orbit. The curve for $C = 38\cdot845$ in Fig. 9 below was constructed in this way, but it is marked as corresponding to $38\cdot85$ because, after all, we only know the value to be somewhat less than $38\cdot9$.

This result gives us the end of the A″ series, considered as directly moving satellites. It is yet more interesting to determine the beginning.

The A″ family of figure-of-8 orbits must arise from a form coalescent with the A′ family of figure-of-8 orbits, [originally] erroneously classified in my previous paper as belonging to the A family of satellites. In that paper I had the A′ family for $C = 38, 38\cdot5, 39, 39\cdot5$. It is therefore possible to construct a figure in which the abscissæ are C and the ordinates $x_0 - 1$. The upper curve in Fig. 7, as far as $C = 39\cdot5$ and $x_0 - 1 = \cdot0650$, shows the result as far as explained as yet; and one single point, viz. $C = 38\cdot85$ and $x_0 - 1 = 0$, has been obtained in the lower curve in Fig. 7.

It then became necessary to find another point on the lower curve, namely that for $C = 39\cdot5$. After some very troublesome calculations I found that if we start the third body at $x_0 - 1 = \cdot0036$, so that it is no longer quite an orbit of ejection, we shall obtain a figure-of-8 orbit which belongs to the family A″.

These two branches for A′ and A″ have to be joined by a curve of rather sharp flexure. I made the junction conjecturally, and guessed that it would occur very nearly when $C = 39\cdot8$. It remained then to draw the two figure-of-8 orbits for this value of C. A great many tedious computations were made, whereby I found that the A′ orbit for this value of C will begin with $x_0 = 1\cdot0418$, and the A″ with $x_0 = 1\cdot0103$. These values give one point more on each of the two branches, and it appeared that the guess of $39\cdot8$ as the critical value was not a very good one. However, it did not seem worth while to proceed to a new conjecture, and the two branches are joined by a dotted curve from which we may conclude with fair accuracy that the critical coalescent figure-of-8 orbit would be found for $C = 39\cdot9$ and $x_0 - 1 = \cdot025$.

The family of orbits called A′ is exhibited in Fig. 8. The four outer curves were originally shown in Fig. 1, Plate IV. of my previous paper and were classified as belonging to the family A. But in the present republication that figure only shows the true A family, and the family A′ is exhibited in this present Fig. 8. This family A′ begins with the conjectural coalescent form when C is $39\cdot9$, and continues until C is 38; the starting-point continually recedes from J.

In Fig. 9 we have the A″ family beginning with the same coalescent form for $C = 39\cdot9$, and continuing until C has fallen to $38\cdot85$, when the orbit has assumed the ejectional form; after this the orbit becomes retrograde. During these changes the starting-point continually approaches J.

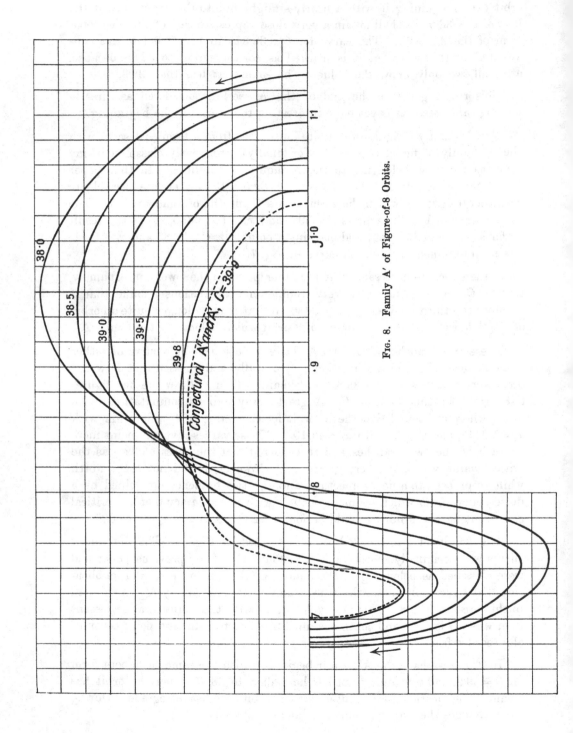

Fig. 8. Family A' of Figure-of-8 Orbits.

In the families A′ and A″ the third body describes circuits alternately about J and about the point of zero force. If it moves exactly on one of those curves it may perform these alternate circuits for ever, but otherwise it will end by moving off towards the neighbourhood of S or towards J. Now, for all starting-points on the line of syzygy between those of A′ and of A″ the body will ultimately or shortly move off towards the neighbourhood of S. Hence these prescribe the limiting forms of the paths of bodies which escape towards the Sun.

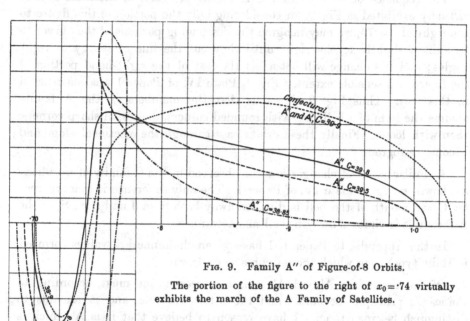

FIG. 9.   Family A″ of Figure-of-8 Orbits.

The portion of the figure to the right of $x_0 = \cdot74$ virtually exhibits the march of the A Family of Satellites.

It is obvious that the third body might describe one circuit round J and then two round the point of zero force, and thus there must be another pair of limiting orbits, say A‴ and Aᴵⱽ, in which the small loop of the 8 is described twice. The range of starting-points between A‴ and Aᴵⱽ will be [narrower] than in the case of A′ and A″. So further there must be other pairs of orbits in which the small loop is described, three, four, and more times, and there will be a decreasing range between the starting-points.

The neck of the hour-glass figure of the curves of zero velocity first opens when $C = 40\cdot1821$, but the A′ and A″ orbits only appear when C has fallen to $39\cdot9$. [As C diminishes in magnitude these are the first figures-of-8 to appear.

When the neck of the hour-glass becomes wider there will be a pair of figure-of-8 orbits which we may call $A^{(2n+1)}$, $A^{(2n+2)}$, in which the body moves once round J and then describes n circuits in the smaller loop of the 8, and when the neck has a considerable width n may be large. Mr Hough has shown (in Paper 2) how each successive pair of multiple figures-of-8 springs into existence as C diminishes, and he has indicated how they furnish an illustration of M. Poincaré's asymptotic orbits*.]

The sequence of changes in the A family of orbits of satellites is also virtually exhibited in Fig. 9, for considering only the portion of this figure to the right of $x = \cdot74$, we may imagine the descending portions of the curves to be shifted slightly to the right until they cut the line of syzygy at right angles; and this change will not affect the rest of the right-hand portion of the figure to a sensible extent. Fig. 1, Plate IV. of Paper 1 was constructed in this way. Thus Fig. 9 shows us that the members of the A family assume the form of a bell, first with rounded corners, then with sharp corners, then with loops. Finally these orbits pass through the ejectional stage and become retrograde.

A similar series of changes is to be observed in the C family of satellites, as shown in Fig. 3, Plate IV. of Paper 1. The only difference is that whereas for C the mouth of the bell is directed away both from J and from S, in the A family it is directed away from J and towards S.

In the Appendix to Paper 1, I have given the numerical values (printed in italic type) from which these figures were drawn.

Of the two orbits A' and A" near coalescence, one must, according to Poincaré's principle, be stable and the other unstable, and it remains to distinguish between them. I have reason to believe that it is A' which is unstable, and therefore A" must be stable.

When the curvature of an orbit is abrupt both methods of discussing stability become very intractable. As the curvature of the A' curve is rather less sharp than that of A" it was chosen for treatment.

It was shown on p. [86] of "Periodic Orbits" that when $C = 39$ the function $\Delta \sin^2 \frac{1}{2}\pi \sqrt{\Phi_0}$ for the A' orbit was negative, and that the orbit was unstable, according to the class called even instability. In that case the representation of Φ by the Fourier series was not very satisfactory, yet sufficiently so, I think, to establish the conclusion. When the determinantal method was applied to the orbit A' for $C = 39\cdot8$, the representation of Φ by the Fourier series was very wild, and the result can hardly be relied on further than as giving the sign of the determinant with some probability of correctness. I found it to be negative, and the instability was therefore probably of the same class as that in the case of $C = 39$.

* [This passage in the original paper was wrong, as Mr Hough has convinced me.]

When Mr Hough's method was applied I found the same negative sign, indicating that c was of the form $\alpha \sqrt{(-1)}$, but the numerical value differed much from that found from the determinant.

It seems, then, almost certain that A' is unstable near coalescence, and if this is so A" must be stable. The margin of stability is, however, probably very small, and indeed it is possible that by the time C has fallen from 39·9 to 39·8 both orbits may have become unstable.

V. REVIEW OF THE PRESENT INVESTIGATION.

§ 9. *Indications as to the Development of Old and the Existence of New Families.*

Several conclusions may be drawn from the results obtained above, of which some are certain and others rather speculative.

The families B and C of satellites were traced from their coalescent origin when C was 39·3 until they came to differ very widely from one another. In Part IV. the close analogy was pointed out between the development of the A family, as now traced, with that which was found before for the C family. The origin of these two families is, however, very different, for the A family begins with an infinitely small circle round J, and as the orbit expands with falling value of the constant C, it gradually loses its circularity; whereas B springs into existence along with the family C when the constant has fallen to 39·3.

In Part IV. the relationship has been traced between the A satellites and the A' and A" figure-of-8 orbits, and we are now able to see that the existence of C will connote two families of figures-of-8, C' and C", with the smaller loop of the 8 directed away from S. Indeed, a fairly accurate drawing of one of these might be constructed from Fig. 1, Plate III. of " Periodic Orbits," where the orbit C is shown alongside of the oscillating satellite (b); for it is only necessary to make a connection crosswise between these two. I believe that this orbit would be of the family C".

It is not as yet at all clear what is the nature of the future development of the family B, when the constant C falls below 38. From the succession of figures in " Periodic Orbits " it is obvious that the orbit B comes to agree more and more closely with the larger loop of the orbit A', and that the right-hand side of the orbit of the oscillating satellite (a) is separated by an ever-diminishing space from the orbit B. May we not presume that B, A', and (a) stretch out until they all draw near to the apex of the equilateral triangle drawn on SJ as a base? But I do not hazard any conjecture as to the changes beyond that stage.

We may feel confident that the family A of planets must pass through changes analogous to those of the A family of satellites, but the constant C

will have to fall considerably below 38 before conspicuous changes set in. Thus the A planets must become retrograde by passing through the ejectional stage with ejection from S towards J. It would seem from an inspection of Fig. 4 that this stage is nearly reached when $C = 26$. There will probably also be a pair of families of figure-of-8 orbits bearing the same relationship to the planetary A that A′ and A″ bear to the satellite A.

Presumably a pair of planetary orbits B and C will spring from coalescence, of which C will at first be stable. The planetary orbit C will lose its stability and will cut SJ nearer and nearer to S, while stretching outwards toward the point of zero force remote from J. This orbit must then become bell-shaped, ejectional, and then retrograde. In correlation with the planetary family C two new planetary figure-of-8 orbits must arise with the smaller loops circling the point of zero force on the side of S remote from J, and they must pass through changes analogous to those of the satellites C′ and C″. The point of zero force remote from J is so far distant from both bodies that all these changes will be much less conspicuous than in the cases already considered.

In this paper three members are traced in Fig. 3 of the family D of superior planets, which move directly in space but are retrograde with respect to the rotating axes. As the value of C falls the size of these orbits diminishes. I do not venture to make any suggestions as to their future development.

In Fig. 3 there is also shown a portion of a non-periodic orbit. The loop described in this case gives a fairly certain indication of the existence of two families of orbits, say D′ and D″, which are analogous to figure-of-8 orbits, but have the smaller loop of the 8 inside the larger one. We may guess that one of these two becomes retrograde by passing through the stage of ejection from J towards S. Then again there must be two others of these inverted figures-of-8 connected with the point of zero force remote from J.

Possibly all these figures stretch themselves out towards the apices of the equilateral triangles erected on SJ. This last suggestion leads me to remark on a possible mode of transition between direct and retrograde orbits. The apex of the equilateral triangle is a point of zero force at which the third body may be at rest, although unstably unless the Sun is more than 25 times as massive as Jove. Now, it would seem that the body may reach the apex, moving asymptotically with an infinitely small velocity either in the direct or retrograde sense. If this is the case, passage through the apex would afford a mode of transition from the direct to the retrograde.

It is clear that much more remains to be discovered even about these simply periodic orbits, but the present investigation has added something to our knowledge, and it may be hoped that the heavy work which has been undertaken to obtain the foregoing results has proved its justification.

APPENDIX.

An example of the method of discussion contained in Part I., p. 143.

Stability of the Orbit $C = 40\cdot25$, $x_0 = 1\cdot1150$ (see p. 79).

The common difference of arc was $\Delta s = \cdot03$, $S = \cdot80844$, $\dfrac{\pi\Delta s}{S} = \cdot11658$.

s	δq_1	$d\delta q_1/dz$	δq_2	$d\delta q_2/dz$
$\cdot00$	$1\cdot00000$	$\cdot00000$	$\cdot00000$	$\cdot11658$
3	$\cdot97994$	$-\cdot04031$	$\cdot11580$	1421
6	$\cdot91873$	8258	$\cdot22666$	0652
9	$\cdot81375$	$\cdot12788$	$\cdot32657$	$\cdot09195$
$\cdot12$	$\cdot66221$	7529	$\cdot40750$	6819
5	$\cdot46386$	$\cdot22047$	5915	$+\cdot03310$
8	$\cdot22502$	5439	6996	$-\cdot01319$
$\cdot21$	$-\cdot03707$	6499	3079	6552
4	$\cdot29448$	4461	$\cdot34060$	$\cdot11299$
7	$\cdot51747$	$\cdot19811$	$\cdot21010$	4487
$\cdot30$	$\cdot68696$	4038	$+\cdot05719$	5803
3	$\cdot79925$	$\cdot08542$	$-\cdot10110$	5668
6	$\cdot86084$	3931	$\cdot25342$	4700
$\cdot39$	$\cdot88046$	$-\cdot00087$	$\cdot39395$	3280
$\cdot42$	$-\cdot86362$	$+\cdot03447$	$-\cdot51794$	$-\cdot11432$

From the last five entries we have :—

s	$-\delta q_2 d\delta q_1/dz$
$\cdot30$	$+\cdot008028$
$\cdot33$	$-\cdot008636$
$\cdot36$	$-\cdot009961$
$\cdot39$	$-\cdot000344$
$\cdot42$	$+\cdot017855$

By interpolation when $s = \frac{1}{2}S = \cdot40422$

$$-\left(\delta q_2 \frac{d\delta q_1}{dz}\right)_s = \cdot007283$$

$$\sin^2 \tfrac{1}{2}\pi c = -\frac{S}{\pi\Delta s}\left(\delta q_2 \frac{d\delta q_1}{dz}\right)_s = \cdot06247$$

$$\tfrac{1}{2}\pi c = \pi \pm 14° \ 2'$$

The fundamental value of c is that which lies nearest to $\sqrt{\Phi_0}$, and is such that

$$\tfrac{1}{2}\pi c = 194° \ 2'$$

$$c = 2\cdot156$$

The method of the infinite determinant (see p. 80) gave

$$\sin^2 \tfrac{1}{2}\pi c = \cdot0630, \quad c = 2\cdot161$$

D. IV.

THE RETROGRADE SATELLITE (see pp. 155—6).

The coordinates from which the stability was computed are as follows :—

C = 40

s	x	y	ϕ
·00	1·08596	·00000	π
3	8081	− ·02939	$\pi - 20°\ \ 4'$
6	6582	5521	$40°\ 14'$
9	4282	7424	$60°\ 36'$
·12	1462	8404	$\pi - 81°\ \ 9'$
5	·98478	8328	$+78°\ 12'$
8	5712	7205	$57°\ 35'$
·21	3516	5182	$37°\ \ 8'$
4	2161	− ·02522	$+10°\ 53'$
·27	·91807	+ ·00443	$-\ \ 3°\ 16'$

When $y_1 = 0$, $x_1 = $ ·91796, $\phi_1 = -$ 23′, $\frac{1}{2}S = $ ·26571, $nT = 27°\ 13'$.

By the method of Part I. I find

$$\tfrac{1}{2}\pi c = 166°\ 49', \quad c = 1·8535$$

The orbit being truly retrograde, according to the formulæ in § 2, I find

$$T\left(n - \frac{d\omega}{dt}\right) = 26°\ 22'$$

$$T\frac{d\omega}{dt} \qquad = 0°\ 51'$$

$$\frac{2nT}{2\pi - nT}\frac{d\omega}{dt} = 0°\ 54'$$

The period is short compared with that of most of the orbits previously traced.

The orbit is very stable, and the motion of the pericentre is small.

FAMILY D OF SUPERIOR PLANETS (see p. 157),

direct in space, but retrograde with respect to moving axes.

C = 39

When $x_0 = 2·34$, $\phi_1 = +$ 9′, when $x_0 = 2·40$, $\phi_1 = -$ 29′.

From interpolation between these the following periodic orbit is determined.

FAMILY D OF SUPERIOR PLANETS *continued*.

s	x	y
0	2·3543	·0000
0·5	·2983	− ·4959
1·	·1335	·9668
1·5	1·8685	1·3896
2·	·5169	1·7437
2·5	·0963	2·0123
3·	0·6272	·1823
3·5	+ ·1322	·2455
4·	− ·3644	·1989
4·5	·8388	·0444
5·	1·2676	1·7894
5·5	·6297	·4463
6·	·9072	·0317
6·5	2·0862	0·5660
7·	− 2·1508	−0·0723
	− 2·154	·000

$$nT = 513°$$

From my notes I find that the orbit is stable, but the computation has been lost.

This conclusion is no doubt correct since this orbit should be more stable than that when $C = 37$, of which details are given below.

C = 38 $x_0 = -1·9675.$

This orbit was computed from the other syzygy. The final value of ϕ is $\phi_1 = \pi - 9'$, and this is accepted as being sufficiently nearly equal to π for periodicity.

s	x	y
0	−1·9675	·0000
0·5	·9068	·4951
1·	·7276	·9606
1·5	·4400	1·3689
2·	·0643	·6959
2·5	0·6201	·9228
3·	− ·1344	2·0368
3·5	+ ·3645	·0317
4·	·8478	1·9082
4·5	1·2880	·6733
5·	·6595	·3403
5·5	·9401	0·9279
6·	2·1121	+ ·4597
6·5	2·1638	− ·0364
6·4643	2·1644	·0000

$$\phi_1 = \pi - 9', \quad nT = 546°$$

According to my notes the orbit is stable, but the computation is lost.

FAMILY D OF SUPERIOR PLANETS *continued.*

C = 37 $x_0 = -1.7503$.

The final value of ϕ is $\phi_1 = \pi - 5'$, and the orbit is accepted as being periodic.

s	x	y
0	-1.7503	·0000
0·5	·6821	·4937
1·	·4816	·9501
1·5	·1650	1·3352
2·	0·7567	·6213
2·5	$-$ ·2872	·7891
3·	$+$ ·2099	·8278
3·5	·7000	·7361
4·	1·1497	·5207
4·5	·5276	·1955
5·	·8056	0·7817
5·5	·9604	$+$ ·3079
6·	1·9757	$-$ ·1903
5·8119	1·9863	·0000

$$\phi_1 = \pi - 5', \quad nT = 602°$$

The stability was determined by the new method, and I find the orbit to be stable, with $c = 1.3674$. Nearly the same value was also found by the determinantal method.

$$2\pi\left(\tfrac{1}{2}c + 1\right) = 606° \quad 8'$$

$$nT - 2\pi\left(\tfrac{1}{2}c + 1\right) = -6° \ 17'$$

$$2\pi\left[1 - \frac{\tfrac{1}{2}c}{nT/2\pi - 1}\right] = -6° \ 23'$$

PERIODIC ORBIT OF EJECTION, $C = 20$.

The initial portion of the orbit was computed from the analytical formulæ of Part III., p. 164.

s	x	y	ϕ	Coordinates referred to fixed axes	
				X	Y
·2	·1998	$-$·00934	$\pi + 83°\ 10'$		
·3	·2983	2664	76°\ 58'		
·4	·3940	5528	69°\ 19'	·3978	$-$·0010
·5	·4845	9749	60°\ 29'		
·6	·5671	·1537	51°\ 0'	·5876	$-$·0009
·7	·6392	·2228	41°\ 29'		
·8	·6990	·3028	32°\ 7'	·7617	$+$·0051
·9	·7449	$-$·3915	$\pi + 22°\ 33'$		

PERIODIC ORBIT OF EJECTION *continued.*

| | | | | Coordinates referred to fixed axes | |
s	x	y	ϕ	X	Y
1·0	·7751	− ·4867	$\pi +$ 12° 33′	·9152	+ ·0213
·1	881	·5857	$\pi +$ 2° 21′		
·2	834	·6855	$\pi -$ 7° 41′	1·1040	·0457
3	617	·7830	17° 19′		
4	243	·8757	26° 31′	1·1340	·0730
·5	·6730	·9614	35° 13′		
·6	096	1·0386	43° 28′	1·2000	·0990
·7	·5360	·1062	51° 20′		
·8	·4540	·1633	58° 53′	1·2425	·1222
·9	·3653	·2094	66° 10′		
2·0	·2716	·2440	73° 16′	1·2653	·1419
·2	+ ·0750	·2780	$\pi -$ 87° 3′	707	575
·4	− ·1241	·2646	+ 79° 24′	593	688
·6	·3145	·2049	65° 44′	327	762
·8	·4850	·1013	51° 36′	·1900	793
3·0	·6240	0·9583	36° 35′	297	779
·2	·7192	·7832	20° 10′	1·0495	712
·4	578	·5879	+ 1° 47′	0·9458	594
·6	291	·3911	− 18° 49′	·8152	·1411
·7	·6881	·3001	29° 46′		
·8	·6303	·2186	40° 57′	·6567	·1172
·9	·5579	·1496	51° 47′		
4·0	·4742	·0951	61° 50′	·4756	·0877
·1	·3827	551	70° 41′		
·2	·2864	284	78° 3′	·2825	·0552
·3	·1877	128	83° 43′		
4·4	− ·0880	− ·0054	− 87° 26′	·0859	·0196

Perihelion $r = $ ·0001, $\theta = - $ 0° 38′.

Thus within limits of error the body reaches S tangentially to the line of syzygies.

The period, as computed somewhat roughly, was found to give $nT = 190°$.

PART II

THE TIDES

THE ENGLISH TEXT OF AN ARTICLE CONTRIBUTED TO
THE GERMAN *ENCYCLOPÆDIA OF MATHEMATICS*

4.

THE TIDES.

[Article "Bewegung der Hydrosphäre," *Encyklopädie der Mathematischen Wissenschaften*, Vol. VI. 1, (1908), pp. 3—83. Chapter B is by Mr S. S. Hough, F.R.S., H. M. Astronomer at the Cape of Good Hope, the rest of the article is by Sir George Darwin.]

TABLE OF CONTENTS.

[The remainder of the article is by G. H. Darwin.]

CHAPTER C.

Practical Applications.

CHAPTER D.

Miscellaneous Investigations.

CHAPTER E.

Tidal Friction and Speculative Astronomy.

Note on § 21.

LITERATURE.

{Separate memoirs are referred to in the text.}

G. B. AIRY. "Tides and Waves," Encyclopedia Metropolitana, 1845 (pp. 241*—396*). {Airy, Tides.}

A. W. BAIRD. Manual of Tidal Observation (pp. 54 and xl, Taylor and Francis, London, 1886). {Baird, Manual.}

A. B. BASSET. Hydrodynamics (Deighton, Cambridge, 1888), Vol. II., Ch. XX., pp. 199—231.

D. BERNOULLI. Traité sur le flux et reflux de la mer. Recueils des pièces, &c., Acad. Roy. des sciences Paris, Vol. IV. (1741), pp. 7, 53; reprinted in Lesueur and Jacquier's edit. of the Principia, Vol. III.

G. H. DARWIN. "Tides," Encyclopædia Britannica (pp. 353—381), and article in Supplementary Volumes. {Darwin, Encycl. Brit.} Scientific Papers, Cambridge, 1907, Vol. I., Oceanic Tides and Lunar Disturbance of Gravity; Vol. II., Tidal Friction.

J. ECCLES, S. G. BURRARD, St G. C. GORE, E. ROBERTS. Details of the tidal observations...from 1872 to 1892, and...methods of reduction. Great Trigonometrical Survey of India, Vol. XVI. (Preface, pp. xv, Part I., pp. 383, Part II., pp. 152, 80 plates and charts, including drawings of the tide-predicting instrument, Dehra Dun, 1901).

W. FERREL. Tidal Researches (pp. 268, Washington Government Printing Office), Appendix to Report of U. S. Coast and Geodetic Survey, 1874. {Ferrel, Tidal Researches.}

S. GÜNTHER. Handbuch der Geophysik, Stuttgart (1899), Vol. II., pp. 456—480.

R. A. HARRIS. Manual of Tides (Washington Government Printing Office), Appendices to Reports of Coast and Geodetic Survey, Part I. (pp. 320—469), 1897; Part II. (pp. 472—618), 1897; Part III. (pp. 123—262), 1894; Part IV. A (pp. 537—675), 1902. {Harris, Manual 1, 2, 3, 4.}

J. C. HOUZEAU and A. LANCASTER. Bibliographie de l'Astronomie. Vol. II. contains a complete bibliography of the theory of tides down to 1881.

H. LAMB. Hydrodynamics (Camb. Univ. Press, 1906), Ch. VIII., pp. 236—338.

P. S. LAPLACE. "Des oscillations de la mer et de l'atmosphère," Mécan. Cél., Book III. "Des oscillations des fluides qui recouvrent les planètes," Mécan. Cél., Book XIII. "Recherches sur quelques points du système du Monde," Œuvres Compl., t. IX., p. 187.

M. LÉVY. Leçons sur la théorie des Marées, première partie (pp. 298, Gauthier-Villars, Paris, 1895).

I. NEWTON. Principia, Lib. I., Prop. 66, Cor. 19; Lib. III., Props. 24, 36, 37.

[H. POINCARÉ. Leçons de Mécanique Céleste, t. III., Théorie des Marées, rédigée par E. Fichot (pp. 472, Gauthier-Villars, Paris, 1910).]

[J. P. VAN DER STOK. "Studiën over Getijden," Parts I. to XIII. (1891—6), Tijdschrift Kon. Inst. van Ingenieurs, Afd. Nederlandsch-Indië (Batavia), Parts XIV., XV. (1896), Tijdschrift Kon. Nat. Vereenigung in Nederlansch-Indië, Dl. LVI. (Batavia). "Études des Phénomènes des Marées sur les côtes néerlandaises," Kon. Nederlandsch Meteorologisch Instituut, No. 90 (1905), also "Elementaire Theorie der Getijden—Getij-Constanten in den Indischen Archipel," ibid., No. 102 (1910).]

W. THOMSON and P. G. TAIT. Natural Philosophy (Camb. Univ. Press, 1883), Vol. I., Part II., § 798 et seq. {Thomson and Tait, Nat. Phil.}

Popular Works.

G. H. DARWIN. Tides and kindred phenomena in the solar system (pp. [20]+346, Murray, London, 1898, 1902 and 1911); Houghton, Boston (1899); German translation, Teubner, Leipzig (1902, 1910 or 1911); Hungarian translation, Buda Pest (1904); Italian translation, Turin (1905). {Darwin, Tides.}

Lord KELVIN {W. THOMSON}. Lectures and Addresses (Macmillan, London, 1891), Vol. III., pp. 139—227.

INTRODUCTION.

THE problems of physical astronomy are divisible into three classes :—in the first, namely the lunar and planetary theories, the celestial bodies are treated as particles; in the theories of precession, geodesy and gravitation, which form the second class, they are treated as rigid bodies; and the third class concerns the relative motions of the parts of which any one of these bodies is composed. To this last class belong the theory of the figure of planets in so far as we suppose the internal strata of equal density to be in hydrostatic equilibrium, and the theory of the tides in which the relative motions of the parts of the elastic or plastic nucleus of the planet, and the oscillations of the superincumbent oceans of water and air are considered.

Since the tidal motions of the nucleus are certainly very small, and since the reaction of the oscillating atmosphere upon the subjacent ocean is excessively minute, by far the larger part of our discussion will treat of the motions of an ocean resting on a rigid nucleus, with no superincumbent atmosphere*. This simplified problem needs further subdivision into the cases where the moving parts are free from internal friction or viscosity, and where they are subject to that influence.

The article is conceived on the following plan :—After a few words on the history of the subject (Chapter A), the oscillations of a frictionless ocean are treated in Chapter B; we then, in Chapter C, consider the practical applications of all kinds to terrestrial oceans†; in Chapter D certain miscellaneous questions are discussed; and finally in Chapter E we consider the influence of frictional resistance to tidal motion, and the various problems of speculative astronomy to which such resistance gives rise.

Within the limits of space at our disposal it is impossible to go into the details either of the methods or of the results of so wide a subject, but we have endeavoured to give such a sketch as will enable the reader to gain some insight into the matter, and will put him into communication with the

---

* The equilibrium and oscillations of a wholly liquid planet are considered by H. Poincaré, G. H. Darwin, and G. H. Bryan. See H. Poincaré, *Acta Math.*, Vol. VII. (1885), p. 259 ; *Phil. Trans. R. S.*, Vol. 198, A (1902), p. 333; G. H. Darwin, *Phil. Trans. R. S.*, Vol. 178, A (1887), p. 242; Vol. 198, A (1902), p. 301; G. H. Bryan, *Phil. Trans. R. S.*, Vol. 180, A (1889), p. 187. See also J. H. Jeans, *Phil. Trans. R. S.*, Vol. 200, A (1902), p. 67; A. Liapounoff, *Annales de la Faculté des Sciences de Toulouse* (1904), (translation of a Russian memoir of 1884); *Acad. Imp. St Pétersbourg Bull.*, Vol. XVII., No. 3, 4 May 1905; *ibid.*, 21 Mars 1906, and *Mémoires*, 1906.

† No attempt is made to give even the outline of results as applicable to the tides at individual ports, and thus an extensive branch of literature is either referred to only incidentally or is left wholly out of account. Amongst the latter we may mention, *inter alia*, the writings of J. P. van der Stok, referred to on p. 187, the *Reports* of the Survey of India, of the Canadian Survey (Dep. of Marine and Fisheries), and of the United States Coast Survey. See also Harris, *Manual* 4 A.

various authors who have discussed the several branches into which the investigations are divided. No attempt is made to give a complete bibliography of the subject, but the reader will find in the second volume of S. Günther's *Handbuch der Geophysik* references to many papers which are not mentioned here.

CHAPTER A.

§ 1. *History*.*

Although both Kepler and Galileo speculated as to the nature and origin of the tides it was Newton who in 1687 laid the foundation of all that has been done since when he brought the theory of gravitation to bear on the subject. In 1738 the Academy of Paris offered the theory of the tides as a subject for a prize. Several remarkable essays were submitted, but the only one which has so immediate a bearing on the subject as to demand our attention here, was that of Daniel Bernoulli. His essay contained a full development of the equilibrium theory, which will be explained below.

The subject was then still in its infancy when Laplace took it up in 1774. He had the insight to perceive, as had Newton before him, that the problem is essentially a dynamical one, and he not only succeeded in determining the oscillations of an ocean covering the whole earth, but he showed how theory and observation were to be combined in the discussion of the tides at any one place.

In the first half of the nineteenth century J. W. Lubbock, W. Whewell and G. B. Airy made important contributions to the subject by improving the methods of treating observations, collecting and coordinating enormous masses of data, constructing tide tables for many ports, discussing the peculiarities of the tides in channels and estuaries, constructing cotidal maps, and considering many other points.

The work of these investigators having shown that the mathematical methods were not wholly appropriate to the problems to be solved, Thomson (Lord Kelvin) suggested about 1870 a new departure in the mathematical treatment, namely that by harmonic analysis; it will be explained below. This method has met with ever increasing approbation from men of science, and may now be fairly considered as an international system with a universally accepted notation.

---

* The most complete account of the history known to me is contained in Harris's *Manual of Tides*, Part I., Chaps. V. to VIII., but the reader may also consult Kelvin's *Popular Lectures*, and Darwin's *Tides*, &c. See also R. Almagià, " La dottrina della marea nell' antichità classica e nel medio evo," *Accad. dei Lincei*, v. (1905), p. 377; *Rivista Geografica Italiana*, Anno X.—XI. (1903—4), p. 2; G. Magrini, Italian translation of Darwin's *Tides* (1905), p. 365.

Our knowledge of the tides of the world is many times greater than was that of Laplace, but it is not too much to say that we are only at the beginning of a general tidal survey, and it may be many generations before it will be possible to make with absolute confidence any wide generalisations as to the nature of the tide wave in the open ocean.

It affords a striking testimony both as to the powers of Laplace and as to the difficulty of the problem that absolutely no advance was made from 1774 to 1897 in the general discussion of the tides of an ocean covering the whole planet, and we are apparently as far off as ever from finding any solution where the ocean is interrupted by barriers of arbitrary shape such as the continents.

The names of many investigators who have made important contributions will be mentioned below, but amongst all these Newton stands out as first, and next to him we must rank Laplace. However original any future work may be it would seem that it must perforce be based on what these two men have written.

CHAPTER B.   DYNAMICAL THEORY.

§ 2.   *Tide-generating Potential.*

The forces acting on the ocean due to the attraction of a disturbing body, say the moon, may be best expressed by a potential function.

The gravitational potential of the moon at the point P is m/R, where m is the moon's mass in gravitational units, and R is the distance of P from the moon's centre.

The forces due to this potential are effective in maintaining the motion of the earth as a whole and also in producing tidal distortion. To evaluate that portion of the whole which is effective in producing tides we must subtract from the moon's potential a function whose differential coefficients at any point give the components of acceleration of the earth's centre.

If we write r for the moon's radius vector and x for the abscissa of the point P measured from the earth towards the moon, the acceleration of the earth's centre due to the moon's attraction is m/r^2, and the function to be subtracted is mx/r^2.

If z denotes the moon's geocentric zenith distance as viewed from P, and ρ the geocentric radius vector of P, we have

$$R^2 = \rho^2 + r^2 - 2\rho r \cos z, \quad x = \rho \cos z$$

The expression for V, the tide-generating potential, is therefore given by

$$V = \frac{m}{(\rho^2 + r^2 - 2\rho r \cos z)^{\frac{1}{2}}} - \frac{m\rho \cos z}{r^2}$$

Expanding in powers of the moon's parallax, namely ρ/r, and suppressing a constant term m/r, we have

$$V = \frac{m\rho^2}{r^3}\left(\tfrac{3}{2}\cos^2 z - \tfrac{1}{2}\right) + \frac{m\rho^3}{r^4}\left(\tfrac{5}{2}\cos^3 z - \tfrac{3}{2}\cos z\right) + \ldots$$

The moon's parallax being only $\frac{1}{60}$, the first of these terms suffices for all practical purposes, and we may put

$$V = \frac{m\rho^2}{r^3}\left(\tfrac{3}{2}\cos^2 z - \tfrac{1}{2}\right) \quad \ldots\ldots\ldots\ldots\ldots\ldots\ldots(1)$$

A corresponding potential will represent the forces due to any other disturbing body, but it is clear that it is only necessary to consider two, namely the sun on account of its great mass, and the moon on account of her small distance.

§ 3. *Equilibrium Theory of the Tides.*

The tidal problem is essentially a dynamical one, and was so regarded by Newton, but our attention is naturally more drawn to the changes of configuration of the ocean at moments separated by considerable intervals of time rather than to the state of motion at any one instant. It was this aspect of the case which probably led Daniel Bernoulli to adopt the hypothesis that the figure of the ocean is at each instant in equilibrium under the action of the disturbing forces at the moment under consideration.

According to this assumption the surface of the ocean must be an equipotential surface under the action of gravitation, centrifugal force, and tide-generating force.

If U denotes the combined potentials of gravitation and centrifugal force, and V the tide-generating potential, U and V must be constant at the surface.

If g be the value of gravity at the earth's surface, and ζ the elevation above the surface of the point to which U applies, it is clear that U is approximately equal to a constant $-g\zeta$. It follows that the equation may be written

$$\zeta = \frac{V}{g} + \text{constant} \quad \ldots\ldots\ldots\ldots\ldots\ldots\ldots(2)$$

The constant in (2) is to be determined by the condition that the volume of the ocean remains constant.   Expressed analytically this gives

$$\iint \zeta dS = 0 \quad\dots\dots\dots\dots\dots\dots\dots\dots\dots(3)$$

where dS is an element of surface, and the integration is extended over the oceanic portion of the globe.

If the ocean covers the whole globe, and if a denotes the earth's mean radius, we have from (1) and (2)

$$\zeta = \frac{ma^2}{gr^3}\left(\tfrac{3}{2}\cos^2 z - \tfrac{1}{2}\right)\dots\dots\dots\dots\dots\dots\dots\dots(4)$$

The function of z being a zonal harmonic, the condition (3) is obviously satisfied.

Since the equilibrium tide as given in (4) is proportional to the disturbing potential, it is convenient in all cases to specify the disturbing forces by means of the equilibrium tide which they would produce.

If the density of the ocean be less than that of the planet an ocean covering the whole nucleus is in stable equilibrium for all possible deformations, but if the nucleus is lighter than the ocean it will float in the ocean with part of its surface dry.   The problem of determining the figure of the surface in the latter case, even in the absence of rotation and of external disturbance, has hitherto remained unsolved*.

§ 4.   Development of the Tide-generating Potential.

The expression (4) for the equilibrium tide shows that the surface is a prolate spheroid of revolution with the major axis of symmetry directed towards the moon, and as the earth rotates the height of water at any point fixed in the earth must undergo periodic changes.

We shall see hereafter in §§ 20 and 23 that $\cos z / r^3$ may be expressed in terms of the latitude and longitude of the place of observation and of the celestial coordinates of the moon.

When this substitution is effected the disturbing potential becomes expressible as the sum of an infinite series of terms, each of which consists of a spherical harmonic of the latitude and longitude, multiplied by a simple harmonic function of the time.

If we limit our consideration of the subject to the single term of V given

---

* Thomson, in Thomson and Tait's *Nat. Phil.*, § 816; Darwin, "Tides," *Encyc. Britan.*, § 19.

in (1), these terms will be harmonics of the second order and of three classes, namely :—

(1)   zonal harmonics, with a time factor of long period; the shortest period being a fortnight.

(2)   tesseral harmonics of unit rank, with a time factor of approximately diurnal period.

(3)   tesseral harmonics of rank 2, with a time factor of approximately semi-diurnal period.

If then μ denotes the sine of the latitude and ϕ the longitude, a typical term of the tide-generating potential will be

$$\gamma_n{}^s P_n{}^s(\mu) \frac{\cos}{\sin}(\lambda t + s\phi) \quad\dots\dots\dots\dots\dots\dots(5)$$

where $P_n(\mu)$ denotes the zonal harmonic of order n, and

$$P_n{}^s = (1-\mu^2)^{\frac{1}{2}s}\frac{d^s P_n}{d\mu^s}$$

In the cases to which we limit our consideration, $n = 2$, and $s = 0, 1, 2$, in the three kinds of harmonics specified above, respectively; also λ is respectively small, or approximately equal to ω, or to 2ω, where ω is the angular velocity of the earth's diurnal rotation.

§ 5.   *Correction to the Equilibrium Theory for the Mutual Attraction of the Water.*

Let us assume that the water covers the whole earth and that the tide-generating potential consists of a single term

$$V = \gamma_n{}^s P_n{}^s(\mu)\cos(\lambda t + s\phi) \quad\dots\dots\dots\dots\dots\dots(5)$$

where n is not necessarily equal to 2.

We may assume also that the corrected equilibrium tide ζ is proportional to V, and is therefore expressible by a harmonic of the same order and rank.

If δ be the density of the water and σ the mean density of the earth, we have $g = \frac{4}{3}\pi\sigma a$; and the ocean may be regarded as surface density $\zeta\delta$ residing on the surface of the globe.

By the theory of spherical harmonic analysis the internal and external potentials of such a layer are

$$\frac{4\pi\delta\rho^n\zeta}{(2n+1)a^{n-1}} \quad\text{and}\quad \frac{4\pi\delta a^{n+2}\zeta}{(2n+1)\rho^{n+1}}$$

At the surface where $\rho = a$, these assume the same value, viz.

$$\frac{3g\delta}{(2n+1)\sigma}\zeta$$

Including this as part of the disturbing potential, (2) becomes

$$-g\zeta + g\,\frac{3\delta}{(2n+1)\,\sigma}\,\zeta + V = \text{constant}$$

It is obvious that the effect of the additional term is to augment the tide in the proportion

$$1 : 1 - \frac{3}{2n+1}\frac{\delta}{\sigma}$$

In the case of spherical harmonics of the second order n is 2, and we know that the mean density of the earth is about $5\frac{1}{2}$ times as great as that of water. Hence the factor of augmentation is $\dfrac{1}{1 - \dfrac{3}{5 \times 5\frac{1}{2}}} = \dfrac{55}{49}$.

§ 6. *Correction for Distribution of Land and Water.*

If the disturbing potential contains only a single term as in (5), we have, as in § 3,

$$\zeta = C + \frac{\gamma_n{}^s}{g}\,P_n{}^s\,(\mu)\cos(\lambda t + s\phi) \quad \dots\dots\dots\dots\dots(6)$$

The condition for the constancy of volume of the ocean is that ζ integrated over the ocean shall vanish. This enables us to determine C in terms of certain definite integrals.

If we write Ω for the area of the ocean and

$$A = \iint P_n{}^s\,(\mu)\cos s\phi\,dS, \qquad B = \iint P_n{}^s\,(\mu)\sin s\phi\,dS$$

where the integrals are taken over the surface of the ocean, the elimination of C from (6) enables us to write it in the form

$$\zeta = \frac{\gamma_n{}^s}{g}\left[\left\{P_n{}^s\,(\mu)\cos s\phi - \frac{A}{\Omega}\right\}\cos\lambda t - \left\{P_n{}^s\,(\mu)\sin s\phi - \frac{B}{\Omega}\right\}\sin\lambda t\right]\dots(7)$$

For any given distribution of ocean on the planet the integrals Ω, A, B may be evaluated by quadratures. It appears that for the existing terrestrial oceans the corrections are insignificant*.

The result (7) shows that in the corrected theory the phases of the tide and of the tidal force are no longer in agreement.

Poincaré has shown how the effects of the mutual gravitation of the water may be included in the corrected theory, but his results are too complex to admit of numerical calculation. He points out, however, that with certain

---

* Darwin and Turner, *Proc. R. S.*, April, 1886 [or Vol. i., p. 328]; Harris, *Manual of Tides*, Part ii., App. 9, Report U. S. Coast Survey, 1898.

distributions of land and water the correction for mutual gravitation may rise in importance to such an extent that (7) will no longer be even approximately correct*.

§ 7. Dynamical Theory—fundamental equations.

We now proceed to the formation of the equations of motion applicable to the ocean.

Let ρ, θ be the radius vector and colatitude of a point with reference to the earth's centre and its axis of rotation, and let ϕ be the longitude of the point with reference to a prime meridian fixed in the earth, and therefore rotating with an angular velocity ω.

Let w, u, v be the relative velocity components of the water in the radial, latitudinal and longitudinal directions.

Let V' be the potential of the field of force to which the liquid is subject, and let

$$\psi = V' - \frac{p}{\delta} + \tfrac{1}{2}\omega^2\rho^2\sin^2\theta + \text{const.}$$

where p is the pressure at ρ, θ, ϕ, and δ the density of the water.

The general equations of motion of liquid referred to such axes, when the squares and products of the components of relative velocity are neglected, have been expressed by Poincaré in a form which is easily reducible to the following† :—

$$\left.\begin{array}{l} \dfrac{\partial w}{\partial t} - 2\omega v\sin\theta = \dfrac{\partial\psi}{\partial\rho} \\[2ex] \dfrac{\partial u}{\partial t} - 2\omega v\cos\theta = \dfrac{\partial\psi}{\rho\,\partial\theta} \\[2ex] \dfrac{\partial v}{\partial t} + 2\omega\,(w\sin\theta + u\cos\theta) = \dfrac{1}{\rho\sin\theta}\dfrac{\partial\psi}{\partial\phi} \end{array}\right\} \quad\ldots\ldots\ldots\ldots(8)$$

The water being confined to a thin, nearly spherical sheet ρ will never differ much from a; and since the form will always be approximately spherical, w will be small compared with u and v. Hence we may replace ρ in the two latter equations by a.

In these equations we may regard ψ as sensibly independent of ρ, since, on account of the small depth, the variations of ψ with ρ can only become important if $\partial\psi/\partial\rho$ is a quantity of higher magnitude than $\partial\psi/a\partial\theta$

---

* *Liouville's Journ.* (5ᵉ série), Vol. II., 1896, p. 7.
† *Acta Math.*, Vol. VII., p. 356.

and $\partial \psi / a \sin \theta \, \partial \phi$. The first of the equations indicates that this will not be the case unless the vibrations are of such rapidity as to render the term $\partial w / \partial t$ of importance. It will be assumed that the tidal oscillations are not of this character. Thus we may drop the first equation and may regard ψ, u, v as having sensibly the same values at all points in the same vertical line.

§ 8. *The Equation of Continuity.*

The formation of the equation expressive of the constancy of the volume of the ocean is materially simplified by the fact that we may regard the horizontal velocity as the same at all points on the same vertical line.

Let h be the depth of the ocean when undisturbed. Then by considering the rates at which the water enters a small rectangular area bounded by constant values of θ, $\theta + \delta\theta$, ϕ and $\phi + \delta\phi$ we may obtain the equation of continuity in the form

$$a^2 \sin \theta \, \delta\theta \, \delta\phi \, \frac{\partial \zeta}{\partial t} = ahu \sin \theta \, \delta\phi - \left[ahu \sin \theta \, \delta\phi + \delta\theta \, \frac{\partial}{\partial \theta} (ahu \sin \theta \, \delta\phi) \right]$$

$$+ ahv \, \delta\theta - \left[ahv \, \delta\theta + \delta\phi \, \frac{\partial}{\partial \phi} (ahv \, \delta\theta) \right]$$

which gives
$$\frac{\partial \zeta}{\partial t} = - \frac{1}{a \sin \theta} \left\{ \frac{\partial}{\partial \theta} (hu \sin \theta) + \frac{\partial}{\partial \phi} (hv) \right\} \quad \dots\dots\dots\dots(9)$$

§ 9. *Pressure Condition at the Free Surface.*

Since the pressure must be zero or constant at the surface,

$$\psi - V' - \tfrac{1}{2} \omega^2 \rho^2 \sin^2 \theta$$

must be constant at the surface.

Hence denoting values at the undisturbed surface by square brackets, and differentiation along the normal to the undisturbed surface by $\dfrac{\partial}{\partial n}$, we must have

$$[\psi - V' - \tfrac{1}{2}\omega^2\rho^2 \sin^2 \theta] + \zeta \left[\frac{\partial}{\partial n} \{ \psi - V' - \tfrac{1}{2}\omega^2\rho^2 \sin^2 \theta \} \right] = \text{const.}$$

If V_0' denotes the value of V' when there is no disturbance, v the disturbing potential, and v' the potential due to the layer of liquid between the mean and actual surfaces

$$V' = V_0' + v' + v$$

Now
$$\left[\frac{\partial}{\partial n}\{V_0' + \tfrac{1}{2}\omega^2\rho^2\sin^2\theta\}\right] = -g$$

and since the surface of reference is a possible free surface when there is no disturbing force $[V_0' + \tfrac{1}{2}\omega^2\rho^2\sin^2\theta] = $ constant.

Hence neglecting squares of small quantities depending on the disturbance, the condition reduces to

$$\psi = v' - g\zeta + v + \text{const.} \quad \dots\dots\dots\dots\dots(10)$$

Since we have in future only to deal with surface values, the square brackets may be omitted, as being no longer necessary.

§ 10.   *Tides in Canals.*

Let χ be the inclination of the direction of a narrow canal of breadth b to the meridian at any point of its length, and U the horizontal velocity in the direction of the canal.   Then

$$u = U\cos\chi, \quad v = U\sin\chi$$

Thus if we multiply the second and third of (8) by $\cos\chi$ and $\sin\chi$ respectively, and add, we obtain

$$\frac{\partial U}{\partial t} = \frac{\partial\psi}{a\partial\theta}\cos\chi + \frac{1}{a\sin\theta}\frac{\partial\psi}{\partial\phi}\sin\chi = \frac{\partial\psi}{\partial s}$$

where s is distance measured along the canal from some fixed point in it.

The equation of continuity now becomes

$$\frac{\partial\zeta}{\partial t} + \frac{1}{b}\frac{\partial}{\partial s}(bhU) = 0$$

whence on eliminating U we have

$$\frac{\partial^2\zeta}{\partial t^2} + \frac{1}{b}\frac{\partial}{\partial s}\left(bh\frac{\partial\psi}{\partial s}\right) = 0$$

On further eliminating ψ by means of (10) the equation becomes

$$\frac{\partial^2\zeta}{\partial t^2} - gh\frac{\partial^2\zeta}{\partial s^2} = -h\frac{\partial^2 v}{\partial s^2}$$

This may be readily integrated when v is expressed as a function of s and t.

If the canal be re-entrant on itself, without any barriers, and of length l, v must be expressible in a Fourier series of sines and cosines of multiples of $2\pi s/l$ in which the coefficients are simple harmonic functions of the time. Thus v may be expressed as the sum of such terms as

$$A\frac{\cos}{\sin}\left(\lambda t + \frac{2r\pi s}{l}\right)$$

where r is a positive or negative integer.

Taking (as is justifiable) a single term of this form and with the cosine, we have

$$\frac{\partial^2 \zeta}{\partial t^2} - gh\frac{\partial^2 \zeta}{\partial s^2} = \frac{4r^2\pi^2}{l^2} Ah \cos\left(\lambda t + \frac{2r\pi s}{l}\right)$$

which admits of the solution

$$\zeta = \frac{\dfrac{4r^2\pi^2}{l^2}h}{\dfrac{4r^2\pi^2}{l^2}gh - \lambda^2} A \cos\left(\lambda t + \frac{2r\pi s}{l}\right)$$

In a more general solution two other terms may be added to this, namely

$$B \cos\left(\lambda t + \frac{s}{\sqrt{gh}} + \epsilon_1\right) + C \cos\left(\lambda t - \frac{s}{\sqrt{gh}} + \epsilon_2\right)$$

For a canal without ends B and C must in general be zero, but if the canal be terminated by barriers B, C, ϵ_1, ϵ_2 must be determined so as to secure zero velocity at the barriers for all values of t.

A detailed development of the theory of tides in canals has been given by Airy*. The problems dealt with by him all resolve themselves into appropriate development of the function v, so as to admit of the application of the method sketched here.

§ 11. *Laplace's Differential Equation for the Tides.*

Replacing the trigonometrical functions by exponentials we consider a single term $\gamma P_n^s(\mu) e^{i(\lambda t + s\phi)}$, whose real and imaginary parts correspond with the real and imaginary parts of the disturbing potential.

If we suppose that u, v, ψ, ζ are each proportional to $e^{i(\lambda t + s\phi)}$, the equations (8), (9) become

$$\left.\begin{aligned} i\lambda u - 2\omega v \cos\theta &= \frac{1}{a}\frac{\partial\psi}{\partial\theta} \\[2mm] i\lambda v + 2\omega u \cos\theta &= \frac{is\psi}{a\sin\theta} \\[2mm] i\lambda\zeta &= -\frac{1}{a\sin\theta}\left\{\frac{\partial}{\partial\theta}(hu\sin\theta) + ishv\right\} \end{aligned}\right\} \quad\cdots\cdots\cdots\cdots(11)$$

Writing $\qquad \cos\theta = \mu, \quad D = (1 - \mu^2)\dfrac{d}{d\mu}, \quad \sigma = \dfrac{2\omega s}{\lambda}$

and substituting from the first two equations in the third, we find

$$a^2(1 - \mu^2)\zeta = (D + \sigma\mu)\left[\frac{h}{(\lambda^2 - 4\omega^2\mu^2)}(D - \sigma\mu)\psi\right] - \frac{s^2}{\lambda^2}h\psi\cdots\cdots(12)$$

We will now examine certain special cases of importance as indicated at the end of § 4.

* " Tides and Waves," *Encyc. Metrop.*

§ 12. *Tides of Long Period—Solution in Power Series.*

For the tides of long period $s = 0$; and putting $\dfrac{\lambda}{2\omega} = f$, the equation (12) reduces to

$$4\omega^2 a^2 \zeta = \frac{d}{d\mu} \left\{ \frac{h(1-\mu^2)}{f^2 - \mu^2} \frac{d\psi}{d\mu} \right\} \quad \dotsc\dotsc\dotsc\dotsc\dotsc(13)$$

Assume that $\dfrac{1}{f^2 - \mu^2} \dfrac{d\psi}{d\mu} = B_1 \mu + B_3 \mu^3 + B_5 \mu^5 + \dots \quad \dotsc\dotsc\dotsc\dotsc(14)$

where the coefficients B_1, B_3, &c. are undetermined.

On multiplying (14) by $f^2 - \mu^2$ and integrating we derive

$$\psi = K + \tfrac{1}{2} B_1 f^2 \mu^2 + \tfrac{1}{4}(B_3 f^2 - B_1)\mu^4 + \tfrac{1}{6}(B_5 f^2 - B_3)\mu^6 + \dots \dotsc(15)$$

Again on substituting from (14) in (13) we find

$$\frac{4\omega^2 a^2}{h}\zeta = B_1 + 3(B_3 - B_1)\mu^2 + 5(B_5 - B_3)\mu^4 + \dots \dotsc\dotsc(16)$$

But denoting the disturbing potential by $\alpha(\mu^2 - \tfrac{1}{3})$, and neglecting the mutual gravitation of the water, equation (10) gives

$$\psi = -g\zeta + \alpha(\mu^2 - \tfrac{1}{3})$$

Replacing ψ, ζ by their expansions and equating coefficients of like powers of μ, we find

$$K = -\tfrac{1}{3}\alpha - \frac{B_1}{\beta}, \quad B_3 - B_1\left(1 - \frac{1}{2.3}f^2\beta\right) - \tfrac{1}{3}\alpha\beta = 0 \left.\vphantom{\begin{array}{c}1\\1\\1\\1\end{array}}\right\}$$

and in general $\dotsc\dotsc(17)$

$$B_{2n+1} - B_{2n-1}\{1 - (2n)'f^2\beta\} - (2n)'\beta B_{2n-3} = 0 \left.\vphantom{\begin{array}{c}1\\1\end{array}}\right.$$

where we have written $(n)' = \dfrac{1}{n(n+1)}$ for brevity, and $\beta = \dfrac{4\omega^2 a^2}{gh}$.

At first sight it might appear that the equations (17) will give B_1, B_3, B_5 &c. in terms of K, α. The quantity α is one of the data of the problem, but K is an arbitrary constant to determine the value of which seems to require the introduction of some other condition than those we have already expressed. The assumed solution is symmetrical with respect to the equator and hence if the ocean is bounded by parallels of latitude also symmetrical to the equator, such a condition would be furnished by the fact that no flow could take place across these barriers.

Provided the solution thus derived be a true solution of the problem the series (14), (15), (16) must admit of a real physical significance and hence must be convergent series for all values of μ which apply to points between the barriers, but for points external to the barriers they need not necessarily converge.

Hence if the ocean extends right up to the poles the series in question must converge for all values of μ between $\mu = \pm 1$ and including these limiting values. The series

$$B_1 + B_3 + B_5 + \ldots$$

must therefore be a convergent series, which cannot be the case unless

$$\underset{m=\infty}{\mathrm{Lt}} \; B_{2m+1} = 0 \quad \ldots\ldots\ldots\ldots\ldots\ldots\ldots\ldots(18)$$

The condition (18) taken in conjunction with equations (17) will serve for the determination of the constant K as we now proceed to show*.

From (17) we derive

$$\frac{B_{2n-1}}{B_{2n-3}} = \frac{-(2n)'\beta}{1 - (2n)'f^2\beta - \dfrac{B_{2m+1}}{B_{2n-1}}} \quad (n > 1)$$

If we suppose that $B_{2m+1} = 0$, $(m > n)$, we derive by successive applications

$$\frac{B_{2n-1}}{B_{2n-3}} = \frac{-(2n)'\beta}{1 - (2n)'f^2\beta +} \; \frac{(2n+2)'\beta}{1 - (2n+2)'f^2\beta} \cdots \frac{+(2m)'\beta}{1 - (2m)'f^2\beta}$$

whence on proceeding to the limit when m is made indefinitely great we derive, in virtue of (18),

$$\frac{B_{2n-1}}{B_{2n-3}} = -N_n$$

where N_n denotes the infinite continued fraction

$$\frac{(2n)'\beta}{1 - (2n)'f^2\beta +} \; \frac{(2n+2)'\beta}{1 - (2n+2)'f^2\beta +} \; \frac{(2n+4)'\beta}{1 - (2n+4)'f^2\beta} + \ldots \text{ad inf.}$$

With this notation, equations (17) yield

$$\frac{B_1}{a} = -2N_1, \quad K = a\left(\frac{2N_1}{\beta} - \tfrac{1}{3}\right), \quad \frac{B_3}{B_1} = -N_2, \quad \&c.$$

whence from (16)

$$\zeta = -\tfrac{1}{2}\frac{ah}{\omega^2 a^2}[N_1 - 3N_1(N_2+1)\mu^2 + 5N_1N_2(N_3+1)\mu^4 - \ldots] \quad \ldots(19)$$

The problem thus resolves itself into the evaluation of the continued fractions N_1, N_2 &c. With numerical data suitable to the conditions existent on the Earth these continued fractions, and the series (19), converge with

---

* The validity of the corresponding procedure in the cases of the semi-diurnal and diurnal tides was impugned by Airy, *Tides and Waves*, and *Phil. Mag.*, Vol. L. (4th series), 1875, p. 277. He was supported by W. Ferrel, *Phil. Mag.*, Vol. I. (5th series), 1876, p. 182, *Gould's Astron. Journ.*, Vols. IX. (1889), p. 41, and X. (1890), p. 121, and *Smithsonian Miscell. Collections*, No. 843, but Lord Kelvin, *Phil. Mag.*, Vol. L. (4th series), 1875, pp. 227, 279, 388, confirmed the correctness of the procedure in question, which is that of Laplace, and it is now universally regarded as correct. Compare also Lord Rayleigh, *Phil. Mag.*, Vol. v. (6th series), 1903, p. 136, a paper which has an important bearing on § 37 below.

great rapidity. For numerical illustrations we may refer to the work of G. H. Darwin* and H. Lamb†.

§ 13.   Tides of Long Period—Solution in Zonal Harmonics.

The use of harmonic functions not only permits us to include the effects of mutual gravitation but also leads to more convergent series.

Assume that

$$\psi = \Sigma \Gamma_n P_n(\mu), \quad \zeta = \Sigma C_n P_n(\mu), \quad v = \Sigma \gamma_n P_n(\mu)$$

where $P_n(\mu)$ denotes the zonal harmonic of order n.

Then the potential of the ocean is given by

$$v' = \Sigma \frac{3g\delta}{(2n+1)\sigma} C_n P_n(\mu)$$

where δ, σ denote respectively the density of the water and the mean density of the Earth.

Since $\psi = v' - g\zeta + v + \text{const.}$

we must have $\Gamma_n = \gamma_n - g_n C_n$ ....(20)

where $g_n = g \left\{ 1 - \dfrac{3\delta}{(2n+1)\sigma} \right\}$ ....(21)

Again on substituting the assumed forms in (13) we find

$$\Sigma h (1-\mu^2) \Gamma_n \frac{dP_n}{d\mu} = - 4\omega^2 a^2 (1-\mu^2) \Sigma C_n \left\{ \frac{(f^2-1)}{n(n+1)} \frac{dP_n}{d\mu} \right.$$

$$\left. - \frac{1}{(2n+1)(2n+3)} \frac{dP_{n+2}}{d\mu} + \frac{2}{(2n-1)(2n+3)} \frac{dP_n}{d\mu} - \frac{1}{(2n-1)(2n+1)} \frac{dP_{n-2}}{d\mu} \right\}$$

no arbitrary constant of integration being necessary since both sides vanish when $\mu = \pm 1$.

We now assume the depth to be uniform and equate the coefficients of each harmonic term separately.

If it be understood that $C_{-1} = 0$, $C_0 = 0$, this leads to

$$\frac{C_{n-2}}{(2n-3)(2n-1)} - L_n C_n + \frac{C_{n+2}}{(2n+3)(2n+5)} = \frac{h\gamma_n}{4\omega^2 a^2} \quad ....(22)$$

where $L_n = \dfrac{f^2-1}{n(n+1)} + \dfrac{2}{(2n-1)(2n+3)} - \dfrac{hg_n}{4\omega^2 a^2}$

Except for the presence of γ_n on the right, these equations present characteristics similar to those of series (17) discussed in § 12.

* Art. "Tides," *Encyclopædia Britannica*, 9th edition, or *Proc. R. S.*, Vol. XLI., 1886, pp. 337—342 [or Vol. I., p. 366].
  † *Hydrodynamics*, § 216.

We now denote by H_n, K_n respectively the continued fractions

$$\left. \begin{array}{cc} \cfrac{1}{\cfrac{(2n+1)(2n+3)^2(2n+5)}{L_n-}} & \cfrac{1}{\cfrac{(2n-3)(2n-1)^2(2n+1)}{L_{n-2}-}} \cdots \\[20pt] \cfrac{1}{\cfrac{(2n-3)(2n-1)^2(2n+1)}{L_n-}} & \cfrac{1}{\cfrac{(2n+1)(2n+3)^2(2n+5)}{L_{n+2}-}} \cdots \end{array} \right\} \ldots(23)$$

in which the former terminates, and the latter proceeds to infinity.

Then replacing γ_n by zero, we obtain

$$\left. \begin{array}{l} C_n/C_{n+2} = (2n+1)(2n+3)\,H_n \\ C_n/C_{n-2} = (2n-1)(2n+1)\,K_n \end{array} \right\} \ldots\ldots\ldots\ldots\ldots(24)$$

The latter of these equations however is dependent on the condition that C_∞ shall vanish, which is necessary in order that the series for ζ may be convergent at the poles.

Where one of the quantities γ, say γ_r, differs from zero, the first of (24) holds when n is less than r, and the latter when n exceeds r. Thus on substituting for the ratios C_{r-2}/C_r, C_{r+2}/C_r in the equation involving γ_r, we obtain

$$C_r (H_{r-2} - L_r + K_{r+2}) = \frac{h\gamma_r}{4\omega^2 a^2}$$

an equation which serves to determine C_r in terms of known quantities.

All the quantities C with suffixes differing from r by an odd integer will be zero, while the ratio of the remaining C's to C_r can be found by (24); we therefore derive as the height of the tide due to a single term $\gamma_n P_n (\mu)$ in the disturbing potential

$$\zeta = \frac{h\gamma_n}{4\omega^2 a^2 (H_{n-2} - L_n + K_{n+2})} \left\{ \begin{array}{l} \ldots + (2n-7)(2n-5)(2n-3)(2n-1)H_{n-4}H_{n-2}P_{n-4} \\ \qquad + (2n-3)(2n-1)H_{n-2}P_{n-2} \\ + P_n + (2n+3)(2n+5)K_{n+2}P_{n+2} \\ + (2n+3)(2n+5)(2n+7)(2n+9)K_{n+2}K_{n+4}P_{n+4} + \ldots \end{array} \right\}$$

$$\ldots\ldots\ldots(25)$$

The tides tend to become infinitely great if the period of the disturbing force is such that

$$H_{n-2} - L_n + K_{n+2} = 0 \ldots\ldots\ldots\ldots\ldots\ldots\ldots(26)$$

The periods which satisfy this condition are evidently the periods of the " free" vibrations of zonal type. For a practical method of solving the equation (26) and determining the free periods, together with numerical applications of the formula (25) we refer to Hough on " The Dynamical Theory of the Tides*."

* *Phil. Trans. R. S.*, Vol. 189, A (1897), p. 201.

§ 14. *Diurnal Tides—Laplace's Solution.*

For the diurnal tides we assume that

$$v = \gamma_2^1 P_2^1(\mu)\, e^{i(\phi+\lambda t)} = 3\gamma_2^1 \mu \sqrt{(1-\mu^2)}\, e^{i(\phi+\lambda t)}$$

and further, following Laplace that $\lambda = \omega$ rigorously.

The following are then appropriate forms for ψ, ζ, viz. :—

$$\psi = \Gamma_2^1 P_2^1(\mu)\, e^{i(\phi+\omega t)}, \quad \zeta = C_2^1 P_2^1(\mu)\, e^{i(\phi+\omega t)}$$

The pressure equation (10) gives, as before,

$$\Gamma_2^1 = \gamma_2^1 - g_2 C_2^1$$

while the equation (12) becomes reducible to

$$\omega^2 a^2 C_2^1(\mu) = \Gamma_2^1 \frac{dh}{d\mu}$$

If therefore the depth of the ocean were a function of the latitude defined by $h = \alpha + \beta\mu^2$, we should find

$$C_2^1 = \frac{2\beta}{\omega^2 a^2}\Gamma_2^1 = \frac{2\beta}{\omega^2 a^2}\{\gamma_2^1 - g_2 C_2^1\}$$

and hence

$$C_2^1 = \frac{2\beta g_2/\omega^2 a^2}{1 + 2\beta g_2/\omega^2 a^2}\gamma_2^1 \quad\dotsfill(27)$$

The rise and fall of the free surface vanish when $\beta = 0$, i.e. when the depth of the water is uniform. In other cases, where the depth can be expressed by the formula $\alpha + \beta\mu^2$, the tide will be "direct" or "inverted" according as the depth at the poles is greater or less than the depth at the equator.

§ 15. *Semi-diurnal Tides—Laplace's Solution.*

For the semi-diurnal tides, we take $\lambda = 2\omega$, $s = 2$, $\sigma = 2$ and, with h constant, (12) becomes

$$\frac{4\omega^2 a^2}{h}\zeta = \left(\frac{d}{d\mu}+\frac{2\mu}{1-\mu^2}\right)\left(\frac{d}{d\mu}-\frac{2\mu}{1-\mu^2}\right)\psi - \frac{4}{1-\mu^2}\psi$$

which reduces to

$$\frac{d^2\psi}{d\mu^2} - \frac{6+2\mu^2}{(1-\mu^2)^2}\psi = \beta g\zeta \quad\dotsfill(28)$$

where

$$\beta = 4\omega^2 a^2/gh \quad\dotsfill(29)$$

But from (10), if we neglect the mutual attraction of the water particles we obtain

$$\psi = -g\zeta + v$$

whence (28) may be written

$$\frac{d^2\psi}{d\mu^2} + \left[\beta - \frac{(6+2\mu^2)}{(1-\mu^2)^2}\right]\psi = \beta v \quad\dotsfill(30)$$

If we write $\nu = \sqrt{(1 - \mu^2)}$, (30) becomes

$$\nu^2(1 - \nu^2)\frac{d^2\psi}{d\nu^2} - \nu\frac{d\psi}{d\nu} - (8 - 2\nu^2 - \beta\nu^4)\psi = \beta\nu\nu^4 = \beta g\,\mathfrak{C}_2{}^2\nu^6 \dots\dots(31)$$

where $\mathfrak{C}_2{}^2(1 - \mu^2)$ represents the height of the equilibrium tide due to the disturbing potential ν.

For an ocean covering the whole Earth, assume

$$\psi = B_0 + B_2\nu^2 + B_4\nu^4 + \dots + B_{2j}\nu^{2j+2} + \dots \dots\dots\dots(32)$$

On substituting for ψ in (31) and equating coefficients of like powers of ν in the two members, we derive

$$\left.\begin{aligned} B_0 = 0,\quad B_2 = 0,\quad 0 \cdot B_4 = 0 \\ 16B_6 - 10B_4 - \beta g\,\mathfrak{C}_2{}^2 = 0 \end{aligned}\right\} \dots\dots\dots(33)$$

and thenceforward

$$2j(2j + 6)B_{2j+4} - 2j(2j + 3)B_{2j+2} + \beta B_{2j} = 0$$

Now since ψ must remain finite when $\nu = 1$, we must have $B_\infty = 0$. Denoting by N_j the continued fraction

$$\frac{\beta}{2j(2j + 3) -} \quad \frac{2j(2j + 6)\beta}{(2j+2)(2j+5) -} \quad \frac{(2j + 2)(2j + 8)\beta}{(2j + 4)(2j + 7)} - \dots \text{ ad inf.}$$

we may derive as in § 12

$$B_4 = -N_1 g\,\mathfrak{C}_2{}^2,\quad B_6 = N_2 B_4$$

and generally

$$B_{2j+2} = N_j B_{2j}$$

Thus substituting in (32), and thence determining ψ, we have

$$\zeta = \frac{\nu - \psi}{g} = \mathfrak{C}_2{}^2\{\nu^2 + N_1\nu^4 + N_1 N_2\nu^6 + N_1 N_2 N_3\nu^8 + \dots\} \dots\dots(34)$$

Numerical applications of this formula have been given by Laplace[*]. The method has been further extended by Lord Kelvin[†] to the case where the ocean is limited by zonal barriers and when the law of depth is more complex.

A discussion of the equation (31), with bibliographical references, will be found in a paper by G. H. Ling on " Laplace's Kinetic Theory of Tides[‡]."

[*] *Mécanique Céleste*, Livre IV., Chap. I.; *Œuvres Compl.*, Vol. II., p. 211.
[†] *Phil. Mag.*, Vol. L. (4th series), 1875, p. 388.
[‡] *Annals of Mathematics*, Vol. X. (1895—1896), p. 95.

§ 16. *Transformation of Laplace's Equations.*

With the view of obtaining a more general solution of our problem, we begin with certain preliminary transformations of the equation (12).

We define a new operator Δ, by

$$\Delta = \frac{d}{d\mu} D - \frac{s^2}{1-\mu^2} = \frac{1}{1-\mu^2} (D^2 - s^2) \quad\ldots\ldots\ldots\ldots(35)$$

so that

$$\Delta^{-1} = \frac{1}{D^2 - s^2} (1 - \mu^2) \quad\ldots\ldots\ldots\ldots\ldots(36)$$

The relationship between the operators D, Δ is

$$(D - \sigma\mu)(D + \sigma\mu) - (s^2 - \sigma^2\mu^2) = (1 - \mu^2)(\Delta + \sigma) \quad\ldots\ldots(37)$$

where, as before $\sigma = \dfrac{2\omega s}{\lambda}$.

Let us now introduce an auxiliary function Ψ, defined by

$$\frac{h\sigma^2}{4\omega^2 a^2 (s^2 - \sigma^2\mu^2)} (D - \sigma\mu)\psi = \Psi - \mu\zeta \quad\ldots\ldots\ \ldots\ldots\ldots(38)$$

The equation (12) then gives

$$(D + \sigma\mu)(\Psi - \mu\zeta) = (1 - \mu^2)\zeta + \frac{h\sigma^2}{4\omega^2 a^2}\psi \ldots\ldots\ldots\ldots(39)$$

and on eliminating ψ by means of equation (38)

$$(D - \sigma\mu)(D + \sigma\mu)(\Psi - \mu\zeta) = (D - \sigma\mu)(1 - \mu^2)\zeta + (s^2 - \sigma^2\mu^2)(\Psi - \mu\zeta)$$

which, in virtue of (37), becomes

$$(\Delta + \sigma)(\Psi - \mu\zeta) = \frac{d}{d\mu}\{(1 - \mu^2)\zeta\} - \sigma\mu\zeta$$

or

$$(\Delta + \sigma)\Psi = \Delta(\mu\zeta) + \frac{d}{d\mu}\{(1 - \mu^2)\zeta\}$$

Applying the operator Δ^{-1}, we obtain in virtue of (36)

$$(1 + \sigma\Delta^{-1})\Psi = [\mu + D\Delta^{-1}]\zeta \quad\ldots\ldots\ldots\ldots\ldots(40)$$

Again, operating on this equation with D, we obtain

$$(D + \sigma D\Delta^{-1})\Psi = D(\mu\zeta) + D^2\Delta^{-1}\zeta$$

$$= D(\mu\zeta) + \{s^2 + (1 - \mu^2)\Delta\}\Delta^{-1}\zeta$$

in virtue of (35).

Combining the last equation with (39) we obtain

$$\sigma(D\Delta^{-1} - \mu)\Psi = -\sigma\mu^2\zeta + s^2\Delta^{-1}\zeta - \frac{h\sigma^2}{4\omega^2 a^2}\psi \quad\ldots\ldots\ldots(41)$$

Again, multiplying the equation (38) by $(s^2 - \sigma^2 \mu^2)$ and applying the operator $D + \sigma\mu$, we find

$$\frac{h\sigma^2}{4\omega^2 a^2}(D + \sigma\mu)(D - \sigma\mu)\,\psi = (s^2 - \sigma^2 \mu^2)(D + \sigma\mu)(\Psi - \mu\zeta)$$

$$- 2\sigma^2 \mu\,(1 - \mu^2)(\Psi - \mu\zeta)$$

which, by means of (39) and (37) reduces to

$$\frac{h\sigma^2}{4\omega^2 a^2}(\Delta - \sigma)\,\psi = (s^2 + \sigma^2 \mu^2)\,\zeta - 2\sigma^2 \mu\Psi$$

Finally, combining this equation with (41) we derive

$$\frac{h\sigma^2}{4\omega^2 a^2}\Delta\psi = s^2\,(1 + \sigma\Delta^{-1})\,\zeta - \sigma^2\,(\mu + D\Delta^{-1})\,\Psi \quad\ldots\ldots\ldots\ldots(42)$$

The equations (40), (42), replacing (12), (38), have the advantage that these operations performed on any one of the functions in question reproduce functions of an exactly similar character.

§ 17.   *Solution in Spherical Surface Harmonics.*

The function $P_n{}^s(\mu)$ is a solution of

$$\frac{d}{d\mu}\left\{(1 - \mu^2)\frac{dP_n{}^s}{d\mu}\right\} + \left\{n(n+1) - \frac{s^2}{1 - \mu^2}\right\}P_n{}^s = 0$$

With the new notation this may be written

$$\Delta P_n{}^s = -n(n+1)P_n{}^s, \quad\text{or}\quad \Delta^{-1}P_n{}^s = -\frac{1}{n(n+1)}P_n{}^s \quad\ldots\ldots(43)$$

Hence, by means of known properties of harmonic functions,

$$(\mu + D\Delta^{-1})P_n{}^s = \frac{(n+2)(n-s+1)}{(n+1)(2n+1)}P^s{}_{n+1} + \frac{(n-1)(n+s)}{n(2n+1)}P^s{}_{n-1}\ldots(44)$$

Assume now that

$$\psi = \sum_{n=s}^{n=\infty} \Gamma_n{}^s P_n{}^s(\mu)\,e^{i(\lambda t + s\phi)}, \qquad \zeta = \sum_{n=s}^{n=\infty} C_n{}^s P_n{}^s(\mu)\,e^{i(\lambda t + s\phi)}$$

$$\Psi = \sum_{n=s}^{n=\infty} D_n{}^s P_n{}^s(\mu)\,e^{i(\lambda t + s\phi)}, \qquad \upsilon = \sum_{n=s}^{n=\infty} \gamma_n{}^s P_n{}^s(\mu)\,e^{i(\lambda t + s\phi)}$$

On substituting in (40), (42) and performing the operations involved by means of (43), (44) we derive

$$\sum_{n=s}^{n=\infty} D_n{}^s\left\{1 - \frac{\sigma}{n(n+1)}\right\}P_n{}^s$$

$$= \sum_{n=s}^{n=\infty} C_n{}^s\left[\frac{(n+2)(n-s+1)}{(n+1)(2n+1)}P^s{}_{n+1} + \frac{(n-1)(n+s)}{n(2n+1)}P^s{}_{n-1}\right]$$

$$\frac{h}{4\omega^2 a^2}\sum_{n=s}^{n=\infty} n(n+1)\Gamma_n{}^s P_n{}^s = -\frac{s^2}{\sigma^2}\sum\left\{1 - \frac{\sigma}{n(n+1)}\right\}C_n{}^s P_n{}^s$$

$$+ \sum D_n{}^s\left[\frac{(n+2)(n-s+1)}{(n+1)(2n+1)}P^s{}_{n+1} + \frac{(n-1)(n+s)}{n(2n+1)}P^s{}_{n-1}\right]$$

On equating coefficients of corresponding functions in the two members of these equations and replacing σ by $2\omega s/\lambda$ we obtain

$$\frac{(n+1)^2(n-s)}{2n-1}C^s{}_{n-1}-\left[n(n+1)-\frac{2\omega s}{\lambda}\right]D_n{}^s+\frac{n^2(n+s+1)}{2n+3}C^s{}_{n+1}=0$$

$$\dots\dots(45)$$

$$\frac{(n+1)^2(n-s)}{2n-1}D^s{}_{n-1}-\frac{\lambda^2}{4\omega^2}\left[n(n+1)-\frac{2\omega s}{\lambda}\right]C_n{}^s$$

$$+\frac{n^2(n+s+1)}{2n+3}D^s{}_{n+1}=\frac{n^2(n+1)^2\,h\Gamma_n{}^s}{4\omega^2a^2}\dots\dots(46)$$

But from the equation (10), as before in the case of the zonal harmonics,

$$\Gamma_n{}^s=\gamma_n{}^s-g_nC_n{}^s$$

whence (46) becomes

$$\frac{(n+1)^2(n-s)}{2n-1}D^s{}_{n-1}-\left[\frac{\lambda^2}{4\omega^2}\left\{n(n+1)-\frac{2\omega s}{\lambda}\right\}-\frac{n^2(n+1)^2\,hg_n}{4\omega^2a^2}\right]C_n{}^s$$

$$+\frac{n^2(n+s+1)}{2n+3}D^s{}_{n+1}=\frac{n^2(n+1)^2}{4\omega^2a^2}h\gamma_n{}^s\dots\dots(47)$$

When all the quantities $\gamma_n{}^s$ but one are zero, we can by means of (45) and (47) express the ratio of any two consecutive terms of the series $\dots C_r{}^s$, $D^s{}_{r+1}$, $C^s{}_{r+2}\dots$ by means of a continued fraction, either infinite or terminating, whose elements involve only the known data of the problem. The equation which involves $\gamma_n{}^s$ will then serve to determine each of these unknown coefficients in terms of $\gamma_n{}^s$. The method of procedure has been already explained in § 13.

For numerical applications to the leading tidal constituents and the determination of the periods of free vibration see Hough on "The Dynamical Theory of the Tides, Part II*."

CHAPTER C.   PRACTICAL APPLICATIONS.

§ 18.   *Tidal Observation.*

A tide-gauge† is an instrument which gives automatically a continuous tide-curve, whence we may read off the height of water at every instant of time. But in fact a large proportion of tidal data consists of observations

---

* *Phil. Trans. R. S.*, Vol. 191, A (1898), p. 139.

† Temporary tide-gauge, Art. Hydrography, *Admiralty Scient. Manual* (London, 1886). Descriptions of various gauges, Baird's *Manual of Tidal Obs.* (Taylor and Francis, London, 1886); Thomson (Lord Kelvin), Tidal Instruments, *Inst. Civ. Eng.*, Vol. LXV., March 1, 1881, and *Popular Lectures*, Vol. III., p. 170 (Macmillan, London, 1891); Harris, *Manual of Tides*, Part II., App. 9, Rep. U.S. Coast Survey, 1897; W. U. Moore, "Richard Tide-gauge" (official paper of H. B. M. Admiralty, 1898); Mensing of the Imperial German Navy patented a pressure-gauge (No. 94007) and current-meter (No. 102874), and a commission is said to have reported favourably on them; also a pamphlet (in English), *The Self-registering Tidal-Gauge for the open Sea* (Julius Spring, Berlin, 1904). [Another form of pressure tide-gauge has been patented (No. 4138/1908) by Admiral Mostyn Field and Captain Purey-Cust, and is found to work satisfactorily.]

made by eye of the times and heights of high and low water*. Mariners are satisfied with a knowledge of such times and heights, and these phenomena were formerly universally chosen as the subject of observation. It is, however, recognised that the tabulation of hourly heights, as read off from a tide-curve, affords far more satisfactory data for the reductions necessary to determine the laws of the tide.

The selection of the site for a tide-gauge is not easy, for the curve as registered at an apparently well-sheltered spot may exhibit many minor irregularities which are not due to the true tide, and a tide-curve should obviously be as smooth as possible†.

C. Lallemand has written a book entitled *Nivellement de Précision*‡ in which he gives a description of his instrument called a "Medimarémétre§." The object of the instrument is to obliterate the tidal oscillations, and thus to derive the mean sea-level without incurring the labour of reducing tidal observations. References to the use of this instrument will be found in C. Lallemand's various reports to the International Geodetic Association.

§ 19.   *Seiches and Vibrations of Lakes and of the Sea.*

The irregularities in a tide-curve are due to certain secondary waves which agitate enclosed areas of water‖. Our knowledge of these waves is principally derived from the work of Dr F. A. Forel¶ of Lausanne, and the

---

* On the precautions to be adopted see Art. "Tides," (Brit.) *Adm. Sci. Man.* The matter is further complicated by "seiches" and "vibrations," see § 19 below.

† On sites see Baird's *Manual*, pp. 8—9.

‡ (1896) Paris.

§ See also *Rivista di Topografia e Catasto*, N. 1, July 1896.

‖ These secondary waves have sometimes been called "periodic tides" (*Nature*, 12 Jan. and 20 Ap. 1899); but the name is very inappropriate.

¶ F. A. Forel, *Le Léman*, Vol. II. (Lausanne, 1895). The author summarizes his own papers published between 1873 and 1885 in the *Bull. Soc. Vaud. Sci. Nat.*; *Arch. Sci. phys. et nat. Genève*; *Schweiz. Naturf. Gesell.* ; *Ann. de Chimie et de Phys.*; *Assoc. Fr. Av. Sci.* (Montpellier).

Forel refers to the observations and the theories of Fatio de Duillier, *Hist. de Genève*, Vol. II. (1730), p. 463 ; J. Jallabert, *Acad. Roy. Sci. Paris* (1742), p. 26 ; H. Saussure, *Voyages dans les Alpes Neuchâtel* (1779), p. 12 ; G. B. E. Vaucher, *Soc. Phys. Genève*, Vol. VI. (1804), p. 35 ; D. Milne, *Roy. Soc. Edinb.*, Vol. I. (1832—44), p. 457 ; B. Studer, *Lehrb. der Phys. Geog.*, Vol. V., p. 78 ; C. O. Meyer, *Physik der Schweiz* (1854), p. 353 ; D. F. J. Arago, *Œuvres Compl.*, Vol. IX. (1857), p. 580 ; E. Favre, *Recherches Géologiques*, Vol. I. (1867), p. 12.

A popular account of Forel's work is given in Chapter 2 of Darwin's *Tides*, etc. ; also in supplementary article, "Tides," *Encycl. Brit.*, 1902. See also Ph. Plantamour, *Arch. Sci. Phys. nat.*, Vol. I. (1879), p. 335 ; G. B. Airy, *Phil. Trans. R. S.*, Vol. 169 (1878), p. 123 ; H. C. Russell ("Seiches of Lake George"), Ann. Address Roy. Soc. N.S. Wales, May 7, 1885, reproduced in *Nature*, Vol. XXXII. (1885), p. 232 ; A. Endrös (Seiches on Chiemsee), Dissertation (Traunstein), 1903.

Of special importance in the theory is G. Chrystal, *Trans. Roy. Soc. Edinburgh*, Vol. XLI., Part III. (1905), p. 599. Also G. Chrystal and E. Maclagan-Wedderburn, in the same volume, p. 823 ; P. White and W. Watson, *Proc. Roy. Soc. Edinburgh*, Vol. XXIV. (1905—6), p. 142 ; K. Honda, T. Terada and D. Isitani, *Phil. Mag.*, Vol. XV. (6th series), 1908, p. 88 ; the same authors, together with Y. Yoshida, on "Secondary Undulations of Oceanic Tides," *Journ. Coll. Sci. Tokyo*, Vol. XXIV. (1908).

waves of longer period are generally known by the Genevan name of Seiche.

Seiches are oscillations of the waters of the whole basin about certain nodal lines. In the lake of Geneva some oscillations were longitudinal, with one, two or more nodes, others were transverse. The period of the uninodal longitudinal seiche is twice the time occupied by a long wave in traversing the length of the lake. Seiches usually begin suddenly with maximum amplitude, and then gradually die away in about six hours. They are almost certainly due to local variations of atmospheric pressure, to cessations of a wind which has been blowing for some time previously, and to minute earthquakes. Forel has suggested other less plausible, or at any rate less frequent causes.

The water of lakes is also agitated by other oscillations of periods so short that they cannot be seiches in which the whole of the lake moves as a single system. Forel called them vibrations, and he estimated their periods at from 20 seconds to two minutes. They seem, at least in part, to be caused by wind, but they may also be produced by steam-boats. Vibrations may be observed to persist with perfect regularity and diminishing amplitude during several hours.

F. N. Denison suggests that vibrations are caused by the variations of barometric pressure which occur with some approach to regularity when a high wind is blowing. He connects this variation of pressure with the instability which, according to H. von Helmholtz, exists when one stratum of fluid streams over another. The atmosphere is not strictly in adiabatic equilibrium, and there is usually an abrupt change in what W. von Bezold calls the "potential temperature*" at the surface which separates two strata of air, of which the upper is moving relatively to the lower. The two strata are then different fluids from a mechanical point of view. H. von Helmholtz has considered the lengths of the waves which must be generated at the surface of separation. F. N. Denison has undertaken an elaborate comparison of the small barometric fluctuations occurring in stormy weather with the contemporaneous vibrations of the Canadian lakes and arms of the sea. He believes that he can prove that the two phenomena are related†, [but subsequent observation hardly seems to confirm his theory‡].

* *Berlin Ber.*, 1888, p. 1189 = *Ges. Abh.*, Braunschweig (1906), p. 128.

† H. von Helmholtz, "Ueber atmosphaerische Bewegungen," *Sitzb. k. Pr. Ak. Wiss.* (1889), pp. 761—780; F. Napier Denison, "Secondary undulations...on the great Lakes," *Proc. Canadian Inst.*, 1897, pp. 28, 55; 1898, p. 134; *Canadian Engineer*, Oct. and Nov. 1897; "Air Barometer," *Ast. Phys. Soc. Toronto*, 1897, p. 1; *Brit. Assoc.* (Dover), 1899, p. 656, and (Glasgow) 1901, p. 577. Some very remarkable secondary undulations are shown in a paper by W. Bell Dawson, *Trans. Roy. Soc. of Canada*, Vol. v., Section III. (1899—1900), pp. 23—26. A general account of the subject is given in G. H. Darwin, *The Tides*, etc. (Murray, London).

‡ [K. Honda and others, *Tokyo Journ. Coll. Sci.*, Vol. xxiv. (1908).]

§ 20. *Tide-generating Forces.*

It has already been shown in § 2 that the potential of the moon's tide-generating force may be written

$$V = \frac{3m}{2r^3} \rho^2 (\cos^2 z - \tfrac{1}{3})$$

where m, r are the moon's mass and radius vector; ρ is the radius vector of the point at which V is the potential; and z is the angle between r and ρ. Since it is only necessary to consider the potential at the earth's surface, and since the earth's small ellipticity of figure is immaterial, we may replace ρ by a, the earth's mean radius.

If h, δ be the moon's Greenwich westward hour-angle and declination at the given time; λ the north latitude and l the west longitude of the place of observation, the above formula leads to

$$V = \frac{3ma^2}{4r^3} [\cos^2 \lambda \cos^2 \delta \cos 2 (h - l) + \sin 2\lambda \sin 2\delta \cos (h - l)$$
$$+ 3 (\tfrac{1}{3} - \sin^2 \delta)(\tfrac{1}{3} - \sin^2 \lambda)]$$

The westward component of the tide-generating force is $dV/a \cos \lambda dl$, and the northward component is $dV/ad\lambda$. The ratios of these to gravity give the apparent deflections of the vertical.

These component forces are periodic, excepting as to a certain small constant portion of the northerly component, which causes a small permanent ellipticity of the earth's figure*.

Under the action of these forces a pendulum describes simultaneously two ellipses, one in a period of a half and the other in the period of a whole lunar day. The centre about which these ellipses are described also oscillates slowly northward and southward, with a very small amplitude and a fortnightly period.

These three motions correspond with Laplace's tides of the third, second and first species respectively referred to in § 4. The mean value of the coefficient $\frac{3}{2} \frac{m}{M} \left(\frac{a}{r}\right)^3$ is $0''\cdot0174$.

---

* See G. H. Darwin, *M. N. R. Ast. Soc.*, Dec. 1899, p. 119 [or Vol. iii., p. 114].

§ 21.   *Deflection of the Vertical.*

If it were possible to measure the actual deflections of a pendulum, a comparison of results with those derived from the above formulæ would tell us the amount by which the solid earth yields to the tide-generating forces.

An account of the various attempts to measure the deflections of the vertical falls outside the scope of this article; we must therefore confine ourselves to but a few words on the subject.

The earlier attempts, although they failed, often led to collateral results of interest[*]. The experiments of E. von Rebeur-Paschwitz, who used and improved the horizontal pendulum[†], were rewarded with a certain amount of success. His work was continued by R. Ehlert and I. T. Kortazzi[‡], and others.

Finally O. Hecker attained far more concordant results by long-continued observations with two horizontal pendulums, hung in azimuths at right angles to one another, in a deep well at the Geodetic Institute at Potsdam[§]. E. Paschwitz, R. Ehlert and I. T. Kortazzi concluded that the oscillations of the pendulums were from one-half to two-thirds of the theoretical amount on a perfectly rigid earth. O. Hecker's observations showed that the fraction is almost exactly two-thirds, and we believe that this result must be very near to the truth. It thus appears that the effective rigidity of the earth's mass must be about the same as if it were made of steel[‖].

One of the difficulties which surround the attempts to make this measurement lies in the fact that the varying load of the oceanic tide must cause

---

[*] G. H. Darwin and H. Darwin, "The Lunar Disturbance of Gravity," *B. A. Reports* (1881), p. 93; (1882), p. 95 [or Vol. I., pp. 389, 430].

[†] F. Zöllner wrote three papers on the instrument and its history in *Pogg. Ann.*, Vol. CL. (1873), pp. 131, 134, 140. These are followed by one by A. Šafařik (*Pogg. Ann.*, Vol. CL., p. 150) on the history. F. Zöllner's papers had been previously published in *Ber. k. Sächs. Gesell.*, Vol. XXI. (1869), pp. 281—4; Vol. XXIII. (1871), pp. 479—575; and Vol. XXIV. (1872), pp. 183—192; and Šafařik's paper in *k. Böhm. Gesell.* (1872). Zöllner attributes priority to A. Perrot (*Paris C. R.*, Vol. LV. (1862), p. 728), but the same idea seems to have occurred previously to J. Gruithuisen (said to be in *Neue Analekten, etc.*, Munich, 1832, Vol. I., Part I.). An account of alleged experiments (doubtless fraudulent) by one L. Hengler is given in *Dingler's Polytech. Journ.*, Vol. XLIII. (1832), pp. 81—92. E. Paschwitz gives a valuable bibliography in "das Horizontalpendel," *Nov. Act. k. Leop. Car. D. Akad.*, Vol. LX. (1892), pp. 1—216. A complete list of Paschwitz's own papers is contained in an obituary notice *Beitr. z. Geophysik*, Vol. II. (1895), pp. 16—18. His most important papers are the above and "Horizontalpendelbeobachtungen zu Strassburg," *Beitr. z. Geophysik*, Vol. II. (1895), pp. 211—535. A summary of his work up to 1893 (and he died in 1895), is contained in *B. A. Rep.* (Nottingham), 1893. This is especially important as showing how he was led to modify some opinions advanced in previous papers.

[‡] R. Ehlert's paper is in *Beitr. z. Geophysik*, Vol. III. (1896), pp. 131—215; I. T. Kortazzi's is in Russian in *Isvestia Russk. Astron. Obshchestva*, Part IV. (1895), pp. 24—56, and Part V. (1896), pp. 301—9. This constitutes a particularly valuable series of observations, and it is unfortunate that it has not (as far as I know) been translated into any of the western languages. A general account of the subject is given in Darwin's *Tides*.

[§] O. Hecker, *k. Preuss. Geodät. Inst.*, Neue Folge, No. 32 (1907).

[‖] See also § 37, p. 238.

flexures of the solid earth at stations near the sea-coast*. This effect seems
to have been actually observed by A. d'Abbadie close to the sea†, and
theoretical investigations seem to point to the possibility that it may still be
sensible at great distances from the coast‡. [W. E. Plummer has shown the
effect very conclusively by observations with a horizontal pendulum at
Bidston Observatory, and has even been able to submit the observed deflections to harmonic analysis so as to render the effects of the several tidal
components evident §.]

Another difficulty arises from a large diurnal oscillation of the pendulum
which has been observed everywhere. This oscillation is undoubtedly thermal,
but it is uncertain how far it is a merely local change, and how far it is a
general terrestrial change affecting the whole hemisphere exposed to sunlight.
R. Ehlert argues in favour of the latter view‖.

§ 22. *Methods of discussing the actual Tides of the Ocean.*

It was Laplace who first showed how theory and observation should be
combined¶. His discussion rests on the principle of forced oscillations;
namely, that—

*The state of oscillation of a system of bodies in which the primitive con-
ditions of movement have disappeared through friction is coperiodic with the
forces acting on the system.*

It follows that if the sea is solicited by a periodic force expressed as
a coefficient multiplied by the cosine of an angle increasing uniformly with
the time, there will result a partial tide, also expressed by the cosine of an
angle increasing at the same rate; but the phase and the coefficient will
differ from those in the corresponding term of the equilibrium theory, and
are only derivable from observation. It is then only the rate of increase of
the angle which is determined by theory.

If the tidal forces due to the sun and moon are expressed in a series of
such terms, the oscillations of the sea will clearly be expressible by a series
of partial tides of unknown amplitudes and phases, but of known periods**.

* G. H. Darwin, "Lunar Disturbance of Gravity," *B. A. Rep.* (1882), p. 106, or reprint of
portion in *Phil. Mag.* (March 1897), p. 177 [or Vol. ɪ., p. 450].

† "Recherches sur la verticale," *Ann. Soc. Sci. Bruxelles*, Vol. v., Part 2 (1881), p. 37. See
also F. Omori, *Earthquake Investigation Committee*, No. 21 (1905), p. 5; *Bull. Imp. Earthq.
Investig. Committee*, Vol. ɪ. (1907), p. 167.

‡ But see E. Paschwitz, *Beiträge z. Geophysik*, Vol. ɪɪ. (1895), pp. 332—8, who argues
forcibly that the theoretical conditions are scarcely applicable, at least in Europe.

§ [See a letter from J. Milne in *Nature*, Vol. ʟxxxɪɪ. (1910), p. 427; and more fully in *Report
of Seismol. Committee of B. A.*, Sheffield (1910). At the end of this paper will be found a note
on this section giving, as an extract from the last-mentioned report, a contribution thereto by
the author.]

‖ *Beiträge z. Geophysik*, Vol. ɪɪɪ. (1896), p. 131.

¶ Chapters ɪɪɪ. and ɪv. of Book ɪv. of *Méc. Cél.*

** Laplace attributes the tidal forces to a number of fictitious satellites. These are specified
in G. H. Darwin, *Phil. Trans.*, Vol. cʟxx., Part ɪɪ. (1879), p. 465 [or Vol. ɪ., p. 55].

In accordance with the principle of the superposition of small motions these may be added together to give the total tide at the place of observation.

The comparison between tidal theory and observation has been carried out in two ways which may be called the "Synthetic" and the "Analytic." The two methods differ only in the completeness with which the tide-generating forces are expressed in a form to which the principle of forced oscillations is strictly applicable.

The semi-diurnal periodicity of the tide, with the weekly alternation of Spring and Neap, suggests the choice of mathematical formulæ of like apparent simplicity; and this indication was followed by Newton, Bernoulli, Maclaurin, Laplace, Whewell and Airy. At places where there is but little diurnal inequality, a single harmonic function, with a slowly and periodically varying amplitude, and with an argument which increases at a rate varying slowly about a mean value, is found to represent the tide in a tolerably satis-factory manner. But where the diurnal inequality is large, as in the Pacific and Indian Oceans, the analytical simplicity of the synthetic method loses its apparent advantage.

We have in § 20 given the expression for the lunar potential in the form appropriate to the synthetic method. The three terms of that expression are approximately harmonic functions of the time, and their coefficients obviously oscillate slowly about mean values. A similar expression would give the solar potential. The mathematical basis of the synthetic method consists in a synthesis of such formulæ. The lunar and solar semidiurnal terms may be combined into one, and a similar fusion may be effected for the diurnal and slowly varying terms. The comparison of the observations at any port with such an analytical expression was however found to exhibit its insufficiency, and it became necessary to add many corrections.

An important new departure in the treatment of the tides was made by Lord Kelvin, who initiated the Harmonic method, in which each term is a true harmonic function of the time*.

In this method the principle of forced oscillations becomes strictly instead of approximately applicable.

Since that principle forms the basis of all the practical treatment of the problem, it seems more convenient to consider first that method in which it has been applied in its more logical form, reserving the less satisfactory method for subsequent consideration. In adopting this course we invert history and consider the more recent before the older method.

---

* G. B. Airy, and after him B. Chazallon (*Paris C. R.*, Vol. XLII. (1856), p. 966), expanded the rise and fall of the tide in a single day in a Fourier series, but this cannot be regarded as a true foreshadowing of the analytic method. See G. B. Airy, "Tides and Waves," *Encyc. Metrop.*, and P. Hatt, *Phénomènes des Marées*, Paris, 1885.

§ 23. *Harmonic Method—Analytical Development**.

The expression for the moon's potential in § 20 involved $h - l$, which is the moon's local hour-angle. Since a/r is the moon's parallax, her position was determined by her local hour-angle, declination and parallax. The position on the earth was fixed by the latitude. But the moon's position may also be specified by means of her mean longitude, the longitudes of her perigee and node, the obliquity of the ecliptic, the inclination and eccentricity of the orbit. If this mode of expression be adopted each one of the three terms of our previous expression will lead to an infinite but rapidly convergent series of terms. Each of the new terms will consist of a constant coefficient multiplied by a simply harmonic function of the time. But it is not found necessary to carry out the process with the same degree of logical completeness as in the lunar and planetary theories. It suffices in fact to make a development which would be rigorously exact if the moon's node and the plane of her orbit were absolutely fixed in space. In consequence of the movement of the node and orbit the coefficients in the expansion of the potential are no longer rigorously constant, but slowly execute small oscillations about mean values; and the arguments of the harmonic functions of the time require small and slowly variable corrections to their phases.

We may in this way regard the potential as expressed by a series of temporarily constant coefficients multiplied by simple harmonic functions of the time, subject to temporarily constant corrections of phase. The principle of forced oscillations then allows us to conclude that each term will correspond with a partial tide of an unknown range and with an unknown phase.

If the equilibrium theory be thus completely developed, the range and phase of each partial tide is exactly determinate, and therefore the range of the corresponding tide at any actual port may be expressed by multiplying the equilibrium range by some factor derivable only from observation. Similarly the phase of the actual tide is derivable by applying some constant correction to the equilibrium phase, also derivable from observation. The object of harmonic analysis is virtually to determine this factor and this correction to the phase for each partial tide. We shall explain below how

* This section is founded on a Report to the Brit. Assoc. (Southport), 1883, by G. H. Darwin and J. C. Adams [or Vol. I., p. 1]. Substantially the same development is contained in C. Börgen, "Harm. Anal. der Gezeitenbeob.," *Ann. d. Hydrographie*, Vol. XII. (1884), pp. 305, 387, 438, 499, 558, 615, 664; P. Hatt, "Analyse harmonique des Marées après les travaux anglais," *Ann. Hydrographiques*, 1893 (Impr. Nat.) ; also without full details, *Explication élémentaire des Marées*, Gauthier-Villars; R. A. Harris, *Manual of Tides*, Part II., App. No. 9 to *Rep. of U. S. Coast Survey*, 1897; J. Eccles, *Great Trig. Survey of India*, Vol. XVI., Dehra Dun, 1901. See also Ferrel, "Tidal Researches," *Rep. U.S. Coast Survey*, 1874. Lord Kelvin's original papers are Reports to Brit. Assoc. 1868, 1870, 1871, 1872, 1876, 1878.

it is found to be practically convenient to give effect to this idea, but for the present we may continue to consider the results of the equilibrium theory as expressed in this mode of development.

In the notation which has now been universally adopted the angular velocity of the earth's rotation is denoted by γ, and the mean motions of the moon, sun and lunar perigee are written σ, η, ϖ ($\gamma\hat{\eta}$, $\sigma\epsilon\lambda\acute{\eta}\nu\eta$, $\H{\eta}\lambda\iota\sigma$). The "speed" of any partial tide is the rate of increase of the "argument" in the simple harmonic function of the time, and all the speeds are functions of γ, σ, η, ϖ.

A nomenclature for the several partial tides becomes a practical necessity, and initial letters, chosen at random, have been assigned to them*. In a few cases where no initial has been adopted it is convenient to specify the tide by means of its speed; for example the evectional monthly tide may be denoted by $\sigma + \varpi - 2\eta$.

In the mathematical expression for the tide it is rather the mean longitudes than the speeds which arise naturally. Hence we write s, h, p for the mean longitudes of the moon, sun and lunar perigee, and t for the mean solar hour angle at the place of observation; so that the argument for any partial tide is a function of t, s, h, p.

In the following schedules various symbols are used, and we must now define them.

The obliquity of the ecliptic is ω, and that of the equator to the lunar orbit is I. It is clear that I is a function of N the longitude of the moon's node, and that it oscillates about the mean value ω. The inclination of the lunar orbit to the ecliptic is i.

The eccentricities of the lunar and solar orbits are e, e_1.

If m, c are the moon's mass and mean distance, τ denotes $\frac{3}{2}m/c^3$, and τ measures the intensity of the lunar tide-generating force. If m_1, c_1 denote the corresponding quantities for the sun and if $\tau_1 = \frac{3}{2}m_1/c_1^3$, it is clear that τ_1/τ (which is equal to ·46035) is the factor required for bringing the solar and lunar tides to a common measure.

It will be observed that the arguments in the schedules below involve ξ and ν; these symbols denote the longitude in the lunar orbit and the right

---

* The principal lunar and solar tides are denoted respectively by M, S (moon, sun). The elliptic lunar tides L, N (connected with M) are analogous to R, T for the sun (connected with S). The diurnal lunar O corresponds with the solar P. The analytical expressions for the evectional tides λ, ν are closely connected with those for the elliptic tides L, N. The variational tide μ has a similar connection with M. Sa, Ssa, Mm, Mf, MSf denote the solar annual, solar semi-annual, lunar monthly and fortnightly, and luni-solar fortnightly tides. There seem to be no obvious reasons for the choice of the other initials.

ascension of the descending node of the equator on the lunar orbit, and are functions of N. The functions of ξ and ν which occur in the arguments are those small corrections to the phases which are necessary to take account of the position of the moon's node. It has been found sufficiently exact to adopt constant values of ξ and ν as applicable to a whole year.

We shall now give the results of the development of the equilibrium theory in the form of schedules.

The general expression for the equilibrium tide is:

$$h = \Sigma \text{ " general coeff." } \times \text{ " coeff." } \times \cos \text{ " argument."}$$

The actual tide is derivable from the equilibrium tide by multiplying each "coefficient" by a factor derived from observation, and by subtracting from the "argument" an angle also so derived.

The numerical values of the coefficients in any of the three groups of tides would indicate the relative importance of the several terms, but they afford no indication of the relative importance of tides belonging to different groups.

A. LUNAR TIDES.

1. Semi-diurnal Tides.

General coefficient $\frac{3}{2}\frac{m}{M}\frac{a^4}{c^3}\cos^2\lambda$.

Name	Initial	Coefficient	Argument $2t+2(h-\nu)$
Principal lunar	M_2	$\frac{1}{2}(1-\frac{5}{2}e^2)\cos^4\frac{1}{2}I$	$-2(s-\xi)$
Luni-solar (lunar part)	K_2	$\frac{1}{2}(1+\frac{3}{2}e^2)\frac{1}{2}\sin^2 I$	
Larger elliptic	N	$\frac{1}{2}\cdot\frac{7}{2}e\cos^4\frac{1}{2}I$	$-2(s-\xi)-(s-p)$
Smaller elliptic	L	$\frac{1}{2}\cdot\frac{1}{2}e\cos^4\frac{1}{2}I$	$-2(s-\xi)+(s-p)+\pi$
Elliptic second order	$2N$	$\frac{1}{2}\cdot\frac{17}{2}e^2\cos^4\frac{1}{2}I$	$-2(s-\xi)-2(s-p)$
Larger evectional	ν	$\frac{1}{2}\cdot\frac{105}{16}me\cos^4\frac{1}{2}I$	$-2(s-\xi)+(s-p)+2(h-s)$
Smaller evectional	λ	$\frac{1}{2}\cdot\frac{15}{16}me\cos^4\frac{1}{2}I$	$-2(s-\xi)-(s-p)-2(h-s)$
Variational	μ	$\frac{1}{2}\cdot\frac{2}{8}m^2\cos^4\frac{1}{2}I$	$-2(s-\xi)+2(h-s)+\pi$

2. Diurnal Tides.

General coefficient $\frac{3}{2}\frac{m}{M}\frac{a^4}{c^3}\sin 2\lambda$.

Name	Initial	Coefficient	Argument $t+(h-\nu)$
Lunar	O	$(1-\frac{5}{2}e^2)\frac{1}{2}\sin I\cos^2\frac{1}{2}I$	$-2(s-\xi)+\frac{1}{2}\pi$
Luni-solar (lunar part)	K_1	$(1+\frac{3}{2}e^2)\frac{1}{2}\sin I\cos I$	$-\frac{1}{2}\pi$
Larger elliptic	Q	$\frac{7}{2}e\cdot\frac{1}{2}\sin I\cos^2\frac{1}{2}I$	$-2(s-\xi)-(s-p)+\frac{1}{2}\pi$
Smaller elliptic	M_1	This tide, being really compounded of two of speeds $\gamma-\sigma+\varpi$ and $\gamma-\sigma-\varpi$, has a complicated coefficient and argument	
Speed $\gamma+\sigma-\varpi$	J	$\frac{1}{2}e\cdot\frac{1}{2}\sin I\cos I$	$+(s-p)+\frac{1}{2}\pi$

3. Long Period Tides.

General coefficient $\frac{3}{2}\frac{m}{M}\frac{a^4}{c^3}(\frac{1}{3}-\frac{1}{2}\sin^2\lambda)$.

Name	Initial	Coefficient	Argument
Change of mean level		$(1+\frac{3}{2}e^2)(\frac{1}{3}-\frac{1}{2}\sin^2 I)$	of variable part is N
Monthly	Mm	$3e(\frac{1}{3}-\frac{1}{2}\sin^2 I)$	$(s-p)$
Evect. monthly		$\frac{4}{5}me(\frac{1}{3}-\frac{1}{2}\sin^2 I)$	$-(s-p)+2(s-h)$
Var. monthly		$3m^2(\frac{1}{3}-\frac{1}{2}\sin^2 I)$	$2(s-h)$
Fortnightly	Mf	$(1-\frac{5}{2}e^2)\frac{1}{2}\sin^2 I$	$2(s-\xi)$

B. SOLAR TIDES.

1. Semi-diurnal Tides.

General coefficient $\frac{3}{2}\frac{m}{M}\frac{a^4}{c^3}\cos^2\lambda$.

Name	Initial	Coefficient	Argument
Principal solar	S_2	$\frac{1}{2}\frac{\tau_1}{\tau}(1-\frac{5}{2}e_1^2)\cos^4\frac{1}{2}\omega$	$2t$
Luni-solar (solar part)	K_2	$\frac{1}{2}\frac{\tau_1}{\tau}(1+\frac{3}{2}e_1^2)\frac{1}{2}\sin^2\omega$	$2t+2h$
Larger elliptic	T	$\frac{1}{2}\frac{\tau_1}{\tau}\cdot\frac{7}{2}e_1\cos^4\frac{1}{2}\omega$	$2t-(h-p_1)$
Smaller elliptic	R	$\frac{1}{2}\frac{\tau_1}{\tau}\cdot\frac{1}{2}e_1\cos^4\frac{1}{2}\omega$	$2t+(h-p_1)+\pi$

2. Diurnal Tides.

General coefficient $\frac{3}{2}\frac{m}{M}\frac{a^4}{c^3}\sin 2\lambda$.

Name	Initial	Coefficient	Argument
Solar	P	$\frac{\tau_1}{\tau}(1-\frac{5}{2}e_1^2)\frac{1}{2}\sin\omega\cos^2\frac{1}{2}\omega$	$t-h+\frac{1}{2}\pi$
Luni-solar (solar part)	K_1	$\frac{\tau_1}{\tau}(1+\frac{3}{2}e_1^2)\frac{1}{2}\sin\omega\cos\omega$	$t+h-\frac{1}{2}\pi$

3. Long Period Tide.

General coefficient $\frac{3}{2}\frac{m}{M}\frac{a^4}{c^3}(\frac{1}{3}-\frac{1}{3}\sin^2\lambda)$.

Name	Initial	Coefficient	Argument
Semi-annual	Ssa	$\frac{\tau_1}{\tau}(1-\frac{5}{2}e_1^2)\frac{1}{2}\sin^2\omega$	$2h$

Notes. The tide L should have compounded with it another of speed $2\gamma-\sigma+\varpi$; this renders the coefficient and the argument more complex than as shown. The coefficients of all the evectional and variational tides should involve in their coefficients the full values of those inequalities; as written above they only show the first terms of the true series.

The solar and lunar K tides fuse together. As to Ssa see section 24.

§ 24. *Meteorological Tides, Over-Tides and Compound or Shallow-water Tides.*

Every tide whose period is an exact multiple or submultiple of a mean solar day or of a tropical year is affected by meteorological conditions, and is obviously liable to some uncertainty.

As a wave progresses in shallow water its contour comes to depart considerably from that of the curve of sines. A Fourier-expansion for the form of wave contains terms of twice and thrice the frequency of the fundamental wave. Such terms may be called over-tides, in analogy with Helmholtz's over-tones in acoustics. It is usual for this reason to introduce the tides M_4, M_6, S_4, S_6 with speeds $4(\gamma - \sigma)$, $6(\gamma - \sigma)$, $4(\gamma - \eta)$, $6(\gamma - \eta)$ respectively. The over-tides are commonly negligible (or are at least neglected) in other cases.

When the heights of two coexistent waves are not small fractions of the depth of the water, compound waves are generated whose frequencies are the sum and difference of the two fundamental waves. The principal tides of this kind are MS, MSf with speeds $2(\gamma - \sigma) \pm 2(\gamma - \eta)$ arising from M_2 and S_2; the latter of these is also an astronomical tide; 2SM with speed $2\gamma + 2\sigma - 4\eta$ arising from S_4 and M_2; and μ with speed $2\gamma - 4\sigma + 2\eta$ arising from S_2 and M_4, and also as the Variational tide. It may be well to mention also M_3 with speed $3(\gamma - \sigma)$ arising from that term in the lunar potential, varying as the fourth power of the moon's parallax, which we omitted in § 2.

In harmonic analysis the following are the tides which are usually evaluated, M_2, M_4, M_6, S_2, S_4, K_2, N, L, ν, μ, 2SM, MS, K_1, O, P, Q, J, Sa, Ssa.

The tides M_1, Mm, Mf, MSf, are also determined, but they generally have merely a theoretical interest.

§ 25. *The results of Harmonic Analysis.*

We consider only a single partial tide as typical of all.

In accordance with the schedule of § 23 the tide M_2 is expressed by

$$k_m \frac{3ma^4}{2Mc^3} \cos^2 \lambda . \tfrac{1}{2}(1 - \tfrac{5}{2}e^2) \cos^4 \tfrac{1}{2}I \cos[2t + 2(h - v) - 2(s - \xi) - \kappa_m]$$

In this expression k_m is the coefficient by which the equilibrium semi-range is to be multiplied in order to obtain the actual semi-range; and κ_m is the alteration of the phase. Both k_m and κ_m are only derivable from observation.

If we write

$$H_m = k_m \frac{3ma^4}{2Mc^3} \cos^2 \lambda \cdot \tfrac{1}{2}(1 - \tfrac{5}{2}e^2) \cos^4 \tfrac{1}{2}\omega \cos^4 \tfrac{1}{2}i$$

$$f_m = \frac{\cos^4 \tfrac{1}{2}I}{\cos^4 \tfrac{1}{2}\omega \cos^4 \tfrac{1}{2}i}, \qquad V_m = 2(t + h - s), \qquad u_m = -2(v - \xi)$$

the expression becomes $f_m H_m \cos(V_m + u_m - \kappa_m)$. H_m consists of two factors of which one involves constants derivable from astronomical data, and the other k_m is a constant to be derived only from observation. Hence H_m may be regarded as a constant appropriate to the tide M_2 and to the place, and is derivable only from observation. If the "tidal constants" H_m, κ_m are known, the height of water as due to M_2 is calculable at any time future or past. f_m is the factor and u_m is the change of phase necessary in order to allow for the actual position of the moon's node*. Their values change slowly, and it is usual to adopt mean annual values.

Every tide has its appropriate f and u. The u for any of the tides is that part of the "argument" which involves v and ξ. Thus in the purely Solar Tides S_2, T, P, u is zero, and f is obviously unity.

The tides K_2, K_1, need special consideration as arising conjointly from Moon and Sun. The tides L and M_1 also require special treatment for other reasons.

The theory of waves in shallow water shows that for the second and third over-tide f is the square and the cube of the fundamental f, and u is twice and thrice the fundamental u. For compound tides f is the product of the constituent f's, and u is the sum or difference (as the case may be) of the constituent u's.

The final result of harmonic analysis is that we obtain H, κ for each tide. If then A_0 denotes the height of the mean sea-level above some datum ashore, the height of the sea at any future or past time is expressed by

$$A_0 + \Sigma fH \cos(V + u - \kappa)$$

We shall below require to indicate the several H, κ separately. This will be done by appending as a subscript the initial letter of the tide. For example H_m, κ_m indicate the relationship to the tide M_2. For the tides K_2, K_1 this notation is not convenient, and therefore we write H'', κ'', and H', κ' respectively in those cases.

---

* See Tables in Baird's *Manual of Tidal Observation*, and Harris, *Manual of Tides*, *U.S. Coast Survey*, Part II., 1897 [or corresponding formulæ, Vol. I., p. 68].

§ 26. *Numerical Harmonic Analysis.*

The twelfth or the twenty-fourth parts of the period of any one of the semidiurnal or diurnal tides may be described as one hour of a special time appropriate to the tide in question.

We might measure the heights on the tide-curve at the exact special hours, but the heights of the water at each exact hour of mean solar time suffice to represent the record. This approximation virtually gives the mean height of the water as estimated over a mean solar hour, of which the middle is at an exact special hour. Allowance may be made for this source of error by augmenting the amplitudes of the several oscillations.

When the mean solar hourly heights are classified according to their incidence amongst the 24 hours of a special time, we obtain by summation 24 heights appertaining to the 24 hours of that special time. Each hour of any other special time will fall indiscriminately in all hours of this time, and thus all the partial tides other than the one under consideration disappear from the mean result. The 24 special hourly heights are then expressible by the Fourier series,

$$A_0 + R_1 \cos(nt - \zeta_1) + R_2 \cos(2nt - \zeta_2) + R_3 \cos(3nt - \zeta_3) + \dots$$

where $2\pi/n$ is the special day.

If the tide for which we are in search is a diurnal one, R_2, R_3 etc. will be very small compared with R_1; if it is semidiurnal R_1, R_3 etc. will be very small compared with R_2, and so forth. It is indeed sufficient to evaluate only the particular R and ζ which theory shows to exist as a partial tide.

Let us suppose that the semi-range R becomes R', when it has been augmented by the factor which is appropriate for the correction due to the fact that the method of computation in reality gives the mean of the height of water estimated over half an hour before and half an hour after the exact hour of special time. Then since the partial tide in question is theoretically expressible by $fH \cos(V + u - \kappa)$, H is equal to R'/f.

If V_0 denotes the value of the "argument" at the instant at which the observations begin, we obviously have $V_0 + u - \kappa = -\zeta$, or $\kappa = \zeta + V_0 + u$. Thus the two tidal constants H, κ are derived from numerical analysis.

The classification or grouping of the mean solar hourly heights has been carried out in several ways. That in use in India involves the recopying of all the observations into a series of schedules*.

---

* There are certain small mistakes in the schedules (prepared for the Government of India by Mr Edward Roberts), but they are not such as to cause any material error. See G. H. Darwin, *Proc. R. S.*, Vol. LII., p. 346 [or Vol. I., p. 217].

C. Börgen uses sheets of tracing paper which are laid on to the tabulated data, and indicate by lines the correct grouping. The United States Coast Survey arrives at the same result by means of cardboard stencil plates.

G. H. Darwin has devised a method wherein observations appertaining to each day are treated as a whole*. Each of these methods appears to possess its own advantages and disadvantages.

C. Börgen has also devised an entirely different method of reduction in which the amount of addition is considerably diminished†. The process is most ingenious, but it has not, so far as I am aware, received the attention it deserves.

The harmonic tidal constants may also be derived from observations of high and low water‡. Each such observation is equivalent to two observations of height only, so that there are virtually nearly eight ($7\frac{43}{59}$) conditions per diem to be fulfilled by the analytical expression for the tide. But these conditions are not convenient for the evaluation in question, and it is possible that the most efficient method of extracting the constants from the observations would be to construct an empirical tide-curve with maxima and minima occurring at the observed times and with the observed heights.

We have hitherto only referred to the semidiurnal and diurnal tides. The tides of long period are clearly derivable from daily means of the height of the water. These daily means are grouped according to rules devised to bring out the amplitudes and phases of the tides Sa, Ssa, Mm, Mf, MSf. Since these tides are generally very small, it is necessary to apply corrections which shall free the daily means from the residual effects of the large semidiurnal and diurnal tides. This process is called the "clearance of the daily means." The reader may refer for details to the various publications specified in a note at the beginning of § 23.

If a function h be expressed as a series of harmonic terms and if one pair of these terms be $A \cos nt + B \sin nt$, then if T be a multiple of the period $2\pi/n$ we have

$$A = \frac{2}{T}\int_0^T h \cos nt\, dt, \quad B = \frac{2}{T}\int_0^T h \sin nt\, dt$$

* Harris, *Manual of Tides*, Part II., Chap. v., App. 9, of *Report of U.S. Coast Survey*, 1897. Instructions are given for preparing templates. "An Apparatus for facilitating, etc.," *Proc. R. S.*, Vol. LII., 1893, pp. 346—389 [or Vol. I., p. 216].

For reduction of a short series of observations see Darwin, (British) *Admiralty Scientific Manual*, 1886 [or Vol. I., p. 119], or Rollet de l'Isle, *Annales Hydrographiques*, 2me Série, Nʳ 780 (1896), deuxième section, p. 196.

† "Eine neue Methode, etc.," *Ann. der Hydrographie* (1894), pp. 219, 256, 295.

‡ G. H. Darwin, "Tidal Observations of H. and L. Water," *Proc. R. S.*, Vol. XLVIII. (1890), pp. 278—340 [or Vol. I., p. 157]. Harris, *Manual*, etc., Part III., Chap. III.

Thus a machine which will effect these integrations will determine A and B. In the reduction of tidal observations we suppose h to be the height of the water at any time during the interval T, and n to be the speed of any one of the partial tides. An attempt has been made to use the ball and table integrator of James Thomson for the reduction of tidal observations[*]. A tide-reducing machine (now in the South Kensington Museum) was actually constructed at a considerable expense under Lord Kelvin's direction. It was designed for evaluating the tides M_2, S_2, K_1, O, P and mean water level A_0. It has never been used, as there seemed reason to suppose that the reductions would be more expensive and less satisfactory than by the numerical process.

It is not within the scope of this article to discuss the tides at particular ports[†].

§ 27. *Explanation of Tidal Terms in common use; Datum Levels.*

We have already indicated at the end of § 25 how the semi-ranges and changes of phase of the several harmonic tides may be indicated by subscript letters attached to the letters H, κ.

The mean height at Spring tide between high and low water is called the "Spring rise" or "Spring range," and is equal to $2(H_m + H_s)$. "Neap range" is equal to $2(H_m - H_s)$; it is usually about a third of Spring range. The height between mean high-water mark at neap-tide and mean low-water mark at spring-tide is called "Neap rise," and is equal to $2H_m$. The mean period between full or change of moon and spring-tide is called "the Age of the tide"; it is equal to $\frac{1}{2}(\kappa_s - \kappa_m)/(\sigma - \eta)$ and is commonly about 36 hours. The period elapsing between the moon's upper or lower transit and high water is called "the Interval" or "luni-tidal interval." Both the interval and the height of the tide are subject to a "fortnightly inequality." The interval at full or change of moon is called "the Establishment of the port" or "the

[*] *Proc. Roy. Soc.*, Vol. xxiv. (1876), p. 262; also Thomson and Tait, *Nat. Phil.*, Vol. i., p. 488; see also "Tidal Instruments," *Inst. Civ. Eng.*, Vol. lxv. (1881), p. 10.

[†] I cannot profess to give a complete list of the places at which the harmonic constants for various places may be found, but the following may be mentioned:—

All the Indian results up to 1892 are collected in Vol. xvi. of the *Great Trigonometrical Survey of India*, Dehra Dun (1901); *Reports of the U.S. Coast Survey*; P. Hatt, publications of the French Service Hydrographique; Börgen, *German Ann. der Hydrographie*; J. P. van der Stok, *k. Instit. van Ingenieurs, Afd. Ned. Indië*, and *Ak. Wet. Amsterdam*; A. Wijkander, *k. Sv. vet. Ak.*, Bd. xv. (1899); a collection of results from various sources, G. H. Darwin and A. W. Baird, "First series of results," *Proc. R. S.*, Vol. xxxix. (1885), pp. 135—207; G. H. Darwin, "Second series," *Proc. R. S.*, Vol. xlv. (1888), pp. 556—611; Wright, *Proc. R. S.*, Vol. lxxi. (1902), p. 91; [Vol. lxxxiii., A (1910), p. 127].

vulgar establishment." The interval at spring-tide is called "the corrected or mean establishment." It is possible to determine the latter from the former when the age of the tide is known.

The French call a quantity, which appears to be equal to $H_m + H_s$, "the unit of height," and define any other inequality by means of a coefficient *.

The practice of the British Admiralty is to refer soundings and tide-tables to "mean low water mark of ordinary spring-tides." The position of this level is not susceptible of exact scientific definition, and where there is a large diurnal tide the uncertainty of meaning may perhaps be considerable. But when the Admiralty Datum has once been fixed with reference to a bench-mark ashore, it serves as well as any other datum level.

A special low-water datum has been adopted for new stations in the Indian Survey: it is called "Indian low-water mark," and is defined as $H_m + H_s + H' + H_0$ below mean sea level. Such a level will not in general differ much from the Admiralty datum, and is low enough to exclude almost all negative entries from a tide-table.

A valuable list of datum levels is given by J. N. Shoolbred, and further information on the subject is contained in a Report by J. J. A. Bouquet de la Grye†.

§ 28.   *Synthetic method—Semi-diurnal Tides*‡.

We might proceed to develop the form of potential as given in § 20, but it is more satisfactory to proceed from the harmonic development. The mean solar time and mean longitude of the moon have to be replaced by hour-angle ψ, right ascension α, and declination δ. We also have to introduce P the ratio of the moon's parallax to her mean parallax, and Δ such a constant declination that $\cos^2 \Delta$ is the mean value of $\cos^2 \delta$. Further, the results can be written more succinctly if we introduce δ' to denote the moon's declination at a time earlier than t by the "age of the declinational inequality," viz., $\frac{1}{2}(\kappa'' - \kappa_m)/\sigma$, and P' for the value of P at a time earlier by the "age of the parallactic inequality," viz., $(\kappa_m - \kappa_n)/(\sigma - \varpi)$. These two ages do not in general differ much from the "age of the tide," viz., $\frac{1}{2}(\kappa_s - \kappa_m)/(\sigma - \eta)$.

---

* P. Hatt, *Phénomènes des Marées* (Paris, 1885), p. 151.

† J. N. Shoolbred, *Brit. Ass. Rep.* (1879), p. 220; J. J. A. Bouquet de la Grye, *International Geodetic Association Report* (1898). [See also reports on Tide-Gauges to the same by G. H. Darwin in 1903 and 1909.]

‡ This section is founded on Darwin, *B. A. Report* (1885), p. 35 [or Vol. I., p. 70]. See a correction by Börgen, "Berechnung, etc.," *Ann. der Hydrog.*, Vol. XVII. (1889), p. 45.

A subscript 1 applied to any symbol will denote that the result refers to the sun; in this case the several "ages" may be regarded as vanishing.

In this method it is necessary to divide the K_2 and K_1 tides into their lunar and solar portions, which are proportional to ·683 and ·317 respectively.

The following expression is then equivalent to the harmonic expansion

$$h_2 = \frac{\cos^2 \Delta}{\cos^2 \Delta_1} H_m \cos(2\psi - \kappa_m) + H_s \cos(2\psi_1 - \kappa_s) \qquad [\text{☽ and ☉}]$$

$$+ \frac{\cos^2 \delta' - \cos^2 \Delta}{\sin^2 \Delta_1} \cdot 683 H'' \cos(2\psi - \kappa'') \qquad [\text{Declination ☽}]$$

$$+ \frac{\cos^2 \delta_1 - \cos^2 \Delta_1}{\sin^2 \Delta_1} \cdot 317 H'' \cos(2\psi_1 - \kappa'') \qquad [\text{Declination ☉}]$$

$$- \frac{\sin 2\delta}{2\sigma \sin^2 \Delta_1} \frac{d\delta}{dt} \left[\frac{\cdot 683 H''}{\cos(\kappa'' - \kappa_m)} - H_m \tan^2 \Delta_1 \right] \sin(2\psi - \kappa_m)$$
$$[\text{Change of declination ☽}]$$

$$+ \frac{\cos^2 \Delta}{\cos^2 \Delta_1} (P' - 1) \frac{H_n \cos \kappa_n - H_l \cos \kappa_l}{e \cos \epsilon} \cos(2\psi - \epsilon) \qquad [\text{Parallax ☽}]$$

$$+ \frac{\cos^2 \Delta}{\cos^2 \Delta_1} \frac{dP/dt}{\sigma - \varpi} \left[4 H_m e - \frac{H_n}{\cos(\kappa_m - \kappa_n)} - \frac{H_l}{\cos(\kappa_l - \kappa_m)} \right] \sin(2\psi - \kappa_m)$$
$$[\text{Change of parallax ☽}]$$

where ϵ is an auxiliary angle defined by

$$\tan \epsilon = \frac{H_n \sin \kappa_n - H_l \sin \kappa_l}{H_n \cos \kappa_n - H_l \cos \kappa_l}$$

The above is in some respects a closer approximation than the expression from which it is derived, since hour-angles, declinations and parallaxes involve all the lunar and solar inequalities.

This formula may clearly be written in the form

$$h_2 = M \cos 2(\psi - \mu) + M_1 \cos 2(\psi_1 - \mu_1)$$

and the reader will easily perceive the forms which must be attributed to M, μ, M_1, μ_1.

In the equilibrium theory each H is proportional to the corresponding term in the harmonically developed potential, and such a proportionality is actually approximately true of tides of nearly the same speed.

If this law be assumed accurate for the purely solar tides S, R, T and solar K_2, we have

$$\frac{\cos^2 \Delta_1}{\sin^2 \Delta_1} \cdot 317 H'' = \frac{H_t - H_r}{3e_1} = H_s$$

whence
$$M_1 = P_1^3 \frac{\cos^2 \delta_1}{\cos^2 \Delta_1} H_s, \quad 2\mu_1 = \kappa_s$$

A similar synthesis of M cannot be legitimately carried out, because the several lunar tides differ considerably in speed, but the following partial synthesis would give fairly good results

$$M = P'^3 \frac{\cos^2 \Delta}{\cos^2 \Delta_1} H_m + \frac{\cos^2 \delta' - \cos^2 \Delta}{\sin^2 \Delta_1} \cdot 683 H'' \cos (\kappa'' - \kappa_m)$$

$$2\mu = \kappa_m + \frac{\cos^2 \delta' - \cos^2 \Delta}{\sin^2 \Delta_1} \cdot 683 \frac{H''}{H_m} \sin (\kappa'' - \kappa_m)^*$$

With whatever degree of accuracy M, μ, M_1, μ_1, be expressed the final semidiurnal synthesis is effected by writing

$$h_2 = H \cos 2 (\psi - \phi)$$

where

$$H = \sqrt{[M^2 + M_1^2 + 2 M M_1 \cos 2 (\alpha - \alpha_1 + \mu - \mu_1)]}$$

$$\tan 2 (\mu - \phi) = \frac{M_1 \sin 2 (\alpha - \alpha_1 + \mu - \mu_1)}{M + M_1 \cos 2 (\alpha - \alpha_1 + \mu - \mu_1)}$$

Since the excess of lunar over solar right ascension, namely, $\alpha - \alpha_1$, goes through its period in a lunation, and since the interval after moon's transit† up to high-water is determined by the value of ϕ, it follows that the "height" H and "the interval" are subject to "fortnightly or semi-menstrual inequalities."

Since spring tide occurs when $\alpha - \alpha_1 = \mu_1 - \mu$, and since the mean value of $\alpha - \alpha_1$ is $s - h$ and the mean value of $\mu_1 - \mu$ is $\frac{1}{2} (\kappa_s - \kappa_m)$, it follows that the mean interval after full and change of moon up to spring tide is $\frac{1}{2} (\kappa_s - \kappa_m)/(\sigma - \eta)$. The association of springs with full and change is so obvious that a fiction has been adopted whereby it is held that springs are generated at those configurations and take a certain time to reach the place of observation. Accordingly $\frac{1}{2} (\kappa_s - \kappa_m)/(\sigma - \eta)$ is called the "age of the tide." The average age of the tide seems to be about 36 hours. By analogy we speak of "the ages of the declinational and parallactic inequalities."

In computing a tide-table it is not practically convenient to use $\alpha - \alpha_1$, which refers to the unknown time of high water, but it is replaced by the difference of right ascensions at the time of moon's transit, which we may denote by A, then

$$\alpha - \alpha_1 = A + \frac{\frac{d}{dt} (\alpha - \alpha_1)}{\gamma - \frac{d\alpha}{dt}} \phi$$

---

* These formulæ have been used in an example of a tide-table in *Admir. Scientific Manual* (1886) [or Vol. I., p. 119].

† Lubbock, Whewell and others consider that the formulæ adapt themselves more closely to actuality when each high-water is associated with the second, third or even fourth preceding transit instead of the immediately preceding one.

With rougher approximation in which $\dfrac{d\alpha}{dt} = \sigma,\ \dfrac{d\alpha_1}{dt} = \eta,\ \phi = \mu$, it may be shown that

$$\alpha - \alpha_1 + \mu - \mu_1 = A - \mu_1 + \tfrac{30}{29}\mu$$

This approximate formula may be used in computing the fortnightly inequality in the height and interval.

We have supposed above that the declinational and parallactic corrections are applied to the lunar and solar tides before their synthesis, but a table of the fortnightly inequality may be based on the mean values H_m and H_s, and the corrections may be applied afterwards. This is the process usually employed, but it involves loss of accuracy, although there is some gain of convenience.

§ 29. *Synthetic method—Diurnal Tides.*

If it be assumed that the lunar portion of K_1 bears the same ratio to O as in the equilibrium theory, and that the solar portion is similarly related to P, it may be shown that the whole diurnal tide is expressed by

$$H_0\,\sqrt{2}\,\frac{\sin 2\delta'}{\sin 2\Delta_1}\cos\left(\psi' + \kappa'\right) + H_p\,\sqrt{2}\,\frac{\sin 2\delta_1}{\sin 2\Delta_1}\cos\left(\psi_1 + \kappa'\right)$$

where δ', ψ' are the declination and hour-angle of an ideal moon which follows the true moon in her orbit at a distance $\tfrac{1}{2}(\kappa' - \kappa_0)$. If the above assumption be not true there is another term also. A synthesis of these two diurnal tides may be carried out as with the semidiurnal tides.

Another form of synthesis together with an example of its use in computing a tide-table is given in a Report made to the British Association in 1885 and in the *Admiralty Scientific Manual* 1886 [Vol. I., pp. 70, 119].

It is perhaps not surprising that the treatment of the diurnal tides by the synthetic method has not been carried out very thoroughly, when we reflect that such an inequality is scarcely perceptible in the North Atlantic ocean. In 1836 W. Whewell wrote[*]:

"The existence of such an inequality (i.e. the diurnal) in the heights of high water has often been noticed by seamen and other observers,.... But its reality has only recently been confirmed by regular and measured observations, and its laws have never been correctly laid down...this inequality had never been obtained in numbers till the recent discussion of the Liverpool tides."

This passage shows in a striking manner how much our knowledge has advanced since the time of Whewell.

[*] *Phil. Trans.* (1836), p. 131.

§ 30.   *Reduction of tidal Observations of High and Low Water.*

Since all the older observations were of this kind, this section must be to some extent historical.

The observations which Laplace treated referred almost entirely to the port of Brest.   He had at his disposal complete observations extending from 1711 to 1716, and again from 1807 to 1822.   The process which he adopted for the determination of the numerical values of the several inequalities consisted in the selection of observations made at times near to certain configurations of the moon and sun.

Lubbock made an important advance in the inclusion of all the observations in the reduction.   His processes (of which a sketch is given by Airy in *Tides and Waves*) consisted in the formation of parcels of observations corresponding to the various elements for which corrections are needed. The greater part of the work relates to British Ports, more especially to London and Liverpool*.

Whewell† made important contributions to science in discussing the theory of Lubbock's numerical results and in adapting them to prediction. He also for the first time made extensive use of curves of fortnightly inequalities in time and height for reduction of observations and for the use of the results in prediction.   His methods remain the standard to which reductions of this kind still conform.

This treatment of the tides is in some respects more accurate than that of harmonic analysis, since it includes various small local peculiarities which can

---

* J. W. Lubbock's papers were as follows:—*Phil. Trans.* 1831, pp. 379—416 (tides of London); 1832, pp. 51—6 ; 1832, pp. 595—600 (tides of London) ; 1833, pp. 19—22 ; 1834, pp. 143—166 ; 1835, pp. 275—300 (Liverpool) ; 1836, pp. 57—75 (Liverpool) ; 1836, pp. 217—266 (London). Also (of less importance) *Phil. Mag.*, Vol. IX. (1831), pp. 333—5 ; Vol. VII. (1835), pp. 457—463 ; Vol. XI. (1837), pp. 195—6.   Also *Reports to Brit. Assoc.*, 1831—2, pp. 189—195 ; 1836, pp. 285—7.

The best account of his treatment of theory is contained in a pamphlet, *An Elementary Treatise on the Tides*, by J. W. Lubbock.   London, Knight, 1839, 54 pp.

† Whewell's papers are as follows :—*Phil. Trans.*, 1833, pp. 147—236 (Cotidal lines, see § 33) ; 1834, pp. 15—46 (London tides); 1835, pp. 83—90 (British Cotidal lines, see § 33 below); 1836, pp. 1—16 (Liverpool tides); 1836, pp. 131—148 (Diurnal inequality at Liverpool evaluated for the first time); 1836, pp. 289—342 (Cotidal lines); 1837, pp. 75—86 (Diurnal inequality at Liverpool and Singapore); 1837, pp. 227—244 (Diurnal inequality; abortive attempt at cotidal lines for this class of tide); 1838, pp. 231—248 (treatment of a short series of observations); 1839, pp. 151—161 (Plymouth tides) ; 1839, pp. 163—166 (Indian tides based on very imperfect observations); 1840, pp. 161—174 (Petropavlovsk tides); 1840, pp. 255—272 (Rise and fall of each day's tide); 1848, pp. 1—30 (Pacific tides) ; 1850, pp. 227—234 (British tides).   Also *Reports to Brit. Assoc.*, Vol. V. (Bristol), 1837, p. 285 ; Vol. VI. (Liverpool), 1838, p. 285 ; Vol. VII. (1839), p. 19 ; Vol. VIII. (1840), p. 18 ; Vol. IX. (1841), p. 436 ; Vol. X. (1842), p. 30 ; *Phil. Mag.*, Vol. XVII. (3rd series), 1840, pp. 321—5; *African Quarterly* (1835), pp. 367—372 (Tides at Cape of Good Hope).

only be represented accurately in the harmonic method by the inclusion of a number of small terms. Nevertheless the advantages of the harmonic method are on the whole so great as far to outweigh those to which we now refer.

Airy discussed* the methods of Laplace, Lubbock and Whewell in his *Tides and Waves.*

The harmonic constants are derivable from tidal observations of high and low-water by means of appropriate arithmetical methods†. But this kind of observation is not well fitted for the evaluation of the harmonic constants, and the computation is necessarily complex if accurate results are to be attained.

§ 31. *Tidal Prediction: Methods of forming Tide-tables.*

Up to the time of the initiation of the harmonic method, prediction was made by reference to the transit of the moon, the interval after transit and the height being derived from curves or tables of the fortnightly inequalities. Corrections were applied subsequently for the effects of the declinations and parallaxes of the moon and sun, and for the diurnal inequality‡. At places where the diurnal inequality is very large this method becomes laborious but it may still be used, if a sufficient number of curves be drawn§.

The older methods are satisfactory for places where the diurnal tide is small, and many tide-tables are still computed in that way. But the use of the harmonic system is continually growing in favour, and methods are needed of computing predictions direct from the harmonic constants.

The limitations of space preclude us from giving an account of the various methods which have been devised for numerical prediction, but references to several papers on the subject are given below‖.

In order to obviate the great labour involved in the arithmetical process, instruments have been devised for drawing a tide-curve and thus furnishing predictions.

The mechanical principle involved in the more important tide-predicters

---

* *Encyclopedia Metropolitana.*

† G. H. Darwin, *Proc. Roy. Soc.*, Vol. xlviii. (1891), pp. 278—340, and Vol. lii. (1893), pp. 388—9 [both in Vol. i., p. 157]; R. A. Harris, *Manual of Tides*, Part iii., App. 7, *U.S. Coast Survey Rep.* (1894), pp. 149—150.

‡ For an account of empirical methods of prediction see Whewell, *History of Inductive Sciences*, Vol. ii. (1873), p. 248, or in abstract, Darwin, *Tides*, p. 80.

§ G. H. Darwin, *Phil. Trans.*, Vol. 182, A (1891), pp. 159—229 [or Vol. i., p. 258].

‖ C. Börgen, *Ann. der Hydrog.*, Vol. xvii. (1889), pp. 1, 43, 89, 131; R. A. Harris, *Manual 3*, pp. 183—7; G. H. Darwin, *Admiralty Manual* (1886), p. 72 [or Vol. i., p. 141]; and *Phil. Trans.*, Vol. 182, A (1891), p. 159 [or Vol. i., p. 258]. [J. P. van der Stok has devised a method by which a fairly accurate tide-table may be computed direct from the harmonic constants. See " Elementaire Theorie der Getijden," *Kon. Nederlandsch Meteorologisch Instituut*, No. 102, 1910.]

is simple. A number of moving pulleys execute harmonic oscillations representing in phase and amplitude the several harmonic tides. A single cord passes round all the pulleys, and a pencil is attached to the end. The pencil sums up all the several harmonic motions and writes its record on a drum, which is geared to all the pulleys*.

Sir William Thomson (Lord Kelvin) appears to have been the first to suggest the application of this principle to tidal prediction in 1872, and the first instrument of the kind was constructed in about 1876 and is now in the South Kensington Museum. The second instrument was designed by E. Roberts for the Government of India. It sums 20 partial tides and it has for many years been in use for the preparation of the Indian tide-tables†; [it is now in 1910 at the National Physical Laboratory].

The Governments of France and of the United States have had instruments of a similar character constructed.

A tide-predicter of another type was invented by W. Ferrel and was successfully used in the United States. It determines only the maxima and minima of the sum of a number of harmonic terms without drawing the complete curve‡.

When a good tide-table has been computed for any place, it is often possible to predict the tides at a neighbouring port by reference to the predictions at the first. The commonest rule is to add or subtract a constant time to all the predictions, and to multiply the heights above and below mean sea level by a constant factor, but the application of such a simple rule often gives very erroneous predictions. W. Bell Dawson has applied this method extensively in making predictions for Canadian waters, and he has shown how corrections, variable according to the position of the moon, may be employed with advantage§.

* This mechanical principle owes its origin to F. Bashforth who read a paper entitled "A description of a machine for finding the numerical roots of equations and tracing a variety of useful curves" before the Brit. Assoc. (Cambridge) in 1845. The author has circulated copies in lithograph of the paper *in extenso*. The same idea also occurred independently to Mr W. H. Russell, *Proc. Roy. Soc.*, Vol. xviii. (1869), pp. 72—3.

† E. Roberts, "A new Tide-predicter," *Proc. Roy. Soc.*, Vol. xxix. (1879), pp. 198—201; W. Thomson, *The Engineer* (London), Dec. 19th, 1879; "Tidal Instruments" (together with discussions as to priority and as to other points), *Inst. Civ. Eng.*, Vol. lxv. (1881), pp. 15—72; also *Popular Lectures*, Vol. iii., p. 185; Thomson and Tait, *Nat. Phil.*, Vol. i. (ed. 1), p. 479; compare also Sir W. Thomson, *Popular Lectures*, Vol. ii., p. 185.

‡ An abridged account of the instrument is contained in Harris's *Manual*, Part iii., and G. H. Darwin, *Tides and Kindred Phenomena, etc.*; W. Ferrel, "A maxima and minima tide-predicting machine," *U.S. Coast Survey*, 1883, Appendix 10, pp. 253—272. Harris gives a short account of the instrument in *Manual 3*, p. 181; see also Darwin, *Tides*, p. 218.

§ *Survey of Tides and Currents in Canadian Waters*, especially for 1902. See also Harris, *Manual of Tides*, Part iii., Chap. v.

§ 32. *Errors of Tide-tables.*

If it were not for the instability of the meteorological elements, tide-tables might doubtless be constructed which would attain to any desired degree of accuracy. Under actual circumstances the amount of error depends on the length of the series of observations from which the tidal constants have been found, on the adequacy of the harmonic expression for the tide, and on the amount of instability in the wind and in the barometric pressure. In estuaries the tide-wave is generally much transformed under the influence of the shallowness of the water, so that in order to secure an accurate representation of the tide it would be necessary to introduce the over-tides of a number of tides for which they are commonly found to be negligible. Thus in some estuaries the predictions based on even 20 harmonic tides are subject to considerable error. It has been found possible, however, to improve such predictions by empirical corrections dependent on the hour of moon's transit, and on the moon's parallax and declination*. These corrections might of course be represented by a sufficient extension of the harmonic method, but the concomitant increase of labour in computing a tide-table would render this plan practically impossible.

Under these circumstances it seems useless to give an example of the actual errors experienced at any ports†.

Attempts have been made with various amounts of success to connect the barometric pressure and the wind with errors in prediction. P. Daussy found that the sea responds to the atmospheric pressure like a barometer, and variations of the height of barometer from the mean corresponded with inverted variations in the height of water in the ratio of 1 to 14·7‡. A table for correction of tidal predictions based on this result is given in the *Annuaire des Marées* of the French Government.

Lubbock and Airy gave results as to the effects of wind and barometric pressure. There is a great discrepancy between their conclusions and those of Daussy§.

Perhaps the most thorough investigation of this subject was carried out by Sir James Ross, when in 1848 his ships *Enterprize* and *Investigator* were embedded in the ice in 71° N. lat. and 91° W. long. The absence of wave-motion as the ice-sheet rose and fell afforded very favourable conditions for investigation during the long Arctic winter. He found that the factor was almost exactly the same as the ratio of the specific gravity of sea water to

---

* E. Roberts, *Great Trigon. Survey of India*, Vol. xvi. Chap. viii. (Dehra Dun, 1901).

† Samples of actual results are given in Chap. xiv. of Darwin, *Tides and Kindred Phenomena*, *etc.* Many comparisons between prediction and observation will be found in the *Reports of the United States Coast Survey* and of the *Survey of India*.

‡ *Conn. des Temps* (1834), pp. 85—7.

§ *Phil. Trans.* (1837), pp. 97—140. Airy (*Tides and Waves*, § 573) however states that both P. Daussy and J. W. Lubbock agree in the belief that the wind has no effect, and refers to *Phil. Trans.* (1832), p. 51; but obviously those observers changed their opinions.

that of mercury, namely, as 1 to 13·2. I imagine that this would be the best value to assume where nothing definite is known as to the law governing the tides at the port in question*.

W. N. Greenwood has attempted a much more elaborate form of correction which seems to have been pretty successful, at least for Glasson Dock, Lancaster†. He claims that this correction will virtually embrace the effects both of wind and pressure.

§ 33.   *Cotidal Charts.*

The motion of a progressive tide-wave in time and space may be exhibited graphically by a series of lines passing through the points at which mean high water at spring-tide occurs at XII, I, II...XI o'clock of Greenwich time. These cotidal lines present to the eye the march of the semidiurnal tide-wave. A similar chart would be needed for the diurnal wave, but the materials are altogether too scanty for even a preliminary sketch‡. Spring tide is chosen as the epoch because the lunar and solar tides are then in the same phase, and the compound tide must travel almost exactly as a purely harmonic semidiurnal tide would do.

In a chart of this kind it is virtually postulated that the tide-wave is single and progressive, for if in any partially closed portion of the ocean the motion consists of an oscillation about fixed nodes, the cotidal lines would merely represent the loops. It is probable that in many cases the tide consists in part of standing oscillations of this kind and in part of a progressive wave, so that the method of cotidal lines may become not merely defective but actually misleading§. The same difficulty arises when there are two sets of progressive waves crossing one another.

* *Phil. Trans.*, Vol. 144, Part II. (1854), pp. 285—296. J. Ross quotes J. W. Lubbock as quoting P. Daussy, and he concludes that P. Daussy found the ratio for Brest, as 1 to 16, and that J. W. Lubbock found it for London as 1 to 10. I am not clear how he arrived at these conclusions.

† W. N. Greenwood, Master Mariner, *Tides...in connection with atmospheric pressure*, Paper No. 30 of the 5th Session of the Shipmasters' Society of London, 1894. A short abstract is contained in *Quart. Journ. R. Meteorolog. Soc.* (London), Vol. XII. (1886), p. 283. See also an important paper by F. L. Ortt [*Nature*, Vol. LVI. (1897), p. 80], who has investigated the subject carefully, and refers to other writers. He seems also to have written in *Tijdschr. Kon. Inst. Ingenieurs* (1896—7), p. 117, and *Ann. der Hydrogr.* (1897), p. 27. See also F. Bubendey, *Zentralblatt der Bauverwaltung* (1895), p. 72; H. Mohn, *Norske Nordhavs Expedition* (1887); H. Hermann, *Marine Rundschau* (1897), p. 798; S. Günther, *Handbuch der Geophysik* (1899), Vol. II., p. 465.

‡ See W. Whewell, *Phil. Trans.* (1837), pp. 227—244, on diurnal cotidal lines. [But J. P. van der Stok has drawn a diurnal cotidal chart, as well as a semi-diurnal one, for the coasts of Java, Sumatra and New Guinea in *Kon. Meteorologisch Instituut*, No. 102, 1910.]

§ An attempt has been made by R. A. Harris (*Manual of Tides*, Part IV. a, Washington, 1902) to divide the ocean into regions in which the oscillations are not progressive. We do not however find his attempted solution of the problem satisfactory. See G. H. Darwin, *Nature*, Vol. LXVI. (1902), p. 444; S. S. Hough and R. A. Harris, Vol. LXXIII. (1905—6), pp. 228, 388. [Harris's theory is discussed by H. Poincaré in Vol. III. (1910) of *Leçons de Mécan. Cél., Théorie des Marées*, p. 364.]

It is only in narrow seas and near the coast that we may confidently regard the tide as progressive, so that it is only in these cases that the interpretation of cotidal lines is free from doubt.

In 1807 Dr Thomas Young discussed the march of the tide-wave round the world, and sketched the cotidal lines round England; but he made no allowance for the refraction of the waves in shallow water*. The subject was then almost untouched when Whewell attacked it about 1830. He collected data from every part of the world, and by aid of the British Admiralty caused special observations to be made at 666 places in Europe, America and Africa. The coordination of this enormous mass of material would alone suffice to place him in the first rank of investigators of the tides, and his conclusions seem to me to deserve more respect than has been accorded to them by some later writers.

W. Whewell's results are contained in two charts. The first of these relates to the world at large, but the Pacific Ocean is nearly a blank†, and the lines of the Indian Ocean are somewhat defective. The second chart relates to the seas surrounding Britain, inclusive of the North Sea; it may be accepted as substantially accurate, and little has been since discovered to improve it‡.

These maps may be studied in Airy's *Tides and Waves* better than in their original forms, since Airy incorporated improvements suggested by Whewell himself in later papers§.

W. Whewell regarded the Atlantic tide as a wave travelling from the south, but G. B. Airy pointed out that the Atlantic basin is large enough for the generation of its own proper tides and that they would be standing oscillations. He does not deny the existence of a progressive wave but considers W. Whewell's cotidal lines for the Atlantic as illusory, and thinks that we cannot as yet at all interpret their meaning.

W. Ferrel, going further than Airy, held that the Atlantic tides would be hardly affected if a barrier were erected across the ocean between Africa and South America. He held that the Atlantic tides are standing oscillations between east and west‖.

---

* T. Young, *Lectures on Nat. Phil.* (edit. of 1845), I., pp. 444—50, Vol. I. and Plate 38, Vol. II.

† C. Börgen seeks to explain the Pacific Tides, as far as they are known, from the consideration of the coexistence of two waves. See *Segelhandbuch für den Stillen Ozean*, Chap. XIII., p. 353.

‡ See however C. Börgen, "Gezeitenerscheinungen im dem Engl. Kanal, etc.," *Ann. Hydrog.*, Vol. XXVI. (1898), p. 414, and *Segelhandbuch der Südküste Irlands*.

§ Whewell, "Essay towards a first approximation to a map of cotidal lines," *Phil. Trans.* (1833), pp. 147—236. The papers on the British cotidal lines are in *Phil. Trans.*, 1835, pp. 83—90; 1836, pp. 289—342; Airy, "Tides and Waves," *Encyclop. Metrop.*, §§ 521—530, §§ 574—584.

‖ Ferrel, *Tidal Researches*, pp. 237 *et seq.*

C. Börgen does not admit that the supposed southern barrier would be without effect, and yet does not deny that oscillations transverse to that ocean may coexist with the progressive wave from south to north. By calculations as to the cotidal lines when two systems of progressive waves cross one another, he concludes that the main system in the Atlantic is a progressive wave from the southward. He thus supports Whewell's view in its main features, and I do not hesitate to express the opinion that of late years insufficient weight has been assigned to Whewell's great work [*].

Amongst later works are two papers by M. S. W. Jefferson on the tides of the coasts of Canada and the United States. The author draws the cotidal lines on these coasts more accurately than Whewell had done, and gives references to papers in the *Reports* of the United States Coast Survey [†].

The Indian Survey has published a cotidal map of the Indian Ocean from Aden to Singapore, in which the deficiencies and errors of Whewell's map are supplemented and corrected [‡].

A cotidal Chart founded on recent data is given in Berghaus's *Physical Atlas*. No attempt is made to join the lines across the main oceans. The explanatory note attached to the chart contains very scanty information as to the manner in which it was drawn up.

§ 34. *Tidal Currents—The Bore.*

These subjects fall under the heads of Hydrography and Hydrodynamics, and must accordingly be treated here very shortly.

The theory of long waves in shallow water shows that the current is in the direction of wave propagation when the water is higher than the mean, and the other way when it is lower than the mean. At the mouth of an estuary the sea is subject to a tidal oscillation, which generates a wave in the estuary and river. The energy of the motion is gradually annulled by friction during the journey of each wave up stream, so that there is a succession of waves travelling in one direction. If it were not for the current proper to the river, the tidal current would change its direction as the water level passes through its mean position; that is to say, the changes of current would occur at a quarter of the complete period later than the changes of level. In consequence of the proper current, however, the portion of the complete period during which the current runs down stream is more than

---

[*] C. Börgen, *Die Gezeiten im Nordlichen Atlantischen Ocean*, § 5; *Segelhandbuch, etc.*, *Deutsche Seewarte.*

[†] *National Geographical Magazine* (1898), Vol. IX., pp. 400, 465, 498.

[‡] The Chart was published in 1895. It bears the marks No. 388—S. 95; Reg. No. 159, S. I. D., April, 95—75.

half of the whole.   The moments at which the currents change in any actual river are of course dependent on the amount of water delivered by the river and on the configuration of the estuary.   In some estuaries, where wide extents of sand or mud are laid bare at low water, the reversal of current from down-stream to up-stream is retarded until extreme low water is reached, and the reversal takes place suddenly and violently in the form of a wall of water which rushes up the estuary.   This kind of tide-wave is called the Bore or Eager (German *Stürmer*, French *Mascaret* or *Barre*). After the passage of the bore the water level continues to rise rapidly but without violence for about half an hour.   There are however sometimes smaller bores which succeed the main one.   I am not aware that this remarkable phenomenon has been subjected to theoretical examination, and indeed it seems to surpass the powers of our present methods of treating fluid motion*.   Although the bore affords a worthy subject of observation, the descriptions of it are for the most part efforts of literature rather than of science.   I only know of two attempts to measure and examine a bore.

As a rule the bore only occurs at spring-tides and with certain winds, but in the Tsien-Tang river in China the bore not only occurs at every tide without exception, but is the largest of which we have knowledge.   This example was carefully examined in 1888 and 1892 by Captain Moore, R.N., when in command of H.M.S. *Rambler*.   His reports contain measurements and a vivid description of the phenomenon.   It may here suffice to say that he found the bore to be from 12 feet to 15 feet in height, and that its rate of progress was about 12 knots.

At a later date W. Bell Dawson made observations at Moncton on the Petitcodiac river in the Bay of Fundy of a bore only slightly less remarkable than the Chinese example.   By means of measurements of the rise of water and of the rate of progress of the bore he was able to draw its profile†.

The tidal currents in narrow seas, such as the English Channel, follow the same law as holds in a stagnant river, the change of current being retarded relatively to the change of water level.   The friction of the water on the shores also causes rotation of the tidal currents in the sea.   The currents past projecting headlands are termed tide-races.   The race of Portland attains a speed of seven knots‡.

F. A. Forel has suggested an explanation of the strange alternation of

---

* The hydrodynamical investigations of Sir W. Thomson in *Phil. Mag.*, Vol. xxii. (1886), p. 353, and Vol. xxiii. (1887), p. 52, have some bearing on this case.

† W. U. Moore, *Bore of the Tsien-Tang-Kiang*, 1888, and *Further Report*, etc., 1893, Potter, London.   An abstract of these Reports, together with certain previously unpublished pictures (presented to the author by Capt. Moore) will be found in G. H. Darwin, *Tides and Kindred Phenomena*, etc., Chap. iii.   W. B. Dawson, *Report of Tidal Department of the Survey of Canada*, 1898; an abstract of the same is given in *Nature*, Vol. lx. (1899), p. 291.

‡ A discussion of the currents of the English Channel and of the North Sea will be found in Airy, *Tides and Waves*, §§ 521—531.

regularity and of apparent caprice in the currents of Euripus which occur between the coast of Eubœa and the mainland of Greece*.

Information as to the currents of territorial waters is generally published in the sailing directions of the Hydrographic Departments of the several nations.

[J. P. van der Stok has applied harmonic analysis to tidal currents with some success. The two components of velocity, say northward and westward, play the same part as the height in the ordinary harmonic analysis; and the phases of the components of oscillation are referred to the time of transit of the fictitious satellite under consideration†.]

§ 35. *Tides of Lakes and of Land-locked Seas.*

If a lake be of such depth and size that a free wave would travel over the length of the lake several times in $12\frac{1}{2}$ hours (the period of the principal lunar tide), its tides may be evaluated from the equilibrium theory. In this case we may assume that the water will always stand at right angles to the instantaneous position of the vertical‡.

As regards the tide the Mediterranean Sea resembles two lakes, because the channel between Sicily and Tunis is too narrow for free tidal inter-communication. The Levantine portion is about 1100 miles in length and of such depth that a free wave in it will travel 300 or 400 miles an hour. Hence in this case the equilibrium theory must be true as a fair approximation. It appears that the ranges of the tide at Malta and on the coast of Syria should be about 6 inches, but this is slightly less than either of the observed ranges. The observed times of high water at the two ends also agree with theory§. But the tide of such a long inlet as the Adriatic cannot be found even approximately from this simple calculation. The range of tide at springs amounts to about 4 feet at Venice‖, and as this oscillation must be regarded as derived from the main Mediterranean tide the factor of kinetic augmentation must be of considerable magnitude. The peculiar tides of the Gulf of Mexico appear to be explicable by the fact that the Yucatan channel affords the only considerable entrance from the sea, so that the Gulf has almost the character of a lake.

* F. A. Forel, *Le Léman*, Vol. II., pp. 163—168.

† [J. P. van der Stok, "Études des Phénomènes de Marée sur les côtes néerlandaises" (Nos. II. and III.), *Kon. Nederlandsch Meteorologisch Instituut*, No. 90, 1905; reviewed by G. H. Darwin in *Nature*, Vol. LXXXIV. (1910), p. 144.]

‡ In applying the equilibrium theory it is necessary to bear in mind that the volume of water must remain constant. This condition is satisfied by choosing the fixed point about which the oscillations take place as the centre of inertia of the lake area. See R. A. Harris, *Manual of Tides*, Part II., p. 535. The corresponding correction to the equilibrium theory for the whole earth is given in Thomson and Tait's *Nat. Phil.*, Vol. I., Part II., § 808; see also § 6 above, p. 194.

§ R. A. Harris, *Manual of Tides*, Part I. (1897), p. 356.

‖ G. B. Airy, *Tides and Waves*, § 521.

The only fresh-water lakes in which (as far as I am aware) the tide has been well determined are Lakes Superior and Michigan, and the results are found to be in close accordance with theory. F. A. Forel failed to detect the tide in the Lake of Geneva*.

Lord Kelvin has determined the oscillations of rotating water in an infinite straight canal, and has suggested that the peculiar nature of the oscillations on the two banks of the canal is such as to throw light on the marked difference between the tides on the English and French shores of the English Channel. If this conjecture is correct that difference is therefore due to the Earth's rotation†.

CHAPTER D.   MISCELLANEOUS INVESTIGATIONS.

§ 36.   *Determination of the Moon's Mass from the Tides.*

The ratio of the mean lunar to the mean solar semidiurnal tides with a proper correction for the different motion in mean longitude of the two bodies, would afford the means of determining the moon's mass.

Laplace determined in this way the moon's mass from the observations at Brest and found it to be 1/74·946 of the Earth's mass‡.

W. Ferrel undertook a much more elaborate determination principally from the harmonic constants at several ports, with the following results for the ratio of the earth to the moon§:

Brest (H. and L.W. obs. from 1812 to 1830 inclusive) 78·0; with a second hypothesis, 77·4; and, following Laplace exactly, 77·1; by another method which follows the principles of Laplace he obtains 62·5. He remarks that this last result, as well as one of Airy's‖, condemns Laplace's principles.

Boston (19 years H. and L.W. observation discussed in U.S. Coast Survey, 1868¶) gives 81·7.

Liverpool (1857–60 and 1866–70, harmonic analysis) 73·3.

Portland, England (1851, 1857, 1866, 1870, harmonic analysis) 80·1.

Fort Point, California (1858 to 1861, harmonic analysis) 61·8.

Karachi (1868–71, harmonic analysis) 76·5; a second determination 78·6; a third determination 77·8.

It is obvious that the general tendency is to attribute too large a value to the moon's mass.

* R. A. Harris, *Manual of Tides*, Part II., p. 535; Ferrel, *Tidal Researches*, pp. 250—255; F. A. Forel, *Le Léman*, Vol. II., pp. 25—29.

† Sir W. Thomson, "On the Gravitational Oscillations of Rotating Water," *Phil. Mag.* (1880), p. 109. Also *Nature*, Vol. XIX. (1878), pp. 152, 571.

‡ *Méc. Cél.*, Liv. XIII., Chap. III., p. 230.

§ W. Ferrel, *Tidal Researches, U. S. Coast Survey*, 1874, pp. 185—236.

‖ *Phil. Trans.*, Vol. 135 (1845), p. 105.

¶ *U. S. Coast and Geodetic Survey*, 1871, p. 86.

§ 37.　*Elastic Tides and the Rigidity of the Earth**.

Lamé and Résal determined the strains of an elastic sphere when subject to any superficial forces, but Lord Kelvin treated the more general problem where the sphere is also under the action of the forces due to a potential expressible as a solid spherical harmonic function. His solution is applicable for finding the effects of tide-generating force or of rotation in an elastic gravitating sphere.

Although Lord Kelvin's solution is interesting, it is not easy to justify its applicability to the earth in the case where the elastic solid is compressible. If we conceive a homogeneous, isotropic, elastic and compressible sphere to become endowed with the power of gravitation, a state of internal strain would be induced. If the sphere had the dimensions and mass of the earth the strains would far exceed the elastic limits of any known material, and the homogeneity, and presumably also the isotropy would disappear in consequence of permanent set†. This consideration shows that homogeneity and isotropy are inadmissible hypotheses for representing the earth, if the material be compressible, and it must be so to some extent. Accordingly it

---

* The following are references to leading authorities on the subject:—Lamé, *Coordonnées curvilignes*, Paris, 1859, Leçons 17—19; H. Résal, abstract in I. Todhunter and K. Pearson's *History of Elasticity*, Vol. II., Part I., pp. 385—394. The original source is said to be "Sur les Équations de l'Élasticité et leur applications à l'équilibre d'une croûte planétaire," Paris, 1855, reproduced in *Traité élémentaire de Mécanique Céleste*, Paris, 1865. It does not seem to be in the edition of 1884 of the last work, and I have not been able to refer to the paper. Thomson and Tait, *Nat. Phil.*, 1883, Vol. I., Part II., § 834 to the end; and Thomson, *Phil. Trans.*, 1863, Part II., p. 583; A. Clebsch, *Crelle's Journ.*, Vol. LXI. (1863), pp. 195—262; C. Chree, an important series of papers in *Camb. Phil. Trans.*, Vol. XIV. (1883—89), pp. 250—369; Vol. XV. (1889—94), pp. 1—36, pp. 139—266, pp. 313—337, pp. 339—390; Vol. XVI. (1894—98), pp. 14—57, pp. 133—151; Vol. XVII. (1897—99), pp. 201—230; also *Quart. Journ. Math.*, Vol. XXI. (1886), pp. 193—208, Vol. XXIII. (1888), pp. 11—32; and a valuable discussion of the applicability of theory to geophysics in *Phil. Mag.*, Vol. XXXII. (5th series), 1891, pp. 233—252, pp. 342—353. A. E. H. Love, *Theory of Elasticity* (1892), Vol. I., Chaps. X., XI.; *Lond. Math. Soc.*, Vol. XIX. (1888), pp. 170—207, and *Camb. Phil. Trans.*, Vol. XV. (1890), pp. 107—118: certain errors in these papers are corrected in the *Elasticity*; [also *Proc. Roy. Soc.*, Vol. LXXXII., A (1909), p. 73]. I. Todhunter and K. Pearson, *History of Elasticity* (1893), Vol. II., Part II. (by Pearson), §§ 1721—6; and Pearson, *Quart. Journ. Math.*, Vol. XVI. (1879), p. 375; P. Järisch (giving reference to L. Henneberg in *Annali Matem.*, Vol. IX. (1879), p. 193), *Crelle's Journ.*, Vol. LXXXVIII. (1880), p. 131; G. H. Darwin, "On Stresses due to Weight of Continents," *Phil. Trans.*, Vol. CLXXIII. (1882), pp. 187—230, and a correction, *Proc. Roy. Soc.*, Vol. XXXVIII. (1885), pp. 322—328 [or Vol. II., p. 459]; P. Rudzki discusses the deformation of the solid earth by the weight of inland ice, *Bull. Acad. Sci. Cracovie*, April, 1899, pp. 171—215, Nov., 1899, pp. 445—468; J. H. Jeans, "The Vibrations and Stability of a Gravitating Planet," *Phil. Trans.*, Vol. 201, A (1903), p. 157; G. Herglotz, "Elasticität der Erde bei Berücksichtigung ihrer variablen Dichte," *Schlömilch's Zeitschrift für Math. und Physik*, Vol. LII. (1905), p. 273.

† Love, *Elasticity*, Vol. I., § 127; see also Chree, *Camb. Phil. Trans.*, Vol. XIV. (1883—89), p. 282, and *Phil. Mag.*, Sept. 1891, pp. 247—9, 250—1.

seems advisable to examine the results which would follow from the hypothesis of compressibility, although the conclusions must necessarily be speculative*. Now J. H. Jeans has pointed out that a spherical and concentric arrangement of the strata of equal density in a planet will be unstable if the compressibility of the material exceeds certain definite limits. He has thus been led to make some important contributions to the theories of cosmogony and he has passed on thence to formulate a theory of the asymmetry of the earth's figure. This investigation has suggested to Lord Rayleigh and to A. E. H. Love to make other very interesting contributions to the subject†.

However notwithstanding the difficulty in justifying the assumption of the incompressibility of the matter forming the earth, it seems to be well worth while to investigate the conclusions which result from the hypothesis that the solid sphere, while retaining its elasticity of form, is incompressible. As far as we can judge, the enormous pressures which must exist in the interior of the earth, at least in its present condition, should tend in the direction of increasing the resistance to further compression, and thus the abolition of the compressibility may not be so wide of the truth as would be the case if the matter possessed the same properties as it has in our laboratories.

We shall, then, henceforward suppose the strained sphere to be absolutely incompressible. In this case the strain due to normal superficial forces expressible by a surface spherical harmonic is exactly proportional to that due to a potential expressible by the corresponding solid harmonic. Under strain the surface of the body is no longer spherical, and the effect of the weight or deficiency of weight of the parts protuberant above or depressed below the true sphere is equivalent to the imposition of surface tractions acting on a true spherical surface. Thus the strains of an incompressible, gravitating, elastic sphere may be determined as though they were due to a potential acting throughout the whole mass, in the absence of any superficial forces.

If the undisturbed sphere of radius a be devoid of the power of gravitation; if the modulus of rigidity be n; and if the potential of the disturbing forces, estimated per unit volume of the sphere of density σ, be $\sigma \tau \rho^2 S_2$, where ρ is radius vector, S_2 a surface harmonic of the second order, and τ a constant; then Lord Kelvin's solution shows that the equation to the surface will be

$$\rho = a \left[1 + \frac{5\sigma a^2}{19n} \tau S_2 \right]$$

* Chree, *Camb. Phil. Trans.*, Vol. XVI. (1894—98), p. 142.

† J. H. Jeans, *Phil. Trans.*, Vol. 199, A (1902), p. 1; *ibid.*, Vol. 201, A (1903), p. 157; Lord Rayleigh, *Proc. Roy. Soc.*, Vol. LXXVII., A (1906), p. 486; A. E. H. Love, *Phil. Trans.*, Vol. 207, A (1907), p. 171 [*Proc. Roy. Soc.*, Vol. LXXX., A (1908), p. 553].

For the case of rotation ω, we have $\tau S_2 = \frac{1}{2}\omega^2(\frac{1}{3} - \sin^2 \text{lat.})$; in the case of tidal disturbance

$$\tau S_2 = \frac{3}{2}\frac{m}{r^3}[\cos^2(\text{moon's z.d.}) - \frac{1}{3}]$$

In the first case the axis of symmetry is that of rotation, in the second it is the moon's radius vector.

By a proper choice of the meaning of S_2 we may (as indicated above) apply this result so as to include the effects of gravitation, but Lord Kelvin obtains the same result synthetically.

Writing $$\mathfrak{r} = \frac{19n}{5\sigma a^2}, \qquad \mathfrak{g} = \frac{2g}{5a}$$

he shows that the ellipticities of the strained figure are $\tau/\mathfrak{r}$ for pure elasticity (as shown above); $\tau/\mathfrak{g}$ for pure gravitation; and $\tau/(\mathfrak{r} + \mathfrak{g})$ for elasticity and gravitation combined.

It follows that the ellipticity for elasticity and gravitation combined is to that for pure gravitation as unity to $1 + \mathfrak{r}/\mathfrak{g}$.

For rigidities equal to those of steel and glass $\mathfrak{r}/\mathfrak{g}$ is 2 and $\frac{2}{3}$ respectively. Hence in these cases the factors of reduction are $\frac{1}{3}$ and $\frac{3}{5}$.

These results are applicable only to the equilibrium tide. Now the gravest free periods of oscillation of an elastic sphere of the size and density of the earth, with moduli of rigidity of glass and of steel, are respectively two hours and one hour. Since also the gravest gravitational oscillation of such a liquid sphere is $1\frac{1}{2}$ hours, it follows that the equilibrium theory will be applicable with fair approximation[*].

One of the most interesting applications of this theory is to determine the amount of effect produced on the tides of an ocean due to the simultaneous yielding of the solid elastic planet. We have seen that the equilibrium tidal ellipticities of globes of glass and steel are respectively $\frac{3}{5}$ and $\frac{1}{3}$ of those of a liquid globe. If then we may neglect the strains produced in the elastic sphere by the varying pressure of the tidally disturbed ocean[†], we see that the tides of the ocean would have amplitudes equal respectively to $\frac{2}{5}$ and $\frac{2}{3}$ of the corresponding tide on an unyielding nucleus.

From this argument Lord Kelvin concludes that it is nearly certain from the observed magnitude of the oceanic tide that the tidal effective rigidity must be much greater than that of glass[‡]. But the researches on oceanic tides made subsequently to the time at which Lord Kelvin wrote tend to

---

[*] H. Lamb, "The Vibrations of Elastic Sphere and Shell," *Lond. Math. Soc.*, Vol. xiii. (1882), pp. 189—212, Vol. xiv. (1883), pp. 50—56; T. J. I'A. Bromwich, "Influence of Gravity on Elastic Waves...of an Elastic Globe," *Lond. Math. Soc.*, Vol. xxx. (1890), pp. 98—120.

[†] C. Chree has shown that the correction on this account may amount to about 10 or 20 per cent., *Camb. Phil. Trans.*, Vol. xvi. (1894—98), p. 141.

[‡] Thomson and Tait, *Nat. Phil.*, § 843.

show that we have but little power of determining the magnitude of the oscillations of oceans, limited by barriers of land, on an unyielding nucleus, and I am not disposed to regard the evidence of the semidiurnal oceanic tides as carrying with it more than a presumption in favour of a high degree of rigidity in the nucleus.

In the passage quoted above Lord Kelvin referred to the semidiurnal oceanic tide, and he then proceeded to argue, in accordance with Laplace's view, that in consequence of friction the oceanic tides of long period must obey the equilibrium law very closely.

Acting on this view I carried out for the second edition of the *Natural Philosophy* (§ 847 *et seq.*) a comparison with the equilibrium theory of the observed lunar fortnightly and monthly oceanic tides at 33 stations, and found them to have amplitudes of about two-thirds of the theoretical amount on an unyielding nucleus.

Subsequently W. Schweydar treated 194 years of observation of the fortnightly and monthly tides in the same way. The factor found by him is ·6*.

From this the conclusion was drawn that the solid earth has an effective rigidity equal to that of steel. But we have shown in § 12 that when the ocean covers the whole earth there is reason to dissent from Laplace's opinion that fluid friction will suffice to bring the tides of long period into accordance with the equilibrium theory. For some years, therefore, it appeared impossible to deduce the rigidity of the earth from tidal theory with any degree of exactness. At length however in 1903 Lord Rayleigh showed that when the ocean is interrupted by land, as in the case of the earth, those modes of fluid motion would be annulled which are the cause of the divergence between the dynamical and the equilibrium theories†. He thus explained how the tides of long period should conform themselves to the equilibrium law, but for a different reason from that assigned by Laplace. In thus reinstating Laplace's conclusion, he has at the same time reestablished the numerical evaluation of the rigidity of the earth as derived from tidal theory.

It is, in this connection, worthy of notice that the prolongation of the free Eulerian nutation, from 305 days to about 430 days as the effect of elastic yielding of the nucleus‡, led Hough to conclude that the earth must be somewhat stiffer than steel§.

---

* W. Schweydar, *Gerland's Beiträge zur Geophysik*, Vol. IX. (1907), p. 41.   See however § 21 above for a different and perhaps more trustworthy evaluation of the rigidity of the earth, agreeing exactly with this estimate.

† Lord Rayleigh, *Phil. Mag.*, Vol. V. (6th series), 1903, p. 136.

‡ S. Newcomb, *M. N. Ast. Soc.*, Vol. LII. (1892), pp. 336—341.

§ S. S. Hough, *Phil. Trans.*, Vol. 187, A (1896), pp. 319—344.

There are thus three lines of research, namely, the direct observation of the lunar attraction (see § 21), the theory of the tides, and the prolongation of the Eulerian nutation, which agree in giving a rigidity nearly equal to that of steel. It is proper to add that the velocity of propagation of earthquake waves points to a much higher rigidity. [A. E. H. Love and also C. Lalle-mand have, however, pointed out that these several lines of investigation really involve the elasticity of the earth in different ways, and they thus explain, at least in great measure, the apparent divergence of results*.]

We have seen above that the general effect of the yielding of the nucleus must be to reduce the oceanic tides, but a superficial yielding of the rocks of the sea bed and of an adjoining coast line may produce an apparent local magnification of the range of tide†.

§ 38. *Tides of the Atmosphere.*

Laplace has shown that the tides of an atmosphere resting on an ocean covering the whole planet are deducible from the tides of a homogeneous incompressible sea. He considers numerically the case where the height of the homogeneous atmosphere is 8·8 kilometres, and the depth of the uninter-rupted ocean is 17·6 kilometres. He finds that the combined lunar and solar semidiurnal tides at springs should give rise at the equator to a barometric oscillation of $0^{mm}·63$ of mercury. Such an oscillation would be within the range of careful observation. He also computes that this inequality of pressure would cause a wind of three-quarters of a centimetre per second, but this would necessarily escape detection by observation‡.

He subsequently discussed eight years (1815–23) of barometric observa-tion (presumably at Brest), and found that the range of the purely lunar semidiurnal tide was only $0^{mm}·0544$ of mercury, with maximum at $3^h 18^m$ of lunar time§.

J. J. A. Bouquet de la Grye has discussed "l'ensemble des observations de hauteurs barométriques, faites à Brest antérieurement à 1879" by grouping the mean solar hourly heights on each day according to lunar time, and he finds a mean range of $1^{mm}·3$ of water, or say $0^{mm}·1$ of mercury; but there are large irregularities, for the amplitude goes as high as $5^{mm}·8$ of water when lunar perigee coincides with maximum southerly declination. He also gives the data of other observers at Singapore, St Helena, Cape Horn and

---

* [*Proc. Roy. Soc.*, Vol. LXXXII., A (1909), p. 73; *Annuaires du Bureau des Longitudes*, 1909 and 1910.]

† G. H. Darwin, "Lunar Disturbance of Gravity," *B. A. Rep.* (1882), p. 472 [or Vol. I., p. 457].

‡ *Méc. Cél.*, Book I., § 37, Book IV., Chap. V. (pp. 117—124, Vol. I., pp. 310—314, Vol. II., *Œuvres Compl.*).

§ *Méc. Cél.*, Book XIII., Chap. VII. (p. 262, Vol. V., *Œuvres Compl.*).

Batavia. It is remarkable that Bouquet finds a retardation of phase of only about half an hour, whereas Laplace found it to be 3^h*.

By far the larger portion of the periodic oscillations of the barometer are of thermal origin; although this subject falls outside the scope of the present article, yet we may mention that A. Angot gives the following formula for the diurnal oscillations of the barometer:

$$0^{mm} \cdot 926 \, \frac{h \cos^2 \delta}{760 r^2} \cos^4 \lambda \cos (2m + 64^\circ) \text{ mercury}$$

where r, δ, m are the sun's radius vector, declination and apparent hour angle, λ the latitude, and h the mean pressure in mm. of mercury at the place of observation†. A small portion of this oscillation must be due to the solar astronomical tide.

The maxima occur at 10 a.m. and 10 p.m., and Lord Kelvin has remarked that the solar attraction on this accelerated tide must cause a minute thermo-dynamic secular acceleration of the earth's rotation‡.

The retarded lunar astronomical tide must cause a still smaller secular retardation of the rotation.

§ 39. *Precession and Nutation.*

If free oscillations exist in an ocean covering a planet the resultant moment of momentum of the system remains unchanged, but the oscillation of the ocean must impart oscillation to the solid nucleus. The mass of our ocean being a very small fraction of that of the whole planet, the disturbance thus produced in the rotation of the planet is negligible. If now an external disturbing force acts on the system the resultant moment of momentum is unchanged by the interaction between sea and earth, and the precessional and nutational couples are the same as though sea and earth were instan-taneously and rigidly connected. The additions to the couples are therefore due to the attractions of the moon and sun on the excess or deficiency of water above or below mean sea level. These additions are obviously excessively small compared with the couples due to the attractions of the disturbing bodies on the equatorial protuberance, and therefore precession and nutation take place sensibly as though the sea were congealed in its mean position§.

* "Ondes Atmosphériques," *Ann. Bureau des Long.* (1895), pp. A. 1—A. 20. The paper contains much matter of interest outside our topic.

† "Marche diurne du Baromètre," *Ann. Bureau Central Météor. France* (1887), pp. B. 237—B. 244.

‡ *Société de Physique*, Sept. 1881, p. 200, or *Edinb. Roy. Soc.*, 1881—2, p. 396; see also Thomson and Tait, *Nat. Phil.*, § 830.

§ G. H. Darwin, "Tides," *Encyc. Brit.*, § 20; *Phil. Trans.*, Vol. 167, Part I. (1877), pp. 271—311 (appendix by Thomson) [or Vol. III., p. 1]; W. Thomson, Address to British Association, 1876; *Trans. Geol. Soc. Glasgow*, 1874, Vol. XIV., p. 312; G. H. Bryan, "Precession and Nutation of a Liquid Planet," *Phil. Trans.*, Vol. 180, A (1889), pp. 187—219.

§ 40.   *The Latitudinal or Eulerian Tide.*

The ocean cannot instantaneously partake of the motion of the solid planet when moving with a free Eulerian nutation, whether the period of that nutation be prolonged by elastic yielding or not. It follows that we have here a source of tidal movement of a wholly different kind from all others considered hitherto. I have called it the latitudinal tide, because the prolonged Eulerian nutation manifests itself as a variation of latitude with a period of 427 days. Since the radius of the circle described by the instantaneous axis of rotation about the pole of the earth is only about $4\frac{1}{2}$ metres, the tide in question must be very small. H. G. Bakhuyzen has sought to detect it from tidal observations made on the coast of Holland, and A. S. Christie from the tides observed in the United States. Their results are fairly consistent with one another and with the theoretical phase and amplitude which would correspond with the variations of latitude. But I do not feel convinced that either author has taken sufficient pains to clear the daily means of the height of the water from the residual effects of the tides of short period, and as there are irregular and unexplained variations of sea level of many times the magnitude of the tide to be detected it seems quite possible that accident may have led to this agreement. In any case we cannot feel any confidence in the reality of the detection of this latitudinal tide *.

CHAPTER E.   TIDAL FRICTION AND SPECULATIVE ASTRONOMY.

§ 41.   *History.*

In 1754 Kant wrote as follows :—

"The water of the ocean covers at least one third (*sic*) of its [the earth's] surface and is kept in continual motion by the attraction of...[the sun and moon]. Moreover this motion is in one direction, exactly opposed to the rotation [of the earth]....Since this flow is opposed to the rotation of the earth, we have here a cause which may be counted upon to retard and diminish rotation continuously to the extent of its capacity....The termination of this change of rotation will occur when the earth's surface, from the point of view of the moon, shall be relatively at rest, i.e., when it rotates in the same time as the moon revolves. If it were fluid throughout the moon's attraction would soon bring its rotation down to this fixed remainder. This immediately reveals the cause which has compelled the moon in its journey round the earth always to expose to it the same aspect....The attraction

* H. G. van de Sande Bakhuyzen, *Ast. Nach.*, Vol. 136 (1894), p. 33 ; A. S. Christie, *Phil. Soc. Washington, Bull.*, Vol. XII. (1895), p. 103.

which the earth exerts on the moon, acting on the satellite while it was still fluid, must have reduced the rotation of the moon (formerly no doubt greater than now) to this fixed residue in the manner just explained *."

This statement of the results of tidal friction is very clear, but there is no allusion to the effects of tidal reaction. One of the sentences in this passage amounts to the argument that if the earth were not solid, internal tidal friction would long ago have reduced its rotation to identity with the moon's orbital motion. In another paper Kant distinctly says that the planets are solid throughout †.

Laplace in his *Système du Monde* refers the present identity of the moon's orbital and rotational motions to the effects of tidal friction‡.

In 1848 J. R. Mayer said in his paper on Celestial Dynamics that " the tidal wave causes a diminution of the velocity of the rotation of the earth§." But the reaction on the moon was not considered by him. Helmholtz in 1854 discussed the tidal retardation both of the moon and of the earth‖.

In 1853 W. Ferrel considered the problem of tidal friction and included the reaction on the moon. As he believed that Laplace had fully explained the secular acceleration of the moon's motion, he supposed the tidal retardation to be compensated by the acceleration due to the contraction of the earth in cooling¶.

W. Ferrel again returned to the subject in 1865, and showed that a libration of the moon would ultimately result from the tidal retardation of the moon's rotation. He pointed out that the effect of the terrestrial tides in the moon must be $562\frac{1}{2}$ times as great as that of the lunar tides in the earth **.

In the same year C. Delaunay adduced tidal friction in explanation of that portion of the secular acceleration which Laplace's lunar theory, as corrected by J. C. Adams, left unexplained. Delaunay did not mention tidal reaction in this first paper, but in a later paper he said he had not

---

* Translation by G. F. Becker in an admirable article entitled "Kant as a Natural Philosopher" (*Amer. Journ. Sci.*, Vol. v., Feb. 1898, p. 97) of Kant's paper, " Ob die Erde in ihrer Umdrehung...einige Veränderung...erlitten habe," Vol. i., pp. 179—186 of collected works (by Hartenstein, 1868). I take other interesting facts from Becker's article.

A translation into French of Kant's " Histoire Naturelle générale et Théorie du ciel " is given in C. Wolf's *Hypothèses Cosmogoniques*, Gauthier-Villars, 1886.

† G. F. Becker, *loc. cit.*, quotes " Ob die Erde veralte, etc.," *Kant*, Vol. i., pp. 187—206.

‡ Note vii., p. 551, Sixth Edit., Bruxelles, 1827. G. F. Becker, *loc. cit.*, says this passage was first inserted in the revised edition of 1824, being absent from earlier ones.

§ Translation of " Dynamik des Himmels," Heilbronn, 1848, in *Phil. Mag.*, Vol. xxv. (4th series), 1863, p. 403.

‖ *Vorträge und Reden*, Vol. i., p. 79 (fifth ed.), Brunswick, 1903. English translation, 1893, p. 167. He incorporates the conclusions of C. Delaunay, J. C. Adams, and W. Thomson.

¶ *Gould's Astron. Journ.*, Vol. iii. (1853), p. 138.

** *Amer. Acad. Arts and Sci.*, Vol. vi. (1862—1865), pp. 379 and 390.

overlooked it, and made a slight correction to a note by J. Bertrand who had in the meantime drawn attention to tidal reaction*.

In 1867 E. J. Stone considered the effect of tidal friction on the obliquity of the ecliptic. His hypothesis as to the nature of the frictional couple is not satisfactory, and he failed to integrate his differential equations. He concluded that the effect was practically insignificant†.

The problem of the actual amount of tidal retardation of the earth's rotation is discussed in Thomson and Tait's *Natural Philosophy*‡.

§ 42. *General Consideration of Tidal Friction.*

The most important effects of tidal friction may be deduced by considering the energy of the system.

If special units of mass, length, and time be chosen, the moment of momentum of the planet's rotation may be taken as equal to n, its angular velocity of rotation; and its kinetic energy of rotation as equal to $\frac{1}{2}n^2$. With the same units, if x^2 be the radius vector of the satellite, the moment of momentum of the motion in a circular orbit is x; the potential energy of the planet and satellite is $-1/x^2$; and the kinetic energy of the orbital motion is $+1/2x^2$.

Thus if h be the moment of momentum, and e the energy of the system

$$h = n + x$$

$$2e = n^2 - \frac{1}{x^2}$$

The meaning of these equations may be discussed graphically, and the fact that tidal friction must reduce the energy enables us to see how the system will change.

The maxima and minima of energy are given by $\dfrac{de}{dx} = 0$, which leads to the quartic

$$x^4 - hx^3 + 1 = 0$$

This equation also represents the fact that the system revolves as a rigid body.

If h is greater than $4/3^{\frac{3}{4}}$ or $1\cdot755$ there are two real roots, one greater and the other less than $\frac{3}{4}h$. The smaller root corresponds to a maximum of energy, the larger to a minimum. Under tidal friction the system will degrade from the maximum to the minimum configuration.

---

* C. Delaunay, *C. R.*, Vol. LXI. (1865), pp. 1023—1032, and *C. R.*, Vol. LXII. (1866), pp. 197—200; J. Bertrand, *C. R.*, Vol. LXII. (1866), p. 162.

† *M. N. R. A. S.*, Vol. XXVII. (1867), p. 192.

‡ In the first edition of the *Natural Philosophy* the details of Adams' calculation were not given, but in the second edition the calculation was verified by G. H. Darwin in Appendix G (a), p. 503, and reference is made to Newcomb's discussion of ancient eclipses in his *Researches on the Moon*, Washington, 1878, Part I., pp. 13 and 280 [see Vol. II., p. 75].

In the course of the change the number of rotations made by the planet during the revolution of the satellite, that is to say the number of days in the month, rises from unity to a maximum and falls again to unity.

In the case of the moon and earth the configuration of maximum energy is when the moon nearly touches the earth, and the maximum number of days in the month is about 29. We have already passed through that maximum *.

The application of this result to the explanation of the origin of the moon is obvious.

§ 43. *The Tides of a Viscous Spheroid*†.

In a complete investigation of the effects of tidal friction it is necessary to adopt some definite theory of frictionally resisted tides. Since the earth was probably once wholly or partially plastic, the tides of a viscous spheroid will be of interest.

The problem for an incompressible viscous spheroid is virtually the same as that of the strain of an incompressible elastic sphere. The latter has been solved by Lamé and Thomson (Lord Kelvin), and merely needs adaptation to give the new result‡.

On neglecting the effects of inertia the solution is as follows :—

Supposing a periodic tide-generating potential $S \cos vt$, of the second order of harmonics, to perturb a homogeneous planet of density w, radius a, modulus of viscosity v and gravity g, the equation to the surface at time t is

$$r = a \left[1 + \frac{5a}{2g} S \cos \epsilon \cos (vt - \epsilon) \right]$$

---

* G. H. Darwin, "Secular Effects of Tidal Friction," *Proc. Roy. Soc.*, Vol. xxix. (1879), pp. 168—181; portions of the same are given in Thomson and Tait, *Nat. Phil.*, Part ii., Vol. i., Appendix G (b), and *Encyc. Brit.*, "Tides." The contours of the energy surface for an elliptic orbit are given in *Phil. Trans.*, Vol. 171, Part ii. (1880), p. 890 [see Vol. ii., pp. 195, 378].

† This and the following sections are founded on the following papers by G. H. Darwin in the *Phil. Trans.* :—"The Bodily Tides of Viscous and Semi-elastic Spheroids, etc.," Part i., Vol. 170 (1879), pp. 1—35. The appendix to this paper is superseded by § 848 of Thomson and Tait's *Nat. Phil.* (1883). {See also H. Lamb, "On the Oscillations of a Viscous Spheroid," *Lond. Math. Soc.*, Vol. xiii. (1881), pp. 51—66.} "The Precession of Viscous Spheroids, etc.," Part ii., Vol. 170 (1879), pp. 447—528. "Problems connected with the Tides of a Viscous Spheroid," Part ii., Vol. 170 (1879), pp. 539—593. "Elements of the Orbit of a Satellite, etc.," Part ii., Vol. 171 (1880), pp. 713—891. The vernal equinox is inadvertently described throughout as the autumnal equinox, but no error is caused thereby. "Planet attended by several satellites, etc.," Part ii., Vol. 172 (1881), pp. 491—535. "Stresses caused by weight of Continents, etc.," Part i., Vol. 173 (1882), pp. 187—230; together with a correction in *Proc. Roy. Soc.*, Vol. xxxviii. (1885), pp. 322—8. "Analytical expressions which give the history of a Fluid Planet attended by a single Satellite," *Proc. Roy. Soc.*, Vol. xxx. (1880), pp. 255—278. In several cases the abstracts of these memoirs (in *Proc. Roy. Soc.*) contain expositions whereby the nature of the results obtained analytically is explained by general reasoning. [These papers are reproduced in Vol. ii.]

‡ See § 37 above.

where
$$\tan \epsilon = \frac{19vv}{2gaw}$$

The height of tide is therefore equal to the equilibrium tide reduced by the factor $\cos \epsilon$, and the phase is retarded by ϵ.

The oceanic tides relatively to the viscous nucleus are equal to those on a rigid nucleus reduced by the factor $\sin \epsilon$, and the time of high-water is apparently accelerated by the time $(\tfrac{1}{2}\pi - \epsilon)/v$.

§ 44. *Nature of the Problem of Tidal Friction and its Subdivision.*

Each constituent partial harmonic tide suffers frictional reduction of amplitude and retardation of phase, differently according to its speed. These are determinate for the viscous spheroid, and as this is the only complete theory of frictionally resisted tidal oscillations, it is well to follow it to its logical conclusion. Any other theory would lead to closely similar results, and accordingly the value of the conclusions will be wider than that attaching to the hypothesis of viscosity.

The problem to be considered may be stated as follows :—

A planet is attended by one or more satellites which generate frictionally resisted tides in it; it is required to find the secular changes in the elements which define the motions of the satellites and of the planet itself.

I shall for brevity speak of the planet as the earth and of the satellites as the moon and sun, although the discussion of the applicability of the results to cosmogony will be omitted. The latter aspect is considered in the memoirs referred to above and elsewhere*.

The problem may be treated either, by means of the perturbing couples and forces which affect the planet's rotation and the motion of the satellite, or otherwise by the method of the disturbing function.

In the method of the disturbing function the differentiation with respect to the elements is an artifice to avoid the determination of the components of disturbing force by differentiation with regard to the coordinates of the body whose motion is disturbed. Hence it follows that although there may actually be only one satellite, yet in the expression for the disturbing function we must suppose that there are two, namely, one a tide-generating satellite and another a disturbed satellite.

If we give to the planet the name of the earth, the tide-raising satellite may be conveniently called Diana, and the satellite whose motion is disturbed

---

* See for examples, G. H. Darwin, *Tides and Kindred Phenomena*, and *Encyc. Brit.*, " Tides."

may be called the moon. When the differentiations have been made, Diana may be made identical with the moon or with the sun at will. Similarly a double name might be adopted for the earth in its two capacities as the subject of the tide and as a rotating·body, but it may suffice to speak of the tidal earth and of the rotating earth.

The first task is to determine the form of the tidal earth under the influence of Diana. This is effected by replacing the pair of "tidal constants" of § 25 for each oceanic tide by constants determined for the whole planet from the theory of the viscous spheroid. The potential of this retarded equilibrium tide at an external point is then expressible by means of spherical harmonic analysis. It involves the three coordinates of Diana relatively to the tidal earth and the three angular coordinates of the tidal earth relatively to a fixed plane and axes, but it also involves the three coordinates of the external point at which the potential is determined relatively to the rotating earth, and the angular coordinates of the rotating earth relatively to a fixed plane and axes. If now the external point be made to coincide with the moon and if the potential be multiplied by the moon's mass, we have a force function for the moon and the rotating earth expressible by the sum of a series of terms, each of which consists of a product of two factors, which may be described as positional and tidal. One of these factors involves the positions of the moon and of the rotating earth, and the other involves the positions of Diana and of the tidal earth; the positional factor involves the coordinates of the moon and of the rotating earth in the same form as the tidal factor involves those of Diana and of the tidal earth, save that the tidal retardation and reduction of amplitude are introduced in the latter factor.

This force function serves as the disturbing function for the motion of the moon as well as for that of the rotating earth. Since it involves the product of the masses of the moon and of Diana and the product of the reciprocals of the cubes of their mean distances, it is said to depend on the square of the tide-generating force.

The equations of variation of elements involve the differentials of this function with respect to the elements of the moon and of the rotating earth, and after differentiation the tidal earth is made identical with the rotating earth, and Diana is identified with the moon or with the sun. It is obvious that the disturbing function will contain a number of terms which are not only periodic in time, but also retain their periodic character after the identifications of Diana with the moon and of the tidal with the rotating earth. These terms give rise to periodic inequalities of negligible magnitude. But there are other terms which cease to be periodic when the tidal and rotating earth are identified, and others which lose their periodicity when, in addition to this first identification, Diana is identified with the moon. These terms give rise to the cumulative secular changes in the elements which it is our object to determine. When the tidal and rotating earth are identified whilst Diana becomes the sun, we obtain the effects due to the action of the moon on the solar tides. As regards the earth's rotation we find that the solar and lunar elements are involved symmetrically, and it becomes obvious that the effect of the sun on the lunar tide is equal to that of the moon on the solar tides; hence we may take this latter effect into account by simply doubling the former. But as regards the moon's motion, the effect of the attraction of the retarded solar tide is *nil*, since this reaction affects only the orbital motion of the earth round the sun.

Again the earth's rotation is affected by the action of the sun on the solar tides; but this need not be specifically evaluated since it may be obtained at once by symmetry from the effect of the moon on the lunar tides. The solar tides react on the sun, but the effect on the earth's orbital motion is negligible.

We thus conclude that the earth's rotation is affected in four ways:—

(1) By the attraction of the moon on the lunar tides.

(2) By that of the sun on the solar tides, determinable by symmetry with (1).

(3) By the attraction of the moon on the solar tides. This effect is equal in amount to

(4) The effect of the attraction of the sun on the lunar tides.

The result may, therefore, be obtained by dropping the fourth mode and doubling the effect of the third.

The moon's orbital motion is only affected in two ways, namely,

(1) By the attraction of the lunar tide on the moon; and

(2) By the attraction of the solar tide on the moon.

Little interest attaches to those elements of the moon's orbit and of the earth's rotation which specify the positions which the bodies would occupy at a given instant of time, for we are merely concerned with those elements which describe the natures of the moon's orbit and of the rotation of the earth.

These latter elements are the mean distance, inclination and eccentricity of the lunar orbit, and the obliquity and diurnal rotation of the earth. Moreover, we only require to trace the secular *changes* in all these five elements whilst we may neglect both those sets of *inequalities* which are usually called secular and periodic. It is, however, unfortunately impossible to direct the investigation strictly according to these considerations. Amongst the ignored elements are the longitudes of the nodes of the orbit and of the equator on the fixed plane, and in one part of the investigation it is found necessary to take into account secular *inequalities* both in the five elements and in the motions of the two nodes.

The whole problem is one of great complexity, but nothing of importance is lost if it be subdivided into two simplified problems. These are:—

(1) Where the moon's orbit is circular, but inclined to the ecliptic.

(2) Where the orbit is eccentric, but always coincident with the ecliptic.

§ 45. *Problem where the Lunar Orbit is circular but inclined.*

It appears that the problem requires further subdivision, for the following reasons:—

It is known that the orbit of a satellite perturbed both by planetary oblateness and by another satellite maintains a constant inclination to a certain plane, which is said to be *proper* to the orbit; and that the nodes of the orbit on the plane revolve with a uniform motion, subject however to periodic inequalities of no present interest.

If then the moon's proper plane be inclined at a very small angle to the ecliptic, the nodes revolve nearly uniformly on the ecliptic, and the orbit is inclined at a nearly constant angle thereto. In this case the equinoctial line also revolves nearly uniformly, and the equator is inclined at nearly a constant angle to the ecliptic.

Here then any inequalities in the motion of the earth and moon, which depend on the longitudes of the node or of the equinox, are periodic in time (although they are secular inequalities), and cannot lead to any cumulative effects which will alter the elements of the earth and moon.

Again suppose that the moon and earth are the only bodies in existence. Here the normal to the invariable plane remains fixed in space, and the earth's axis and the normal to the lunar orbit must always be coplanar with the normal to the invariable plane, so that the orbit and equator must have a common node on the invariable plane. This node revolves with a uniform precessional motion, and, so long as the earth is rigid, the inclinations of the orbit and equator to the invariable plane remain constant. Here also inequalities which depend on the longitude of the common node are periodic in time, and can lead to no cumulative effects.

But if the lunar proper plane be not inclined at a small angle to the ecliptic, the nodes of the orbit may either revolve irregularly, or may oscillate about a mean position, which will itself be subject to a slow precessional motion. In this case the inclinations of the orbit and equator to the ecliptic may oscillate considerably. Here then inequalities which depend on the longitudes of the node and of the equinoctial line are not simply periodic in time, and may and will lead to cumulative effects.

This explains what was stated above, namely, that we cannot entirely ignore the motion of the two nodes.

The problem is thus divisible into three cases :—

(i)   Where the nodes revolve uniformly on the ecliptic, and where there is a second disturbing satellite, namely the sun.

(ii)   Where the earth and moon are the only bodies in existence.

(iii)   Where the nodes either oscillate, or do not revolve uniformly.

The cases (i) and (ii) are distinguished by our being able to ignore the nodes. In none of the three cases can the tides generated by any one satellite produce directly a secular change in the mean distance of any other satellite.

In cases (i) and (ii) the tides generated by any one satellite cannot produce any secular change in the inclination of the orbit of any other satellite to the plane of reference. This is not true of case (iii).

Before proceeding further it may be well to remark that we shall have occasion to describe several of the elements as possessing stability or instability. But these terms are not in strictness applicable, because, if any element is momentarily stable or unstable, the simultaneous changes in all the other elements will upset the permanence of that characteristic. The expression is however convenient and sufficiently explanatory.

We will now consider case (i) where the moon's node revolves uniformly on the ecliptic, and will state the results as to the rates of variation of the several elements.

The obliquity of the ecliptic in general increases with the time. But if, with the month of its present length, the obliquity of the ecliptic were about 87°, it would diminish. There is therefore a position of stability for a very large obliquity. The angle of obliquity which corresponds with the position of stability diminishes as the moon's periodic time diminishes, and so long as the month is greater than twice the day zero obliquity is still a position of instability. But if the month is less than twice the day zero obliquity becomes stable. These results refer to the hypothesis that the amount of viscosity is small.

Reverting to the case of the two disturbing bodies, it appears that the combined effect of sun and moon operates adversely to the separate effects of each.

The tidal friction will always cause the earth's diurnal rotation to diminish. The separate and combined effects cooperate, but the tides of long period have no effect. All the results which we are now stating depend on tidal friction, but it is convenient to reserve this name especially for the effect on the diurnal rotation.

The tidal reaction on the moon is expressed by the rate of increase of the square root of the moon's distance. The change in this element is in general positive, although the retarded tides of long period tend to make the moon approach the earth. The combined effect of sun and moon is evanescent, and accordingly no satellite can change the mean distance of any other satellite by means of its tides.

The inclination of the lunar orbit to the ecliptic will diminish unless the obliquity of the ecliptic be very large, but if this were the case we could no longer assume the lunar nodes to revolve uniformly on the ecliptic. This increase of inclination will always hold good so long as the month is a considerable multiple of the day, but the tendency may be reversed if that condition is not fulfilled.

The analytical results expressive of case (ii) where the moon and earth are the only bodies are easily derivable from those of case (i) by putting the sun equal to zero and by referring the motion to the invariable plane instead of to the ecliptic.

§ 46. *Problem where the Lunar Nodes oscillate or revolve irregularly.*

This is the case denoted by (iii) above. The geometrical relationships of the several parts of the system are very complex, even when we regard the obliquity of the ecliptic and the inclination of the orbit as small, and we cannot undertake to explain the peculiar difficulties of the case within the limits of space at our disposal.

It must suffice to state that it is necessary to refer the lunar orbit to a "proper plane," and the earth's motion to another "proper plane." There are thus four inclinations and four nodes, namely the inclination and node of each plane on the ecliptic and the node and inclination of the orbit and of the equator on their respective proper planes. Of these eight angles only four are independent, two of the inclinations and two of the nodes being deducible from the other pairs. It may be shown that this system of planes and nodes reduces on the one hand to the case with which we are familiar in the case of the earth and moon, where the effect of the planetary oblateness is small; and on the other hand serves to represent such motion as we observe in the satellites of Mars when the effect of oblateness is large.

When the moon was near the earth the effect of planetary oblateness was large, because the earth was very oblate and the satellite very near. Thus in tracing the moon's history in retrospect we have to pass from the one extreme case to the other. It is known that the kind of approximation requisite for the lunar theory differs from that appropriate to the theory of the other satellites, and the peculiar difficulty of the present problem resides in the fact that we have to adopt the same treatment for all such cases.

The four elements which define the system sufficiently for solving the problem in hand are (1) the angular velocity of rotation of the earth, (2) the square root of the moon's radius vector, (3) the inclination of the earth's proper plane to the ecliptic, and (4) the inclination of the lunar orbit to its proper plane. All the other elements may be set aside, since they are either deducible from these, or else are merely required for determining the position of either body at a given moment of time.

The differential equations specifying the rates of change of the four elements under the influence of tidal friction, may be determined by the method of the disturbing function. It is then possible to find an approximate form of the equation of conservation of moment of momentum, as modified by the effects of solar tidal friction. By means of this we determine the length of day for any assumed length of month, or (as comes to the same thing) for any assumed value of the square root of the moon's radius vector. This equation virtually replaces two of the differential equations, and it then becomes possible to eliminate the time from the other two equations which give the rates of change of the inclinations, and to adopt the square root of the moon's radius vector as independent variable.

The equations which remain are so complex that it is only possible to treat them by quadratures, but the results obtained in that laborious way seem to throw light on the present configuration of the moon and earth.

§ 47. *Problem where the Orbit is eccentric, but not inclined.*

I now return to the second of the two problems, where the moon moves in an eccentric orbit coincident with the ecliptic.

It appears that the tides raised by any one satellite can produce no secular change in the eccentricity of the orbit of any other satellite.

In the present configuration of the moon and earth, the eccentricity in general increases with the time, but if the obliquity to the ecliptic were nearly 90° or if the viscosity were very great the eccentricity would diminish.

If the viscosity be small it appears that, if the obliquity of the ecliptic be zero, the eccentricity of the orbit will either increase or diminish according as 18 rotations of the planet take a shorter or a longer time than 11 revolutions of the satellite. From this it follows that in the history of a satellite revolving about a planet of small viscosity the circular orbit is dynamically stable until 11 months of the satellite have become longer than 18 days of the planet. Since the day and month start from equality and end in equality, it follows that the eccentricity of orbit will rise to a maximum and ultimately diminish again.

It also appears that if a satellite be started to move in a circular orbit with the same periodic time as that of the planet's rotation (with maximum energy for given moment of momentum), then if infinitesimal eccentricity be given to the orbit, the satellite will ultimately fall into the planet; and if, the orbit being circular, infinitesimal decrease of distance be given the satellite will fall in, whilst if infinitesimal increase of distance be given the satellite will recede from the planet. Thus this configuration, in which the planet and satellite move as parts of a rigid body, has a complex instability, for there are two sorts of disturbance which cause the satellite to fall in, and one which causes it to recede from the planet.

A retrospective integration by quadratures for the eccentricity of the lunar orbit shows it diminishing from its present value of $\frac{1}{18}$ to about $\frac{1}{10800}$.

T. J. J. See has applied the results of this investigation to the hypothesis of two stars mutually disturbing one another, and seeks to explain thereby the high degree of eccentricity of the orbits of many double stars*.

§ 48. *Analytical Solution for two Bodies.*

In the special case where there are only two bodies, where the viscosity is small, and where the obliquity, the inclination and eccentricity of orbit are small, it is possible to find an analytical solution of the system of differential equations. The results are instructive, but cannot safely be applied to the case of the moon and earth, since the solar influence plays a very important part in the result†.

§ 49. *A Speculation as to the Time and as to the Origin of the Moon.*

The amount of time needed to produce any given amount of change in the system is directly dependent on the degree of viscosity postulated. The most serious obstacle to the full applicability of this theory to the history of the moon and earth resides in the length of time required, for it appears that a period of at least 50 to 60 million years is needed; but we do not propose to discuss questions of cosmogony in these pages.

In the memoirs of which I am now giving an account a suggestion was made as to a possible explanation of the manner in which the moon may have detached itself from the primeval planet.

The primeval planet must have been wholly or in great part liquid or gaseous. The fundamental free period of oscillation of such a body would

---

* T. J. J. See, "Die Entwickelung der Doppelsternsysteme." Inaugural Dissertation, 1872. Schade, Berlin. *Evolution of the Stellar Systems*, Vol. I., Lynn, Mass., U.S.A., 1896.
† *Proc. Roy. Soc.*, Vol. XXX. (1880), p. 255 [or Vol. II., p. 383].

probably be three or four hours*. If then the planet were rotating with a period of six to eight hours the solar semidiurnal tide would be of enormous magnitude on account of its unison with the free period. Hence the solar attraction might induce instability when the mere rapidity of rotation would prove insufficient to cause it.

§ 50.   *Tidal Friction where there are several Satellites.*

The problem presented by the tidal friction of a planet attended by two satellites moving in circular orbits may be solved by a graphical method analogous to that referred to in § 42; and a similar treatment is applicable where there are more satellites, although the surface of energy cannot actually be constructed if we have to postulate space of more than three dimensions. A paper†, in which this subject is considered, contains also a discussion of the applicability of the theory of tidal friction to the cosmogony of the solar system.

§ 51.   *Collateral Problems.*

The changes in the rotation of the planet and in the orbit of the satellite are produced by tangential force between the surface of the mean sphere and the tidal protuberance. In the case of a viscous or plastic spheroid this action would raise wrinkles on the surface perpendicular to the axis of greatest pressure. These wrinkles would run meridionally at the equator and would trend eastward in northerly and southerly latitudes. The intensity of the force, varying as the square of the cosine of the latitude, would be greatest at the equator. It is possible that the general configuration of our continents is in part due to this cause‡.

The apparent torsion of the planets has also been considered from another point of view by W. Prinz§.

The energy converted into heat whilst the day is prolonged by tidal friction from $5\frac{1}{2}$ hours to 24 hours is sufficient to raise the whole earth, with specific heat equal to that of iron, to 1000° cent. On the hypothesis of the viscosity of the whole planet by far the larger portion of the heat is generated in the central parts. It is improbable that we can attribute to this cause more than a very small part of the observed gradient of temperature in mines and borings. But if the tidal friction affected only the more superficial portions of the mass, this conclusion might be modified to some extent‡.

---

* A. E. H. Love, "Oscillations of a Rotating Liquid Spheroid and the Genesis of the Moon," *Phil. Mag.*, Vol. XXVII. (5th series), 1889, pp. 254—264.

† *Phil. Trans.*, Part II. (1881). On lines 4 and 7 of p. 524 for 5000 and $\frac{1}{2500}$ read 50,000 and $\frac{1}{25000}$. [The mistake is corrected in Vol. II.]

‡ G. H. Darwin, *Phil. Trans.*, Vol. 170 (1879), pp. 539—593 [or Vol. II., p. 140].

§ W. Prinz, *Annuaire de l'Obs. Roy. Bruxelles*, 1891.

NOTE ON § 21, p. 212.

[An extract from a proof sheet of the Fifteenth Report of the Committee of the British Association on Seismological Investigations (Sheffield, 1910).]

OBSERVATIONS OF DEFLECTION OF THE VERTICAL DUE TO TIDAL LOAD.

Towards the latter end of last year it occurred to Professor Milne that the conditions under which the earthquake records were made at Bidston might be utilised to determine the amount of deformation of the earth's surface due to the accumulation and removal of a heavy load of tidal water.

A few years ago, in the basement of the Victoria Club at Ryde, Professor Milne made some observations with this in view. Contrary to expectations, it was found that when the tide rose the strand rose also. This was attributed to the banking up of drainage from the land and the consequent bulging up of the same. It was, however, pointed out by Sir George Darwin that the greater quantity of water in the English Channel might more than counterbalance the effect of the smaller volume in the Solent.

In the Mersey, as shown by the tide gauges on the Liverpool Landing Stage, the variation in the height of the tide can considerably exceed 10 feet, and in the Dee, at Helbre Island, the oscillation is practically the same. The difference in the time of high water at these two stations is about half an hour. Consequently, as a glance at the rough map of the coast-line will show (Fig. 1), there is a tendency for the load to balance on the east and west sides, while on the north and south, apparently, the difference would be most marked. Under these circumstances it is a little difficult to determine what would be the most appropriate azimuth to mount the pendulum, but as the boom in the original seismometer was placed north and south, in the new instrument the direction was made east and west. The seismometer can, however, be turned through any angle if it be felt desirable to continue the investigations.

The instrument used was designed by Professor Milne and his assistant, Mr S. Hirota. All the observations were made and discussed by Mr W. E. Plummer, Director of the Bidston Observatory.

The boom differs in some essential particulars from the type ordinarily used in the Milne seismometer. It is divided into two parts, one, nearer to the stand, consists of a stout brass rod, carrying a weight of about 7 lb. At

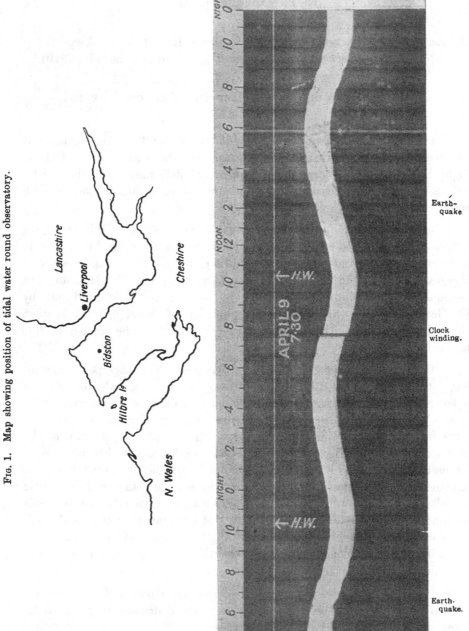

Fig. 1. Map showing position of tidal water round observatory.

Fig. 2. Deflections recorded at Bidston due to Tidal Load.

the extremity of this rod, which is only about 30 inches in length, is placed
a light magnifying style, independently carried, and attached to the boom
proper by means of a magnetised needle, capable of moving between a
slender iron fork. The sensitiveness of the instrument can be increased at
will by reducing the distance between the pivot on which the magnifying
style works and the end of the boom. In the original construction the
multiplying arm was 10 inches long, rotating about a centre 1 inch from the
end of the boom, consequently the displacement was magnified ten times.
The arrangements for photographing the movement were of the ordinary
character. The sensitised paper was paid out at the rate of 5 millimetres an
hour, so as to make the small amplitude of the oscillation apparent. The
tidal displacements were sufficiently noticeable, and the accordance with the
time of high water was satisfactory. To increase the sensitiveness of the
instrument so as to make the motion more distinct and easily measured, and
since the needle sometimes failed to engage the steel fork, it was felt
desirable to adopt a different method of connection. With this view,
Professor Milne suggested that the magnetised needle should be removed
and the multiplying piece mounted as a bifilar pendulum, an arrangement
which allowed the centre of motion to be brought much nearer to the end of
the boom and gave a multiplication of about forty times. The method of
photographing the point of light was changed, and a thin strip of black
paper substituted. This apparatus has been in use since last March (1910),
and generally worked satisfactorily. The diagram (Fig. 2) shows the character
of the photographs that are now being taken. The instrument is not well
adapted for the record of earthquake waves, but two small tremors are shown,
the second one is not to be found on the ordinary earthquake film.

Some difficulties were introduced by the greater sensitiveness, and some
have been made more apparent, but Mr Plummer was not able to eliminate
them satisfactorily. One difficulty was to determine the linear displacement of the boom due to an angular tilt of the instrument; for the smallest
angular motion which it was possible to make with accuracy moved the
multiplying style off the scale. It seemed necessary to reduce the sensitiveness by a known factor; this was done by increasing the distance between
the supports of the bifilar portion. There are objections to this plan, and
up to the present Mr Plummer has left the results in the form of the actual
measured displacement. Another difficulty arose from a long slow movement
of a very minute order in one direction, probably masked in the less sensitive
instrument, but now distinctly noticeable in a continued series of observations.
The whole seismometer was mounted on a slate slab on the top of a drain
pipe, 2 feet in diameter. This form of stand was preferred by Professor Milne,
because it avoided the drying of mortar or cement, which, in a brick-built
pier, would take a very considerable time. The observed creeping may be
due to some motion in the hill on which the Observatory is built, akin to the

annual variation in the azimuthal error of the transit instrument. While the instrument has been in use the temperature has been increasing. Observations in the second half of the year may clear up this point.

It must not, however, be overlooked that one possible cause for this creeping may be found in the seasonal shift in the direction of the north-south barometrical gradient, accompanied by a seasonal change in the mean sea-level. In summer time the region of high barometrical pressure lies to the north of Great Britain, whilst in winter it lies considerably to the south.

The amplitudes on the diagrams seem sufficiently large to warrant an attempt to determine the tidal constants by means of harmonic analysis in the same way that the records of a tidal gauge are treated. It may be said here that it was hoped originally to determine from the residuals between the computed and observed curves the direct effect of the moon's tide-generating force. At the present moment such an inquiry is no doubt rendered difficult owing to the slow creeping of the pendulum towards the north. Mr Plummer remarks that he feels himself to be in the position of a man who endeavours to find the height of the tides from readings on a scale that is continually sinking into the ground, and at a rate which he cannot determine and which he does not know to be uniform. There are also other practical difficulties connected with the winding of the clock, attending to the illumination, &c. It is by no means certain that after a disturbance the boom returns to the position originally occupied with no greater error than the small quantity sought. The discussion of the results, so far as they have gone, is useful as emphasising these difficulties, and with that view they are printed here. The observations from April 14 to April 28 seemed as free from objection as any that have been made, and as a first attempt it was arranged to derive the several tides in the manner described by Professor Sir G. H. Darwin. Clearly, if the main tides could not be recognised, it was hopeless to look for more recondite effects. There is a slight want of definiteness in the edge of the photograph; but this defect has been to some extent removed, it is hoped, by measuring both sides and using the mean. The curve was read off to a tenth of a millimetre, and that unit has been used throughout.

The results of the harmonic analysis are given in the following table. About these Sir George Darwin writes as follows:

"Since the oscillations of the pendulum are due to the weight of sea-water, it seems best to compare them with the tidal constants, as derived from ten years of observation at Helbre Island*. This place being near the mouth of the Dee seems to afford a better means of comparison than does Liverpool. The constants for Liverpool, however, differ but slightly

* See Baird and Darwin, *Proc. Roy. Soc.*, Vol. xxxix. (1885), p. 196, col. 33.

from those at Helbre Island. It is further desirable to compare the results with those derived from the equilibrium theory of tides for a place in lat. 53° 24′, approximately that of Bidston. I gave in Table E of the Report on Tides to the Brit. Assoc. for 1883 (*Scientific Papers*, Vol. I. p. 25) a theoretical scale of importance of the several tides expressed in terms of the principal lunar semi-diurnal tide M_2 as unity. But this table takes no account of the latitude of the place of observation, merely giving the relative importance of the several 'coefficients.' What we require is to know what would be the deflections of the pendulum at Bidston if it were erected on an absolutely unyielding soil, and were only affected by the tide-generating forces due to moon and sun. The values given in that table for the semi-diurnal tides may be quoted directly therefrom, and give the results in terms of M_2 as unity. But to reduce the diurnal tides to the same measure for this latitude, we must multiply the tabular values by $2 \tan \lambda$, where λ is latitude. In this way we obtain a scale of relative importance for the luni-solar tide-generating force at Bidston.

		Helbre Island	Tide-generating force at Bidston
Lunar semi-diurnal M_2	$\begin{cases} H = 17{\cdot}52 \frac{1}{10} \text{ mm.} \\ \kappa = 318° \end{cases}$	9·758 ft. 319°	1·000 0°
Solar semi-diurnal S_2	$\begin{cases} H = 7{\cdot}45 \\ \kappa = 327° \end{cases}$	3·128 ft. 3°	·465 0°
Luni-solar semi-diurnal K_2	$\begin{cases} H = 2{\cdot}03 \\ \kappa = 327° \end{cases}$	·890 ft. 358°	·127 0°
Luni-solar diurnal K_1	$\begin{cases} H = 5{\cdot}64 \\ \kappa = 346° \end{cases}$	·391 188°	1·572 0°
Solar diurnal P	$\begin{cases} H = 1{\cdot}88 \\ \kappa = 346° \end{cases}$	·146 174°	·520 0°
Lunar diurnal O	$\begin{cases} H = 1{\cdot}86 \\ \kappa = 237° \end{cases}$	·370 41°	1·118 0°

Since the series of observations only extended over a fortnight, it was necessary to *assume* that the phase of K_2 was the same as that of S_2, and the amplitude about $\frac{3}{11}$ths. Similarly the phase of P is assumed to be identical with that of K_1, and the amplitude one-third. Hence in the case of the pendulum there are really only four independent evaluations, and the values of K_2 and of P might have been omitted as far as concerns the provision of a means of comparison between the pendulum and the tide.

"A fortnight is much too short a period of observation to afford trustworthy values for the deflections of the pendulum, and therefore we should not place implicit reliance on the exact numerical values obtained.

"The phase of M_2 for the pendulum is virtually identical with that of the tide, but this exactness of coincidence is probably to some extent accidental. The high tide, so to say, for the solar tide S_2, differs in phase from that of the water by 36° or 1 h. 12 m., and the amplitude is considerably greater relatively to M_2 than is the corresponding ratio for the sea.

"The phases of the diurnal sea-tides at Helbre Island are very abnormal, for whereas it might have been expected that they should all come out nearly the same, the phases of K_1 and O differ by 147°. The result is, however, derived from so many years of observation that it is certainly correct and is, moreover, confirmed by the tidal constants for Liverpool. In the case of the pendulum we observe a similar abnormality, for the phases of K_1 and O differ by 109°. It is, however, remarkable that these tides are almost inverted with reference to the sea-tides. One may conjecture that there are perhaps nodal lines for these tides at some short distance out to sea, and that the bulk of the sea which produces the flexure is in the opposite phase from that which gives the visible tide at Helbre Island and Liverpool. The amplitudes of K_1 and O are also very discordant, both in absolute amount and between themselves. In the sea K_1 and O have nearly the same amplitude, but with the pendulum that of K_1 is three times as great as that of O. This would result if the supposed node of K_1 were nearer the shore than that for O, because if this were so there would be a larger weight of water, oscillating in a phase opposite to that of the sea in shore, to produce flexure in the case of K_1 than in that of O. However, the series is much too short to justify any confidence in such conjectures.

"The last column gives the relative importance of the tide-generating forces for the several tides, and it will be seen that the force for K_1 is much larger and that for O somewhat larger than that for M_2. We see that both in the sea and in the case of the pendulum there is an enormous reduction of amplitude for diurnal tides as compared with the semi-diurnal ones, but the reduction is markedly less for the pendulum. If these values should be confirmed, we may perhaps suspect that the direct luni-solar tide-generating force is rendering itself evident in the K_1 tide, and such a conjecture would accord with the phase of K_1 approaching 360° without the intervention of the nodal line at sea suggested above. However, as already pointed out, it is too soon to draw any conclusions with confidence."

Whatever may yet come from this new departure in observations bearing upon Earth Physics, the work already accomplished is suggestive of certain conclusions.

We see that an observatory near to a shore line, in consequence of the diurnal tilting to which it may be subjected, is unsuitable for certain investigations. This, however, was pointed out by Sir G. H. Darwin in his Report to the British Association in 1882. The discussion suggests precautions in the determination of the nadir at an observatory on the sea-coast, and probably the deepest mine in central Britain is still unsuitable as a place in which to measure the effects of lunar gravitation.

The deflections accompanying tidal loads observed at Bidston indicate a relationship between the yielding of areas represented by rocks and other materials and loads which are fairly well measurable.

These deflections which accompany a 10-feet tide amount at Bidston to approximately 0″·2. This yielding may be truly elastic, or it may possibly be partly due to the sagging of a surface like that of a raft under the influence of load. This latter idea falls in line with seismological observations, which show day after day that the large waves of earthquakes, whether passing beneath the alluvial plains of Siberia or beneath the crystalline rocks of North America, do so at a uniform speed. Seismology suggests that we live on a congealed surface, which, whether it is thick or thin, light or dense, apparently responds in a uniform manner to undulations which pass beneath it.

PART III

MISCELLANEOUS PAPERS
IN CHRONOLOGICAL ORDER

5.

ON SOME PROPOSED FORMS OF SLIDE-RULE.

[*Proceedings of the London Mathematical Society*, Vol. VI. (1875), p. 113.]

THE object of the author was to devise a form of slide-rule which should be small enough for the pocket, and yet be a powerful instrument.

The first proposed form was to have a pair of watch-spring tapes graduated logarithmically, and coiled on spring bobbins side by side. There was to be an arrangement for clipping the tapes together and unwinding them simultaneously. Two modifications of this idea were given.

The second form was the logarithmic graduation of several coils of a helix engraved on a brass cylinder. On the brass cylinder was to fit a glass one, similarly graduated.

To avoid the parallax due to the elevation of the glass above the other scale, the author proposed that the glass cylinder might be replaced by a metal corkscrew sliding in a deep worm, by which means the two scales might be brought flush with one another.

6.

AN APPLICATION OF PEAUCELLIER'S CELL.

[*Proceedings of the London Mathematical Society*, Vol. VI. (1875), pp. 113, 114.]

THE object was to devise a mechanical method of making a force which shall vary inversely as the square of the distance from a fixed point. This may be done by means of a Peaucellier's cell. Let O be the fixed pivot of a cell, in equilibrium under the action of two forces P and P' acting at B and D.

Then, by the principle of virtual velocities,

$$P' \cdot \delta OD + P \cdot \delta OB = 0$$

Now

$$OD \cdot OB = OA^2 - AD^2$$

therefore

$$\frac{\delta OD}{OD} = -\frac{\delta OB}{OB}$$

therefore

$$P' \cdot OD = P \cdot OB$$

whence

$$P = P' \frac{OA^2 - AD^2}{OB^2}$$

If, then, P' is a constant force acting away from O, P is an attractive force varying as OB^{-2}.

The idea was the joint production of the author and of his brother, Horace Darwin.

7.

THE MECHANICAL DESCRIPTION OF EQUIPOTENTIAL LINES.

[*Proceedings of the London Mathematical Society*, Vol. VI. (1875), pp. 115—117.]

Fig. 1 represents two Peaucellier's cells, of which O_1, O_2 are the respective fulcra, which are fixed; A, P and B, P the poles. Let the moduli of the cells be respectively proportional to two charges of positive electricity at O_1 and O_2. Thus $O_1A = m_1/O_1P$, $O_2B = m_2/O_2P$, so that, if $O_1A + O_2B$ is constant, P traces the equipotential line of m_1 and m_2 placed at O_1 and O_2 respectively. The constancy of $O_1A + O_2B$ may be attained thus: Let the pivots at O_1, O_2, about which the cells turn, be two needles; fasten a piece of pack-thread to A, pass it round O_1, round another needle E (driven into the drawing-board), round O_2, and fasten the other end to B. The broken line in Fig. 1 represents this thread. Then, if P is moved so as to keep the thread taut, it describes one of the equipotential lines. If the needle E is shifted to F and G, we get other equipotential lines.

Fig. 2 represents the arrangement of the thread where one of the charges is negative. In this and the succeeding figures, fixed points are marked with crosses, and the needles are exaggerated so as to show the disposition of the strings, and the bars of the cells are omitted, leaving only the tracing point P and the other poles marked. In this second case the requisite is that $OA \sim OB$ should be constant. This is obviously attained by making a knot K, or a loop in the string which passes round the fixed needle E. This knot or loop is held in one hand and pulled, whilst P, the tracing point, is released with the other, so as always to keep the strings taut. By shifting the knot or loop from point to point along the string, we get successive equipotentials.

Some figures submitted to the Society were drawn with rough cells, the tracing point being the blunt end of a needle; and they appeared to the author sufficiently good to show that a carefully made instrument would perform very well.   Pack-thread was found to be but very slightly extensible.

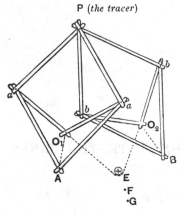

(Two positive charges.)

FIG. 1.

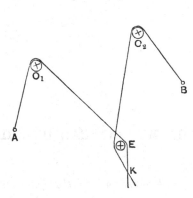

(One positive charge, and one negative.)

FIG. 2.

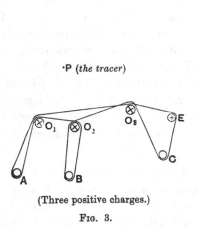

(Three positive charges.)

FIG. 3.

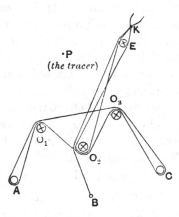

(Two positive charges, one negative.)

FIG. 4.

In an attempt to draw with one of Mr Hart's 4-bar reciprocators, the bars interfered much more with the fixed pivots than in the case of Peaucellier's cells; and with positive Peaucellier's cells but a small part of the field could be drawn.

Professor Sylvester has suggested to the author that it should be possible to do without string; and it would doubtless be possible, but it is to be doubted whether any linkage would practically allow of the easy alteration of the constant which is permitted by the shifting of needle E, so as to draw successive lines quickly.

Figs. 3 and 4 show the arrangement of the string requisite for 3 points, either all positive, or two positive and one negative. No endeavour has been made to construct such an instrument, as it would require very careful workmanship; the string would have to be endless, and therefore smooth pegs would have to be fixed to the non-tracing poles of the cells. The field throughout which tracing would be possible would also be much restricted.

A similar kind of arrangement of strings is clearly theoretically applicable to any number of points, whatever are the signs of their charges.

ON A MECHANICAL REPRESENTATION OF THE SECOND ELLIPTIC INTEGRAL.

[*Messenger of Mathematics*, Vol. IV. (1875), pp. 113—115.]

LET $\dfrac{x^2}{a^2} + \dfrac{y^2}{b^2} = 1$ be the equation to an elliptic cylinder $BADE$ (Fig. 1), then the equations to its circular sections through the origin are

$$y \sqrt{\left(\frac{1}{b^2} - \frac{1}{a^2}\right)} = \pm \frac{z}{a}$$

Let BCD be a circular section of the cylinder; its equation is

$$(y - b)\sqrt{(a^2 - b^2)} + bz = 0$$

whence $\qquad PN = \dfrac{b-y}{b}\sqrt{(a^2 - b^2)}$

now $\qquad CO = \sqrt{(a^2 - b^2)}$

therefore $\qquad PM = \dfrac{y}{b}\sqrt{(a^2 - b^2)}$

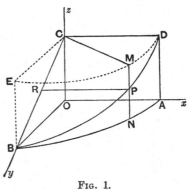

FIG. 1.

Supposing the elliptic cylinder to roll along a plane, the circular section BPD will trace out a curve on the plane. For determining this curve, take in the plane as axis of ξ, the straight line on which the dotted ellipse EMD rolls, as origin the point at which EB when in contact with the plane intersects the axis of ξ. And measure η perpendicular to ξ. Then it is clear that

$$\eta = PM = \frac{y}{b}\sqrt{(a^2 - b^2)}$$

$$\xi = \text{arc } ME = \int_y^b \sqrt{\left(1 + \frac{a^4 y^2}{b^4 x^2}\right)}\, dy = \frac{b}{\sqrt{(a^2 - b^2)}} \int_\eta^{\sqrt{(a^2 - b^2)}} \sqrt{\left(1 + \frac{a^4 y^2}{b^4 x^2}\right)}\, d\eta$$

therefore
$$\frac{d\xi}{d\eta} = \frac{-b}{\sqrt{(a^2 - b^2)}}\sqrt{\left(1 + \frac{a^4 y^2}{b^4 x^2}\right)}$$

Whence
$$x^2(a^2 - b^2)\left[\left(\frac{d\xi}{d\eta}\right)^2 + 1\right] = a^4$$

therefore
$$\frac{x^2}{a^2} = \frac{a^2}{a^2 - b^2}\left[\left(\frac{d\xi}{d\eta}\right)^2 + 1\right]^{-1}$$

also
$$\frac{y^2}{b^2} = \frac{\eta^2}{a^2 - b^2}$$

therefore
$$\frac{a^2}{\left(\frac{d\xi}{d\eta}\right)^2 + 1} + \eta^2 = a^2 - b^2$$

therefore
$$\frac{d\xi}{d\eta} = \pm\sqrt{\left(\frac{b^2 + \eta^2}{a^2 - b^2 - \eta^2}\right)}$$

$$\xi = \pm\int\sqrt{\left(\frac{b^2 + \eta^2}{a^2 - b^2 - \eta^2}\right)}\,d\eta$$

Put
$$\eta = \sqrt{(a^2 - b^2)}\cos\theta = ae\cos\theta$$

Then taking the lower sign

$$\xi = a\int\sqrt{(1 - e^2\sin^2\theta)}\,d\theta$$

Now, if two circles of radius a be fastened together with their planes parallel, and inclined to the line joining their centres at an angle α, they clearly form the circular sections of an elliptic cylinder, of which the major axis is a, and the eccentricity $\cos\alpha$.

Therefore, if two such circles be rolled along a sheet of paper, the equation to their track is
$$\eta = a\cos\alpha\cos\theta$$

$$\xi = a\int\sqrt{(1 - \cos^2\alpha\sin^2\theta)}\,.\,d\theta$$

ξ is the second elliptic integral, and therefore this contrivance is a mechanical method of drawing a curve in which the abscissæ are simple elliptic integrals of the second kind.

If the circles be fixed together like an ordinary pair of parallel rulers (Fig. 2), they may be so arranged as to give the value of any elliptic integral of this kind, where e is less than unity.

In what follows, I will, for brevity, call the line joining the centres of the two discs L (see Fig. 3).

Let it be required to evaluate $a\int_0^\theta \sqrt{(1 - e^2\sin^2\theta)}\,d\theta$ for some given value of θ.

First set the discs so that $\cos \alpha = e$.

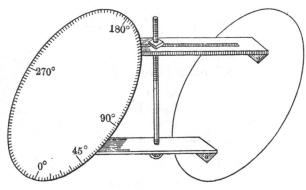

FIG. 2.

It is clear from spherical trigonometry that θ is the angle BCP in Fig. 1, that is to say, θ is the angle between the projection of L on the disc, and the radius of the disc which goes to the point of contact with the plane; see Fig. 3. Suppose both discs graduated about their centres like a protractor, to degrees, the two projections of L being the two lines of 0° on the two discs respectively.

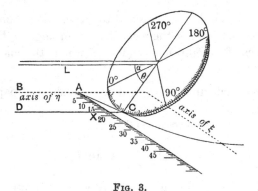

FIG. 3.

Draw on a sheet of paper two parallel lines distant L apart, and intersect them by a perpendicular to either, AB (Fig. 3). Place the discs so that the two 0° are at A and B respectively, and roll them until the two points of contact come to θ on the angle-scale of each disc; let C, D be then the points of contact. Join CXD, intersecting one of our parallel lines (say A) in X. Then AX is the required elliptic integral.

It is clear that our two parallel lines will be the same for all values of e; so that if they were drawn once for all, and were graduated, the result might be read off at once.

The complete integral $a \int_0^{\frac{1}{2}\pi} \sqrt{(1 - e^2 \sin^2 \theta)}\, d\theta$ is half the distance from node to node of the sinuous curve.

When e is small the curve approximates to a straight line, when e is nearly unity it approximates to a succession of semicircles.

If the circular wheels were replaced by elliptic wheels, so arranged as to be sections of a circular cylinder, we should obtain the simple harmonic curve

$$\eta = \overset{\circ}{a} \sin \frac{\xi}{a}$$

9.

ON MAPS OF THE WORLD.

[*Philosophical Magazine*, Vol. L. (1875), pp. 431—444.]

THE ordinary stereographic projection of the world in two hemispheres is utterly worthless as giving a true impression of the whole; for the linear scale at the margins of the circles is twice that at their centres. Its only merit is that there is no angular distortion. Mercator's projection gives a still more fallacious impression, except as regards the equatorial regions.

It appears to me therefore that there is a want, in the school-room and lecture-room, of some map which shall give a more truthful representation of the globe than the above, and which yet shall not be so expensive and cumbrous as a globe.

A gnomonic projection on to the faces of a regular icosahedron is but very slightly distorted, although a slight amount of angular distortion is here introduced. I have been told that at the recent Geographical Congress at Paris, some such projections as this were exhibited, and that they were of old date. Mr Proctor has also made star-maps by projection on to the faces of a regular dodecahedron; but in 1873, when the idea occurred to me of using this projection, I was not aware of the fact.

If the icosahedral projection be developed and arranged as a band of ten triangles round the equator, with saw-like edges of five triangles in the north and five in the south, a very fair representation of the globe is given. And the interstices between the teeth of the saws may be arranged so as not to damage the continents very severely.

In this map the meridians are straight lines, but are broken in direction at the junction of two triangles. The parallels of latitude become ellipses, which may be easily laid out by aid of a property of conic sections; viz. if a circular cone be placed with its vertex at the centre of a sphere, and a section made by a tangent plane to the sphere, the radius of curvature at the vertices of this conic section is constant for all tangent planes, and varies as the tangent of the semi-angle of the cone.

Now in our map the ellipses are represented with sufficient accuracy by the circles of curvature at their vertices; and the radii of these circles may be taken direct with the compasses from a sector, as the cotangents of the corresponding latitudes.

Besides a map of this kind, I have also constructed a portable quasi-globe with this method of projection. The faces of the icosahedron are made to hinge together, so that the whole can be packed flat in the form of a half-hexagon. Such a globe was exhibited at the British Association Meeting at Bradford. When mounted, the icosahedron circumscribed a sphere of 25 inches diameter. This form of globe might doubtless be constructed much cheaper than a truly spherical one, because the framework would be ordinary carpentry, and the twenty map-sheets might be printed flat like ordinary maps.

In 1872 I showed the above described maps and globe to General (afterwards Sir Richard) Strachey; and he suggested that by cutting down the icosahedron in some way, a still more satisfactory projection might be attained. It then occurred to us that by truncating the solid angles of the icosahedron, a solid figure of 32 faces would be obtained, viz. 20 hexagons and 12 pentagons.

If the truncation be carried on by slices until the truncating planes touch the sphere enclosed in the icosahedron, these hexagons are not regular, but have two sets of three sides equal to one another; a long side is always opposite to a short side. If unity is the radius of the sphere, the long sides and short sides are respectively ·4913 and ·3401. The pentagons are always regular; and at this particular degree of truncation the side of the pentagon is ·4913, and a pentagon is therefore always contiguous to the long side of a hexagon; whilst hexagons are always contiguous along their short sides*.

This projection was utilized by having a sort of umbrella-like stand, with a pentagonal face in the middle, surrounded by five hexagons; or else with a hexagon in the middle, surrounded by three pentagons and by three hexagons. The maps were drawn on 32 separate sheets; and the sheets required to represent any part of the world were mounted on the umbrella.

By these means about one-fifth of the globe is shown at once; and thus the equivalent of a very large globe might be used in a room of ordinary size. The sheets may also be conveniently kept, since they are all flat, and will lie one on another.

---

* This leads me to observe that if the angles of any one of the regular solids be truncated in this way, another one is ultimately produced. The 20-hedron and 12-hedron, the 8-hedron and cube, and the tetrahedron and tetrahedron are thus correlated. This property is of course due to the fact that the polar reciprocal of any regular solid is itself a regular solid. It is curious to observe the transitional forms as the slices are cut off the angles.

The figures 1, 2, 3, 4 show the forms of the various map-sheets, together with the figures required for laying out the meridians and parallels of latitude.

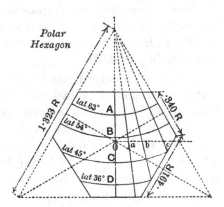

Polar Hexagon

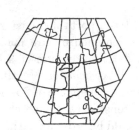

Diameter of inscribed sphere is 3 inches.

Fig. 1.

The radii of the circles of lat. given by R cot l.
The distances OA, OB, OC, OD by
$$R \cot (l + 37° \ 23').$$
The distances Oa, Ob, Oc by
R sin 37° 23' tan y where $y = 9°$, 18°, 27°.
R is radius of inscribed sphere and l latitude.

Fig. 2.

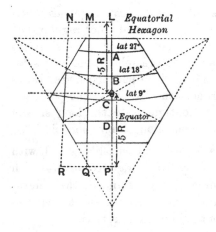

Equatorial Hexagon

The radii of the circles of lat. given by R cot l.
The distances OA, OB, OC, OD by
$$R \cot (l + 79° \ 11').$$
The intercepts LM, LN, PQ, PR given by
R tan y (sin 79° 11' $\pm \frac{1}{2}$ cos 79° 11')
where $y = 9°$ and 18°.

Fig. 3.

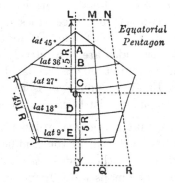

Equatorial Pentagon

The radii of the circles of lat. given by R cot l.
The distances OA, OB, OC, OD, OE by
$$R \cot (l + 63° \ 26').$$
The distances LM, LN, PQ, PR, by
R (sin 63° 26' $\pm \frac{1}{2}$ cos 63° 26') tan y
where $y = 9°$ and 18°.

Fig. 4.

Besides those kinds shown in the figures, there are two pentagons which close in the two poles; but it is so easy to lay them out, that it does not seem worth while to give a figure. The meridians on the equatorial faces converge so little that it is more convenient to set them out by finding two points through which they pass. The broken lines in the figures are merely constructional.

In order that the meridians and lines of latitude may fall symmetrically on each face, it is better to set them every 9° or 6°, instead of every 10° as is usually done. For the whole globe, there are required 10 equatorial hexagons, 10 equatorial pentagons (5 in N. and 5 in S.), and 2 polar pentagons.

This 32-faced figure is a very close approximation to the globe.

The Murchison Fund of the Geographical Society (£40) has been granted for carrying this scheme out practically; and a Committee has been appointed, of which General Strachey and Mr (now Sir) Francis Galton are members. The scale is large, the polyhedron being designed to circumscribe a sphere of 10 feet diameter. The various sheets of the map are stretched on light wooden frames; and they can be hasped on to a kind of umbrella, of which the handle is held horizontal. It is expected that it will be finished shortly; and it will, I believe, be placed in the rooms of the Society.

Another somewhat similar plan has occurred to me, and seems to me preferable, at any rate for somewhat smaller globes than the one above referred to.

Suppose ABC to be one face of a regular icosahedron *inscribed* in a sphere (see Fig. 5), and that we bisect the arcs of great circles subtended by the sides BC, CA, AB respectively in D, E, F. Then pass a plane through DEF, and three others through AEF, BDF, CDE respectively. The face ABC may be replaced by the equilateral triangle DEF and the three isosceles triangles AEF, BDF, CDE. If this be done with every face of the icosahedron, we have a solid figure of 80 faces—20 equilateral triangles, and 60 isosceles (but nearly equilateral) triangles—inscribed in the sphere. If we project the globe on to this surface, with the vertex of projection at the centre, we obtain an excellent approximation to the true globe.

Now this plan would be very complicated if it were necessary to have 80 different map-sheets. Fortunately, however, the form of the triangles makes it advantageous to have four sheets united together, viz. the equilateral triangle and the three isosceles ones which have replaced the face of the original icosahedron.

Fig. 6 represents one of these sets of four sheets when spread out flat. These four sheets may be printed from a single plate, and may be pasted on to quasi-triangles, such as ABC (Fig. 6), which are hinged or creased along the lines DE, EF, FD. If the scale on which it is carried out is sufficiently small to permit of the faces being made of cardboard, it would, I think, answer very well. We should then have to select the five appropriate sheets (each comprising four faces) and mount them on the umbrella stand; the five sheets would then represent one-quarter of the globe.

The map-sheets may be kept in a very small compass, because the isosceles triangles may be folded down over the equilateral triangles, as shown in Fig. 7.

In the other scheme it requires six or seven sheets to represent nearly one-fifth of the globe: but it has the countervailing advantage of permitting a greater choice of the region which is to be in the middle; for, by having two umbrella stands, we can either place a pentagon or a hexagon in the middle.

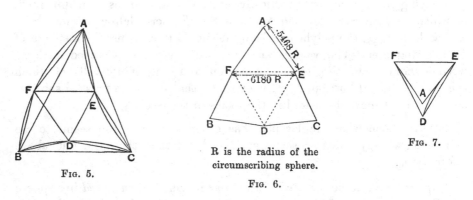

Fig. 5.

R is the radius of the circumscribing sphere.

Fig. 6.

Fig. 7.

In the instrument as made for the Geographical Society, the same general framework serves for both umbrellas, which may be shifted with great ease. It was this advantage in choice of the central region displayed which induced the Committee to prefer the original 32-faced polyhedron.

A similar construction may of course be applied to a dodecahedron inscribed in a sphere; and we thereby obtain a 72-faced surface, viz. 12 pentagons, and 60 obtuse-angled isosceles triangles. Here, as before, the map-sheets might be printed in sets of six, viz. a pentagon surrounded by five triangles; three map-sheets will then give one-quarter of the globe. In this figure the pentagons are so large compared with the triangles that the approximation to the sphere is not very close.

Models of the various plans explained were exhibited at Bradford; but, by an oversight, no abstract of this paper appeared in the British Association *Report*.

10.

A GEOMETRICAL PUZZLE.

[*Messenger of Mathematics*, Vol. VI. (1877), p. 87.]

THE following puzzle was shown to me by a friend. It consists of a square (Fig. 1) made of cardboard, ruled into 64 equal squares, and divided into four pieces by cuts along the thick lines. These four pieces can apparently be put together (Fig. 2) so as to form a rectangle of 5×13 squares, and the puzzle

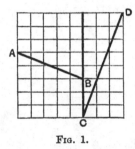

FIG. 1.

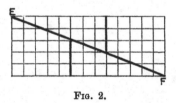

FIG. 2.

consists in explaining where the error lies, as we thus seem to transform 64 into 65 equal squares. The explanation is of course simple, as AB descends two squares in five, and CD (when turned round) three squares in eight, while EF, the diagonal of the rectangle in Fig. 2, descends five squares in thirteen. Since

$$\tfrac{5}{13} - \tfrac{3}{8} = \tfrac{1}{104}, \quad \tfrac{2}{5} - \tfrac{5}{13} = \tfrac{1}{65}, \quad \tfrac{2}{5} - \tfrac{3}{8} = \tfrac{1}{40}$$

we see that the two lines AB and CD are so nearly in a straight line, and deviate so little from the diagonal EF of the rectangle, that the eye does not detect the differences, and a very slight want of fitting is all that can be noticed. Any number of similar puzzles could no doubt be easily made on the same principle, it being requisite to find two fractions differing but little from the ratio of the sides of the rectangle, which would be best effected by expressing this ratio as a continued fraction (thus $\dfrac{5}{13} = \dfrac{1}{2+}\dfrac{1}{1+}\dfrac{1}{1+}\dfrac{1}{2}$, the convergents to which are $\frac{1}{2}$, $\frac{1}{3}$, $\frac{2}{5}$, while we obtain $\frac{3}{8}$ by omitting the final denominator), but generally it would be necessary to dissect the figure into more parts. Exactly the same dissection applies for the square 169 and the rectangle $168 = 8 \times 21$, for

$$\tfrac{5}{13} - \tfrac{8}{21} = \tfrac{1}{273}, \quad \tfrac{5}{13} - \tfrac{3}{8} = \tfrac{1}{104}, \quad \tfrac{8}{21} - \tfrac{3}{8} = \tfrac{1}{168}$$

and AB would descend three squares in eight, and CD five in thirteen. Probably, however, the 64 and 65 transposition would be the best as a puzzle. I have been informed that Hutton in his edition of Ozanam (*Mathematical Recreations*, 1803, Vol. I., p. 298), gives a transformation of a rectangle 3×11 squares into two rectangles of 4×5 and 2×7 squares, but that it is not nearly so good.

11.

A GEOMETRICAL ILLUSTRATION OF THE POTENTIAL OF A DISTANT CENTRE OF FORCE.

[*Messenger of Mathematics*, Vol. VI. (1877), pp. 97, 98.]

LET A, B, C be the principal moments of inertia of a body of mass M, at its centre of gravity, and let ρ be the distance of a point from the centre of gravity. Then, if the latter point be very distant, the potential V of the body at that point is approximately $\dfrac{M}{\rho} + \dfrac{A+B+C-3I}{2\rho^3}$, where I is the moment of inertia of the body about the radius vector ρ. Conversely, if the distant point be an attractive centre, the above expression gives the resultant potential of the point on the body. If the centre of force and the centre of gravity be fixed, the body will not in general be in equilibrium without some further external constraint. It is proposed to illustrate geometrically the nature of this external constraint, or, in other words, of the resultant couple exercised by the centre of force on the body. For convenience, the body is considered as fixed and the centre of force as moveable on the surface of a sphere of radius ρ.

Conceive the body replaced by its momental ellipsoid. Then the position of the centre of force may be indicated by P, its projection on the surface of the ellipsoid. It is clear that V is constant for all positions of P which make I constant, and since I varies inversely as the square of the radius vector of the momental ellipsoid, therefore when P lies on the intersection of a sphere with the ellipsoid, V is constant. Thus no work is done in moving the body so that P moves along any such sphero-conic. The central circular sections of the momental ellipsoid constitute one such intersection of a sphere and the ellipsoid.

Let $Ax^2 + By^2 + Cz^2 = K^4$ be the equation to the momental ellipsoid. Construct a series of spheres, the reciprocals of the squares of whose radii proceed in arithmetical progression with a common difference $\dfrac{2\rho^3}{3K^4}$, and a given term $\dfrac{B}{K^4}$, so that $\dfrac{1}{r_n{}^2} = \dfrac{B}{K^4} + n\left(\dfrac{2\rho^3}{3K^4}\right)$, where n is a positive integer.

Then corresponding to the sphero-conic formed by the sphere r_n, we have

$I = \dfrac{K^4}{r_n{}^2}$; and, therefore $V_n = $ a constant $+\, n$, so that for successive sphero-conics V differs by unity.

Project all these sphero-conics from the centre on to the sphere containing the circular sections of the ellipsoid, the radius of which is $\dfrac{K^2}{\sqrt{B}}$. Then clearly the circular sections themselves form one term of the series of sphero-conics.

The figure shows the general appearance of the sphere after the projection.

Now it clearly requires no work to move P along any of these lines, and it requires unit work to move P from any line to the next, by whatever path it is moved. But when P is at any point, the resultant couple clearly tends to move P perpendicular to the line on which it lies, and, therefore, the axis of the couple is a tangent to the line in question. Also P tends to move from places of higher to places of lower potential. Hence, taking the ordinary convention, when P is at any point the axis of the couple must be measured in the direction of the arrow-heads along the equipotential lines.

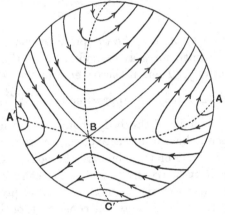

If the system of orthogonal tra-jectories of the equipotential lines be drawn, P will always tend to move along them from A to A' towards C or C'. It appears that there are six positions of equilibrium, but only two are stable, viz. those at C and C'. The equilibrium at B and B' is peculiar, because it is seen to arise from the counterbalancing of four equal and opposite couples. There are an infinite number of paths by which P may pass from A to C, but only one from any other point to C.

If the two principal moments A and B are equal to one another, the two circular sections shut up together, and the equipotential lines become a series of small circles round C as pole, and the arrows run round the sphere in a direction opposed to the motion of the hands of a watch. The resultant couple, therefore, has its axis always in the equator of the body and is at right angles to the meridian passing through P, it is also a function only of the latitude of P and not also of the longitude.

This is exactly the case of the action of the sun or moon on the protube-rant parts of the earth, and the above geometrical manner of looking at the matter enables one to see clearly why the precessional couple acts about an equatorial line 90° from the disturbing body's meridian.

12.

ON GRAPHICAL INTERPOLATION AND INTEGRATION.

[*Messenger of Mathematics*, Vol. VI. (1877), pp. 134—136.]

Interpolation.

IN graphical methods of calculation it is often requisite to draw a curve through the extremities of a number of equidistant ordinates. It is hard to draw a good curve with the free hand, and in the attempt to improve on the first trial the line becomes thickened and the accuracy of the figure is spoilt. A 'French curve' gives somewhat better results, but it is more satisfactory to find intermediate points through which the circle of curvature passes.

The following method gives an easy rule for finding approximately intermediate points on the ordinates which lie half-way between the given ones.

Let A', B', C', D' be the given ordinates, and a', b', c' the intermediate ones, on which it is desired to interpolate.

Join BA, BC, BD; CA, CB, CD and so on along the whole curve.

Let f, g, h be the required interpolated points.

Draw BK, gk perpendicular to AC and BC respectively.

Then the circle of curvature at B may be taken to pass through A and C, and we require to find the points f, g through which it also passes.

By the properties of the circle $\dfrac{BC^2}{BK}$ is nearly equal to $\dfrac{gC^2}{gk}$.

Now if the curvature be not very abrupt $BC = 2gC$ nearly; hence $gk = \tfrac{1}{4}BK$ nearly.

If BC be bisected in k, and kg erected on it and made equal to $\tfrac{1}{4}BK$, the required interpolated point is found. This gives a method of interpolation independent of any axes of reference, and might be used for interpolating between any points which are placed at nearly equal intervals along the arc of the required curve.

This construction would, however, give some little trouble, and where there are ordinates to refer to, and the curve is not inclined at too small an angle to these ordinates or where the curvature is not very abrupt, the rule may with advantage be modified yet further.

The angles kgb, KBF are nearly equal to one another, hence $\dfrac{gb}{gk} = \dfrac{BF}{BK}$, and, therefore, bg is nearly $= \frac{1}{4}FB$.

Hence, we get the following simple rule for interpolation on the intermediate ordinates. Take $af = \frac{1}{4}BF$, $bg = \frac{1}{4}CG$, &c., and the required points are found. It must be observed however, that if we were to work along the curve in the opposite direction, we should get different values for the increments to $a'a$, $b'b$, $c'c$. For example $bg = \frac{1}{4}BF$ or $\frac{1}{4}CG$. Where this discrepancy is perceptible to the eye, it will be best to take a point half-way between the two points given by the rule; that is to say, we take $bg = \frac{1}{8}(BF + CG)$.

To make a good figure this process needs only to be carried out in the parts of the curve, which depart notably from a straight line.

It is interesting to consider the analytical meaning of this process.

Let u_0, u_2, u_4, &c. be the given ordinates.   Then
$$BF = u_2 - \tfrac{1}{2}(u_0 + u_4)$$
therefore
$$u_1 = \tfrac{1}{2}(u_0 + u_2) + \tfrac{1}{4}\{u_2 - \tfrac{1}{2}(u_0 + u_4)\}$$
$$= \tfrac{1}{8}(3u_0 + 6u_2 - u_4)$$

Now since this equation may be written in the form
$$u_1 = u_0 + x(u_2 - u_0) + \frac{x(x-1)}{1.2}(u_4 - 2u_2 + u_0)$$

where $x = \frac{1}{4}$, it follows that the geometrical rule is the same as the rule for interpolation in the calculus of finite differences, as far as the second differences.

If we have four ordinates,
$$u_3 = \tfrac{1}{8}(3u_2 + 6u_4 - u_6)$$
or
$$= \tfrac{1}{8}(3u_4 + 6u_2 - u_0)$$

according as we work along the curve in one direction or the other, and the mean between these two is
$$u_3 = \tfrac{1}{16}\{-u_0 + 9u_2 + 9u_4 - u_6\}$$

and this may be written in the form
$$u_3 = u_0 + x(u_2 - u_0) + \frac{x(x-1)}{1.2}(u_4 - 2u_2 + u_0)$$
$$+ \frac{x(x-1)(x-2)}{1.2.3}(u_6 - 3u_4 + 3u_2 - u_0)$$

where $x = \frac{3}{2}$.

Therefore the rule given with respect to amending the interpolation by working from both ends of the curve, and taking the points half-way between the two points so given on each ordinate, is correct as far as the third differences.

It was stated above that these rules give good results unless the curve is inclined to the ordinates at a very small angle, or unless the curvature is very abrupt; this is precisely equivalent to saying in the language of finite differences, unless the first and second differences are large.

Integration.

It is often advantageous in obtaining an approximation to the value of a definite integral, to find a geometrical construction for a line proportional to the function to be integrated, and then to find about half-a-dozen equidistant values of the function.

But, having found them, the question arises as to the rule by which it is best to combine these values so as to give the integral.

At first sight it would appear best to take the rules for approximate quadrature given by the Calculus of Finite Differences, such as Weddle's rule for areas with seven equidistant ordinates. This view is however fallacious, because every ordinate is affected by error, and the probable error of the result will not be the same whatever rule is taken.

For example take Weddle's rule, that the integral is equal to

$$\frac{3h}{10} \{u_0 + u_2 + u_4 + u_6 + 5u_1 + 6u_3 + 5u_5\}$$

Now if each ordinate has a probable error c, the result has a probable error

$$\frac{3hc}{10} \sqrt{(4 + 25 + 36 + 25)} = \tfrac{3}{10} \sqrt{(90)} \, hc = 2 \cdot 846 hc$$

If, however, we take the rule that the integral is equal to

$$h \{\tfrac{1}{2} u_0 + u_1 + u_2 + u_3 + u_4 + u_5 + \tfrac{1}{2} u_6\}$$

the result has a probable error $hc \sqrt{(\tfrac{1}{4} + 5 + \tfrac{1}{4})} = 2 \cdot 345 hc$, and this is less than if we had taken the better rule.

It must remain an indeterminate question whether we are likely to lose more by taking a bad rule like the latter, or gain more by having a smaller probable error. It can only be decided by a knowledge of the kind of curve of which the area is being found, and by some estimate of the amount of error to which each ordinate is liable, and even then it would seem that there is no certain criterion.

13.

ON A THEOREM IN SPHERICAL HARMONIC ANALYSIS.

[*Messenger of Mathematics*, Vol. VI. (1877), pp. 165—168.]

THE following note is merely an extension to spherical harmonics of a paper by Prof. Toepler on Fourier's theorem*.

Suppose we have a series

$$Y = S_0 + S_1 + S_2 + \ldots + S_i$$

each term of which is a complete spherical harmonic of the order indicated by the suffixes; and that we wish to choose all the constants of the harmonics in such a manner that Y may differ by as little as possible from $F(\theta, \phi)$ all over the surface of a sphere of unit radius.

Following the analogy of the method of least squares, we must make

$$\int_0^{2\pi} \int_0^{\pi} \{F(\theta, \phi) - Y\}^2 \sin \theta \, d\theta \, d\phi$$

a minimum.

It is immaterial whether Y does or does not contain all the harmonics from 0 to i; it might even consist of only a single term.

This last case is the one which I will consider first, so that $Y = S_i$, and we must make

$$\int_0^{2\pi} \int_0^{\pi} \{F(\theta, \phi) - S_i\}^2 \sin \theta \, d\theta \, d\phi$$

a minimum.

---

* "Notiz über eine bemerkenswerthe Eigenschaft der periodischen Reihen," *K. Akad. der Wiss. in Wien*, Sitzung vom 14 December, 1876.

Now if we indicate by $[i, s]$ the series

$$\sin^s \theta \left\{ \cos^{i-s} \theta - \frac{(i-s)(i-s-1)}{4(s+1)1} \cos^{i-s-2} \theta \sin^2 \theta \right.$$

$$+ \frac{(i-s)(i-s-1)(i-s-2)(i-s-3)}{4^2(s+1)(s+2)1.2} \cos^{i-s-4} \theta \sin^4 \theta - \&c. \right\}$$

it is well-known that

$$S_i = \sum_{s=0}^{s=i} [i, s][A_s \cos s\phi + B_s \sin s\phi]^*$$

The problem then consists in the determination of the various constants A_s, B_s so as to satisfy the minimum condition.

By differentiating the integral with respect to each of the constants involved in S_i, we shall get an equation for each of those constants. For example, the equation resulting from the differentiation with respect to A_s is

$$\iint \{F(\theta, \phi) - S_i\} [i, s] \cos s\phi \sin \theta \, d\theta \, d\phi = 0$$

Now since

$$\int_0^{2\pi} \sin s'\phi \sin s\phi \, d\phi = \int_0^{2\pi} \cos s'\phi \cos s\phi \, d\phi = \int_0^{2\pi} \sin s'\phi \cos s\phi \, d\phi = 0$$

except when $s' = s$, and then the two former are not zero, therefore the equation becomes,

$$\iint F(\theta, \phi) [i, s] \cos s\phi \sin \theta \, d\theta \, d\phi = A_s \iint \{[i, s]\}^2 \cos^2 s\phi \sin \theta \, d\theta \, d\phi$$

and, therefore,

$$\pi A_s \int_0^{\pi} [i, s]^2 \sin \theta \, d\theta = \iint F(\theta, \phi) [i, s] \cos s\phi \sin \theta \, d\theta \, d\phi$$

The equation for B_s will only differ from this by having $\sin s\phi$ instead of $\cos s\phi$ under the double integral on the right-hand side.

Let $C = \pi \int_0^{\pi} [i, s]^2 \sin \theta \, d\theta$ for brevity, and let $[i, s]'$ indicate the same series as $[i, s]$, but with θ' written for θ.

Now since A_s, B_s depend only on definite integrals, we may in those integrals write θ' and ϕ' for θ and ϕ.

* See Thomson and Tait's *Natural Philosophy*, pp. 159, 160.

Then the term of S_i

$$[i, s]\{A_s \cos s\phi + B_s \sin s\phi\}$$

$$= \frac{[i, s]}{C}\left\{\cos s\phi \iint F(\theta', \phi')[i, s]' \cos s\phi' \sin \theta' d\theta' d\phi'\right.$$

$$\left. + \sin s\phi \iint F(\theta', \phi')[i, s]' \sin s\phi' \sin \theta' d\theta' d\phi'\right\}$$

$$= \frac{1}{C}\iint F(\theta', \phi')[i, s][i, s]' \cos s(\phi - \phi') \sin \theta' d\theta' d\phi'$$

and all the other terms of S_i may be treated in the same way.

Now it is proved in Thomson and Tait's *Nat. Phil.* (p. 160) that

$$\int_0^\pi \{[i, s]\}^2 \sin \theta \, d\theta = \frac{2}{2i+1} 2^{2s} (s!)^2 \frac{(i-s)!}{(i+s)!}$$

Hence

$$\frac{1}{C}[i, s][i, s]' \cos s(\phi - \phi') = \frac{2i+1}{2\pi} \cdot \frac{1}{2^{2s}(s!)^2} \cdot \frac{(i+s)!}{(i-s)!}[i, s][i, s]' \cos s(\phi - \phi')$$

Also the general term of the biaxal spherical harmonic Q_i is

$$2 \cdot \frac{1}{2^{2s}(s!)^2} \frac{(i+s)!}{(i-s)!}[i, s][i, s]' \cos s(\phi - \phi')$$

Hence, it follows that

$$S_i = \frac{2i+1}{4\pi}\iint F(\theta', \phi') Q_i \sin \theta' d\theta' d\phi'$$

This form of S_i is therefore the best harmonic representation of $F(\theta, \phi)$ of the order i.

But initially we were not constrained to take S_i as a complete harmonic, and any term, or all the terms but one might have been wanting; thus each individual term in this form of S_i is the best harmonic representative of the order i and class s.

Now suppose that there had been more than one term in Y initially; then since the integral of the product of two harmonics of different orders over the sphere is zero, the determination of the constants in each term will be precisely the same as when there was only one term. Thus the best representation of $F(\theta, \phi)$ is

$$\Sigma \frac{2i+1}{4\pi}\iint F(\theta', \phi') Q_i \sin \theta' d\theta' d\phi'$$

The summation being made for all the values of i which exist in Y, with the omission of any terms in each integral which were wanting initially from the completeness of each constituent harmonic.

If all the harmonics are complete, and if there are an infinite number of them, the representation of $F(\theta, \phi)$ is perfect, for it is clear that the higher orders of harmonics must represent less important features in $F(\theta, \phi)$ than the lower, or, in other words, the series is convergent.

We thus obtain the well-known expansion of a function in a series of spherical harmonics; but the present way of looking at the matter shows that it is not merely the series as a whole which represents the function, but that each individual term and sub-term acts independently and contributes the best it can, as though it alone existed.

In the foregoing proof the value of the integral

$$\int_0^\pi \{[i, s]\}^2 \sin \theta \, d\theta$$

is quoted from Thomson and Tait; the proof, however, given in their work does in fact involve this expansion of a function in spherical harmonics, but since $[i, s]$ is merely a finite trigonometrical series the same result might be obtained by simple algebra, though (as remarked by them) possibly it might not be an easy piece of analysis.

Where the space, throughout which the best representation of $F(\theta, \phi)$ is to be found, does not comprise the whole sphere, there is no doubt but that by the choice of proper harmonics of fractional or imaginary orders, a parallel theorem might be proved, but I am unable at present to undertake the investigation.

14.

ON FALLIBLE MEASURES OF VARIABLE QUANTITIES, AND ON THE TREATMENT OF METEOROLOGICAL OBSERVATIONS.

[*Philosophical Magazine*, Vol. IV. (1877), pp. 1—14.]

IF we make any observation, for example the transit of a star, a definite numerical result is obtained. To say that that result is liable to errors of observation, is only correct from one point of view. It is true that the result does not really correspond with the time at which the star crossed the meridian, yet it *is* an accurate representation of a certain very complex event. Undoubtedly, the principal feature in that event is the time of crossing the meridian; but there is also involved in it various properties of the instruments, the atmosphere, and the observer himself, &c. The object of the observation is, of course, to get a result which shall represent that principal feature, after the elimination of the minor features. The comparative simplicity of this, and of many other observations, permits us to unravel the complex event into its constituent parts, and to estimate each numerically. One part consists of corrections of all sorts. But when all these have been made, there still remains a result representing a complex event, viz. the transit of the star, together with unknown properties of the circumstances of the observation. These unknown properties form the subject-matter of the theory of errors of observation; but it is only because there is a consistent theory of the principal phenomenon, that the most probable line of demarcation can be drawn between the two parts of the complex event. The final result is given as a definite value with a margin of uncertainty in either direction.

But in the case of astronomical observations there is complete certainty that we cannot have affected the principal event in any way by the method of observing. In experiments, however, it is impossible to imitate exactly the proposed conditions; so that, even when corrections have been applied, and when we can estimate the degree of uncertainty in the method of observing, there remains another sort of uncertainty, viz. as to the closeness with which the proposed conditions were imitated; that is to say, the principal event is rendered uncertain and complex.

The case of experiments graduates into that of observations of natural phenomena, where we have no control over the disturbing causes, and have no opportunity of slightly altering the conditions.

The line of demarcation between the principal event, whose laws are to be determined, and the disturbances, here becomes still more undefined. And where we are still groping after a law of the phenomena (as in the case of meteorology) it is unknown what is to be classed as the principal event and what as disturbances. It is like looking at a series of irregular waves with ripples of various sizes on their surface; until some law in the formation of the waves is discovered, it is unknown how large a ripple may be neglected in the discovery of that law. Nevertheless the only chance of discovery seems to be to neglect the ripples by some arbitrary rule, and to examine the main features of the series of waves.

The problem of how best to combine a number of discontinuous observations into a continuous law, so as to give a general representation of them after disturbances due to fallibility of measures, runs of chance (as in statistics), &c. have been set aside, is one that constantly presents itself for solution; and a rational and methodical treatment can hardly fail to be of value.

The most frequent occasion for the solution of the problem arises from the necessity of drawing a curve passing close to the extremities of a number of ordinates; and the usual way of solving it is to draw a curve without abrupt changes of curvature as close to the points as possible. If the changes of curvature are abrupt enough, the curve may be made to pass exactly through the points; but then each observation is treated as exact, and we have exactly the case of the series of waves with ripples on them. But, by what precedes, it appears that we had better omit the ripples; and the question remains as to how far we are justified in smoothing down the curve. This process of smoothing is often done by the free hand; but it will probably be done better by a system; and it will be an additional advantage if the system admits of arithmetical as well as graphical application.

In those cases in which an algebraic law can be assigned, to which the ordinates ought to conform, the best method of treating the problem is to determine the constants involved in the function by the method of least squares; but it might often be not worth while to carry out this process, as, for example, where the deviations of the various observations from the law are large, and where it would accordingly be pedantic to assign values to the constants with precision.

Where the law is unknown and the observations are equidistant, a method of treatment might, perhaps, be devised by the assumption of some form of function containing fewer constants than the number of given points, and consisting of a number of simple harmonic terms, none of which go through a large fraction of their period in passing from one ordinate to the next. The constants involved might then be determined by the method of least squares, so that the function should give the best representation of the observations. But the assumption of the form of function would be arbitrary, and the process very laborious.

On the whole it will be more convenient and equally satisfactory, as far as the result is concerned, to proceed empirically from the first, remembering that the main object is to exclude ripples of short period.

The method here suggested is one which I believe is used in some form or other by meteorologists; but I am not aware that its merits have been discussed, or that it has been extended to the smoothing of surfaces and of functions of three or more independent variables, as I here propose to do*.

Empirical Rule.—The observations are supposed to be equidistant and to be functions of only one independent variable. The method may be most easily explained geometrically, and the transition afterwards made to its arithmetical equivalent. It will be convenient also to speak of the deviations of the several observations from the principal part of the complex event which those observations represent as *errors*.

It is proposed to substitute for each pair of consecutive points A, B, a point P which bisects the straight line AB. The points P then lie on a series of ordinates halfway between the original ones.

If the errors of A and B are of opposite sign, P is a better point than either of them; if they have errors of the same sign, P will be better or worse according to the direction of the curvature of the curve. But if the rate of change of curvature of the curve is small (as it must be assumed to be to justify the smoothing process), P is very little better or worse. Now, as on the average the series of points deviate as often to one side as to the other of the curve, there clearly will be on the average an improvement in accuracy from the substitution of P, whilst there will certainly be less abrupt changes of curvature in a curve passing through P than through AB. Where the points already lie on a fair curve with no contrary flexure, the chance is rather more than even that there will be a loss of accuracy of representation, because the substituted points all lie on the same side of the given ones, and the only case where there is improvement is where all the errors have one particular sign, and are not very small. The process of smoothing must then be applied cautiously, and especially at maximum- and minimum-points.

---

* Since this paper has been in the hands of the printer, I have learnt that M. Schiaparelli has written a work entitled *Sul modo di ricavare la vera espressione delle leggi della natura dalle curve empiriche* (Milan, 1867), and that M. De Forest has written on the subject in the *Annual Reports of the Smithsonian Institution* for 1871 and 1873, and in the *Analyst* (Iowa) for May 1877.

If the points P do not lie on a fair curve, the process may be applied again in part of the series or along the whole line; but when once our judgment leads us to think that the curve is smooth enough, every succeeding operation tends to spoil the representation.

Analytically the process may be stated as follows:—

If y_0, y_1, y_2, ... be the successive given ordinates, and if ϕ indicates a single smoothing operation, so that ϕy_x indicates the substituted ordinate corresponding to the abscissa x; then clearly

$$\phi y_x = \tfrac{1}{2}\left(y_{x+\frac{1}{2}} + y_{x-\frac{1}{2}}\right)$$

and generally

$$\phi^n y_x = \left(\frac{E^{\frac{1}{2}} + E^{-\frac{1}{2}}}{2}\right)^n y_x$$

It is clear that an odd number of operations will leave us with points on ordinates halfway between the original ones, whilst an even number will leave us on the original ones. There is a practical advantage in proceeding by two operations at a time, because it is not then necessary to draw the intermediate ordinates, and because a double operation has a very simple geometrical and analytical meaning.

From the above formula,

$$\phi^2 y_x = \tfrac{1}{2}\left\{y_x + \frac{y_{x+1} + y_{x-1}}{2}\right\}$$

If in the figure MA, NB, QC are the three ordinates y_{x-1}, y_x, y_{x+1}, then $Nb = \tfrac{1}{2}(y_{x+1} + y_{x-1})$, and P, which bisects Bb, is the point to be substituted for B.

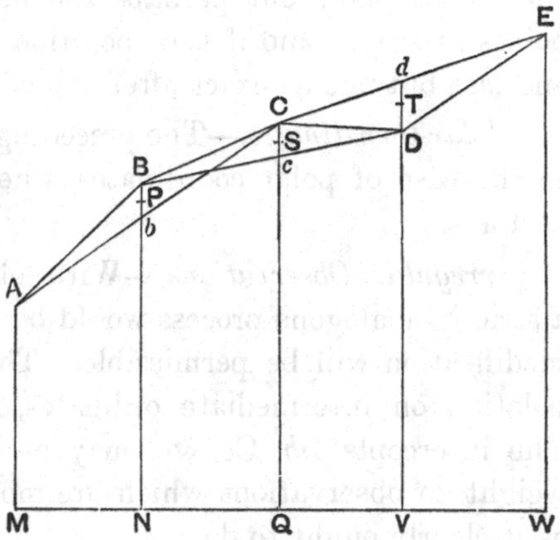

The practical rule of construction may be stated thus:—

Let A, B, C, &c. be the given points; join every point to that next it and next but one to it. Then the points to be substituted bisect the intercepts Bb, Cc, &c. in P, S, &c. If the curve which may be drawn through P, S, &c. still seems too sinuous, repeat the operation.

Fig. 1.

The arithmetical application of this process is obviously very simple; for

$$\phi^2 y_x = y_x + \tfrac{1}{4}\left\{y_{x-1} - 2y_x + y_{x+1}\right\}$$

$$= y_x + \tfrac{1}{4}\Delta^2 y_{x-1}$$

and therefore the correction to be applied to any ordinate y_x is $\tfrac{1}{4}\Delta^2 y_{x-1}$.

We can see how it is that this process tends to improve the curve. The observed or given values of the function consist of two parts, the first representing the principal event or wave whose law of variation is to be found, and the second the errors or ripples which are to be eliminated. Now the observations are supposed to be so close as not to admit of very large differences between the successive values of the *principal event*; and therefore their second differences will be small. On the other hand, the errors will be some positive and some negative; and therefore their second differences will be very irregular, and probably much larger on the whole than the errors themselves. The second differences of the observed values are the sums formed by the addition of these two sets of second differences; and the justifiability of the process depends on the assumption that the increase of the latter will be sufficient to render the diminished values of the former insignificant by comparison. We thus obtain a series of quantities which depend principally on the errors, except in case the errors are small, or in case of a run of luck in the signs and magnitude of the errors, such as to make them apparently conform to law and thus present small second differences. Now the proposed corrections to the various observed values are the quarters of this series of quantities; and thus in all probability our corrections depend principally on the errors. The process is therefore justifiable unless the points already lie in a smooth curve. The rough criterion of the applicability of the smoothing process is that the second differences of the observed values should not appear to conform to any law.

Every double operation causes the loss of one point at the beginning and one at the end; but perhaps the best course is to treat the first and last points as exact; and if the operation is repeated more than twice, the second and last but one as exact after one of these double operations, and so on.

Polar Coordinates.—The preceding method is applicable with equal justice to the case of polar coordinates, where the ordinates are replaced by radii vectores.

Irregular Observations.—With observations which are not equidistant, a strictly analogous process would be complex; but as it is empirical, a slight modification will be permissible. Thus we may omit the analogue of interpolation on intermediate ordinates, and only retain the double operation. The intercepts Bb, Cc, &c. may be bisected as before; for this gives less weight to observations which are more remote than to those which are near, as it clearly ought to do.

The corresponding numerical rule is to substitute for the ordinate y_r, the value

$$\frac{(x_{r+1} - x_r)(y_{r-1} + y_r) + (x_r - x_{r-1})(y_r + y_{r+1})}{2(x_{r+1} - x_{r-1})}$$

where
$$x_{r-1},\ y_{r-1};\ \ x_r,\ y_r;\ \ x_{r+1},\ y_{r+1}$$
are the coordinates of the three successive points.

The merits of an empirical rule like this must of course depend on how it seems to work practically. I therefore devised the following scheme for testing it. A circular piece of card was graduated radially, so that a graduation marked x was $\dfrac{720}{\sqrt{\pi}} \displaystyle\int_0^x e^{-x^2} dx$ degrees distant from a fixed radius. The card was made to spin round its centre close to a fixed index. It was then spun a number of times, and on stopping it the number opposite the index was read off*. From the nature of the graduation the numbers thus obtained will occur in exactly the same way as errors of observation occur in practice; but they have no signs of addition or subtraction prefixed. Then by tossing up a coin over and over again and calling heads + and tails −, the signs + or − are assigned by chance to this series of errors. About a dozen equidistant values of some function (say sine or cosine) were next taken from a Table, and the errors added to or subtracted from them in order. The errors may be made either small or large by multiplying them by any constant. The falsified values may then be fairly taken to represent a series of observations; but we here know what are the true ones. The corrections were then applied, in some cases arithmetically and in others graphically, and the deviations of the corrected values from the true were observed.

In other cases a series of equidistant ordinates were taken, and a sweeping free-hand curve was drawn to represent the true curve, and the several ordinates of this curve were falsified by the roulette and then corrected by a graphical application of the rule. The general result of a good many trials was such as to justify the smoothing process. When the errors were considerable the mean error was much reduced, although the actual error of some ordinates was increased; where the errors were very small the mean error was even slightly increased. Although the danger of over-smoothing was obvious, and the sharpness of the features of the curve was generally diminished, yet I think it was clear that the method might generally be employed with advantage, especially in such cases as the attempt to deduce some law from statistics or a series of barometric oscillations of considerable periods. The errors must be very large to justify a quadruple operation. This method of trial could not be so well applied to testing the case of an odd number of smoothing operations, where we are left finally at intermediate ordinates.

On the whole, I think the process is justifiable if applied with caution. Nevertheless it undoubtedly tends to spoil the results if applied to a series of points which are already in a sweeping curve; and therefore I have tried to find some other process which should not have this disadvantage. This can only be done by taking more than three points of the curve into consideration; and therefore the process must be more cumbrous.

---

* It is better to stop the disk when it is spinning so fast that the graduations are invisible, rather than to let it run out its course.

The method pursued was as follows :—

Let $-2,\ y''$; $-1,\ y'$; $0,\ y$; $1,\ y_1$; $2,\ y_2$ be the coordinates of five consecutive points on the curve. Suppose them to be represented by a curve whose equation is $y = a + bx + cx^2 + dx^3$; and make the following expression a minimum, viz.

$$\Sigma\,(a + bx + cx^2 + dx^3 - y)^2$$

where the summation is made for the values of $x = -2, -1, 0, 1, 2$, and the corresponding values of y. In other words, the values of a, b, c, d are to be determined by the method of least squares, so that this curve shall give the best representation of the five points.

The equations for finding a, b, c, d are therefore

$$5a\quad + b\Sigma x + c\Sigma x^2 + d\Sigma x^3 = \Sigma y$$
$$a\Sigma x + b\Sigma x^2 + c\Sigma x^3 + d\Sigma x^4 = \Sigma xy$$
$$a\Sigma x^2 + b\Sigma x^3 + c\Sigma x^4 + d\Sigma x^5 = \Sigma x^2 y$$
$$a\Sigma x^3 + b\Sigma x^4 + c\Sigma x^5 + d\Sigma x^6 = \Sigma x^3 y$$

From the manner in which the origin has been chosen the sums of the odd powers of x are all zero, and $\Sigma x^2 = 10,\ \Sigma x^4 = 34,\ \Sigma x^6 = 130$.

Thus the first and third equations are

$$5a + 10c = y'' + y' + y + y_1 + y_2$$
$$10a + 34c = 4y'' + y' + y_1 + 4y_2$$

and the second and fourth may be easily written down. It will be noticed that the first and third equations would be exactly the same if we assumed as the form of the equation $y = a + bx + cx^2$.

Now the proposed method is to substitute for every point of the series of given points the intersection with the ordinate of that point of the curve of the form $y = a + bx + cx^2$ or $y = a + bx + cx^2 + dx^3$ which best represents that point and the two preceding and two succeeding points. In the case we have been considering we are, therefore, to substitute for the point $0, y$ the intersection of this curve with the axis of y ; that is to say, we are to substitute the point $0, a$, because when $x = 0, y = a$.

Now $\qquad\qquad 35a = -3y'' + 12y' + 17y + 12y_1 + 3y_2$

or $\qquad\qquad\quad a = y + \tfrac{3}{35}\{-y'' + 4y' - 6y + 4y_1 - y_2\}$

$$= y - \tfrac{3}{35}\Delta^4 y''$$

Thus the correction δy, to be applied to y, is $-\tfrac{3}{35}\Delta^4 y''$.

Hence generally, since the process is supposed to be applied all along the series,

$$\delta y_x = -\tfrac{3}{35}\Delta^4 y_{x-2}$$

To give a geometrical meaning to the rule, it may be observed that

$$-y'' + 4y' - 6y + 4y_1 - y_2 = -(y'' - 2y + y_2) + 4(y' - 2y + y_1)$$

and therefore if Δ' be a symbol denoting the operation of differencing with the omission of alternate ordinates,

$$\delta y_x = \tfrac{12}{35}\{\Delta^2 y_{x-1} - \tfrac{1}{4}\Delta'^2 y_{x-2}\}^*$$

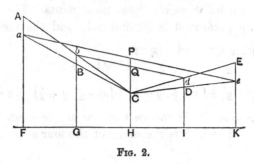

FIG. 2.

Now in the figure let AF, BG, CH, DI, EK be any five consecutive ordinates, and suppose that it is proposed to correct the ordinate CH. Then, if the construction shown in the figure be carried out, it is clear that

$$CP = \Delta^2(BG), \text{ and } CQ = \tfrac{1}{4}\Delta'^2(AF)$$

and therefore $$PQ = \Delta^2(BG) - \tfrac{1}{4}\Delta'^2(AF)$$

Thus the correction to be applied to the ordinate HC is $\tfrac{12}{35}$ (or very nearly $\tfrac{1}{3}$) of PQ. The same process must be applied all along the series for each set of five points.

Four points are lost out of the series, two at each end. For example, if A, B are the first two points, the rule gives no substituted points on those ordinates. The results obtained from the use of this rule do not seem markedly superior to those given by the empirical method, except where the points lie in a fair curve; and as the rule is more cumbrous to apply, it does not seem likely to be of much practical value.

The construction of a fair surface near a number of points.

The preceding process may be extended to the case where the function involves two independent variables. The observed values may, for the sake of clearness, be considered as consisting of a number of ordinates standing on the intersections of the lines of a chess-board, of which two intersecting edges are the axes of x and y.

---

* If the points lie in a fair curve, $\Delta^2 y_{x-1} - \tfrac{1}{4}\Delta'^2 y_{x-2}$ is very small, a property which I have used to give a rule of graphical interpolation on intermediate ordinates (see Paper 12, p. 285). Thus in this case the correction applied is very small.

Let $[x, y]$ indicate the given ordinate which stands on the point x, y; let E, Δ be the operations of writing $x + 1$ for x, and of differencing with respect to x; and let E, D be the like operations with respect to y. Let ϕ represent a single smoothing operation on any series of ordinates which are in a plane parallel to x; and ψ the same with respect to y.

Now apply a smoothing operation ϕ to all the points parallel to x, and then apply the operation ψ to all these new points. The order in which these operations are performed is immaterial; and the result is

$$\phi\psi\,[x,\,y] = \frac{E^{\frac{1}{2}} + E^{-\frac{1}{2}}}{2} \cdot \frac{E^{\frac{1}{2}} + E^{-\frac{1}{2}}}{2}\,[x,\,y]$$

$$= \tfrac{1}{4}\,\{[x+\tfrac{1}{2},\,y+\tfrac{1}{2}]+[x+\tfrac{1}{2},\,y-\tfrac{1}{2}]+[x-\tfrac{1}{2},\,y+\tfrac{1}{2}]+[x-\tfrac{1}{2},\,y-\tfrac{1}{2}]\}$$

This, interpreted geometrically, means that we are to erect in the middle of each square an ordinate which is the mean of the four surrounding ordinates. Again,

$$\phi^2\psi^2\,[x,\,y] = \tfrac{1}{16}\,\{[x+1,\,y+1]+[x+1,\,y-1]+[x-1,\,y+1]$$
$$+\,[x-1,\,y-1]+2\,([x+1,\,y]+[x-1,\,y]+[x,\,y+1]$$
$$+\,[x,\,y-1])+4\,[x,\,y]\}$$

If the figure represents any four squares of the chess-board (in which observe that nine ordinates stand on the intersections), the rule given by a double operation is to substitute for the ordinate at K,

$\tfrac{1}{16}$ of sum of ordinates at A, B, C, D

$+ \tfrac{1}{8}$ of sum of ordinates at E, F, G, H

$+ \tfrac{1}{4}$ of ordinate at K.

It is clear that the operations of smoothing parallel to the two axes are quite independent, and that there is no necessity to smooth the same number of times in each direction. The symbolical way of writing the operation makes it perfectly easy to construct any desired modification of this formula, where ϕ and ψ are each performed any number of times.

The process may also be extended with equal justice to the case where there are three independent variables, although that case no longer admits of geometrical interpretation.

Let t be a third variable; then, if the former notation be extended, and if only a type of each form of term be written down preceded by the sign of summation, it will be found that

$$\phi^2\psi^2\chi^2\,[x,\,y,\,t] = \tfrac{1}{64}\,\{\Sigma\,[x\pm1,\,y\pm1,\,t\pm1]+2\Sigma\,[x,\,y\pm1,\,t\pm1]$$
$$+\,4\Sigma\,[x,\,y,\,t\pm1]+8\,[x,\,y,\,t]\}$$

There are eight terms of the first kind, such as $[x+1, y-1, t+1]$; twelve of the second, such as $[x+1, y, t-1]$; six of the third, such as $[x+1, y, t]$; and only one of the last, viz. $[x, y, t]$.

Application to Ocean Meteorology.—This last process appears to me to be applicable here. Meteorologists have divided the ocean up into squares of 5° of latitude and 5° of longitude. The logs of ships sailing over those squares are consulted for meteorological observations; and the results are classified by months and squares; and the mean result for any one element, such as the height of the barometer, is taken to be the average for the middle of that square and the middle of that month*. Now it seems as though this were a case where smoothing is justifiable, and that it is allowable to make the result for each month depend in some degree on its neighbours both in space and time. There are three independent variables, viz. latitude, longitude, and time; and in the previous formula we may take these to be represented by x, y, and t respectively.

Suppose, for example, we want to modify the mean height of the barometer for any square e for, say, February, both with reference to surrounding squares and to the heights for January and March. Then the rule for finding the amended height is:—Take the sum of the heights for a, c, g, k for January and for March + twice the sum of the heights for b, d, h, f for January and March + twice the sum of the heights for a, g, c, k for February + four times the sum of the heights for b, d, h, f for February + four times the sum of the heights for e for January and March + eight times the height for e for February, and divide the result by 64.

a	b	c
d	e	f
g	h	k

It must be observed that it is not necessary that the smoothing should be carried to the same extent for x, y, and t. If, for example, we wish to smooth only once for time, the formula will be different, and the result will be applicable to the beginnings of the months instead of the middles. A knowledge of the particular requirements of the case is the only guide to the amount of smoothing which is expedient; but the formulæ are so easy to construct that it does not seem worth while to give any other forms.

In concluding this part of the subject, I may mention that the proposed processes may be extended so as to allow various weights to the various observations.

Terrestrial Meteorology.—There are a number of observatories on the land at which observations are taken at the same hours of the day all over the country; from these results, maps are drawn showing the form of the

---

* I owe this explanation to Mr (now Sir) Francis Galton.

"isobars" for each day. After the observations have been reduced to the sea-level and corrected in other ways, they may be considered as correct, and the isobars give a graphical illustration of the successive deformations of the barometric surface. Land meteorology serves, then, to give quite a different kind of result from those of ocean meteorology. In the latter the result is the mean heights of the barometer at stated places and times. The oceanic barometric surface, as far as we know it, is ideal, and does not correspond with its real form at any one time. In oceanic meteorology the smoothing process seems justifiable; for we only seek to study the main features of the changes. In land meteorology this is not the case; for we seek to discover the details of the changes. To return to the former metaphor—in one case the law of the waves is sought, in the other the law of the ripples.

The observatories are scattered irregularly over the country; and it seems probable that the results would be more useful and more easily interpreted if they could be distributed at regular intervals of space. They are already regularly distributed as regards time. My present object is, then, to give a formula (which is, as far as I am aware, new) for the reduction of observations scattered irregularly, to regular stations equidistant in latitude and longitude. It is a problem in interpolation of the ordinary kind where the ordinates are not fallible.

The problem is to find a continuous surface passing through the tops of a number of irregularly spaced ordinates; and it may be solved by an extension of Lagrange's well-known formula for interpolation in two dimensions.

Let x_0, y_0; x_1, y_1; ...; x_n, y_n be the coordinates (latitude and longitude) of a number of points, and let $z_0, z_1, ..., z_n$ be ordinates (barometric heights) corresponding to these points. Lagrange's formula suggests the following as the equation to a surface passing through the tops of $z_0, z_1, ..., z_n$:

$$z = z_0 \frac{(x - x_1)(y - y_1)(x - x_2)(y - y_2) \ldots (x - x_n)(y - y_n)}{(x_0 - x_1)(y_0 - y_1)(x_0 - x_2)(y_0 - y_2) \ldots (x_0 - x_n)(y_0 - y_n)}$$

$$+ z_1 \frac{(x - x_0)(y - y_0)(x - x_2)(y - y_2) \ldots (x - x_n)(y - y_n)}{(x_1 - x_0)(y_1 - y_0)(x_1 - x_2)(y_1 - y_2) \ldots (x_1 - x_n)(y_1 - y_n)}$$

$$+ \ldots\ldots$$

$$+ z_n \frac{(x - x_0)(y - y_0)(x - x_1)(y - y_1) \ldots (x - x_{n-1})(y - y_{n-1})}{(x_n - x_0)(y_n - y_0)(x_n - x_1)(y_n - y_1) \ldots (x_n - x_{n-1})(y_n - y_{n-1})}$$

Then this formula will give the height z of the barometer at any station whose latitude and longitude are x, y, as deduced from the heights at the several observing-stations. The applicability of this interpolation depends, of course, on the assumption that the surface is not contorted between the given ordinates; and if the observatories are numerous enough, this assumption is probably justifiable.

The application of the formula would in general entail a great detail of arithmetic; but in the case of the reduction from irregular to regular stations, the great mass of the work might be done once for all. In this case the co-ordinates of the observing stations x_0, y_0; x_1, y_1; ... are the same day after day, and the coordinates of the fixed stations x, y are constant for each of them. Hence the coefficients of z_0, z_1, ... in the formula may be calculated once for all.

It would be very laborious and unnecessary to make the heights of the barometer at the equidistant stations depend on all the observatories in the country; and it would be probably quite sufficient to make each one depend on the five or six nearest observatories. The practical rule would then run somewhat in this fashion (the numbers being purely hypothetical):—

$$\left.\begin{array}{l}\text{Height of bar at lat. } 51° \\ \text{long. } 1° \text{ W.}\end{array}\right\} = \cdot705 \text{ Oxford} + \cdot20 \text{ Kew} \\ + \cdot092 \text{ Southampton} + \cdot002 \text{ Cambridge.}$$

Every separate point to which the reductions were to be made would require a different set of coefficients, which would depend on the four, five, or six nearest actual observing-stations.

If the heights of the barometer were taken as the excess above 28 inches, the various heights need not be given to more than three figures; and as the coefficients would probably have also three figures, the multiplications might be very easily made by means of Crelle's *Rechentafeln*. By these means the daily observations might be very quickly reduced, and the results of each day's observations would be given by a series of numbers on a map spaced out at regular intervals of latitude and longitude. This would, I think, facilitate the drawing of the "isobars," and it would also be more intelligible than are the results as given at irregularly dispersed stations.

It may be noticed that the same set of coefficients would also be proper for the reduction of any other meteorological element which could be fairly represented by a surface. The calculation of the coefficients would be rather laborious; but if there is any real advantage in thus classifying the observations, this would be of slight consequence, as the work would be performed once for all.

In conclusion, I will add one other rule—namely, for interpolation between the oceanic meteorological observations when smoothed, as before suggested. This is a formula for interpolation in the case of a function of three independent variables, the values of which are given at equal intervals, as is the case in the mean barometer-heights in latitude, longitude, and time.

Let Δ, D be the differences between successive barometer-heights in latitude and longitude respectively, and δ the difference in time (that is to say, between the values for successive months). Then, following the former notation,

$$[x + \xi,\ y + \eta,\ t + \tau] = [x,\ y,\ t] + \xi\Delta + \eta\mathrm{D} + \tau\delta$$

$$+ \frac{1}{1 \cdot 2} \{\xi\,(\xi - 1)\,\Delta^2 + \eta\,(\eta - 1)\,\mathrm{D}^2 + \tau\,(\tau - 1)\,\delta^2$$

$$+ 2\xi\eta\Delta\mathrm{D} + 2\eta\tau\mathrm{D}\delta + 2\tau\xi\delta\Delta\}$$

$$+ \dots\dots$$

The proof of this will be obvious to those acquainted with the Calculus of Finite Differences. No doubt it has been given before, although I do not happen to have met with it. This formula enables us to pass from the regular equidistant values for the middles of squares and months to those for any other neighbouring time and place.

15.

ON THE HORIZONTAL THRUST OF A MASS OF SAND.

[*Proceedings of the Institution of Civil Engineers*, Vol. LXXI. (1883), pp. 350—378.]

§ 1. *Account of Experiments.*

THE pressure of loose earth against revetment walls is a subject which is frequently being brought in a practical manner under the notice of engineers. A considerable number of theoretical investigations have been published by Coulomb, Rankine, Lévy, Boussinesq, and others, on this subject, but it appears that there is a singular deficiency of experimental data for testing the accuracy of the results of theory. The following Paper contains an account of some experiments made with that view in the summer of 1877. Circumstances prevented the Author from carrying out his original intention of experimenting with various forms of apparatus and various materials. Although, then, the investigation is somewhat incomplete, yet it seemed to have been carried far enough to establish several propositions of interest with regard to the horizontal thrust of sand. As will be seen below, the nature of those conclusions was such that the renewal of the experiments seemed inexpedient. An important paper by Mr [afterwards Sir] Benjamin Baker*, M. Inst. C.E., on the lateral pressure of earthwork, was read before the Institution of Civil Engineers in 1878, and excited a discussion of much interest. That paper, which has only recently been brought under the Author's notice, afforded the immediate inducement for bringing forward his results.

It is certain that, unless the theory agrees well with the facts, when the subject of experiment is of the finest and most uniform material, it will be very unlikely to give good results when the material is of the kind actually

---

* "The Actual Lateral Pressure of Earthwork," *Minutes of Proceedings Inst. C.E.*, 1879—80, Vol. LXV., p. 140.

occurring in embankments. Several specimens were accordingly obtained of the sand which is used in sawing marble, and, as the apparatus was to be on a small scale, the finest grained of the samples was chosen. It was the carefully washed and sieved road-scrapings from a flinty country, and consisted almost entirely of fine fragments of flint. When dried it formed a fine powdery sand with no large fragments in it. According to the received theory, the mechanical properties of a granular substance are completely determined when its specific gravity, and its angle of repose, are known, that is to say the greatest inclination to the horizon at which a talus will stand.

The angle of repose of the sand was determined by several experiments to be 35°, but this did not appear to be a constant which it was possible to determine within less than 1°, because in forming a talus there occurs some automatic sorting of the sand into finer and coarser particles. By the weighing of a vessel, into which the sand was poured as lightly as possible, the specific gravity in that condition was determined as 1·40. But when the sand was jarred, shaken, and thoroughly stirred up with a stick, the specific gravity rose to 1·55.

The effect of the stirring was peculiar. When the stick was first introduced, it easily penetrated to the bottom of the vessel, but after a little stirring the substance could be felt to stiffen at the bottom, and the stiffening then crept upwards, so that after a time the sand became impenetrable to the stick. This is not very surprising, seeing that after stirring the sand occupied a tenth less volume than before.

Here already there is a certain deficiency in the mathematical theories, since this very large variability of specific gravity is not taken into account; and whilst actual embankments no doubt correspond much more nearly to the sand in close order than in open order, the angle of repose of shaken sand is a phrase without a meaning. When the Author was beginning these experiments he had the advantage of discussing the subject with the late Professor Clerk Maxwell, who remarked that he supposed that the "historical element" would enter largely into the nature of the limiting equilibrium of sand. By this he meant that sand when put together in different ways would exercise different thrusts, although presenting visibly the same external appearance. The Author kept this valuable remark before him throughout, and found that Maxwell's conjecture was correct. The historical element is one which essentially eludes mathematical treatment.

For the purpose of experimenting a box was made, one end of which was a door turning on horizontal hinges at the bottom; the box was to be filled with sand, and it was then proposed to determine the least force capable of sustaining the door in position. Fig. A is a diagram of the apparatus in section. The supports for the various pulleys and accessories are omitted for the sake of simplicity.

The box *ABCD* is of the dimensions shown in the figure in centimètres and inches, and the width perpendicular to the section is 30·5 centimètres, or a foot, inside measure. The side *BC* was made to slide in and out, in order to facilitate the operation of filling the box with sand. It was originally

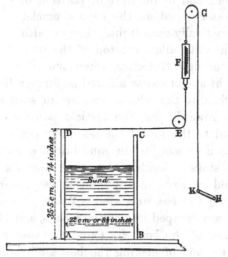

Fig. A. (Scale $\frac{1}{12}$ inch to 1 inch.)

more remote from *AD*, but the operation of filling the box was found so laborious that it was moved up closer, to the distance shown. The sides *BC*, *AD* had films of sand glued to their surfaces. The two vertical sides parallel to the section (which are not, of course, shown) were prolonged for some way beyond *A*. The side *AD*, which is here called the door, turns on hinges at *A*, and is about 30 centimètres broad. The bottom of the box *AB* is of double thickness as far as *A*, to admit of the attachment of the hinges at *A*. A strip of soft silk was glued over the hinges *A*, to prevent their becoming clogged with sand. It is obvious that the door turned between a pair of vertical walls. As the door was to be easily movable, its edges could not be in absolute contact with the vertical sides of the box, and thus there were necessarily small cracks by which the sand might escape, or perhaps cause the door to jam against the sides. After some trouble this difficulty was overcome in a manner which seemed quite satisfactory. Each edge of the door would have left a crack of something less than $\frac{1}{4}$ millimètre open between it and the vertical side of the box, had the sides been flat all along. Fig. B exhibits one edge of the door and part of the vertical side of the box, as seen vertically from above when the door is upright. The flat side of the box is shown as scalloped out with a vertical groove running parallel with the edge of the door; in consequence of this arrangement the crack referred to was narrow when the door was upright, but became broader the moment the door yielded from its upright position. A narrow triangular piece of soft

20—2

silk was glued by one of its edges along the inner face of the door, and along
its other edge to the flat side of the box.   When the
door was upright, as shown in Fig. B, the silk formed a
conical tube, marked as " silk bag," lying in the groove
just outside the door; and in the upright position of
the door the silk just stopped up the narrow cracks,
and made the box practically sand-tight.   The breadth
of the silk was such as to allow the top of the door
to move outwards through 2 inches, when the silk

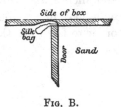

Fig. B.

became stretched tight and of course allowed no further displacement.   There
was a pair of stops fixed to the sides of the box in such a position that the
door could not turn inwards past the vertical position.   There was also a
counterpoise attached to the top of the door, but not shown in Fig. A, and
so arranged that the door was just in equilibrium when it was resting very
lightly against these stops.   Besides these stops there was a pair of bolts by
which the door could be fixed in any position very nearly vertical, during
the operation of filling the box with sand.   When sand was poured into the
box it was rare that any escaped into the silk bags, and the light contact of
the silk with the sides of the box was found to cause no sensible friction.
On the whole this method of making the box sand-tight, and at the same
time avoiding the friction of the door against the sides, was found to be far
more satisfactory than was expected, and it may be accepted that the results
have not been sensibly vitiated by friction.   And even if there were this
cause of error, the greater part of the conclusions would be unaffected by it.
The inner face of the door had a thin coating of sand glued to it, so that the
friction of the sand against the door might be equal to the internal friction
of the loose sand.   The door was also graduated with a scale of centimètres,
for the purpose of reading off the depth of the sand.

The box having been now described, the method must be explained for
recording the couple acting on the door, when the sand was just on the point
of slipping.

To the centre of the top of the door there was hooked on a silk cord DE
(Fig. A), which ran horizontally to the pulley E, and then vertically to the
spring-balance F—one of Salter's graduated on the metric system.   From
the upper end of the balance F there ran another cord which passed over the
pulley G, and then ran vertically down to the spindle K, round which it was
wound several times.   The spindle K was formed of stout brass wire, and H
was a crank-handle, by means of which the silk cord could be wound upon or
unwound from the spindle K.   On unwinding the silk slowly from the
spindle, the tension on DE could be gradually relaxed, and the tension could
be noted at each instant by watching the index of the balance.

The method of observation was to place the box flat on the floor; bolt
the door in the vertical position; fill the box carefully, in the required

manner, to any required depth ; hook on the cord to the middle of the door ; raise the tension of the cord to such an amount as would certainly hold the door in position ; gently unbolt the door, and then gradually relax the tension by unwinding the spindle, whilst attentively watching the index of the spring-balance. When the door yielded there was a sudden motion of the index, and the position from which it yielded was noted down. The Author also found after some time that, by attentive listening, he could detect a yielding of the sand, so small as scarcely to move the index perceptibly ; the ear was thus made to confirm the eye in the observations.

Suppose now that the box has been filled with sand of specific gravity w to a depth l, and that at the moment at which the tension of the cord is relaxed down to the point at which the door yields, that tension is T grammes. At the instant of yielding, the couple tending to hold up the door is $T \times 35\cdot5$ gramme-centimètres.

Then let L be the couple required to hold up a strip of the door of 1 centimètre in breadth, on the supposition that the box is infinitely wide, and let b be the effective breadth of the box. Then clearly $Lb = T \times 35\cdot5$.

There is reason to believe, from the experiments of Series VII. below, that the sides of the box exercised but little influence in supporting the mass of sand, and for the reasons assigned below b may be taken as 29 centimètres, being 1 centimètre less than the actual width of the door.

Let ϕ be the angle of repose of the sand.

Then, according to Rankine's formula,

$$T \times 35\cdot5 = Lb = \tfrac{1}{6}wl^3b \tan^2\left(\tfrac{1}{4}\pi - \tfrac{1}{2}\phi\right)$$

And according to Boussinesq's formula, given in the discussion on Mr Baker's paper, above referred to, viz., equation (16), with b zero, because the wall turns about the lower edge in contact with the mass of sand, and with $\phi_1 = \phi$, because the friction against the wall is equal to the internal friction of the sand,

$$T \times 35\cdot5 = Lb = \tfrac{1}{6}wl^3b \tan^2\left(\tfrac{1}{4}\pi - \tfrac{1}{2}\phi\right) \frac{\cos\left(\tfrac{1}{4}\pi - \tfrac{1}{2}\phi\right)}{\cos\left(\tfrac{3}{2}\phi - \tfrac{1}{4}\pi\right)} \cos\phi$$

These formulæ will be considered below in detail, but at present it is only necessary to state that one important object of the experiments is to test their truth. At the time, indeed, of making the experiments the Author was not aware of Boussinesq's formula, and only made the comparison with Rankine's. It was intended to test the following points, viz., whether the oversetting couple L varies as the cube of the depth of the sand, and whether the constant introduced in order to express the actual amount of that couple corresponds in magnitude with either of the theoretical functions of the angle of repose of the sand. Experiments were also to be undertaken to find the couple L when the surface of the sand was not horizontal ; the theoretical formulæ for these cases will be given in the second part of this paper. Lastly, it was necessary to determine what amount of supporting influence was exercised by the sides of the box.

It will be seen below, in discussing the experiments, that these questions are scarcely susceptible of perfectly rigorous answers; but at present it is only necessary to explain the method by which the answers, as far as they are given, are to be extracted from the experimental results.

The graphical method of record was found to be well suited for the purpose; for the record of a great number of experiments could be made on a single figure, and the result to be deduced could be most easily extracted from the figure itself. The records will accordingly be here reproduced in the graphical form.

The horizontal axis is taken to indicate the depth l of the sand (Figs. 1—7 below), the numbers along the l-axis being centimètres of depth from the bottom of the box. The vertical ordinates represent T, the tension of the cord in grammes at the moment of slipping.

The graphical records are all drawn to the following scale:—The abscissæ, representing the centimètres of depth (l) of the sand, are drawn to the scale of $\frac{1}{8}$ inch to the centimètre, and the numbers represent centimètres. The ordinates are drawn to the scale of $\frac{2}{3}$ centimètre to 100 grams. In each Figure there is introduced a little diagram, to explain the method in which the box was filled with sand; that is to say, to record "the historical element." There are also drawn one or more cubical parabolas, of which the equation referred to the coordinate-axes is $T = \mu l^3$; the value of μ corresponding to each parabola is given. Then by mere inspection it was easy to see which parabola, if any, best agreed with the observations, and by a little simple arithmetic the numerical value of the function of the effective breadth of the door, and of the angle of repose of the sand, could be evaluated.

In fact, each set of the experiments gave the value of $Lb \div \frac{1}{6}wl^3$, which for the Series I. to IV. should, according to Rankine, be equal to $b \tan^2 (\frac{1}{4}\pi - \frac{1}{2}\phi)$, and according to Boussinesq should be

$$b \tan^2 (\tfrac{1}{4}\pi - \tfrac{1}{2}\phi) \cos (\tfrac{1}{4}\pi - \tfrac{1}{2}\phi) \sec (\tfrac{3}{2}\phi - \tfrac{1}{4}\pi) \cos \phi$$

On dividing these values by b (which is taken as 29 centimètres) the numerical values of the function of ϕ are obtained.

After these explanations, the records of the experiments themselves may be given.

It may be worth mentioning that there was at first an index attached to the door, by which any displacement of the door was much magnified. In Series I. and II. this index was still in use, but it was ultimately dispensed with, after some practice had shown it to be unnecessary.

Series I., Fig. 1.—The box was filled with sand with a shovel, or with a tin canister, and at each stage in the filling the strata were maintained nearly level, but it was not possible to make the strata truly level in this way.

The most prominent feature in this series of experiments, recorded in Fig. 1, was the difficulty of asserting that the sand slipped at any definite time. As soon as the tension T had fallen to within 100 or 200 grammes of the point at which equilibrium quite broke down, it was observable that there was an almost continuous yielding of the door as the tension was gradually relaxed. After the upper edge of the door had yielded by perhaps $\frac{1}{20}$ inch the motion began to be much more marked, and the diminution of tension was accompanied by obvious motions of the door, which yielded by little jerks. It was this marked yielding which was taken as the phenomenon

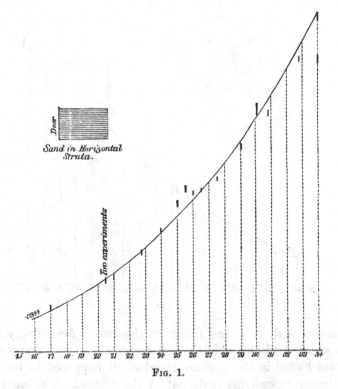

Fig. 1.

to be observed, and it is accordingly only possible to state that this took place with approximately a certain tension on the cord. This fact is noted in Fig. 1 by the record of each observation consisting of a vertical line of greater or less length, according as the jerking motion of the door ranged through a greater or less variation of tension. The curve inserted in the figure is the cubical parabola which, as far as may be judged, best satisfies the observations. Considering how much the element of judgment entered in this record, the experiments seem to accord tolerably well with the law that the oversetting couple varies as the cube of the depth of the sand.

This parabola shows that

$$T = 0\cdot0344 \; l^3 \text{ gramme}$$

$$Lb = 1\cdot221 \; l^3 \text{ gramme-centimètre}$$

$$= \tfrac{1}{6}wl^3 \times 5\cdot234, \text{ with } w = 1\cdot40$$

and

$$L = 0\cdot180 \times \tfrac{1}{6}wl^3 \text{ gramme-centimètre}$$

with

$$b = 29 \text{ centimètres, and } w = 1\cdot40$$

Series II., Fig. 2.—As it was found that there was no sensible efflux of sand into the silk bags, and that the door had no tendency to jam after it had yielded by a small distance, the Author attempted to combine two sets

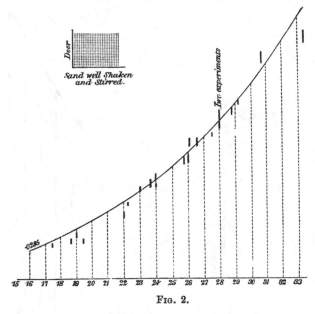

FIG. 2.

of observations with only one filling of the box. Accordingly, after the tension at which the jerky yielding became prominent had been determined, the door was bolted in the position in which it happened to be, that is to say, in all cases exceedingly little out of the vertical. The sand was then thoroughly stirred up with a stick, and the sides of the box hammered. This caused the sand to pack closer by fully 10 per cent., and the level in the box to fall by nearly a centimètre. The door was unbolted and a fresh observation was made. The difficulty of determining what was the tension of the cord, when limiting equilibrium was attained, was considerably greater than in Series I. Nevertheless, observations were taken of the tensions at which the yielding by jerks first became clearly marked, and the results are recorded in Fig. 2. It seems probable that some part of the difficulty arose from the door being slightly elastic, and being put into a state of strain by the ramming down of the sand. This strain would be gradually relaxed as the tension on the cord diminished.

The cubical parabola which seems best to fit the facts is such that

$$T = 0{\cdot}0285\ l^3$$

$$Lb = 1{\cdot}0118\ l^3$$

$$= 3{\cdot}916 \times \tfrac{1}{6}wl^3,\ \text{with}\ w = 1{\cdot}55$$

and $L = 0{\cdot}132 \times \tfrac{1}{6}wl^3,\ \text{with}\ w = 1{\cdot}55,\ b = 29$

This series of experiments cannot be regarded as satisfactory, but as it is the only one in which the earth was stirred and shaken, it is given for what it is worth.

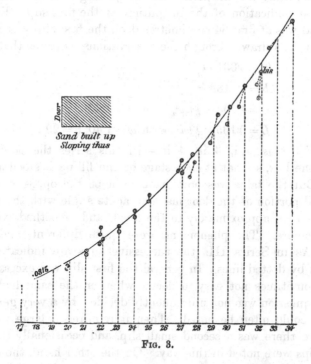

FIG. 3.

Series III., Fig. 3.—In this series the sand was poured in at the side of the box next to the door, so that at each stage of filling it lay at the angle of repose, with the uncovered portion of the door making an obtuse angle with the talus. When the sand was filled up to the desired height the part of the box lying over the talus was filled in, the strata being kept parallel to the talus, until an approximately level surface to the mass of sand was formed. The surface was finally levelled by scraping it with a straight-edge of wood.

Judging by the experience gained in the first two series, the Author fully expected to observe a similar mode of breakdown of equilibrium. He was therefore surprised, in making the first experiment, not to note any perceptible preliminary yielding before the equilibrium completely broke down, and the door yielded to the full extent permitted to it by the silk bags. Indeed, the record of this first experiment was lost through surprise. In subsequent experiments the breakdown of equilibrium was nearly always very marked, although not generally so complete as in the first experiment. After the first sand-slip the tension of the cord was further relaxed until a second slip took place. It will be noted in Fig. 3 that the observations are recorded by two small circles joined by dotted lines. The centre of the upper circle records the tension on the cord when the first slip took place; the lower records the second slip. If the first slip was large the sand fell considerably in the box, so that the second slip corresponds to a smaller depth of sand; thus the inclination of the joining line between a pair of marks gives an indication of the magnitude of the first slip. For instance, when the sand was at first 34 centimètres deep the first slip was a very large one. The parabola drawn through the observations indicates that

$$T = 0.0315 \, l^3$$

$$Lb = 1.1183 \, l^3$$

$$= 4.793 \times \tfrac{1}{6} wl^3, \text{ with } w = 1.40$$

and $\qquad L = 0.165 \times \tfrac{1}{6} wl^3, \text{ with } w = 1.40, b = 29$

Series IV., Figs. 4, i, 4, ii, 4, iii.—In this series the sand was again poured into the box, so that at each stage of the filling it stood at the angle of repose. But the talus was now made to slope the opposite way, so that the uncovered portion of the door made an acute angle with the talus. The sand was filled up approximately to the level, and smoothed with a piece of wood as before. The phenomena were quite different, and somewhat capricious. As in Series III., the successive slips are indicated by small circles joined by dotted lines. In general the first slip was excessively small, and it was sometimes not easy to decide whether the sand had slipped or not, but the question was not unfrequently decided by a very gentle hissing noise, which could often be noted. Then it required a large relaxation of tension before there was a second small slip, and occasionally three or four successive slips were noted in this way. On the other hand, the equilibrium would occasionally break down suddenly by a large slip, as in Series III. The record of these experiments is given in Fig. 4, i. In order that the figure may not be confused the dotted line joining some of the pairs of dots is drawn curved. There are notes added with regard to some of the experiments. The results are so irregular that it is hardly possible to say which cubical parabola suits them best. The parabola $T = 0.036 \times l^3$ seems as near as any of them, but some of the observations are very discordant from it.   This would give $L = 0.189 \times \frac{1}{6} wl^3$, when $w = 1.40$, $b = 29$.   A new series of experiments (Fig. 4, ii) of the same kind was then begun, especially close attention being paid to the first small slips, and to the manner of filling the box, so that it should not be jolted at all before the experiment began.   A similar capriciousness is observable in the results, and the cubical

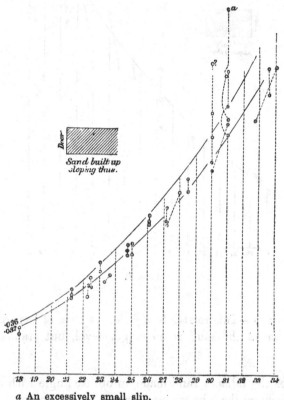

Sand built up sloping thus.

Door

a An excessively small slip.
? Some doubt as to whether there was really a slip.

Fig. 4, i.

parabola which suits the observations perhaps better than any other is $T = 0.036 \times l^3$.   In Fig. 4, iii are collected together all the first slips of both the above series.   The observations marked B were large slips, those marked b were moderately large, and those marked ? were somewhat doubtful on account of the excessive smallness of the slip.   It will be seen that no parabola really satisfies these observations well, but they group themselves approximately about

$$T = 0.036 \times l^3$$
$$Lb = 1.278 \times l^3$$
$$= 5.477 \times \frac{1}{6} wl^3, \text{ with } w = 1.40$$
and
$$L = 0.189 \times \frac{1}{6} wl^3, \text{ with } w = 1.40, b = 29$$

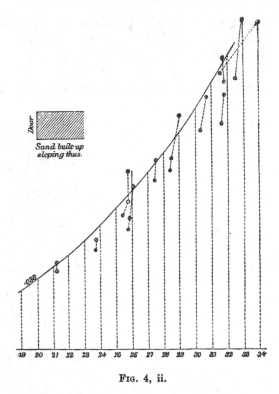

Fɪɢ. 4, ii.

Series V., Fig. 5.—This relates to a different class of experiment, namely, when the surface of the sand is no longer level, but stands at the angle of repose.

In this series the talus of sand made an obtuse angle with the uncovered portion of the door, so that the grains of sand tended to roll away from the door. Throughout the operation of filling the box this arrangement of strata was maintained. The depth l of the sand in this case means the depth at the movable door.

The observations are recorded in the same way as before, and are very fairly consistent amongst themselves. The parabola shows

$$T = 0\cdot028 \times l^3$$

$$Lb = 0\cdot994 \times l^3$$

$$= 4\cdot26 \ \times \tfrac{1}{6}wl^3, \text{ with } w = 1\cdot40$$

and $\qquad L = 0\cdot147 \times \tfrac{1}{6}wl^3, \text{ with } w = 1\cdot40, b = 29$

The couple is, of course, notably less than when the sand was level.

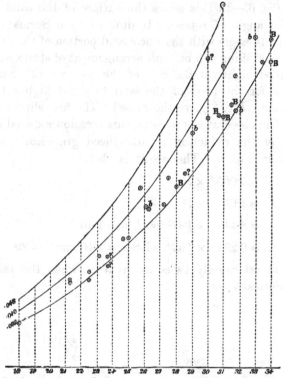

In cases marked ? there was some doubt, generally because of the smallness of the slip.　The cases marked B were large slips, those b were moderate.

Fig. 4, iii.

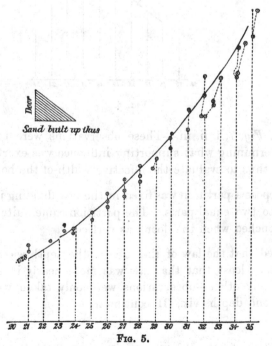

Sand built up thus

Fig. 5.

Series VI., Fig. 6.—In this series the surface of the sand was not level, but stood at the angle of repose. It differed from Series V. in that the talus made an acute angle with the uncovered portion of the door. Throughout the operation of filling the box this arrangement of strata was maintained. As the sand stood higher at the side of the box remote from the door, it was not possible for the level of the sand to stand higher than 23 centimètres at the door. Fig. 6 gives the record. The first slips only are entered in this figure. The observations were on this occasion reduced arithmetically, but the result of the reduction is introduced graphically in the cubical parabola shown in Fig. 6. The result is that

$$T = 0.0555 \times l^3$$
$$Lb = 1.970 \ \times l^3$$
$$= 8.44 \ \ \times \tfrac{1}{6}wl^3, \text{ with } w = 1.40$$

and
$$L = 0.291 \ \times \tfrac{1}{6}wl^3, \text{ with } w = 1.40 \text{ and } b = 29$$

The couple is almost exactly twice as great as when the talus sloped the other way in Series V.

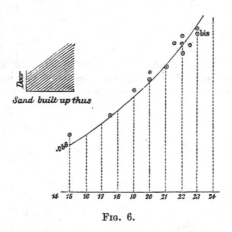

Fɪɢ. 6.

Series VII., Fig. 7, i, ii, iii.—These observations were undertaken with the view of determining what supporting influence was exerted by the sides of the box, and thus to evaluate the effective width of the box.

For this purpose a partition was fixed in the box dividing it, perpendicular to the door, into two equal parts. The partition came quite up to the door, which it just touched when the door was vertical.

As it seemed that the law of the cube of the depth was sufficiently well established—or, at least, that that law was more nearly in accordance with the facts than any other—observations were only taken with the sand at approximately one depth, viz., 31 centimètres.

In the first set of experiments with the partition (Fig. 7, i) the sand was built up as in Series III. The results were reduced arithmetically, but are recorded graphically. They give

$$T = 0.0296\ l^3$$

$$Lb = 1.049\ l^3 = 4.50 \times \tfrac{1}{6}wl^3$$

when $w = 1.40$

and $L = 0.164 \times \tfrac{1}{6}wl^3$

when $w = 1.40,\ b = 27\tfrac{1}{2}$

Series III. had given

$$L = 0.165 \times \tfrac{1}{6}wl^3, \text{ when } b = 29$$

In the next set the sand was built up as in Series IV. The results were reduced arithmetically, but are recorded graphically in Fig. 7, ii. They give

$$T = 0.0328\ l^3$$

$$Lb = 1.165\ l^3 = 4.99 \times \tfrac{1}{6}wl^3$$

when $w = 1.40$

and $L = 0.182 \times \tfrac{1}{6}wl^3$

when $w = 1.40,\ b = 27\tfrac{1}{2}$

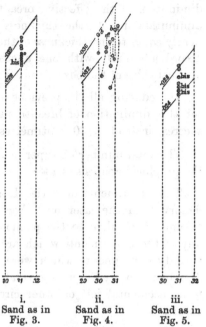

i.
Sand as in
Fig. 3.

ii.
Sand as in
Fig. 4.

iii.
Sand as in
Fig. 5.

FIG. 7.

Series IV. had given $L = 0.189 \times \tfrac{1}{6}wl^3$, when $b = 29$.

In the next set the sand was built up as in Series V. The results were reduced arithmetically, but are recorded graphically in Fig. 7, iii. They give

$$T = 0.0261\ l^3$$

$$Lb = 0.926\ l^3 = 3.967 \times \tfrac{1}{6}wl^3, \text{ when } w = 1.40$$

and $L = 0.144 \times \tfrac{1}{6}wl^3, \text{ when } w = 1.40,\ b = 27\tfrac{1}{2}$

Series V. had given $L = 0.147 \times \tfrac{1}{6}wl^3$, when $b = 29$.

Thus the twenty-nine experiments with the partition gave sensibly the same results as when the partition was absent, provided b (the effective breadth of the box) is taken as $27\tfrac{1}{2}$ centimètres instead of 29 centimètres.

The actual thickness of the partition was 1·3 centimètre, and therefore the actual diminution of the width of the box was 1·3 centimètre. But when the partition was in, there were four vertical surfaces tending to support the mass of sand, instead of only two, as there were when the partition was absent*. If, therefore, these vertical surfaces had exercised

---

* If the experiments had been accurate enough to admit of rigorous treatment, the following would have been the process. The actual width of the door is 30 centimètres. Let x be the

much supporting influence, the experiments of Series VII. would not have given results in accordance with those of the preceding series, without a diminution of the effective breadth of the box over and above the actual diminution due to the thickness of the partition.   As a diminution of b nearly equal to the thickness of the board makes the two sets of experiments accord admirably with one another, it may safely be concluded that the vertical sides of the box exercised very little supporting effect on the sand.

In reducing all the previous experiments, a centimètre has been allowed for this diminution of breadth, and b has always been taken as 29 centimètres, instead of 30 centimètres.

The total number of experiments recorded in Figs. 1—7, and included in the graphical analysis is 143.

The experiments of Series I. and II. are probably somewhat less satisfactory than the later ones, because after much experience the Author became more alive to the phenomena to be watched.   But before making any of the experiments which are here analysed, he had already made over seventy experiments which were all rejected for defects in the apparatus. Moreover, in each series several experiments are excluded on account of some accidental jolt or other circumstance, which might have vitiated the result.

After completing the work of which an account has now been given, the Author was interrupted by other occupations, and has not resumed the subject.

§ 2.   Discussion of the Experiments.

The results of the experiments must now be compared with the theoretical formulæ.   In the Series I. to IV. the surface of the sand was flat.   According to Rankine the couple per unit breadth of the door should be expressed by

$$L = \tfrac{1}{6} wl^3 \tan^2 (\tfrac{1}{4}\pi - \tfrac{1}{2}\phi)$$

and according to Boussinesq by

$$L = \tfrac{1}{6} wl^3 \tan^2 (\tfrac{1}{4}\pi - \tfrac{1}{2}\phi) \frac{\cos (\tfrac{1}{4}\pi - \tfrac{1}{2}\phi)}{\cos (\tfrac{3}{2}\phi - \tfrac{1}{4}\pi)} \cos \phi$$

---

diminution of width due to the support afforded by two sides, as in Series I.—VI.   Then $2x$ is the diminution of breadth due to the support of four sides, as in Series VII.   In Series VII. the breadth of door in actual contact with the sand was $30 - 1 \cdot 3 = 28 \cdot 7$ centimètres.   Thus, comparing III. with VII., i, in order that the two sets of experiments may give identical results, the equation $\frac{4 \cdot 793}{30 - x} = \frac{4 \cdot 493}{28 \cdot 7 - 2x}$ must be satisfied.   This gives $x = 0 \cdot 65$.   Similarly, the comparisons of IV. with VII., ii, and of V. with VII., iii, give respectively $x = 1 \cdot 24$, and $x = 0 \cdot 70$.   It may thus be concluded that the diminution of breadth of the box due to lateral support must be about a centimètre, and accordingly b has been taken as 29 in the reduction of the experiments of Series I. to VI.

The angle of repose of the sand was determined, as nearly as might be, at 35°. In order to show the influence of error in this determination it may be stated that when $\phi = 34°, 35°, 36°$

$$\tan^2 (\tfrac{1}{4}\pi - \tfrac{1}{2}\phi) = 0{\cdot}283, \; 0{\cdot}271, \; 0{\cdot}260 \text{ respectively}$$

and

$$\tan^2 (\tfrac{1}{4}\pi - \tfrac{1}{2}\phi) \frac{\cos (\tfrac{1}{4}\pi - \tfrac{1}{2}\phi)}{\cos (\tfrac{3}{2}\phi - \tfrac{1}{4}\pi)} \cos \phi = 0{\cdot}208, \; 0{\cdot}199, \; 0{\cdot}189 \text{ respectively}$$

Thus the two theories should give, for the values of the coefficient, for this sand, about 0·27 and 0·20 respectively. Now the following were found as the values of the coefficient by experiment: from Series I. 0·180, from Series II. 0·132, from Series III. 0·165, from Series IV. 0·189. The discrepancy from Rankine's value is enormous, and that theory may be safely neglected. The difference from Boussinesq's value in three out of four of the series is not large, but it is (except in IV.) larger than can be fairly put down to errors of observation and an erroneous estimate of the effective width of the box. The disagreement of the various series amongst themselves is considerable, and in Series IV. especially there was much uncertainty as to the proper value to be taken. It will be pointed out below in what way it appears likely that the *à priori* assumptions of the theories are deficient.

The experiments of Series V. and VI. in which the sand formed a talus at the angle of repose will now be considered. It appears from M. Boussinesq's paper* that, in the case of the ascending talus (Series VI.) supported by a vertical wall, the formula for the oversetting couple is

$$L = \tfrac{1}{6}wl^3 \cos^2 \phi$$

---

* The memoir is entitled "Essai théorique sur l'équilibre d'élasticité des massifs pulvérulents, &c. Mémoire présenté à la classe des sciences dans la séance du 6 Juin 1874." Société Royale des Sciences de Belgique. It has also been published by Gauthier-Villars of Paris.

On p. 93 (§ 41 of Sec. VIII.) it is stated that the moment of the thrust about the exterior base of the wall is

$$\tfrac{1}{2}\rho g h^2 K \left(\tfrac{1}{3}h \cos \omega - b \sin \omega\right)$$

Here his ρg is w, the density of the sand; his h is l, the depth of sand against the wall; b is the thickness of the wall, which must be taken as zero; ω is the inclination of the talus to the horizon, estimated as positive when the talus makes an acute angle with the uncovered portion of the wall; and K a certain coefficient. Thus so far his result is in the notation of the present paper $\tfrac{1}{6}wl^3 K \cos \omega$.

On p. 126 (§ 47 of Sec. IX.) it will be found that when the sand is on the point of slipping (*état ébouleux*),

$$K = \tan (\tfrac{1}{4}\pi - \tfrac{1}{2}\phi) \frac{\cos \psi \cos (\phi + \delta) \cos (\omega - i)}{\cos (\phi_1 - \delta) \cos (\omega + \psi)}$$

Here ϕ is the angle of repose of the sand.

On p. 108 (§ 44) i is defined as the inclination of the sustaining wall to the vertical; in the present case i is of course zero.

M. Boussinesq has not been able to evaluate a formula for the descending talus, which formed the subject of experiment in Series V.

Rankine's theory, as applied to a talus, shows that the resultant action parallel to the line of greatest slope of the talus per unit of area across a vertical interface, whose plane is parallel to a horizontal line drawn across the talus, is expressed by

$$wz \cos \omega \, \frac{\cos \omega - \sqrt{(\cos^2 \omega - \cos^2 \phi)}}{\cos \omega + \sqrt{(\cos^2 \omega - \cos^2 \phi)}}$$

where z is the depth of the interface vertically below the surface, ω is the slope of the talus, and ϕ the angle of repose. He supposes that the action on a supporting wall is identical with that across an ideal surface cutting a

---

On p. 109 ψ is defined by the equation

$$\sin (\omega + 2\psi) = \frac{\sin \omega}{\sin \phi}$$

And on p. 126, it appears that

$$\delta = \tfrac{1}{4} \pi - \tfrac{1}{2} \phi - \psi - i$$

where, as already remarked, i is to be put as zero.

The coefficient of friction between the wall and the sand is $\tan \phi_1$. And as in the above experiments the door was sprinkled with sand $\phi_1 = \phi$.

When the surface is level, by putting $\omega = 0$, $i = 0$, $\phi_1 = \phi$, the formula is obtained which has been quoted above for the case of the horizontal surface.

Next, where the surface forms a talus at the angle of repose, and where ω is positive and equal to ϕ, the formula quoted in the text is obtained.

Finally, where the surface forms a talus at the angle of repose, and where ω is equal to $-\phi$, the formula for K gives the value zero. The formula in fact ceases to be even approximately true, and no other is investigated.

In a letter to the Author, dated January 6th, 1883, M. Boussinesq says (translation) :—" In the memoir of 1874 I only attach importance, at least from the practical point of view, to § ix. The preceding paragraphs have only a theoretical value, because I had at that time allowed myself to be led into error by following Rankine, who, very erroneously, did not take into account the perturbations caused by the neighbourhood of the wall. . . . . My present ideas on the thrust of earth, from a practical point of view, are presented in the short paper which you have seen in the discussion on Mr Baker's work. . . . {A new extended edition is published in the *Annales des Ponts et Chaussées*.} . . . I find the results of your useful observations, 0·180, 0·132, 0·165, 0·189, as concordant with my theoretical result 0·20, as could be expected (except perhaps the second); for the theoretical formula supposes the mass of earth and the wall infinitely long in the horizontal direction, and it neglects the disturbing effect of the solid soil which supports the mass of earth, treating the subject as though the depth of the earth were also infinite. To take into account the influence of the subjacent soil would complicate the problem inextricably. It must be observed, moreover, that the horizontal force which you had to measure must have been smaller than the horizontal component of the thrust itself, however little that force be aided by accessory friction in keeping the thrust (*sic*, query sand) in equilibrium. {The Author fails to comprehend this remark.} . . . . As far as relates to the case where ω is negative . . . . I do not think that one can rely on my formula (especially when $i = 0$), because it is founded on the hypothesis that the angle δ is small. Now for negative values of ω, this angle δ soon becomes considerable, and it has begun to be even somewhat too large when $\omega = 0$. . . . The cases where δ is large, which are those where ω is negative, would necessitate integrations which appear to me to pass the power of analysis."

complete talus. In the case where the talus has the slope ϕ, the force becomes simply $wz \cos \phi$, and the moment of all the forces about the base of the wall is

$$\int_0^l wz \cos^2 \phi \, (l - z) \, dz$$

Completing the integration there results, as before from Boussinesq,

$$L = \tfrac{1}{6} wl^3 \cos^2 \phi$$

Since action and reaction across an ideal interface are equal and opposite, it is obvious that moments of the action and reaction about a point vertically under the interface are equal and opposite. If the interface be solidified into a wall, equal and opposite moments are got about the base of the wall, whether it be supposed that all the sand is removed from higher up than the wall, or lower down. Thus the two cases of the ascending and descending talus should give the same oversetting couple.

When $\phi = 35°$, $\cos^2 \phi = 0\cdot671$, and to accord with theory the experiments of Series V. and VI. should both, according to Rankine, and the latter only according to Boussinesq, have given

$$L = \tfrac{1}{6} wl^3 \times 0\cdot671$$

Series V. gave, however, for the numerical factor $0\cdot147$, and Series VI. gave $0\cdot291$. This shows that the theory in this case is utterly at fault.

It will now be shown how, by means of some very simple assumptions with regard to the equilibrium of sand—assumptions which seem to have as good grounds for acceptance as the more complex views which form the basis of the theories of Rankine and Boussinesq—formulæ can be arrived at which present some sort of concordance with the results of experiment. The Author was, in fact, surprised to find that the method suggested actually gives Boussinesq's formula for the case of the flat surface, and moreover gives formulæ very fairly concordant with the results of the Series V. and VI.

Only the case of the vertical wall will be considered, for it is here alone that there is a datum for comparison. It will be clear, however, that the same principles might be applied in almost any case.

It is first assumed that the centre of pressure on a flat vertical surface immersed in the sand obeys the hydrostatic law, and is at two-thirds of the depth below the surface. This is confirmed experimentally by the fact that the oversetting couple varies, with some degree of accuracy, as the cube of the depth. It is next assumed that a certain wedge of earth in the rear of the wall is simply supported by the pressures and frictions acting across the face of the

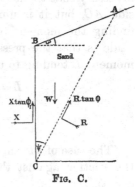

FIG. C.

wall and the internal face in the mass of sand. In Fig. C, let BC be the wall; AB the talus supported by it; let ω be the slope of the talus; and let $BC = l$ the depth of sand. Let ABC be the wedge which is assumed to be supported by the wall; and let the angle at C be called ψ. Let $\tan \phi$ be the coefficient of friction within the sand, and $\tan \phi_1$ that between the sand and the wall; w the density of the sand.

Then the wedge, whose weight is W, is supported by forces as indicated by arrows in the figure.

The angle $\qquad A = \pi - (\tfrac{1}{2}\pi + \omega + \psi) = \tfrac{1}{2}\pi - (\omega + \psi)$

Then $\qquad AB = l\,\dfrac{\sin \psi}{\cos (\omega + \psi)}, \qquad AC = l\,\dfrac{\cos \omega}{\cos (\omega + \psi)}$

and $\qquad W = \tfrac{1}{2} w\,.\,AB\,.\,AC \sin A = \tfrac{1}{2} wl^2 \dfrac{\sin \psi \cos \omega}{\cos (\omega + \psi)}$

Resolving horizontally and vertically,

$$X = R\,(\cos \psi - \tan \phi \sin \psi) = R\,\frac{\cos (\psi + \phi)}{\cos \phi}$$

$$X \tan \phi_1 + R\,(\sin \psi + \tan \phi \cos \psi) = W$$

or $\qquad X \tan \phi_1 + R\,\dfrac{\sin (\psi + \phi)}{\cos \phi} = \tfrac{1}{2} wl^2 \dfrac{\sin \psi \cos \omega}{\cos (\omega + \psi)}$

Substituting for X in terms of R,

$$R\,\{\cos (\psi + \phi) \tan \phi_1 + \sin (\psi + \phi)\} = \tfrac{1}{2} wl^2 \frac{\sin \psi \cos \omega \cos \phi}{\cos (\omega + \psi)}$$

Whence $\qquad R = \tfrac{1}{2} wl^2 \dfrac{\sin \psi \cos \omega \cos \phi \cos \phi_1}{\cos (\omega + \psi) \sin (\psi + \phi + \phi_1)} \left.\begin{array}{c} \\ \\ \\ \\ \end{array}\right)$

and $\qquad X = \tfrac{1}{2} wl^2 \dfrac{\sin \psi \cos \omega \cos (\psi + \phi) \cos \phi_1}{\cos (\omega + \psi) \sin (\psi + \phi + \phi_1)}$

In order that the equation of moments about C may be satisfied, it is necessary that the forces R and $R \tan \phi$ should act at a certain point in the side AC, but it is unnecessary to determine the point for the purpose of finding the moment tending to overset the wall. Part of the assumption made is that the pressure X acts at two-thirds of the depth l. Hence the moment L tending to upset BC about C is given by

$$L = \tfrac{1}{3} l X$$

$$= \tfrac{1}{6} wl^3 \frac{\sin \psi \cos \omega \cos (\psi + \phi) \cos \phi_1}{\cos (\omega + \psi) \sin (\psi + \phi + \phi_1)}$$

The case of the smooth wall, in which $\phi_1 = 0$, will be passed over. When the friction against the wall is equal to the internal friction of the sand $\phi_1 = \phi$.

First, suppose the upper surface of the soil is level (Series I.—IV.), so that $\omega = 0$, and $\phi = \phi_1$.

Then $$L = \tfrac{1}{6} w l^3 \tan \psi \; \frac{\cos (\psi + \phi)}{\sin (\psi + 2\phi)} \cos \phi$$

Following Coulomb and others, and supposing the wedge to be bounded by a plane bisecting the angle between a free talus and the vertical, and assuming that ϕ is the angle of such a talus, $\psi = \tfrac{1}{4}\pi - \tfrac{1}{2}\phi$, it results that

$$L = \tfrac{1}{6} w l^3 \tan^2 (\tfrac{1}{4}\pi - \tfrac{1}{2}\phi) \; \frac{\cos (\tfrac{1}{4}\pi - \tfrac{1}{2}\phi)}{\cos (\tfrac{3}{2}\phi - \tfrac{1}{4}\pi)} \cos \phi$$

and this is Boussinesq's formula for this case. It has already been seen that it presents a fair approximation to some of the experimental results.

Secondly, suppose the upper surface stands at the angle of repose, and that $\omega = + \phi$ (Series VI.), then with $\phi = \phi_1$,

$$L = \tfrac{1}{6} w l^3 \sin \psi \; \frac{\cos^2 \phi}{\sin (\psi + 2\phi)}$$

If, as in the first case, $\psi = \tfrac{1}{4}\pi - \tfrac{1}{2}\phi$, this becomes

$$L = \tfrac{1}{6} w l^3 \sin (\tfrac{1}{4}\pi - \tfrac{1}{2}\phi) \frac{\cos^2 \phi}{\cos (\tfrac{3}{2}\phi - \tfrac{1}{4}\pi)}$$

Thirdly, suppose the upper surface stands at the angle of repose, but with the talus in the opposite direction, so that $\omega = - \phi$ (Series V.); then with $\phi = \phi_1$,

$$L = \tfrac{1}{6} w l^3 \sin \psi \; \frac{\cos^2 \phi}{\cos (\psi + 2\phi)} \; \frac{\cos (\psi + \phi)}{\cos (\psi - \phi)}$$

Putting $\psi = \tfrac{1}{4}\pi - \tfrac{1}{2}\phi$ as before, this becomes

$$L = \tfrac{1}{6} w l^3 \sin (\tfrac{1}{4}\pi - \tfrac{1}{2}\phi) \frac{\cos^2 \phi}{\cos (\tfrac{3}{2}\phi - \tfrac{1}{4}\pi)} \frac{\sin (\tfrac{1}{4}\pi - \tfrac{1}{2}\phi)}{\cos (\tfrac{3}{2}\phi - \tfrac{1}{4}\pi)}$$

or $$L = \tfrac{1}{6} w l^3 \left[\frac{\sin (\tfrac{1}{4}\pi - \tfrac{1}{2}\phi) \cos \phi}{\cos (\tfrac{3}{2}\phi - \tfrac{1}{4}\pi)} \right]^2$$

With $\phi = 35°$ to correspond with the sand, it will be found that

$$\left[\frac{\sin (\tfrac{1}{4}\pi - \tfrac{1}{2}\phi) \cos \phi}{\cos (\tfrac{3}{2}\phi - \tfrac{1}{4}\pi)} \right]^2 = 0\cdot 3125$$

and $$\sin (\tfrac{1}{4}\pi - \tfrac{1}{2}\phi) \frac{\cos^2 \phi}{\cos (\tfrac{3}{2}\phi - \tfrac{1}{4}\pi)} = 0\cdot 1455$$

From Series V. it appeared that the latter coefficient was $0\cdot 147$, and from Series VI. that the former was $0\cdot 291$. The agreement therefore of this rough and semi-empirical rule with experiment is far better than might have been expected.

The reasons for the discrepancies from the theory and for the want of agreement of the different series of experiments amongst one another will now be considered.

It has always been assumed by previous writers that the tangential action across an ideal interface in a mass of loose earth is of the same nature as the statical friction between solids, and that when the tangential stress has attained in magnitude a certain fraction of the normal stress, the equilibrium is on the point of breaking down. That fraction is the coefficient of internal friction of sand, and is supposed to be equal to $\tan \phi$, where ϕ is the angle of a talus of the greatest possible slope. A little consideration will show that the hypothesis cannot be exact, even with an ideal sand with incompressible grains, and absolutely devoid of coherence. For imagine a mass of sand thrown loosely together; then if the grains are of irregular shape a certain portion of them will be resting on points and angles, thus occupying more space than they might do.

If the sand be now compressed, many of the grains will slip and rotate, and fall into interstices; in fact a considerable amount of rearrangement will take place, and the density of the mass will rise considerably—by quite 10 per cent. if the rearrangement be thorough, as found experimentally.

Even if all the grains were spherical a considerable amount of change would take place, and when they are angular of course much more. After a certain amount of pressure or shaking has been applied the grains could not be made to pack closer. This movement of the grains amongst themselves may be described as "settling."

Now if the maximum tangential stress across an interface, which is compatible with equilibrium, be compared before and after settling, it can be seen that they will be very different. For a grain of sand which is well embedded amongst its neighbours will require much more force to displace it than a grain resting partly on points and angles.

Hence it is clear that the coefficient of internal friction of sand is a function of the pressure, and not merely of the pressure then existing, but also of the pressure and shaking to which at some previous period that portion of the mass of sand has been subjected.

No mass of sand can be put together without some history, and that history will determine the nature of its limiting equilibrium. It is quite impossible to say how much these causes will vitiate any mathematical theory of the equilibrium of sand, but experience seems to show that the vitiation is extensive.

On considering these views, it seems reasonable to suppose that, when in Series I. the sand was built up in approximately parallel layers, it underwent a partial settlement, and on the average the grains of sand were not in positions which predisposed them to fall more one way than another. In Series II. the settlement was pretty completely accomplished by shaking and stirring. In both these cases it was difficult to state that there was any exact epoch of instability, and it is a highly plausible supposition that the first process, before a definite sand-slip, was a gradual "unsettlement" of the sand, in which one grain after another partially rotated, assumed a more open order, and caused the whole mass to occupy a larger volume. This gradual unsettlement would almost certainly take place along certain surfaces or narrow regions, which ultimately formed the seat of slipping. After the unsettlement had proceeded to a certain extent a visible slip would take place; probably this slip caused a partial resettlement, and then must follow repetition of the unsettling process and a fresh slip, and so on.

Now from these à priori considerations it seems as though the rough theory above developed, in which the equilibrium is considered of a wedge of sand, treated as a rigid body, should give results more conformable to facts than theories which treat the sand as a continuous medium. For there seems reason to suppose that the unsettling process, preparatory to slipping, only takes place over certain surfaces, and that in other regions the sand behaves like a rigid body. In all modes of building a mass of sand, something of this settling and unsettling must take place.

Consider now the experiments of Series III., where the sand was built up so as to form a talus sloping away from the door. The sand was lying in loose order, and each grain must have stopped whilst falling away from the door. On this occasion when the first sand-slip took place there was often a complete breakdown of the equilibrium.

The following may be offered as a plausible explanation of this fact: The grains having attained a more or less stable configuration for movements away from the door, had probably become jammed together so as to be arranged in arches with the convexity directed upwards and towards the door; displacements towards the door would therefore break up these arches, and equilibrium would not be again attained until the arch-like arrangement was re-established with the concavity towards the door. The sand being in loose order scarcely any preliminary unsettlement was necessary before the equilibrium broke down. Now in the experiments of Series IV. the sand was built up with the talus sloping towards the door. The sand was in loose order, and each grain must have stopped as it was rolling towards the door.

A fact which was noticed during the operation of building up the sand in a talus has not previously been adverted to, viz. that during the process the Author frequently saw or heard a very small partial slip, which occurred a few seconds after a dose of sand had been poured on to the talus, and when equilibrium had apparently been attained. He noticed in this series that, when one of these partial slips occurred near the end of the operation of filling the box, the tension on the cord had to be relaxed, below what might have been expected from the other experiments, before equilibrium broke down, and then the first slip was in general an excessively small one. The sand had probably settled into a configuration of more stability for displacements towards the door than if this partial slip had not taken place.

If the theory of the arch-like arrangement of the sand, suggested to explain the phenomena of Series III. be adopted, it seems that the reason of the anomalous results of Series IV. and of the smallness of the first slips can be explained. For according to this view the sand was built in arches adapted to resist motions towards the door, and when a grain or two was displaced from an arch, only a very small motion was requisite before the arch was re-formed. Moreover, if a partial slip had just occurred at the end of the operation of building up the sand, the arches were stronger because there had been more complete settlement, and thus a greater relaxation of tension of the cord was possible before there was displacement than if the partial slip had not occurred.

Whatever may be thought of these suggested explanations, the fact remains that Series III. gave large slips and fairly consistent results, and Series IV. small slips and inconsistent results.

It will not be necessary to advert here to the experiments of Series V. and VI., since they only differed from those which have just been discussed in the form of the surface of the sand.

In these experiments a material was purposely chosen as much like the ideal loose earth of the mathematician as possible: and although the experiments have been by no means so various and complete as is desirable, yet they have been sufficient to convince the Author, at least, that even with such materials the ideal laws of equilibrium are by no means obeyed. The coefficient of maximum internal friction is probably very different in various parts of a mass of sand, and is not in fact a constant at all; and the process of settlement and unsettlement is of such importance that it is not possible to regard the mass as incompressible; in many cases there is no definite phenomenon which can be called the breakdown of equilibrium; and, lastly, a rough empirical rule for finding the equilibrium of a wedge of earth in the rear of a revetment will probably give results more conformable to fact than the elaborate methods which treat the material as homogeneous. The Author wishes to ask engineers whether there is not in general some preliminary mark of the failing of a wall by cracks appearing first in the earth at the back; and whether, when a wall has begun to bulge out in places, that process does not frequently stop, and the wall remain sound for ever afterwards?

This is what would be expected from the above experiments, for the unsettlement of the earth appeared to be a preliminary in those cases which bore the most resemblance to actual earthworks, and the bulging of the wall may frequently give the earth an opportunity of settling into more stable equilibrium, and the pressure on the wall may thus be relieved. Now if these conclusions are sound with regard to an almost ideally perfect material like dry sand, with how much more force may they be applied to such materials as those with which engineers actually have to deal?

Imagine, for example, a revetment wall, and that it is to be filled up at the rear with a substance like pitch. This is at first friable and powdery, and may lie in a talus, like sand. When the embankment is first made the pressure on the wall will be somewhat the same as though the substance were loose earth. But after a time the pitch will settle and bind, and the pressure on the wall will become the same as though the material were fluid. Now while there is certainly no kind of earth with such perfect viscosity as that of pitch, yet some clays approximate to it. In these cases the pressure on the wall will probably rise largely. But in the case of ordinary loose earth it is probable that the pressure is greatest at first, and that as the earth settles the pressure falls. If the earth is liable to become sodden with water and then dry again, the variations in pressure will probably be very large.

The Author believes that the incomplete experiments of which he has given an account are the first in which a record of the "historical element" has been carefully kept, and in this respect they seem to him to be worthy of record.

It is to be hoped that other experimenters may be induced to take up the subject, and to carry out experiments with other materials, and by other methods. But until this is done, and the conclusions stated above are shown to be faulty, the soundest view seems to be that engineers have no better practical course open to them than, neglecting the elaborate formulæ which have been suggested, to work with semi-empirical rules such as those of Coulomb, and to allow a large coefficient of safety.

16.

ON THE FORMATION OF RIPPLE-MARK IN SAND.

[*Proceedings of the Royal Society*, Vol. XXXVI. (1884), pp. 18—43.]

THE following paper contains an account of experiments and observations on the formation of ripple-mark in sand. The first section is devoted to experiments on the general conditions under which ripple-mark is formed, and especially on the mode of formation and maintenance of irregular ripples by currents. In the second section it is shown that regular ripple-mark in sand is due to a complex arrangement of vortices in oscillating water; and the last section gives some account of the views of certain recent observers in this field, and a discussion of some phenomena in the vortex motion of air and water.

§ 1. *First Series of Experiments.*

A cylindrical zinc vessel, like a flat bath, with upright sides, 2 feet 8 inches in diameter and 9 inches deep, was placed on a table, which was free to turn about a vertical axis. Some fine sand was strewn over the bottom to a depth of about an inch, and water was poured in until it stood three inches deep over the sand. After some trials of simply whirling the bath, in which no regular ripple-mark was formed, I found that rotational oscillation with a jerking motion of small amplitude gave rise almost immediately to beautiful radial ripples all round the bath. If the jerks were of small amplitude the ripples were small, and if larger they were larger. On one occasion having made large ripple-marks, I oscillated the bath much more rapidly, and a second set of ripples sprang into existence in the furrows of the first set. Another time, when in consequence of irregularity in the motion, a set of radiating waves were generated in the water, a second set of transverse ripples were formed, which produced by interference a beautifully mamellated structure, arranged like a chess-board. In all these experiments the radiating ripples began first to appear at the outer margin of the bath and grew inwards; but the growth stopped after they had extended to a certain distance. If the jerking motion was violent, ripples were not formed near the circumference, and they only began at some distance inwards. After these preliminary trials, arrangements were made for regularising both the frequency and amplitude of oscillation of the bath. An attempt was then made to formulate the laws which govern the generation of ripple-marks. In the following notes of experiments the expression "octave" is used to denote a ripple-length which is one-half of the main or fundamental ripple, and the amplitudes are measured by the displacement of the edge of the bath.

The water stood 1 inch deep in the bath.

1. Amplitude 1 inch; frequency 52 per minute (complete oscillations).
No ripples formed after four minutes.

2. Amplitude $2\frac{1}{2}$ inches; frequency 52.
55 ripples round the circumference, extending about 4 inches inwards; somewhat irregular and with a tendency to break into the octave.

3. Amplitude $6\frac{1}{2}$ inches; frequency 52.
Motion very violent. About 5 large irregular ripples in the circumference, breaking at about 9 inches from the outside into about 40 ripples.

4. Amplitude $1\frac{3}{4}$ inches; frequency 52.
In one part 7 ripples to 9 inches; at another part 12 ripples to 1 foot; some tendency to break into the octave. The ripples only extended an inch or two from the edge.

More water was then poured in until it stood about $1\frac{3}{4}$ inches deep.

5. Repetition of No. 4.
There were 33 ripples in a half circumference (which is 22 inches); the ripples were more regular than in No. 4, with not so much tendency to break into the octave.

6. Water $2\frac{1}{2}$ inches deep; frequency 59; amplitude 3 or 4 inches.
43 ripples to the circumference (which is 44 inches), extending inwards 8 inches.

Hereafter the amplitudes were marked by a pointer projecting 3 inches from the edge of the bath.

7. Water $2\frac{1}{2}$ inches; frequency 75; amplitude $2\frac{1}{2}$ inches.
66 ripples to circumference, very regular, and extending inwards 4 or 5 inches.

8. Water $2\frac{1}{2}$ inches; frequency 74; amplitude 3 inches.
63 or 64 ripples, extending 6 or 7 inches; broken in two or three places.

9.   Water $2\frac{1}{2}$; frequency 75; amplitude 4 inches.

53 ripples, extending 8 or 9 inches; not so regular.

10.   Water $2\frac{1}{2}$; frequency 78; amplitude 5 inches.

47 ripples, extending 10 or 11 inches.

11.   Water $2\frac{1}{2}$; frequency 75; amplitude 6 inches.

Agitation violent; all the coarser sand collected round the margin, without ripple-mark for 4 inches inwards; from 4 to 11 inches inwards, rather irregular ripples about 37 to circumference; the usual flat centre.

12.   Water $2\frac{1}{2}$; frequency about 80; amplitude 7 inches.

The water churned up the sand with violence; margin the same as No. 11; from 8 to 12 inches from margin rather irregular ripples, about 34 to circumference.

13.   (Bad observation.)   Water $2\frac{1}{2}$; frequency about 85; amplitude about $1\frac{1}{2}$ to 2 inches.

80 ripples to circumference.

14.   (Bad observation.)   Water $2\frac{1}{2}$; frequency 57; amplitude 2 inches.

No ripples raised.

An analysis of the observations marked 7 to 14 was made on the hypothesis that the water remained still, when the bath oscillated with a simple harmonic motion. I endeavoured to find whether λ, the wavelength of ripple (in inches) was directly proportional to v, the maximum velocity of the water relatively to the bottom (in inches per minute) during the oscillatory motion; also the values of v_1 and v_2, the least and greatest velocities of the water compatible with the formation of ripple-mark.

The following are the results:—

$\lambda \div v$		Feet per sec. v_1		Feet per sec. v_2
·0031	....	·51 to ·56	....	—
·0027 to ·0028	....	·46 to ·51	....	—
·0024	....	·43 to ·49	....	—
·0021	....	·43 to ·52	....	—
·0023	....	·50	....	1·2
·002	....	·56	....	1·12
—	....	more than ·42	....	—
Mean ·00245 min. or ·147 second	....	·503	....	1·2

It appears therefore that ripples are not formed if the maximum velocity of the water relatively to this particular sand, estimated on the above hypothesis, is less than half a foot or greater than a foot per second; and that if v be that maximum velocity in inches per minute, the wave-length of ripples generated is ·00245v, or ·147v when v is measured in inches per second. The results seem as fairly consistent with one another as could be expected. It will appear from § 2 that the maximum velocity of the water, as estimated on the hypothesis that the water as a whole executes a simple harmonic oscillation relatively to the bottom, does not give the maximum velocity of the water in contact with the sand relatively thereto. The quantity called v is not in reality the maximum velocity of the water in contact with the bottom relatively thereto, but it is in fact 6·283 times the amplitude multiplied by the frequency. Thus we cannot conclude that a current of half a foot per second is just sufficient to stir the sand. In the state of oscillation corresponding to v_1, it is probable that part of the water at the bottom is moving with a velocity much greater than half a foot per second relatively to the sand. The number of the experiments analysed is insufficient for the accurate determination of the law connecting wave-length, amplitude, and frequency; but this branch of the subject was not pursued further because other observers, whose work is referred to in § 3, have made a number of experiments with this object.

It was after making this set of experiments that I hit on what appears to be the key-note of the whole phenomenon.

A series of ripples extending inwards for some distance having been made by oscillation, and the water having come to rest, the bath was turned slowly and nearly uniformly round. The ripples were then observed to prolong themselves towards the centre; this shows that a uniform current is competent to prolong existing ripples. The uniform current flattened the tops of the ripples, but made the lee-side steeper. After being exposed to a prolonged current, the ripples were not only not in course of obliteration, but became somewhat more pronounced.

The sand was then smoothed with the edge of a board; after the exposure of the sand to a current, marks made with the edge, which at first were too faint to be seen, became by a course of development well-defined ripples. The whole surface became gradually mottled with irregular chains of ripples of which the weather-side was a very gradual slope, and the lee-side was steep. The appearance was strikingly like that of drifted snow. As it might be conjectured that there would be eddies or vortices on the lee-side, I made some regular ripple-marks by oscillation, and then exposed them to a current. I shortly observed minute particles lying on the surface of the sand climbing up the lee-slope of the ripples apparently *against* stream. This proved conclusively the existence of the suspected vortices.

If when the bath was at rest, a sudden motion was given in one direction, the sand on the lee-side of each ripple was observed to be churned up by a vortex. By giving a short and sudden motion, I was able to see the direct stream pile up the sand on the weather-side and the vortex on the lee-side. Fig. 1 shows the effect of a single short jerk. Two little parallel ridges of sand were formed, namely (a) by the direct stream, and (b) on the lee-side by the vortex, a little below the crest of the ripple-mark.

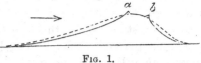

FIG. 1.

It is thus clear that casual surface inequalities are accentuated by the combined action of the direct stream and of the vortex.

For the purpose of examining the vortices a glass tube was drawn out to a fine point, and fitted at the other end with a short piece of india-rubber tube. With this a drop of ink could be squirted out at the bottom of the water. This method was adopted in all subsequent observations, and it proved very valuable. It may be worth mentioning that common ink, which is heavier than water, was better than aniline dye; and the addition of some sulphate of iron to the ink improved its action.

A drop of ink was placed in the furrow between two ripples; as soon as the continuous stream passed, the ink was parted into two portions, one being sucked back apparently against stream up the lee-side of the ripple-mark, and the other being carried by the direct stream towards the crest. By observing the limits of the transport of the ink, I concluded that the stream lines were as shown in Fig. 2. These points being settled, it remained to

FIG. 2.

discover how the vortices were arranged which undoubtedly must exist in the oscillatory formation of regular ripples. The rapidity of the necessary oscillations made this a task of some difficulty.

§ 2. *On the Formation of Ripple-mark by Oscillation.*

The observations were made in two different ways.

In the first of these, which also ultimately proved to be the most successful, the ripple-mark was made in a glass trough about 1 foot long, 5 inches wide, and 6 inches deep. In order to observe the formation of the ripples absolutely in profile, a sheet of glass fitting the trough was placed to stand on four short corks at the bottom. The trough was then put to rock on two corks, one at each side, on the line which bisects its length. Two other slightly shorter corks were put at the ends; these served as stops, and only allowed it to rock through a very small angle. The trough was placed on a window-sill with a strong light outside, and was gently rocked by hand.

When the trough is half filled with water, and sand is sprinkled on the glass plate, it is easy to obtain admirable ripple-marks by gently rocking the trough.

When a very small quantity of sand is sprinkled in and the rocking begins, the sand dances backwards and forwards on the bottom, the grains rolling as they go.

Very shortly the sand begins to aggregate into irregular little flocculent masses, the appearance being something like that of curdling milk. The position of the masses is, I believe, solely determined by the friction of the sand on the bottom, and as soon as a grain sticks, it thereby increases the friction at that place.

The aggregations gradually become elongated and rearrange themselves. As soon as the formation is definite enough to make the measurement of the wave-length possible, it is found that the wave-length is about half of what it becomes in the ultimate formation.

Some of the elongated patches disappear, and others fuse together and form ridges, the ridges then become straighter, and finally a regular ripple-mark is formed with the wave-length double that in the initial stage.

When a drop of ink is put on the glass without any sand, it simply slides to and fro, with perhaps a faint tendency to curdle, but it cannot be caused to form ripple-mark. This shows that the initial stage when the sand is beginning to curdle is due simply to friction.

When the ink is put upon a flocculent mass it betrays some kind of dance in the water, but the layer of water disturbed is so thin that it would hardly have been possible to detect the law of the motion from this case alone. When, however, the nature of that motion, as described below, has been discovered, the same kind of motion may be recognised in the dance of the ink over these flocculent aggregations.

I found in the later experiments that it was advantageous to have a very regular ripple-mark. I therefore sprinkled sand on the sheet of glass, and, before beginning the rocking, I traced regular furrows in it with the point of my finger. A few oscillations of the trough soon effaced all signs of the artificial origin, and the ripple crests with the bare glass in the furrows, were absolutely indistinguishable, except by perfect regularity, from those pro-duced naturally. Most of the observations were, however, made with the natural ripples, and it was only towards the end that I adopted this plan in order to save time and to obtain perfect regularity.

In the rocking trough, the water moves whilst the bottom of the vessel is still, save for the small rocking motion. A second arrangement was, however, made in which the converse is true.   A sheet of plate glass is caused to oscillate in the bottom of a trough with glass sides.   The oscillator is moved by a connecting rod and crank driven by a small water-motor, the throw of the crank is small, and the rapidity of oscillation can be varied within considerable limits.

When sand is sprinkled on the oscillating sheet of glass, phenomena such as described above are again observed, and good regular ripple-mark is formed in the sand.   Although much was learned from this instrument, still the rocking trough was on the whole more useful.

It appeared to be certain from the first set of experiments that ripple-mark was due to eddies or vortices, and the question remained as to how the vortices were arranged in oscillatory motion.   It required some practice, and many hours of watching, to establish the conclusions explained below, indeed the phenomena next described were only detected long after that which follows them in this paper.

If a very gentle oscillation be started, the layer of ink on the crest of a ripple-mark becomes thicker and thinner alternately, swaying backwards and forwards; then a little tail of ink rises from the crest, and the point of growth oscillates on each side of the crest; the end of the tail flips backwards and forwards.   Next the end of the tail spreads out laterally on each side, so that a sort of mushroom of ink is formed, with the stalk dancing to and fro. The height of the mushroom is generally less than a millimètre.

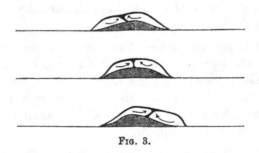

Fig. 3.

Fig. 3 is the best representation I can make of this appearance, which I shall call an ink mushroom.   The first of these figures gives the extreme of excursion on one side, the second the mean position, and the third the extreme on the other side.   The figures show the state of affairs when the oscillation is very gentle, so that the amplitude of oscillation of the main body of water is small compared with the wave-length of the ripple-mark. The elongated hollows under the mushroom are the vortices, and the stem is the upward current.   If the ink be thick these spaces are clouded, and the appearance is simply that of an alternate thickening and thinning of the ink on the crest.   When one is familiar with this motion, after examining it carefully with gentle oscillation over ripple-mark of some size, the same kind of dance may, I think, be detected in the stage of ripple manufacture after the sand has curdled into elongated flocculent masses.

The oscillations being still gentle, but not so gentle as at first, streams of ink from the two mushrooms on adjacent crests creep down the two slopes into the furrow between the adjacent ridges, and where they meet a column of ink begins to rise from the part of the water whose mean position is in the centre of the furrow. The column is wavy, and the appearance is strikingly like that of smoke rising from a fire in still air.

The column ascends to a height of some 5, 10, or perhaps 20 times the height of the ripple-marks, according to the violence of the agitation. It broadens out at the top on each side and spreads out into a cloud, until the appearance is exactly like pictures of a volcano in violent eruption; but the broad flat cloud dances to and fro relatively to the ascending column. The ink continues to spread out laterally and begins to fall on each side. In this stage if the ink is not thick it is often very like a palm-tree, and for the sake of a name I call this appearance an ink tree. The branches (as it were) then fall on each side, and the appearance becomes like that of a beech-tree, or sometimes of an umbrella. The branches reach the ground, and then creep inwards towards the stem, and the ink, which formed the branches, is sometimes seen ascending again in a wavy stream parallel to the stem.

Perhaps a dozen or twenty oscillations are requisite for making the ink go through the changes from the first growth of the tree.

The descending column of a pair of trees comes down on to the top of the mushroom. I have occasionally, when the oscillations are allowed to die, seen both tree and mushroom, but the successful manufacture of the tree necessitates an oscillation of sufficient violence to render the observation of the mushroom very difficult.

The alternate thickening and thinning of the ink on the crests seems to render it probable that with moderate oscillation the mushroom vortices are still in existence, or at any rate that alternately one and the other is there. With violent oscillation, when the stem of the tree is much convoluted, as described below, it cannot be asserted that the mushroom vortices exist, and I am somewhat inclined to believe them to be then evanescent.

Each side of the ink tree is clearly a vortex, and the stem is the dividing line between a pair, along which each vortex contributes its share to the ascending column of fluid. The vortex in half the tree is clearly in the first place generated by friction of the vortex in its correlated mushroom, and is of course endued with the opposite rotation. The ascending stem of the tree is a swift current, but over the mushroom the descending current is slow until close to the mushroom, when it is seen to be impelled by pulses.

I was on one occasion fortunate enough to observe a mote in the water which was floating nearly in the centre of a tree vortex, and counted twelve revolutions which it made before it was caught away from its fortunate position.

If the adjoining crests are of unequal height the stem of the tree is thrown over sideways away from the higher crest; and indeed it requires care to make the growth quite straight. The ink in the stem ascends with a series of pulses, and it is clear that there is a pumping action going on which renders the motion of each vortex somewhat intermittent, the two halves of the tree being pumped alternately.

The amount of curvature in the stem of the tree depends on the amplitude of the oscillation of the water. Figs. 4, 5, 6, give fair representations of ink trees.

Fig. 4.   Fig. 5.

Fig. 4 is the palm-tree stage with gentle oscillation, and Figs. 5 and 6 represent the appearance when the amplitude is greater.

Fig. 6.   Fig. 7.

Fig. 7 exhibits a tree in which the growth is one-sided on account of inequality between the heights of the bounding crests.

Fig. 8 represents a mushroom and a tree which I have occasionally succeeded in observing simultaneously.

The ink is propagated along the convolutions of the stem of the ink tree, but the convolutions are themselves propagated upwards, and each convolution corresponds to one oscillation. The motion of the ink along the convolutions soon becomes slow, but the convolutions become broader and closer. Thus the upper part of the tree is often seen to be most delicately shaded by a series of nearly equidistant black lines. A perfectly normal ink tree, made by a very thin stream of ink, would be like Fig. 9, in which the whole is formed by a single line; but it is not possible to represent the extreme closeness of the lines adequately.

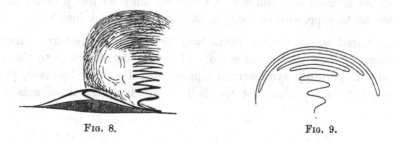

FIG. 8.   FIG. 9.

In the transition from the mushroom stage to the tree stage it appeared to me that it was very frequent that only half the ink tree was formed. At any rate I have frequently noted the mushrooms and half the tree vortices lasting during many oscillations, and then the other halves of the trees gradually appeared. This might, of course, be due to an accidental deficiency of ink in an invisible tree vortex, but I have observed this appearance frequently when there is ink at the stem of the tree, and when there seemed no reason why it should only be carried up in one ascending stream and not in the other.

If the agitation is very gentle the sand on the crests of the ripple-marks is just moved to and fro; with slightly more amplitude, the dance is larger, and particles or visible objects, such as minute air-bubbles, in the furrows, also dance, but with less amplitude than those on the crests. When the rocking is gentle the oscillation in the furrow appears to be in a different phase from that on the crest, with more violent rocking I did not observe the difference of phase. The dance is not a simple harmonic motion like that of the main body of the water relatively to the bottom, but the particles dash from one elongation to the other, pause there, and then dash back again.

As the amplitude further increases the furrows are completely scoured out, and the sand on the crests is dashed to and fro, forming a spray dancing between two limits. With violent agitation this dance must have an amplitude of more than half a wave-length. If the agitation be allowed to subside the dance subsides, and when the water is still the ripple-mark is left symmetrical on both sides. With extremely violent oscillation all the water becomes filled with flying dust, and it is no longer possible to see what is happening. This seems to be the condition when the agitation is too strong for the formation of ripple-mark. It is probable that the rush of water sweeps away the existing ripple-mark, and there is then no longer anything to produce a systematic arrangement of vortices.

In Fig. 10 I have tried to exhibit the dance of the vortices by the succession of Figs. I to VII.   When the amplitude of oscillation is the same as
when the ripple-mark is generated, the series of changes is of the kind shown.
The figures succeed one another in time, but they do not pretend to such
accuracy as to represent the stages at rigorously equal intervals.

The dotted waves show a mean contour of the ripple-mark; they are
introduced to show the displacements of the crests relatively to the mean
position.   A perfectly symmetrical ripple-mark does not, however, present
a simple harmonic outline, for the hollows are flat and the crests rather
sharp.

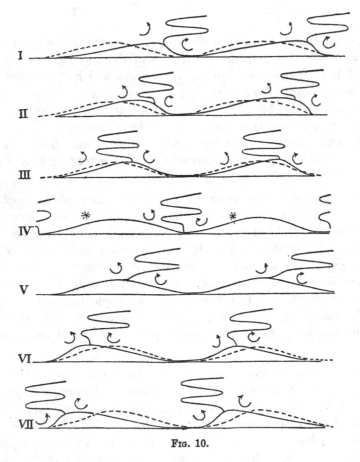

FIG. 10.

The convoluted line is the stem of the ink tree, which forms the
dividing line of the two vortices; the curved arrows show the direction
of rotation.

In I the water is at its elongation to the right.   The crest of the ripple-
mark is also at its extreme to the right, and the right hand slopes are steep,

whilst the left are gentle.   Here the water is at rest except for the vortices, which both tend to carry sand up to the crest.

In II the general mass of water is beginning its movement to the left; it carries with it the upper convolutions of the ink tree, but leaves the root very nearly in the same position as in I. The crest of the ripple-mark is but little displaced.   In III the crest has begun its displacement, so that although the root of the ink tree has begun moving to the left, it is still over the crest. The convoluted stem continues to move to the left with increasing velocity, leaving the root behind it over the crest.

Just before the convolutions reach the position over the middle of the furrow, the root leaves its crest and moves with very great speed to the left. In IV the root is just passing under the convolutions. The whole system is moving with its maximum velocity, but the root outstrips the stem.   The two slopes of the crest are nearly symmetrical.   In V the root has gained so far on the convolutions as to have again reached a crest.

In VI the convolutions have caught up the root, and the crests are being displaced.

Finally VII is a repetition of I in the opposite direction, and the half oscillation is completed.

If in these figures I to VII we take the wave-length of ripple-mark as unity, the amplitude of oscillation of the main body of water is 2·1, that of the crests is ·7, and the breadth of the convolutions of the tree is ·3.   The sum of the amplitude of oscillation of the crest, with the breadth of the convolutions, and the wave-length of ripple-mark is equal to 2, and this is very nearly equal to the amplitude of oscillation of the water, as it ought to be.

The law which governs the intensity of the vortices must be a matter of inference, since I found the motion too rapid to be sure of anything save that the vortices are driven alternately by pulses, and that the motion was most energetic near the elongations.

In I the right hand vortex of a pair must be at its maximum of intensity, and it seems probable that the left hand vortex has a sub-maximum in consequence of the friction of the water along the dividing line.   During the return motion from I to VII, the left hand vortex must be increasing in intensity, so that it is at its maximum in VII.   Probably the right hand vortex diminishes in intensity from I to V, and then increases to its sub-maximum in VII.

I am not able to say from observation that the vortices which have been described as giving rise to the ink mushrooms actually exist in this state of oscillation, but if they are there, one of them should be found at the point marked with an asterisk in IV.

The figures tell better than words the mechanism by which the ripple-mark is made and maintained, and the cause of the dance of the crests. The only difficulty is in stage IV, where the root of the tree is in the state of transference from one crest to the next. In this stage the vortices would seem to be in the act of degrading the ripple-mark, but they are not then either of them at their maximum of intensity, and the time during which this holds good is exceedingly short compared with the whole semi-period of oscillation. It seems somewhat likely that small vortices are called into existence at the points marked with asterisks in IV, which serve to protect the ripple-marks from degradation during the transference.

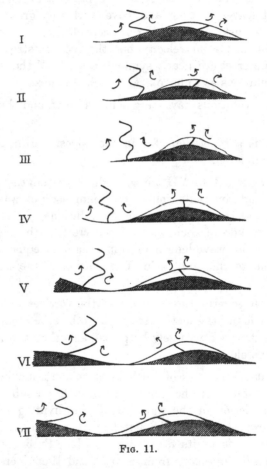

I

II

III

IV

V

VI

VII

FIG. 11.

Fig. 11, I to VII, exhibits the dance of the vortices when the oscillation of the water is considerably less in amplitude than the wave-length of ripple-mark.

Here the crests of the ripple-mark are scarcely sensibly disturbed. Above the crest is drawn the pair of mushroom vortices, the curved arrows showing the direction of rotation being placed outside of the mushrooms; but I am not able to satisfy myself that they are both in existence during the whole oscillation. Fig. 8 above exhibits an appearance which I have sometimes seen, which seems to show that they may both exist together with an ink tree.

We must now draw attention to the manner in which the convolutions are added to the ink tree, and thus show the continuity of this Fig. 11 for gentle oscillation with Fig. 10 for violent oscillation.

In Fig. 10, in II, III, IV, a convolution is added, which is unwrapped again in V. It is again formed in VI and VII, and then becomes permanent, and is transmitted up the stem of the tree.

In Fig. 11 the convolution is added in II, and then remains permanently part of the tree; but a partial convolution is added in III and IV, which is unwrapped again in V, VI, VII.

Thus in violent oscillation the convolution is permanently added just before an elongation, and in gentle oscillation afterwards. It would be easy to construct a figure for an intermediate amplitude in which the convolution is added just at elongation. If the oscillation gradually increases, the convolutions are permanently added sooner and sooner, and at the same time the formation of convolutions and subsequent unwrapping assumes more and more prominence.

It must be understood that these figures are drawn from the results of long watching of the process. My attention was at one time directed to one part of the phenomenon, and at another to a different part, and the amplitudes were constantly varied. I do not pretend to be able to see all these changes in a single half oscillation, lasting barely half a second. It may appear that I am incorrect in some parts of the construction, but I would ask any one who repeats the experiments not to condemn me hastily, for the constructions which I have given are the results of frequent trials and errors in the attempt to represent the changes observed.

I have not been able to determine exactly the mode of motion in the initial stages of ripple-making, when the oscillation has large amplitude, but when the ripple-marks are still in what I have described as the curdling stage.

If a current be passed over existing ripple-mark a vortex is established on the lee of each ripple; if the current be reversed the vortex is on the other side. Thus intermittent opposite currents will form ripple-mark, but probably without giving it a very regular wave-length.

If the intermittence is rapid, the vortex established on the lee-side, when the current is in one direction, is not annulled when the current is reversed, but it will be carried over the crest of the ripple-mark, and will diminish in intensity, whilst the new vortex with opposite rotation is established.

The study of a very gentle oscillation over existing ripple-mark, by means of a drop of ink placed on the ridge, enables us to observe these vortices (see Fig. 3). I think it depends on the amplitude of oscillation whether both vortices are always in existence and simply vary in intensity, or whether the vortex due to motion to the right is quite annulled during the motion to the left, and *vice versâ*.

It may be suspected, therefore, that, in the early stages of ripple-making, when the amplitude of oscillation is large, vortices are set up in the lee of each aggregation of sand, in the same way as if the current were permanent, and that when the current is reversed these vortices are speedily annulled, and a new set on the other side of the aggregations is established. When a drop of ink is put on an aggregation, and the oscillation is started, the ink forms a layer of not more than half a millimetre in thickness. It is easy to see that there takes place some kind of rapid oscillation which is not simply harmonic. It appears to present all the characters of the motion when gentle oscillation is established over ripple-mark of some height, and therefore it is probable that the motion is of the same kind in both cases.

When the aggregations are more pronounced, small correlated tree vortices are set up. As above stated, it has seemed to me that frequently only half of each tree vortex is set up at first.

I am disposed to regard this as the transitional state from the mode of oscillation, which produces the octave with small height of ripple-crest, to the fundamental with considerable height.

In gentle oscillation over high ripple-mark the tree vortices are, in the first instance, seen to be started by the mushroom vortices, and the same is probably true of the condition we are considering.

If the suggested view as to the mode of transition be correct, then we must suppose that at first every alternate tree vortex is started by its correlated mushroom vortex. If there be no tree vortices, or if there be only every alternate one, the vortices can pack twice as close as if the trees are symmetrical; but the existence of a half tree vortex tends to generate its other half, and this half cannot exist normally unless every alternate ripple-mark is removed. The degradation of the alternate ripple-mark must arise then from the existence of the second half of the ink tree. In these early stages the phenomenon is not highly regular, and therefore, besides the smallness of the scale and the rapidity of the motion, we have the difficulty of irregularity to contend with.

Other observers have endeavoured to determine the laws connecting the wave-length in the ultimate formation with the various concomitant circumstances, and I shall leave this subject to the following section, where some account of their work will be given.

We may summarise the results of these observations as follows:—

The formation of irregular ripple-marks or dunes by a current is due to the vortex which exists on the lee of any superficial inequality of the bottom; the direct current carries the sand up the weather slope and the vortex up the lee slope. Thus any existing inequalities are increased, and the surface of sand becomes mottled over with irregular dunes. The velocity of the water must be greater than one limit and less than another, the limiting velocities being dependent on the average size and density of the particles. Existing regular ripple-mark is maintained by a current passing over it perpendicular to the ridges. A slight change in form ensues, the weather slope becoming less steep and the lee slope steeper. The ridges are also slowly displaced to leeward. The regular ripple-mark may also thus be somewhat prolonged, so that although a uniform current cannot, as I believe, form regular ripple-mark, yet it may increase the area over which it is to be found.

Regular ripple-mark is formed by water which oscillates relatively to the bottom. A pair of vortices, or in some cases four vortices, are established in the water; each set of vortices corresponds to a single ripple-crest and the vortices oscillate about a mean position, changing their shapes and intensities periodically, but not with a simple harmonic motion.

The successive changes in the vortex motion, whilst ripple-mark is being established, and when the amplitude of oscillation over existing ripple-mark varies, are complex. As far as I have been able to determine, the following is an account of the phenomena:—

We begin with variation in amplitude of oscillation over existing regular ripple-mark, where the height of the undulations is not a very small fraction of the wave-length.

When the amplitude of oscillation is small compared with the wave-length, a pair of small vortices are established above the crest of each ripple-mark, rotating in opposite directions. In the mean position the upward current is over the crest, and the current of water tends to carry up sand from each furrow to the crest. The dividing line of the vortices oscillates, but the bottom of the line has much less amplitude of oscillation than the top, so that the dividing line is alternately inclined to one and the other side of the vertical. The vortices are thus carried backwards and forwards over the crest of the ripple, but the current always tends to maintain the crest, merely displacing very slightly the position of the highest point. The vortex which is on the lee-side is more intense than the other. We will call these the primary vortices. (See Fig. 3.)

Suppose now the amplitude of oscillation to be somewhat larger; then the primary vortices by their friction on the adjacent water generate two other vortices. The upward current in these secondary vortices has its mean position over the middle of a furrow, and the current comes down immediately over the upward current of the primary pair of vortices. It appears that sometimes only every alternate one of the secondary vortices is established. The upward current of the secondary vortices oscillates with a motion which is very far from being harmonic. It remains at its elongation for a long time and then darts across to the other elongation. (See Fig. 11.) During this mode of oscillation the primary vortices are carried much further backwards and forwards over the crests.

With still larger amplitude of oscillation it is no longer possible to distinguish the primary vortices, and the secondary vortices increase in intensity. It seems probable that the primary vortices are no longer both in existence during the whole oscillation, but that they are alternately created and annulled, so that when one exists the other does not. If this be so the vortex which exists is that which is on the lee side of the ripple in the state of motion at the instant.

With strong oscillation the secondary vortices apparently do all the work, and the primary vortex, if it exists, only exists for a short time, whilst it may serve as a protecting vortex to the ripple-crest, during the rapid transference of the dividing line of the secondary vortices from one crest to the next. Each secondary vortex is alternately a vortex under the lee of a ripple-mark as exhibited in Fig. 10. Mere description is hardly sufficient to explain the motion.

With very violent oscillation the ripple-marks are obliterated, and the water is filled with flying dust.

We now revert to the initiation of ripple-mark.

If the surface is very even, as when sand is sprinkled on glass, when a uniform oscillation of considerable amplitude is established, the sand is carried backwards and forwards and some of the particles stick in places of greater friction. As soon as there is any superficial inequality, it is probable that a vortex is set up in the lee of the inequality which tends to establish a dune there. Such vortices are, however, too small to be seen. The return current in the second half of an oscillation maintains the dune, a vortex being established on the other side, now the lee-side. As the sand tends to stick by friction in a great number of positions, the sand agglomerates into elongated patches, and the patches are so near to one another that the vortex on either side of one patch just fails to interfere with the next patch. As the patches elongate and regularise themselves the vortices increase in intensity, and the vortex established on one lee is not obliterated in the return current. The two vortices are then the primary vortices described above. As the ripple increases in height by the obliteration of some of the elongated patches, the primary vortices set up the secondary vortices. Perhaps the normal state of transition is that only one of the secondary vortices is established at first, and that when the other secondary vortex is set up it tends to obliterate every alternate ripple-mark, and thus to generate a ripple of double wave-length. As the ripples increase in height the secondary vortices become more and more important, and the primary less important. The final or stationary condition is that described above as the case of strong oscillation.

It is to be admitted that this history of the successive stages of the formation of ripple-mark is to some extent speculative, but it is the only method of formation which appears to accord with the various phenomena observed and described above.

It is important to note that when once a fairly regular ripple-mark is established, a wide variability of amplitude in the oscillation is consistent with its maintenance or increase. No explanation of ripple-making can be deemed satisfactory which does not satisfy this condition.

In this summary no attempt has been made to go over again the various peculiarities of the motion, which have been noted above, such as the dance of the crests and of the convolutions of the dividing line of the vortices. We must refer the reader back for the consideration of these points.

§ 3. The Work of previous Observers and Discussion.

Some valuable papers on ripple-mark have been lately published. The first of these is by Mr A. R. Hunt[*]. In it he makes an extensive collection of observations on the natural history of ripple-mark. As, however, he does not touch at any length on the mode of formation, I have but little to say on his work. He remarks that regular ripples are due to alternating currents, and that the irregular marks due to currents ought to be distinguished by another name from the regular marks formed by oscillating water. M. Forel, whose paper is referred to below, takes the same view, and describes these irregular marks as dunes. My own observations seem to accord well with the facts collected by Mr Hunt.

The second paper is by M. Casimir de Candolle[†]. His experiments have led him to enounce (p. 245) the following general law :—

"When a viscous material in contact with a less viscous liquid experiences an oscillatory or intermittent friction, arising from the relative motion of the liquid layer, 1st, the surface of the viscous material is rippled perpendicularly

---

[*] " On the Formation of Ripple-mark," *Proc. Roy. Soc.*, April 20, 1882, Vol. xxxiv., p. 1.

[†] "Rides formées, &c.," *Archives des Sciences Physiques et Naturelles*, Genève, No. 3, Vol. ix., 15th March, 1883.

to the direction of motion ; and 2nd, the wave-length is directly proportional to the amplitude of the oscillation."

The word viscous cannot here have its usual meaning, for sand cannot be called viscous. The epithet seems to denote that the constituent parts of the material are mobile, and that there is a considerable amount of internal friction.

When oscillations are set up in a vessel containing two fluids of very unequal viscosity, such as tar and water, ripples are formed on the more viscous fluid. But if the two fluids do not differ widely in viscosity, as mercury and water, water and turpentine, essence of cinnamon and water, ripple-mark is not generated. If, however, a layer of powder be introduced at the surface of separation, ripple-mark is easily formed.

Ripples were made in sand with a variety of fluids, but with olive oil it was found impossible. According to the views maintained in the present paper, the viscosity of oil is too great to permit the generation of the ripple-making vortices.

At p. 257, M. de Candolle writes :—

"Chaque ride se termine à la partie supérieure par une crête composée des particules les plus légères. Tant que dure le balancement du liquide, les particules sont animées d'un mouvement pendulaire qui les transporte alternativement de part et d'autre de la crête. Aussi longtemps que l'amplitude de ce balancement est égale à celle qui a donné naissance aux rides, les particules mobiles parcourent à chaque demi-oscillation* toute la distance qui sépare l'une de l'autre deux crêtes consécutives. Ce va-et-vient des particules s'étend jusqu'à une certaine distance au-dessous du sommet de chaque crête, mais son amplitude va en diminuant de haut en bas, en raison du poids plus considérable des particules inférieures. Il en résulte que chacune de ces crêtes mobiles a l'apparence d'une lamelle qui oscille sur le sommet de la ride qu'elle termine et s'étire en même temps dans le sens du mouvement de l'eau, ce qui donne tout à fait l'apparence d'un corps visqueux.

"Lorsque l'amplitude du balancement du liquide diminue, il en est naturellement de même des excursions de ces lamelles, et si l'on vient à arrêter subitement le balancement, les particules composant les crêtes mobiles peuvent se déposer entre les rides où elles forment un système de rides secondaires plus minces, intercalées entre celles qui correspondent au maximum d'amplitude du balancement."

In this passage the dance of the particles is in the first place described as being from one side to the other of the ridge, and this, I believe, is the fact.

---

* In a letter M. de Candolle tells me that "demi-oscillation" should read "quart d'oscillation"; but I still do not see how the ambiguity pointed out below is removed by this correction.

This statement is, however, apparently contradicted by what follows, viz., that the dance is from crest to crest. I have very rarely seen the inter-calated ripple-marks to which M. de Candolle refers, but I venture to think that his explanation is not sound, and that they are formed by the particles of sand which, in violent oscillation, have been caught up by the secondary or tree vortices, carried quite round and dropped at the root of the tree, when the oscillations of the water are dying out.

M. de Candolle arrives at the interesting conclusion that the wave-length of ripple-mark is independent of the nature of the oscillating fluid. His suggestion that cirrus clouds are ripple-marks between two aerial currents will be referred to below. The whole paper forms a valuable contribution, and should be read by those who are interested in the subject.

The last paper to which I shall refer is by M. Forel*. He has made extensive observations on ripple-mark, formed both naturally and artificially. He distinguishes between dunes formed by continuous currents either of air or water and ripple-marks formed by oscillation. His view accords with the experiments in § 1 above, but he has not apprehended the importance of the vortex in the lee of the dune, considering that region merely as slack water. I feel some doubt as to the view that a regular series of dunes may be formed by uniform current; at any rate, in my experiments the dunes were irregular, and had no definite wave-length.

His observations on the circumstances which govern the wave-length of ripple-mark are important. He finds that the factors which enter are the amplitude and period of oscillation of the water; and a third factor is the maximum velocity of the water, which he takes as identical with the ratio between the two others. If the water moves as a whole with a simple harmonic oscillation it is undoubtedly true that the ratio of amplitude to period is proportional to the maximum velocity, but the vortices quite disturb this relation. The maximum velocity of the water relatively to the bottom must depend upon the intensity of the vortices, and this depends upon the height of the ripple-mark.

M. Forel finds that the length and breadth of the vessel have no influence on wave-length, but that it diminishes with increasing depth of water. This he attributes to a diminution both of the period and of the amplitude of the oscillation of the water which is in contact with the bottom. The wave-length increases with the coarseness of the sand.

He remarks that when the ripple-mark is once made, the amplitude of

---

* "Les Rides de Fond," *Archives des Sciences Physiques et Naturelles*, Genève, 15 Juillet, 1883. M. Forel quotes an article of mine in *Nature* as attributing the formation of ripple-mark to the action of currents in the sea. My statement was intended merely to imply that shallow-ness of water is favourable to the formation of ripple-mark. I had already made a great part of these experiments when that article was written in 1882.

oscillation is without influence on its wave-length. He draws attention to the two limiting velocities, one too great, and the other too small for the formation of ripples, for which values were found in the experiments of § 1.

M. Forel explains ripple-mark as the confluence of two dunes, formed alternately by the oscillating currents. This theory is undoubtedly correct, if somewhat incomplete.

The wave-length, he says, is the amplitude of oscillation of a grain of sand "librement transportée par l'eau." This expression requires further explanation; if it means the amplitude of oscillation of a particle of water at the bottom, when the oscillation is started, and before the ripples have risen, I am disposed to doubt it. It may mean the distance which an average grain of sand is transported when lying on the surface of other sand, under the like circumstances*; if so, ripple-marks formed with a thin layer of sand on a sheet of glass should have a longer wave-length than if the sand be thick; this I am also disposed to doubt. However this may be, M. Forel considers that the wave-length should vary directly as the amplitude of oscillation, directly as the velocity of the current, inversely as the density of the sand, and inversely as the size of the grains. Considering with M. Forel that the velocity of the current is the ratio of amplitude to period, we should have the wave-length for any one sand varying as the frequency multiplied by the square of the amplitude. The few fairly consistent experiments recorded in § 1, do not accord with this view, for it seemed that wave-length varied as v, which is proportional to frequency multiplied by amplitude. This I understand to accord with M. de Candolle's law. As to the law that the finer the sand the longer the wave-length, M. Forel justly observes that it is in contradiction with the fact that small ripples are formed by fine sand, and large ripples by coarse sand. But he endeavours to remove the apparent inconsistency by remarking in effect that the larger limiting velocity for fine sand is smaller than the smaller limiting velocity for coarse sand. There must undoubtedly be truth in this view, but I hesitate to accept it as the whole truth.

Noticing that, in the same sites in the Lake of Geneva, the ripples are always of the same length, he says, " de ces observations il semblerait résulter que l'intensité des vagues a bien peu d'influence sur la largeur des rides; que la nature du sol est le seul facteur important."

It appears to me that M. Forel's view as to the wave-length of ripple-mark cannot be accepted as final, but he has certainly thrown much light on the subject in his interesting paper.

The following considerations bear upon the laws of wave-length :—

It appeared that in the initial stages of ripple-making, the wave-length

---

* I learn by a letter from M. Forel that this is his meaning.

is at first only half as long as it becomes ultimately, and that when the layer of sand is thin, the wave-length always remains shorter than if it is thick. Hence if a little sand is dusted on to the oscillating sheet of glass, it is found that the wave-length of ripple is long in the middle of the patch of sand, and short near the margins. Thus the patch when ripple-marked presents such an appearance as Fig. 12. If the sand is thin, this appearance often persists however long the oscillation is maintained. This shows that wave-length is a function of the height of the existing undulations; that is to say, not only of the amplitude of oscillation of the upper part of the vortices, but also of their intensity. On the parts of the plate where the sand is thick, a continual rearrangement of ripple-mark goes on; the wave-length extends by the excision of short patches of intercalated ripple-mark, and by

FIG. 12.

general rearrangement. Finally the sand reaches an ultimate condition as regards wave-length, although rearrangement of ripple-mark still appears to go on for a long time. Then we find in this final condition most of the sand arranged with a certain fundamental wave-length, but where the sand is thin, patches remain with the octave or half wave-length.

It is not easy to understand precisely the mode in which the oscillation of the water over the undulating bottom gives rise to vortices, but there are familiar instances in which nearly the same kind of fluid motion must occur.

In the mode of boat propulsion called sculling, the sailor places an oar with a flat blade through a rowlock in the stern of the boat, and, keeping the handle high above the rowlock, waves the oar backwards and forwards with an alternate inclination of the blade in one direction and the other. This action generates a stream of water sternwards. The manner in which the blade meets the water is closely similar to that in which the slopes of two ripple-marks alternately meet the oscillating water; the sternward current in one case, and the upward current in the other are due to similar causes. We may feel confident that in sculling, a pair of vortices are formed with axes vertical, and that the dividing line between them is sinuous. The motion of a fish's tail gives rise to a similar rearward current in almost the same way. These instances may help us to realise the formation of the ripple-making vortices.

Lord Rayleigh has considered the problem involved in the oscillations of a layer of vortically moving fluid separating two uniform streams*. At the meeting of the British Association at Swansea in 1880, Sir William Thomson read a paper discussing Lord Rayleigh's problem†. He showed that, in

---

* "On the Stability or Instability of certain Fluid Motions," *Proc. Lond. Math. Soc.* (Feb. 12, 1880), Vol. XI., p. 57.

† *Nature*, Nov. 11, 1880, pp. 45—6, and see correction on p. 70.

a certain case in which the analytical solution leads to an infinite value, there are waves in the continuous streams in diametrically opposite phases, and that the vortical stratum consists of a series of oval vortices. Fig. 13 illustrates this mode of motion. The uniform current flowing over existing ripple-mark exhibits almost a realisation of this mode of motion, one of the streams of fluid being replaced by the sandy undulations. The same kind of motion must exist in air when a gust of wind blows a shallow puddle into standing ripples.

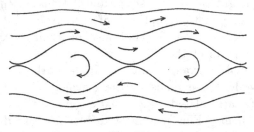

FIG. 13.

It seems probable that what is called a mackerel sky is an evidence of a closely similar mode of motion; and this agrees with M. de Candolle's suggestion that cirrus is aerial ripple-mark. The layer of transition between two currents of fluid is dynamically unstable, but if a series of vortices be interpolated, so as to form friction rollers as it were, it probably becomes stable. It is likely that in air a mode of motion would be set up by friction, which in frictionless fluid would be stable. If one of the currents of air be colder than the other, a precipitation of vapour will be caused in the vortices, and their shapes will be rendered evident by clouds*.

The direction of striation and velocity of translation of mackerel clouds require consideration according to this theory. If the velocity of the upper current be u, and of the lower current be $-u$, the interposed vortices have a velocity zero, and have their axes perpendicular to the velocity u. Hence if relatively to the earth the rectangular components of the upper current are $u + w$, v, and of the lower are $-u + w$, v, the component velocities of the vortices are w, v, and their axes are parallel to the component v.

Therefore the striations should be parallel to that direction in which the two currents have equal components, and the component velocity of the

---

* I had suggested that the centrifugal force of the vortical rotation must produce rarefaction, fall of temperature, and precipitation of vapour. I have to thank Professor Stokes for pointing out that the fall of temperature would necessarily be so small as only to cause precipitation if the air were almost completely saturated. I think the fall of temperature at the centre of the vortex might be between a hundredth and a fiftieth of a degree Centigrade. {Sir W. Thomson tells me that he had given this vortical explanation before the Brit. Assoc. in 1876. The volume, however, gives no abstract. He explains the formation of cloud as due to the upward motion in the vortices, and consequent rarefaction.—Jan. 4, 1884.}

clouds parallel to the striations should be equal to that of either current in
the same direction.   The resultant velocity of the clouds is clearly equal to
a half of that of the two currents, and the component velocity of the striations
perpendicular to themselves is the mean of the components of the two
currents in the same direction.

If one of the currents veers the axes of the vortices are shifted.   The
existing clouds will be furrowed obliquely, and the vortical stratum will be
cut up into diamond-shaped spaces, determined by the intersection of the
old and new vortex-axes.   This would explain the patchwork arrangement
commonly observed in mackerel sky.   May not the lengths of the patches
give a measure of the rate of veering of one of the currents ?

The above account of the formation of ripple-marks shows it to be due to
a complex arrangement of vortices.   The difficulty of observation is considerable, and perhaps some of the conclusions may require modification.
I hope that other experimenters may be induced to examine the question*.

Lord Rayleigh has shown me a mathematical paper†, in which he has
considered the formation of aerial vortices over a vibrating plate.   It seems
possible that an application might be made of similar modes of approximation
to the question of water oscillating over a corrugated bottom.   Even a very
rough solution would probably throw much light on the exact changes
which the ripple-making vortices undergo, and any guidance from theory
would much facilitate observation.

* [The following are references to some other papers on the subject :
A. R. Hunt, *Royal Dublin Soc.* Mar. 17, 1884.
R. D. Oldham, *Geolog. Survey of India*, Vol. xxxiv., Pt. 3.
Vaughan Cornish, *Geographical Journal*, Mar. 1897, May and June 1898, June 1899,
Jan. 1900, Aug. 1901, Aug. 1902 ; also *Dorset Nat. Hist. Club*, Vol. xix., 1898, and
Scottish Geograph. Mag. Jan. 1901.
Bertololy, *Münchener Geograph. Studien*, 9th pt., 1900.
Arctowski, *Soc. Belge de Géologie*, Vol. xv., 1901.
Forchheimer, *k. k. Akad. Wien*, Dec. 1903.
Mrs Ayrton, *Proc. Roy. Soc.* Vol. A, lxxxiv., 1910, p. 285.]
† [*Phil. Mag.* Vol. xv. (5th series), p. 229 ; Vol. xvi. (1883), p. 50.]

17.

NOTE ON MR DAVISON'S PAPER ON THE STRAINING OF THE EARTH'S CRUST IN COOLING.

[*Philosophical Transactions of the Royal Society*, Vol. 178 (1887), A, pp. 242—249.]

MR DAVISON'S interesting paper was, he says, suggested by a letter of mine published in *Nature* on February 6, 1879. In that letter it is pointed out that the stratum of the Earth where the rate of cooling is most rapid lies some miles below the Earth's surface. Commenting on this, I wrote:—

"The Rev. O. Fisher very justly remarks that the more rapid contraction of the internal than the external strata would cause a wrinkling of the surface, although he does not admit that this can be the sole cause of geological distortion. The fact that the region of maximum rate of cooling is so near to the surface recalls the interesting series of experiments recently made by M. Favre (*Nature*, Vol. XIX., p. 108), where all the phenomena of geological contortion were reproduced in a layer of clay placed on a stretched india-rubber membrane, which was afterwards allowed to contract. Does it not seem possible that Mr Fisher may have under-estimated the contractibility of rock in cooling, and that this is the sole cause of geological contortion?"

Mr Davison works out the suggestion, and gives precision to the general idea contained in the letter. He shows, however, that there is a layer of zero strain in the Earth's surface, and that this layer, instead of that of greatest cooling, must be taken to represent Favre's elastic membrane.

It appears that the mathematical discussion of the problem in his paper is unnecessarily laborious, and he has not made various important deductions as to the integral results of distortion and as to the magnitude of the effects to be expected. I, therefore, offer the following note with the intention of rendering more complete an important chapter in the mechanics of geology.

When a spherical shell expands with rise of temperature, it may be said to stretch, in one sense of the word. If the shell were one of the layers of the Earth, such a stretching would have no geological effect, for it would merely involve a change of density. The term "stretching" then requires an explanation in connection with Mr Davison's paper. The stretching which we have to consider is, in fact, simply the excess of the actual stretching above that due to rise of temperature. The negative of such stretching is a contraction, and it would actually be shown by a crumpling of strata.

If ρ be the density of a body, and ϵ its modulus of linear expansion for temperature, then it is obvious that when the temperature is raised θ degrees the density becomes $\rho(1 - 3\epsilon\theta)$. Now suppose that a spherical shell of radius r expands so that its radius becomes $r(1 + \alpha)$, and suppose that at the same time its temperature is raised θ degrees. Then, if k be the modulus of stretching, so that unit length is stretched by a length k, we have, in consequence of the above explanation of stretching,

$$\alpha = k + \epsilon\theta \dots\dots\dots\dots\dots\dots\dots\dots\dots\dots\dots(1)$$

Now let us consider the geometry of changes in a sphere such that a shell of internal radius r, thickness δr, and density ρ, expands until its internal radius r becomes $r(1 + \alpha)$ and its density ρ becomes $\rho(1 - 3\epsilon\theta)$.

The external radius $r + \delta r$ clearly then becomes

$$r(1 + \alpha) + \delta r \left[1 + \frac{d}{dr}(r\alpha)\right]$$

Thus the mass of the shell $4\pi r^2 \rho \delta r$ becomes

$$4\pi r^2 \rho \delta r \left[1 + 2\alpha + \frac{d}{dr}(r\alpha) - 3\epsilon\theta\right]$$

Since the mass remains unchanged, we must have

$$2\alpha + \frac{d}{dr}(r\alpha) - 3\epsilon\theta = 0$$

This equation of continuity may clearly be written

$$\frac{d}{dr}(r^3\alpha) = 3\epsilon\theta r^2$$

Substituting for α in terms of the modulus of stretching as given by (1), we have

$$\frac{d}{dr}(kr^3 + \epsilon\theta r^3) = 3\epsilon\theta r^2$$

Hence

$$\frac{d}{dr}(kr^3) = -\epsilon r^3 \frac{d\theta}{dr}$$

and therefore

$$k = -\frac{\epsilon}{r^3}\int r^3 \frac{d\theta}{dr}\, dr \quad\text{.............................(2)}$$

where the integral is taken from r down to such a depth that there is no change of temperature.

If, now, θ represents the rise of temperature per unit time, and if we replace k by dK/dt, the rate of stretching per unit time, and if we make application of (2) to the case of the Earth, and write v for the temperature of the Earth at a depth x below the surface, and c for the Earth's radius, we have

$$\frac{dv}{dt} = \theta$$

$$c - x = r$$

and (2) becomes

$$\frac{dK}{dt} = -\frac{\epsilon}{(c-x)^3}\int_c^x (c-x)^3 \frac{d^2v}{dx\,dt}\, dx \quad\text{....................(3)}$$

In inserting the limits to the integral, it is assumed that the temperature at the Earth's centre is sensibly constant.

The amount by which a great circle of radius $(c-x)$ is being stretched per unit time is

$$2\pi\,(c-x)\,\frac{dK}{dt}$$

This expression, with the above value (3) for dK/dt, is Mr Davison's result.

We know from Thomson's solution for the cooling of the Earth that, when x exceeds a small fraction of the Earth's radius the temperature gradient and its rate of variation in time are very small. Hence, when x is not very small, $d^2v/dx\,dt$ is very small; therefore we may with sufficient approximation replace $(c-x)^3$ under the integral sign by $c^3 - 3c^2x$. Outside the integral we may simply neglect x. Also the limit c of the integral may be replaced by infinity, and this is desirable because Thomson's solution is really applicable to an infinite slab and not to a sphere. With these approximations we get

$$\frac{dK}{dt} = \epsilon \int_x^\infty \left(1 - \frac{3x}{c}\right)\frac{d^2v}{dx\,dt}\, dx \quad\text{......................(4)}$$

Integrating, with regard to time, from the time t to zero, we shall get the total amount of stretching in the layer x from the epoch of consolidation down to time t. Thus

$$K = \epsilon \int_x^\infty \left(1 - \frac{3x}{c}\right)\frac{dv}{dx}\, dx \quad\text{......................(5)}$$

Now with Thomson's notation

$$\int_x^\infty \frac{dv}{dx}\, dx = V - v$$

Since
$$\frac{dv}{dx} = \frac{V}{(\pi\kappa t)^{\frac{1}{2}}} e^{-x^2/4\kappa t}$$

$$\int_x^\infty x\,\frac{dv}{dx}\,dx = \frac{V}{(\pi\kappa t)^{\frac{1}{2}}}\int_x^\infty xe^{-x^2/4\kappa t}dx = \frac{2V\kappa t}{(\pi\kappa t)^{\frac{1}{2}}}\cdot e^{-x^2/4\kappa t} = 2\kappa t\,\frac{dv}{dx}$$

Hence (5) becomes
$$K = \epsilon\left[V - v - \frac{6\kappa t}{c^2}\cdot c\,\frac{dv}{dx}\right]^* \quad\text{...............(6)}$$

This expression gives us the total amount of contraction since consolidation in terms of the temperature and the temperature gradient. I shall return to this expression later.

Differentiating (6) with respect to the time, we have

$$\frac{dK}{dt} = \epsilon\left[-\frac{dv}{dt} - \frac{6\kappa}{c}\frac{dv}{dx} - \frac{6\kappa t}{c}\frac{d^2v}{dx\,dt}\right]$$

But
$$\frac{dv}{dt} = -\frac{x}{2t}\frac{dv}{dx}, \quad\text{and}\quad \frac{d^2v}{dx\,dt} = \frac{1}{2t}\left(\frac{x^2}{2\kappa t} - 1\right)\frac{dv}{dx}$$

Hence
$$\frac{dK}{dt} = \frac{\epsilon}{2t}c\,\frac{dv}{dx}\left[\frac{x}{c} - \frac{12\kappa t}{c^2} - \frac{6\kappa t}{c^2}\left(\frac{x^2}{2\kappa t} - 1\right)\right]$$

or
$$\frac{dK}{dt} = \frac{\epsilon}{2t}c\,\frac{dv}{dx}\left[\frac{x}{c} - \frac{3x^2}{c^2} - \frac{6\kappa t}{c^2}\right] \quad\text{...............(7)}$$

When x exceeds more than a small fraction of the Earth's radius dv/dx is very small; hence in (7) we may regard x/c as small, and write the equation

$$\frac{dK}{dt} = \frac{\epsilon}{2t}c\,\frac{dv}{dx}\left[\frac{x}{c} - \frac{6\kappa t}{c^2}\right] \quad\text{...............(8)}$$

When $x = 0$, dK/dt is negative, and hence at the surface there is contraction or crumpling. The crumpling continues for some depth downwards, and vanishes when

$$\frac{x}{c} = \frac{6\kappa t}{c^2}$$

Taking, with Thomson, the foot and year as units, κ appears to be about 400; if, therefore, t is τ million years, $\kappa t = 4 \times 10^8\tau$. Now, c being 21×10^6 feet, $\kappa\tau/c^2 = \tau/10^6$ approximately, and

$$x = \frac{24 \times 10^8}{21 \times 10^6}\tau \text{ feet}$$

$$= 114\tau \text{ feet}$$

If τ be 100, $x = 2$ miles.

---

* If x had not been treated as small, we should have got

$$\epsilon(V - v)\frac{c(c^2 + 6\kappa t)}{(c-x)^3} - \frac{2\kappa t}{(c-x)^3}\left[\frac{c^3 - (c-x)^3}{x} + 4\kappa t\right]\epsilon\frac{dv}{dx}$$

and it is easy to see that this leads to the same result as the above for all practical purposes.

Thus, if the time since consolidation be 100 million years, the present depth of the stratum of no strain is 2 miles, and the depth is proportional to the time since consolidation. With a greater value of κ the depth is greater.

With regard to the value of dK/dt at greater depths, we observe that at a few miles below the stratum of zero strain $6\kappa t/c^2$ becomes negligible compared with x/c. Hence dK/dt is approximately proportional to $x\,dv/dx$. Now, if we take the figure drawn in Thomson and Tait's *Natural Philosophy*, Appendix D, and augment or diminish each ordinate NP' proportionately to the corresponding abscissa, it is clear that we shall get just such a curve as that drawn by Mr Davison. His curve might thus have been drawn without any computations at all. The function $x\,dv/dx$ is proportional to dv/dt, and thus his dotted curve is of just the same kind as the other, excepting close to the surface.

Now let us return to the expression for the integral stretching, viz. :—

$$K = \epsilon \left[(V - v) - \frac{6\kappa t}{c^2}\, c\, \frac{dv}{dx} \right] \quad\dots\dots\dots\dots\dots\dots(9)$$

We have
$$\frac{dv}{dx} = \frac{V}{(\pi\kappa t)^{\frac{1}{2}}}\, e^{-x^2/4\kappa t}$$

Hence, if x be small and t large (both of which conditions apply to the present time near the Earth's surface),

$$\frac{dv}{dx} = \frac{V}{(\pi\kappa t)^{\frac{1}{2}}}$$

Hence
$$K = \epsilon \left[V - v - \frac{6\,(\kappa t)^{\frac{1}{2}}}{c\pi^{\frac{1}{2}}}\, V \right]\dots\dots\dots\dots\dots\dots(10)$$

Now, as we have seen above, with such values as those with which we have to deal, $6\kappa t/c^2$ is a small fraction, notwithstanding that it increases with the time.

Hence, for the upper layers, we have approximately

$$K = \epsilon (V - v) \quad\dots\dots\dots\dots\dots\dots\dots(11)$$

Thus it appears that the integral effect is always a stretching, and that it is the same in amount at whatever speed the globe cools. The fact that, if the globe cools suddenly, the integral effect must be stretching has been pointed out by Mr Davison.

If we differentiate (10), we have

$$\frac{dK}{dt} = \epsilon \left[-\frac{dv}{dt} - \frac{3\kappa}{c} \cdot \frac{V}{(\pi\kappa t)^{\frac{1}{2}}} \right]$$

But
$$-\frac{dv}{dt} = \frac{x}{2t}\frac{dv}{dx}$$

and, near the surface,
$$\frac{dv}{dx} = \frac{V}{(\pi\kappa t)^{\frac{1}{2}}}$$

Hence
$$\frac{dK}{dt} = \epsilon\frac{dv}{dx}\left(\frac{x}{2t} - \frac{3\kappa}{c}\right)$$

$$= \frac{\epsilon}{2t}c\frac{dv}{dx}\left(\frac{x}{c} - \frac{6\kappa t}{c^2}\right)$$

and thus we find equation (8) again, as ought to be the case. This may also be written

$$\frac{dK}{dt} = \frac{\epsilon}{2t}\frac{V}{(\pi\kappa t)^{\frac{1}{2}}}\left(x - \frac{6\kappa t}{c}\right)$$

We must now see whether the amount of crumpling of the surface strata can be such as to explain the contortions of geological strata.

It must be borne in mind that, from a geological point of view, contraction is not the negative of stretching. When a stratum is stretched it may perhaps be ruptured, and rock may be squeezed up into the crack, at least for the strata which are very near the surface, and therefore not under great pressure; but when compressed the stratum is no doubt crumpled. Hence it is insufficient to know that the integral effect from the time of consolidation is a stretching; for that stretching may be merely the excess of a stretching over a crumpling. Now we have found above that the depth of the stratum of no strain is given by

$$x = \frac{6\kappa t}{c}$$

Hence, at the time t', given by $t' = cx/6\kappa$, the surface of no strain was at depth x; and at all later times than t' the surface of no strain lies deeper. Therefore, to find the total amount of crumpling at any depth, we require to find the integral effect taken from t' to t, which is greater than t'.

The integral stretching from consolidation to time t is

$$K = \epsilon\left[V - v - \frac{6\kappa t}{c}\frac{dv}{dx}\right]$$

Now near the surface v is nearly equal to $v_0 + x\frac{dv}{dx}$;

hence
$$K = \epsilon(V - v_0) - \epsilon\frac{dv}{dx}\left[x + \frac{6\kappa t}{c}\right]$$

But
$$\frac{dv}{dx} = \frac{V}{(\pi\kappa t)^{\frac{1}{2}}}e^{-x^2/4\kappa t} = \frac{V}{(\pi\kappa t)^{\frac{1}{2}}} \text{ near the surface.}$$

Hence
$$K = \epsilon(V - v_0) - \frac{\epsilon V}{(\pi\kappa)^{\frac{1}{2}}}\left[\frac{x}{t^{\frac{1}{2}}} + \frac{6\kappa t^{\frac{1}{2}}}{c}\right] \quad\dots\dots\dots\dots(12)$$

At the time t', given by $t' = cx/6\kappa$,

$$K = \epsilon\,(V - v_0) - \frac{\epsilon V}{(\pi\kappa)^{\frac{1}{2}}}\left[x^{\frac{1}{2}}\left(\frac{6\kappa}{c}\right)^{\frac{1}{2}} + x^{\frac{1}{2}}\left(\frac{6\kappa}{c}\right)^{\frac{1}{2}}\right] = \epsilon\,(V - v_0) - \frac{\epsilon V}{(\pi\kappa)^{\frac{1}{2}}}\cdot 2\left(\frac{6\kappa}{c}\right)^{\frac{1}{2}} x^{\frac{1}{2}}$$

$$\dots\dots(13)$$

This gives the total stretching from the time of consolidation until the surface of no strain has got down to x.

If, therefore, we subtract (13) from (12), we get the total stretching between the time t' and the time t, and the result is obviously

$$K = -\frac{\epsilon V}{(\pi\kappa t)^{\frac{1}{2}}}\left[x - 2\left(\frac{6\kappa t}{c}\right)^{\frac{1}{2}} x^{\frac{1}{2}} + \frac{6\kappa t}{c}\right]$$

This is clearly
$$K = -\epsilon\frac{dv}{dx}\left[x^{\frac{1}{2}} - \left(\frac{6\kappa t}{c}\right)^{\frac{1}{2}}\right]^2 \dots\dots\dots\dots\dots(14)$$

This expression is essentially negative, and therefore the total effect from t' to t is a crumpling, as was foreseen.

This integral crumpling vanishes at the same place as does dK/dt, that is to say, when

$$x = \frac{6\kappa t}{c}$$

This, we have shown above, will be at a depth of 2 or 3 miles. The amount of crumpling at the surface is given by putting $x = 0$, and

$$K = -\epsilon\frac{6\kappa t}{c}\frac{dv}{dx}$$

Now we have seen that, if τ be the number of millions of years since consolidation,

$$\frac{6\kappa t}{c} = 114\tau \text{ feet}$$

and
$$\frac{dv}{dx} = 0°\cdot02 \text{ Fahr. per foot}$$

Hence
$$K = -\tau\epsilon \times 2\cdot28$$

The total amount by which a great circle is contracted is 25,000 K miles; and, judging by the coefficient of expansion of metals, ϵ may be about 5×10^{-5}.

Thus, with these rough data, the amount of crumpling of a great circle of radius c is

$$2\pi cK = \tau \times \left(\frac{5}{10^5}\right) \times (2\cdot5 \times 10^4) \times 2\cdot28 = 2\cdot85\tau \text{ miles}$$

Thus, in 10,000,000 years, $28\frac{1}{2}$ miles of rock would be crumpled up.

The area of rock crumpled is $4\pi c^2 \cdot 2K$, and with these numerical data

$$4\pi c^2 \cdot 2K = 4c\,(2\pi cK)$$

Now $c = 4000$ miles, and therefore

$$4\pi c^2 . 2K = 22,800\tau \text{ square miles}$$

Thus, in 10,000,000 years, 228,000 square miles of rock will be crumpled up and piled on the top of the subjacent rocks.

The numerical data with which we have to deal are all of them subject to wide limits of uncertainty, but the result just found, although rather small in amount, is such as to appear of the same order of magnitude as the crumpling observed geologically.

The stretching and probable fracture of the strata at some miles below the surface will have allowed the injection of the lower rocks amongst the upper ones, and the phenomena which we should expect to find according to Mr Davison's theory are eminently in accordance with observation. It therefore appears to me that his view has a strong claim to acceptance.

18.

ON THE MECHANICAL CONDITIONS OF A SWARM OF METEORITES, AND ON THEORIES OF COSMOGONY.

[*Philosophical Transactions of the Royal Society*, Vol. 180 (1889), A, pp. 1—69.]

MR [now Sir NORMAN] LOCKYER writes in his interesting paper on Meteorites* as follows:—

"The brighter lines in spiral nebulæ, and in those in which a rotation has been set up, are in all probability due to streams of meteorites with irregular motions out of the main streams, in which the collisions would be almost *nil*. It has already been suggested by Professor G. Darwin (*Nature*, Vol. XXXI., 1884–5, p. 25)—using the gaseous hypothesis—that in such nebulæ 'the great mass of the gas is non-luminous, the luminosity being an evidence of condensation along lines of low velocity, according to a well known hydrodynamical law. From this point of view, the visible nebula may be regarded as a luminous diagram of its own stream-lines'."

The whole of Mr Lockyer's paper, and especially this passage in it, leads me to make a suggestion for the reconciliation of two apparently divergent theories of the origin of planetary systems.

The nebular hypothesis depends essentially on the idea that the primitive nebula is a rotating mass of fluid, which at successive epochs becomes unstable from excess of rotation, and sheds a ring from the equatorial region.

The researches of Roche † (apparently but little known in this country) have imparted to this theory a precision which was wanting in Laplace's original exposition, and have rendered the explanation of the origin of the planets more perfect.

---

* *Nature*, Nov. 17, 1887. The paper itself is in *Roy. Soc. Proc.*, Vol. XLIII., 1887, p. 117.
† Montpellier, *Acad. Sci. Mém.*

But notwithstanding the high probability that some theory of the kind is true*, the acceptance of the nebular hypothesis presents great difficulties.

Sir William Thomson long ago expressed to me his opinion that the most probable origin of the planets was through a gradual accretion of meteoric matter, and the researches of Mr Lockyer afford actual evidence in favour of the abundance of meteorites in space.

But the very essence of the nebular hypothesis is the conception of fluid pressure, since without it the idea of a figure of equilibrium becomes inapplicable. Now, at first sight, the meteoric condition of matter seems absolutely inconsistent with a fluid pressure exercised by one part of the system on another. We thus seem driven either to the absolute rejection of the nebular hypothesis, or to deny that the meteoric condition was the immediate antecedent of the sun and planets. M. Faye has taken the former course, and accepts as a necessary consequence the formulation of a succession of events quite different from that of the nebular hypothesis†. I cannot myself find that his theory is an improvement on that of Laplace, except in regard to the adoption of meteorites, for he has lost the conception of the figure of equilibrium of a rotating mass of fluid.

The object of this paper is to point out that by a certain interpretation of the meteoric theory we may obtain a reconciliation of these two orders of ideas, and may hold that the origin of stellar and planetary systems is meteoric, whilst retaining the conception of fluid pressure.

According to the kinetic theory of gases, fluid pressure is the average result of the impacts of molecules. If we imagine the molecules magnified until of the size of meteorites, their impacts will still, on a coarser scale, give a quasi-fluid pressure. I suggest then that the fluid pressure essential to the nebular hypothesis is, in fact, the resultant of countless impacts of meteorites.

The problems of hydrodynamics could hardly be attacked with success, if we were forced to start from the beginning and to consider the cannonade of molecules. But when once satisfied that the kinetic theory will give us a gas, which, in a space containing some millions of molecules, obeys all the laws of an ideal non-molecular gas filling all space, we may put the molecules out of sight and treat the gas as a plenum.

In the same way, the difficulty of tracing the impacts of meteorites in detail is insuperable; but, if we can find that such impacts give rise to a quasi-fluid pressure on a large scale, we may be able to trace out many

* The very remarkable photograph of the nebula in Andromeda, exhibited to the Royal Astronomical Society by Mr Isaac Roberts on December 6, 1888 [seemed to me when this paper was written to afford] something like a proof of the substantial truth of the nebular hypothesis.

† *Sur l'Origine du Monde*, Paris, Gauthier-Villars, 1884 ; *Annuaire pour l'an* 1885, *Bureau des Longitudes*, p. 757.

results by treating an ideal plenum. Laplace's hypothesis implies such a plenum, and it is here maintained that this plenum is merely the idealisation of the impacts of meteorites.

As a bare suggestion this view is worth but little, for its acceptance or rejection must turn entirely on numerical values, which can only be obtained by the consideration of some actual system. It is obvious that the solar system is the only one about which we have sufficient knowledge to afford a basis for discussion. This paper is accordingly devoted to a consideration of the mechanics of a swarm of meteorites, with special numerical application to the solar system.

The investigation has entailed a considerable amount of mathematical analysis; there is, however, no analysis in §§ 1 and 2. The reader who only wishes to know the arguments and results, without a consideration of the mathematical details, is therefore recommended, after reading §§ 1 and 2, to pass on to the Summary.

§ 1. On the Effective Elasticity of Meteorites in Collision.

When two meteoric stones meet with planetary velocity, the stress between them during impact must generally be such that the limits of true elasticity are exceeded; and it may be urged that a kinetic theory is inapplicable unless the colliding particles are highly elastic. It may, however, I think, be shown that the very greatness of the velocities will impart what virtually amounts to an elasticity of a high order of perfection.

It appears, a priori, probable that, when two meteorites clash, a portion of the solid matter of each is volatilised, and Mr Lockyer considers the spectroscopic evidence conclusive that it is so. There is, no doubt, enough energy liberated on impact to volatilise the whole of both bodies, but only a small portion of each stone will undergo this change.

A rough numerical example will show the kind of quantities with which we are here dealing.

It will appear hereafter that the mean velocity of a meteorite may be at the least about 5 kilometres a second; and, accordingly, the mean relative velocity of a pair would then be about 7 kilometres a second*. Hence, if two stones, weighing a kilogramme, move each with a velocity of $3\frac{1}{2}$ kilometres per second directly towards one another, the energy liberated at the moment of impact is $2 \times \frac{1}{2} \times 10^3 (3\frac{1}{2} \times 10^5)^2$ or 12×10^{13} ergs.

Now Joule's equivalent is $4\cdot2 \times 10^7$ ergs; hence, the energy liberated is about 3 million calories.

---

* If v be the velocity of mean square, $v\sqrt{2}$ is the square root of the mean square of relative velocity.

It is quite uncertain how much of each stone would be volatilised; but, if it were 3 grammes, there would be a million calories of energy applied to each gramme.

The melting temperature of iron is about 1500 degrees Centigrade, and the mean specific heat of iron may be about $\frac{1}{7}$*. Hence, about 300 calories are required to raise a gramme of iron from absolute zero to melting point. I do not know the latent heat of the melting of iron, but for platinum it is 27, and the latent heat of volatilisation of mercury is 62. Hence, about 400 or 500 calories suffice to raise a gramme of iron from absolute zero to volatilisation. It is clear, then, that there is energy enough, not only to volatilise the iron, but also to render the gas incandescent; and the same would be true if the mass operated on by the energy were 30 grammes instead of 3.

It must necessarily be obscure as to how a small mass of solid matter *can* take up a very large amount of energy in a small fraction of a second, but spectroscopic evidence seems to show that it does so; and, if so, we have what is virtually a violent explosive introduced between the two stones.

In a direct collision each stone is probably shattered into fragments, like the splashes of lead when a bullet hits an iron target. But direct collision must be a comparatively rare event. In glancing collisions the velocity of neither body is wholly arrested, the concentration of energy is not so enormous (although probably still sufficient to effect volatilisation), and, since the stones rub past one another, more time is allowed for the matter round the point of contact to take up the energy; thus, the whole process of collision is much more intelligible. The nearest terrestrial analogy is when a cannon-ball bounds off the sea. In glancing collisions fracture will probably not be very frequent.

From these arguments, it is probable that, when two meteorites meet, they attain an effective elasticity of a high order of perfection; but there is, of course, some loss of energy at each collision. It must, however, be admitted that on collision the deflection of path is rarely through a very large angle. But a succession of glancing collisions would be capable of reversing the path; and, thus, the kinetic theory of meteorites may be taken as not differing materially from that of gases.

Perhaps the most serious difficulty in the whole theory arises from the fractures which must often occur. If they happen with great frequency, it would seem as if the whole swarm of meteorites would degrade into dust. We know, however, that meteorites of considerable size fall upon the earth; and, unless Mr Lockyer has misinterpreted the spectroscopic evidence, the

---

* *Physikalisch-Chemische Tabellen*, Landolt and Börnstein.

nebulæ do now consist of meteorites. Hence, it would seem as if fracture was not of very frequent occurrence. It is easy to see that, if two bodies meet with a given velocity, the chance of fracture is much greater if they are large, and it is possible that the process of breaking up will go on only until a certain size, dependent on the velocity of agitation, is reached, and will then become comparatively unimportant.

When the volatilised gases cool, they will condense into a metallic rain, and this may fuse with old meteorites whose surfaces are molten. A meteorite in that condition will certainly also pick up dust. Thus, there are processes in action tending to counteract subdivision by fracture and volatilisation. The mean size of meteorites probably depends on the balance between these opposite tendencies. If this is so, there will be some fractures, and some fusions, but the mean mass will change very slowly with the mean kinetic energy of agitation. This view is, at any rate, adopted in the paper as a working hypothesis. It was not, however, possible to take account of fracture and fusion in the mathematical investigation, but the meteorites are treated as being of invariable mass.

§ 2.　On the Velocity of Agitation of Meteorites, and on its Secular Change.

The velocity with which the meteorites move is derived from their fall from a great distance towards a centre of aggregation. In other words, the potential energy of their mutual attraction when widely dispersed becomes converted, at least partially, into kinetic energy. When the condensation of a swarm is just beginning, the mass of the aggregation towards which the meteorites fall is small; and, thus, the new bodies arrive at the aggregation with small velocity. Hence, initially, the kinetic energy is small, and the volume of the sphere within which hydrostatic ideas are (if anywhere) applicable is also small.

As more and more meteorites fall in, that volume is enlarged, and the velocity with which they reach the aggregation is increased. Finally, the supply of meteorites in that part of space begins to fail, and the imperfect elasticity of the colliding bodies brings about a gradual contraction of the swarm.

I do not now attempt to trace the whole history of a swarm, but the object of the paper is to examine its mechanical condition at an epoch when the supply of meteorites from outside has stopped, and when the velocities of agitation and distribution of meteorites in space have arranged themselves into a sub-permanent condition, only affected by secular changes. This examination will enable us to understand, at least roughly, the secular change in the velocity and in the distribution of the meteorites as the swarm contracts, and will throw light on other questions.

§ 3.   *Formulæ for Mean Square of Velocity, Mean Free Path,
and Interval between Collisions.*

We have to investigate whether, when the solar system consisted of a swarm of meteorites, the velocities and encounters could have been such that the mechanics of the system can be treated as subject to the laws of hydrodynamics. The formulæ which form the basis of this discussion will now be considered.

For the sake of simplicity, the meteorites will, in the first instance, be treated as spheres of uniform size.

The sum of the masses of the meteorites is equal to that of the sun, for the planets only contribute a negligible mass. If M_0 be the sun's mass, and m that of a meteorite, their number is M_0/m.

If, at each encounter between two meteorites, there were no loss of energy, the sum of the kinetic energies of all the meteorites would be equal to the potential energy lost in the concentration of the swarm from a condition of infinite dispersion, until it possessed its actual arrangement. In such a computation the rotational energy of the system is negligible.

Suppose the sun's mass to be concentrated from infinite dispersion until it is arranged in the form of a homogeneous sphere of radius a and density ρ. Then let the sphere be cut up into as many equal spaces as there are meteorites, and let the matter in each space be concentrated into a meteorite. When the number of meteorites is large, the potential energy lost in the first process is very great compared with that lost in the subsequent partial condensation into meteorites*. Thus, the energy lost in the partial condensation is negligible.

If μ be the attractional constant, the lost energy of condensation is well known to be $\frac{3}{5}\mu M_0^2/a$. But on the hypothesis that there is no loss of energy at each encounter, this must be equal to the sum of the kinetic energies of all the meteorites. If, therefore, v^2 be the mean square of velocity of a meteorite, we must have $\frac{1}{2}M_0v^2 = \frac{3}{5}\mu M_0^2/a$, so that $v^2 = \frac{6}{5}\mu M_0/a$.

But homogeneity of density and uniformity of kinetic energy of agitation are impossible; for the meteor-swarm must be much condensed towards its centre, so that we have largely under-estimated the lost potential energy of the system. Also, the velocity of agitation must decrease towards the outside, or else the swarm would extend to infinity. Besides this, the partial conversion of molar into molecular energy, which must take place on each encounter, has been neglected.

---

* It depends, in fact, on the square of the ratio of the diameter $2a$ to the linear dimension of one of the equal spaces.

We shall see below reason for believing that throughout a large central volume the mean square of velocity of agitation is nearly uniform, and that outside this region it falls off.

Suppose, then, that M is the mass and a the radius of that portion of the swarm in which the square of velocity of agitation is uniform; let v_0^2 be that square of velocity, and let it be defined by reference to the potential of M at distance a, so that

$$v_0^2 = \beta^2 \frac{\mu M}{a} \quad \dots\dots\dots\dots\dots\dots\dots(1)$$

where β is a coefficient for which a numerical value will be found below.

The square of velocity of agitation outside the radius a is to be denoted by v^2, and subsequent investigation will be necessary to evaluate v^2 in terms of v_0^2.

If we denote by a_0 the earth's distance from the sun, and by u_0 the earth's velocity in its orbit, we have

$$u_0^2 = \mu \frac{M_0}{a_0} \dots\dots\dots\dots\dots\dots\dots(2)$$

Whence,
$$v_0 = \beta u_0 \left(\frac{M a_0}{M_0 a} \right)^{\frac{1}{2}} \quad \dots\dots\dots\dots\dots\dots(3)$$

If in any distribution of meteorites w is the sum of the masses of all the meteorites in unit volume, or the density of the swarm at any point, and if λ be that distance which is called in the kinetic theory of gases " the mean distance between neighbouring molecules," we have

$$\lambda^3 = \frac{m}{w} \quad \dots\dots\dots\dots\dots\dots \dots\dots(4)$$

Now, the mean density of that part of the swarm in which the kinetic energy of agitation is constant being ρ, we have

$$\rho = \frac{3M}{4\pi a^3} \dots\dots\dots\dots\dots\dots\dots(5)$$

and
$$\lambda^3 = 4\pi a^3 \cdot \frac{m}{M} \cdot \frac{\frac{1}{3}\rho}{w} \quad \dots\dots\dots\dots\dots\dots(6)$$

Suppose that s is " the radius of the sphere of action " of a meteorite, so that when two of them approach so that the distance between their centres is s there is a collision.

Let L and T be the mean free path and mean interval between collisions. Then, since the mean velocity is $v \sqrt{(8/3\pi)}$, we have, according to the kinetic theory of gases*,

$$L = \frac{\lambda^3}{\pi s^2 \sqrt{2}}, \qquad T = \frac{L}{v} \sqrt{\tfrac{3}{8}\pi} \quad \dots\dots\dots\dots\dots(7)$$

---

* Meyer, *Kinetische Theorie der Gase.*

Then, on substitution from (4), (5), and (6), we have

$$L = \left[\frac{a_0{}^3\sqrt{8}}{M_0}\right]\left(\frac{a}{a_0}\right)^3 \frac{M_0}{M} \cdot \frac{\frac{1}{3}\rho}{w} \cdot \frac{m}{s^2}, \qquad T = \left[\frac{a_0{}^3\sqrt{3\pi}}{M_0 u_0}\right]\left(\frac{a}{a_0}\right)^{\frac{7}{2}} \cdot \frac{1}{\beta}\left(\frac{M_0}{M}\right)^{\frac{3}{2}} \frac{v_0}{v} \cdot \frac{\frac{1}{3}\rho}{w} \cdot \frac{m}{s^2}$$

$$\ldots\ldots(8)$$

Now let

$$u_0 = \left(\frac{\mu M_0}{a_0}\right)^{\frac{1}{2}}, \qquad u = u_0\left(\frac{a_0}{a}\right)^{\frac{1}{2}}$$

$$l_0 = \frac{a_0{}^3\sqrt{8}}{M_0}, \qquad l = l_0\left(\frac{a}{a_0}\right)^3 \qquad\qquad \Bigg\} \ldots\ldots\ldots\ldots\ldots\ldots(9)$$

$$\tau_0 = \frac{a_0{}^3\sqrt{3\pi}}{M_0 u_0}, \qquad \tau = \tau_0\left(\frac{a}{a_0}\right)^{\frac{7}{2}}$$

and we have

$$v = \beta u\left(\frac{M}{M_0}\right)^{\frac{1}{2}}$$

$$L = l \cdot \left(\frac{M_0}{M}\right) \cdot \frac{\frac{1}{3}\rho}{w} \cdot \frac{m}{s^2} \qquad\qquad \Bigg\} \ldots\ldots\ldots\ldots(10)$$

$$T = \tau \cdot \frac{1}{\beta}\left(\frac{M_0}{M}\right)^{\frac{3}{2}} \cdot \frac{v_0}{v} \cdot \frac{\frac{1}{3}\rho}{w} \cdot \frac{m}{s^2}$$

We now proceed to calculate u_0, l_0, τ_0, and also $2a_0/l_0$, using the centimetre-gramme-second system of units.

The sun's mass may be taken as 315,511 times that of the earth, and the earth as $6\cdot14 \times 10^{27}$ grammes*; hence

$$M_0 = 10^{33\cdot28718} = 1\cdot9372 \times 10^{33} \text{ grammes.}$$

The attractional constant and the earth's mean distance from the sun are

$$\mu = \frac{648}{10^{10}}, \qquad a_0 = 1\cdot487 \times 10^{13} \text{ cm.}$$

With these values

$$u_0 = 10^{6\cdot46323} = 2{,}905{,}600 \text{ cm. per sec.}$$
$$l_0 = 10^{6\cdot68129} = 4{,}800{,}600 \text{ cm.}$$
$$\tau_0 = 10^{0\cdot25366} = 1\cdot79334 \text{ sec.} \qquad\qquad \Bigg\} \ldots\ldots\ldots\ldots(11)$$
$$\frac{2a_0}{l_0} = 10^{6\cdot79204} = 6{,}195{,}000$$

The dimensions of l_0 and τ_0 are not those of length and time; but, if meteorites of 1 gramme mass, with sphere of action 1 centimetre, and "velocity of mean square" of agitation equal to the earth's velocity in its orbit, have density of distribution equal to one-third of the mean density of the sphere M, then l_0, τ_0 will be the mean free path and time, as stated in centimetres and seconds. We may thus regard l_0, τ_0 as a length and time, provided care be taken in the subsequent use of the symbols to adhere to the centimetre-gramme-second system of units.

* Here and elsewhere I generally use Everett's *Units and Physical Constants*.

§ 4.   *On the Equilibrium of a Gas at Uniform Temperature in Concentric Spherical Layers under its own Gravitation.*

It is assumed provisionally that the conditions are satisfied which permit us to regard the swarm of meteorites as a quasi-gas, subject to the laws of hydrostatics.

The solution of this problem, then, becomes a necessary preliminary to the discussion of the kinetic theory of meteorites.   The equilibrium of a gas under its own gravitation has been ably discussed by Professor Ritter in one of his series of papers on gaseous planets*.   The intrinsic interest of the problem renders an independent solution valuable.   Suppose, then, that a mass M_1 of gas is enclosed in a spherical envelope of radius a_1, and is in equilibrium in concentric spherical layers.   Let $v_1{}^2$, the mean square of the velocity of agitation of the gaseous molecules, be defined by reference to the potential of the mass M_1 at the radius a_1, so that

$$v_1{}^2 = \beta_1{}^2 \frac{\mu M_1}{a_1}$$

where $\beta_1{}^2$ is a numerical coefficient, and μ is the attractional constant.

Let p and w be the pressure and density of the gas at radius r, and k the modulus of elasticity, so that

$$p = kw$$

$$k = \tfrac{1}{3} v_1{}^2 = \tfrac{1}{3} \beta_1{}^2 \frac{\mu M_1}{a_1}$$

Then the equation for the hydrostatic equilibrium of the gas is

$$\frac{r^2}{w} \frac{dp}{dr} + 4\pi\mu \int_0^r wr^2 dr = 0 \quad \dots\dots\dots\dots\dots\dots(12)$$

It is obvious that $\dfrac{-r^2}{\mu w}\dfrac{dp}{dr}$ is equal to the whole mass enclosed inside radius r, and this relation will hold however the equation be transformed, provided we do not multiply the equation by any factor.

In consequence of the relation between p and w this may be written

$$k \left[r^2 \frac{d}{dr} \log w + \frac{4\pi\mu}{k} \int_0^r wr^2 dr \right] = 0$$

---

* "Untersuchungen über die Höhe der Atmosphäre und die Constitution gasförmiger Welt-körper," *Wiedemann's Annalen* (New Series), Vol. xvi., 1882, p. 166.   A very elegant solution of part of my problem has also been given by Mr G. W. Hill in the *Annals of Mathematics*, Vol. iv., No. 1, p. 19 (February, 1888).   Mr Hill's paper only reached my hands after my own calcu-lations had been completed, and I therefore adhere to my own less elegant method.   Mr Hill has obviously not seen M. Ritter's papers.

If ρ_1 be the mean density of the mass M_1, we have

$$4\pi\mu = \frac{3\mu M_1}{\rho_1 a_1{}^3} = \frac{9k}{\beta_1{}^2 a_1{}^2 \rho_1}$$

Hence, we may write the equation (12) in the form

$$ka_1\left[\frac{r^2}{a_1}\frac{d}{dr}\log w + \frac{9}{\beta_1{}^2}\int_0^r \frac{w}{\rho_1}\frac{r^2}{a_1{}^3}\,dr\right] = 0$$

Now, let

$$x_1 = \frac{a_1}{r}, \qquad \frac{w}{\rho_1} = \tfrac{1}{9}\beta_1{}^2 e^{y_1}$$

and the equation becomes

$$\tfrac{1}{9}\beta_1{}^2 \mu M_1\left[-\frac{dy_1}{dx_1} + \int_{x_1}^{\infty}\frac{e^{y_1}}{x_1{}^4}\,dx_1\right] = 0 \quad\ldots\ldots\ldots\ldots(13)$$

By differentiation we obtain the equation

$$\frac{d^2 y_1}{dx_1{}^2} + \frac{e^{y_1}}{x_1{}^4} = 0 \quad\ldots\ldots\ldots\ldots\ldots\ldots(14)$$

It is obvious from (13) that $\tfrac{1}{9}\beta_1{}^2 M_1 dy_1/dx_1$ is the mass enclosed inside radius a/x_1, and therefore $\tfrac{1}{9}\beta_1{}^2 dy_1/dx_1$ is equal to unity when $x = 1$.

A general analytical solution of (14) does not seem to be attainable, and recourse must be had to numerical processes. Although this is an equation of the second degree, and its general solution must involve two arbitrary constants, we shall see (as pointed out by M. Ritter) that the general solution, as applicable to our problem, may be deduced from one single numerical solution. M. Ritter proceeds by a graphical method, which he has worked with surprising accuracy. I shall therefore adopt an analytical and numerical method, which, although laborious, is susceptible of greater accuracy.

Whatever be the arrangement of the gas, the density at the centre must have some value. I therefore start with a central density ω, corresponding to the value η of y_1, so that

$$\frac{\omega}{\rho_1} = \tfrac{1}{9}\beta_1{}^2 e^{\eta} \quad\ldots\ldots\ldots\ldots\ldots\ldots\ldots(15)$$

For the sake of brevity the suffix 1 will be now omitted from the various symbols, to be reaffixed later when it is required.

At the centre, where x is infinite, dy/dx, d^2y/dx^2, &c., are all zero, and we put $y = \eta$.

Let $\xi = e^{\eta}/x^2$, and let us assume

$$y = \eta + v$$
$$= \eta - A_1\xi + A_2\xi^2 - A_3\xi^3 + \ldots$$

Now, the differential equation (14) to be satisfied is

$$x^2\frac{d^2y}{dx^2} = -\frac{e^y}{x^2} = -\xi e^v$$

But
$$x^2 \frac{d^2y}{dx^2} = -2.3A_1\xi + 4.5A_2\xi^2 - 6.7A_3\xi^3 + \ldots$$

and by expanding e^v we obtain

$$-\xi e^v = -\xi + A_1\xi^2 - \left(A_2 + \frac{1}{2}A_1^2\right)\xi^3 + \left(A_3 + A_1A_2 + \frac{1}{2.3}A_1^3\right)\xi^4$$

$$-\left(A_4 + A_1A_3 + \frac{1}{2}A_2^2 + \frac{1}{2}A_1^2A_2 + \frac{1}{2.3.4}A_1^4\right)\xi^5$$

$$+\left(A_5 + A_1A_4 + A_2A_3 + \frac{1}{2}A_1^2A_3 + \frac{1}{2}A_1A_2^2 + \frac{1}{2.3}A_1^3A_2 + \frac{1}{2.3.4.5}A_1^5\right)\xi^6 - \ldots$$

By equating coefficients in these two series, I find

$$A_1 = \frac{1}{6}, \quad A_2 = \frac{1}{120}, \quad A_3 = \frac{1}{1890}, \quad A_4 = \frac{61}{1,632,960}$$

$$A_5 = \frac{629}{224,532,000}, \quad A_6 = \frac{34\cdot07383\ldots}{156 \times 10^6}, \quad \&c.$$

and

$$\log A_1 = 9\cdot2218487, \quad \log A_2 = 7\cdot9208188, \quad \log A_3 = 6\cdot7235382$$

$$\log A_4 = 5\cdot5723543, \quad \log A_5 = 4\cdot4473723, \quad \log A_6 = 3\cdot3392964$$

whence, by extrapolation,

$$\log A_7 = 2\cdot243, \quad \log A_8 = 1\cdot13$$

In M. Ritter's paper, already referred to, he takes a certain function u as equal to 1·031, when the radius is unity. Now, Ritter's function u is equal in my notation to $\frac{1}{2}(w/\rho) \div \frac{1}{3}\beta^2 a^2/r^2$ or $\frac{1}{2}e^y/x^2$. It follows, therefore, that Ritter takes the surface value of $y = \log_e 2\cdot062$. But he intends the central density to be 100 times the surface density; hence, to take the same solution, we must have e^η equal to 100 times 2·062. Therefore, his value of η would be

$$\eta = \log_e 100 + \log_e 2\cdot062$$

or
$$\eta = 5\cdot3288465$$

Now, as I want to make a comparison between my solution and his, I start with this value of η. The only object attained by the choice of this particular value is that the two solutions become easily comparable. It will be seen below that the value of η does not make the central density exactly 100 times the surface density, but only satisfies that condition approximately. In Ritter's graphical treatment of the problem this value 100 is the exact datum, whereas in my method we start with an exact value of η, and proceed to find the ratio of central to surface density.

With this value of η (whence $\log_{10} e^\eta = 2\cdot3142888$) I find the following series for y:—

$$y = 5\cdot3288465 - \frac{34\cdot3667}{x^2} + \frac{354\cdot321}{x^4} - \frac{4638\cdot79}{x^6} + \frac{67,532\cdot0}{x^8} - \frac{1,044,280}{x^{10}}$$

$$+ \frac{16,789,000}{x^{12}} - \frac{2\cdot77 \times 10^8}{x^{14}} + \frac{4\cdot4 \times 10^9}{x^{16}} - \cdots \quad \cdots\cdots(16)$$

and, by differentiation, the series for dy/dx is obvious.

This series will be very accurate from $x = \infty$ to about $x = 8$. Thus, when $r/a = \cdot1$, or $x = 10$, we have

$$y = 5\cdot016558, \qquad \frac{dy}{dx} = + \cdot0568910$$

and even the far less convergent series for d^2y/dx^2 gives $- \cdot0150891$, agreeing with $- e^y/x^4$ to the last place of decimals. When $r/a = \cdot125$, or $x = 8$, we have *

$$y = 4\cdot863925, \qquad \frac{dy}{dx} = \cdot101168$$

whence, $$\frac{d^2y}{dx^2} = - \cdot031624$$

with y correct to four, and probably to five, places of decimals, and dy/dx probably correct to four places of decimals. This is amply sufficient for our purpose. Indeed, accuracy of this order would be altogether pedantic, were it not that the errors accumulate.

We cannot, then, rely on this method of procedure beyond the region included between $x = \infty$ and $x = 8$, and must now make a new departure.

Since $$\frac{d^2y}{dx^2} = - \frac{e^y}{x^4}$$

$$\log\left(- \frac{d^2y}{dx^2}\right) = y - 4 \log x$$

therefore, $$\frac{d^3y}{dx^3} = \frac{d^2y}{dx^2}\left(\frac{dy}{dx} - \frac{4}{x}\right) \quad \cdots\cdots \cdots\cdots\cdots(17)$$

Now, let $$A_n = \frac{1}{n!}\frac{d^ny}{dx^n}$$

where, after differentiation, x is put equal to c, a constant.

Then (17) may be written

$$A_3 = \frac{2!}{3!}A_2\left(A_1 - \frac{4}{c}\right) \quad \cdots\cdots\cdots\cdots\cdots(18)$$

---

* Even when $x = 5$, I find from this series $y = 4\cdot342$, which lies very near to $y = 4\cdot332$, found below. But the series for dy/dx is useless.

It is clear that
$$\frac{d^p A_n}{dc^p} = \frac{n+p\,!}{n\,!} A_{n+p}$$

Hence, differentiating (18) $n-3$ times, we have

$$\frac{n\,!}{3\,!} A_n = \sum_{q=0}^{q=n-3} \frac{n-3\,!}{n-3-q\,!\,q\,!} \frac{2\,!\,n-q-1\,!}{3\,!} A_{n-q-1} \left\{ \frac{q+1\,!}{1\,!} A_{q+1} + (-)^q \frac{4\,(q\,!)}{c^{q+1}} \right\}$$

or

$$A_n = \frac{1}{n\,.\,n-1\,.\,n-3} \sum_{q=0}^{q=n-3} (n-q-1)(n-q-2) A_{n-q-1}$$

$$\times \left\{ (q+1)\, A_{q+1} + (-)^q \frac{4}{c^{q+1}} \right\}$$

or

$$A_n = \frac{1}{n\,.\,n-1\,.\,n-3} \left\{ 2\,.\,1 A_2 \left[(n-2)\, A_{n-2} + (-)^n \frac{4}{c^{n-2}} \right] \right.$$

$$\left. + 3\,.\,2 A_3 \left[(n-3)\, A_{n-3} - (-)^n \frac{4}{c^{n-3}} \right] + \dots \right\} \quad \dots(19)$$

Now, if, for a given value of x, viz., c, we know y, or A_0, and dy/dx, or A_1, then we can compute A_2 from the formula $-\frac{1}{2} e^{A_0} c^{-4}$; and, by the formula (18), viz. :—

$$A_3 = \frac{1}{3\,.\,2\,.\,1} \left\{ 2\,.\,1 A_2 \left[A_1 - \frac{4}{c} \right] \right\}$$

A_3 may be computed.

Afterwards, A_4, A_5, &c., may be computed by successive applications of (19). This being so,

$$y = A_0 + A_1\,(x-c) + A_2\,(x-c)^2 + \dots \quad \dots\dots\dots\dots(20)$$

$$\frac{dy}{dx} = A_1 + 2 A_2\,(x-c) + 3 A_3\,(x-c)^2 + \dots$$

In these series x may have any value, provided the series converges adequately. The convergence may be much improved by an artifice, which, however, I unfortunately did not discover until most of the computations were completed. Let us add and subtract $\log 2x^2$ on the right-hand side of (20).

Now $$\log_e 2x^2 = \log_e 2c^2 + 2 \log_e \left[1 + \frac{x-c}{c} \right]$$

$$= \log_e 2c^2 + 2\,\frac{x-c}{c} - \frac{2}{2}\frac{(x-c)^2}{c^2} + \frac{2}{3}\frac{(x-c)^3}{c^3} - \dots$$

If, then, we write

$$B_0 = A_0 - \log_e 2c^2, \quad B_1 = A_1 - \frac{2}{c}, \quad B_2 = A_2 + \frac{2}{2c^2}, \quad B_3 = A_3 - \frac{2}{3c^3}, \text{ &c.}$$

we have $$y = \log_e 2x^2 + B_0 + B_1\,(x-c) + B_2\,(x-c)^2 + \dots \quad \dots\dots\dots(21)$$

a more convergent series than that with the A's.

The simplest way of computing the B's appears to be by first computing the A's.

The process for obtaining the numerical solution is then as follows:—

We have the values of y, dy/dx, $\frac{1}{2}dy^2/dx^2$ when $x = 8$, that is to say, of A_0, A_1, A_2 when $c = 8$. From these the successive A's and B's are computed, and the resulting series gives the values of y and dy/dx when x is 5 or $r = \cdot2$. Starting from this point a new series gives the result when $r = \cdot3$, another series gives the values for $r = \cdot4$, and so on. Later in the calculation several values may be computed from one formula*.

When the computation has been carried out to $r = a$, we have reached the spherical envelope, but that envelope may be replaced by another at any more remote distance from the centre. Thus, the integration may be pursued for values of x less than unity, and when the lower limit is zero the envelope is at infinity.

If we write $\log u = B_0 + B_1(x - c) + B_2(x - c)^2 + \ldots$

we have $e^y = u \cdot 2x^2$

and $$\frac{w}{\rho} = \tfrac{2}{9}\frac{\beta^2 a^2}{r^2} \cdot u$$

But it may easily be seen that $2\beta^2 a^2/9r^2$ is a particular solution of the problem; hence, u is a factor by which the particular solution is to be multiplied to obtain the general solution. The function u is given by

$$u = \tfrac{1}{2}\frac{e^y}{x^2} = -\tfrac{1}{2}x^2\frac{d^2y}{dx^2} \quad\ldots\ldots\ldots\ldots\ldots\ldots\ldots\ldots(22)$$

A table of the values of u is given below, showing how the general solution shades off into the particular solution. This function, u, is also tabulated by Ritter, and I made use of its value, when $x = 1$, to determine the value of η, with which the integration is to begin. I find, however (see Table I.), that, when $x = 1$, $u = 1\cdot0063$, in place of $1\cdot031$, as given by him.

The last row in Table I. gives the ratio of the central density ω to w, the density at the distance r; this ratio is equal to $e^{\eta - y_1}$.

* If the series be carried as far as B_8, several steps may be included in one series. For example, the first series, when $c = 8$, may be pushed even as far as $r = \cdot4$ without serious error; for it gives $y = 2\cdot960$ instead of the true value $2\cdot965$, and $dy/dx = \cdot957$, instead of the true value $\cdot944$. I have not, however, been satisfied with this degree of accuracy.

The following Table gives the results of the computation, and the suffix 1 is reintroduced in the several symbols.

I. TABLE OF RESULTS.

$\dfrac{1}{x_1} = \dfrac{r}{a_1} =$	0	·1	·125	·2	·3	·4	·5	·6	·7	·8	·9	1·0
$y_1 =$	5·328845	5·01656	4·8639	4·3317	3·6064	2·9653	2·4235	1·9672	1·5794	1·2458	·9553	·6995
$\dfrac{dy_1}{dx_1} =$	0·0	·05689	·10117	·2967	·6217	·9442	1·2404	1·5098	1·7569	1·9867	2·2030	2·4087
$\dfrac{d^2y_1}{dx_1^2} =$	0·0	-·01509	-·03162	-·1217	-·2984	-·4967	-·7053	-·9267	-1·1647	-1·4237	-1·7055	-2·0126
$u = -\tfrac12 x_1^2 \dfrac{d^2y_1}{dx_1^2} =$	0·0	·75445	1·0120	1·5215	1·6577	1·5521	1·4107	1·2871	1·1888	1·1123	1·0528	1·0063
$\dfrac{\omega}{w} = e^{\eta - y_1} =$	1·0000	1·3666	...	2·7105	5·598	10·63	18·29	28·84	42·50	59·32	79·33	102·45

$\dfrac{1}{x_1} = \dfrac{r}{a_1} =$	1·0	1·1	1·2	1·5	2·0	2·5	3·0	∞
$y_1 =$	·6995	·4717	·2672	-·241	-·864	-1·328	-1·699	$-\infty \,(\log 2 x_1^2)$
$\dfrac{dy_1}{dx_1} =$	2·4087	2·6062	2·7971	3·342	4·202	5·139	6·038	$+\infty \left(\dfrac{2}{x_1}\right)$
$\dfrac{d^2y_1}{dx_1^2} =$	-2·0126	-2·3466	-2·7089	-3·976	-6·744	-10·35	-14·82	$-\infty \left(-\dfrac{2}{x_1^2}\right)$
$u = -\tfrac12 x_1^2 \dfrac{d^2y_1}{dx_1^2} =$	1·0063	·9697	·9406	·884	·843	·828	·823	1·0
$\dfrac{\omega}{w} = e^{\eta - y_1} =$	102·45	128·7	157·8	263	489	778	1127	$\infty \left(\dfrac{e^\eta}{\tfrac12 x_1^2}\right)$

It will be noticed that u rises from zero to a maximum of about $1\cdot66$, falls to a minimum of about $\cdot82$, and then rises to unity.

Since $\frac{1}{3}\beta_1^2 dy_1/dx_1 = 1$ when $x_1 = 1$, we have $\frac{1}{3}\beta_1^2 = 1/2\cdot4087 = \cdot4152$.

M. Ritter has $\cdot4143$ for this constant, which he calls m.

It appears from the Table that the density at the centre is $102\frac{1}{2}$ times as great as that where $r = a_1$. M. Ritter's solution is intended to make that ratio exactly 100, but this solution shows that we ought to have started with a slightly different value of η to obtain that result.

In the general solution of the differential equation $d^2y/dx^2 = -e^y/x^4$ the two arbitrary constants may be taken to be the values of y and dy/dx when x is infinite. Now, we have taken arbitrarily $y = 5\cdot329$ when x is infinite, and the physical conditions of the problem imply that dy/dx is zero when x is infinite. For if dy/dx had any positive or negative value different from zero, it would mean that at the centre there was a nucleus of infinitely small dimensions, but of finite positive or negative mass. Now, a_1 is that distance from the centre at which the density is $1/102\cdot45$ of the central density; hence, we may regard a_1 as the arbitrary constant of the solution. Whatever be the elasticity of the gas, we may always take as our unit of length that distance from the centre of the nebula at which the density has fallen to $1/102\cdot45$ of its central value. Hence, the above table gives the general solution of the problem, subject, however, to the condition that there is no central nucleus.

If we view the nebula from a very great distance, a_1 appears very small, and thus the solution of the problem becomes $y = \log 2x^2$. It is easy to verify that this is a particular algebraic solution of the differential equation, as is pointed out by Ritter in his paper[*]. I found this solution very useful in a preliminary consideration of the problem treated in this paper.

The next point which we have to consider is the form which the solution will take, if, instead of taking a_1 as the unit of length, we take any other value.

The density at any distance and the elasticity are to remain unchanged, but are to be referred to new constants.

Thus, w, r, v^2 remain unchanged, but are to be referred to M, ρ, β^2, a, instead of to M_1, ρ_1, β_1^2, a_1.

Now, since w remains unchanged,

$$\tfrac{1}{3}\beta^2\rho e^y = \tfrac{1}{3}\beta_1^2\rho_1 e^{y_1}$$

---

[*] I have made use of this solution in a paper in the *Proceedings of the Royal Society*, Vol. xxxvi. (1884), p. 158, or Vol. iii. of this work p. 69, and it has also been referred to in a paper by Sir W. Thomson, *Phil. Mag.*, Vol. xxiii., p. 287. Sir W. Thomson's paper covers much the same ground as some of M. Ritter's earlier papers, but was written by him independently and in ignorance of them.

and, since v^2 remains unchanged,

$$\beta^2 \rho a^2 = \beta_1^2 \rho_1 a_1^2$$

Also

$$x = \frac{a}{r} = x_1 \frac{a}{a_1}$$

From these relations it is clear that

$$y = y_1 - 2 \log \frac{a_1}{a}$$

and

$$M = \tfrac{1}{3}\beta_1^2 M_1 \frac{dy_1}{dx_1} \left(x_1 = \frac{a_1}{a} \right)$$

Then, since $\tfrac{1}{3}\beta^2 dy/dx = 1$, when $x = 1$, and since $dy = dy_1$ and $dx = dx_1 a/a_1$, it follows that

$$\tfrac{1}{3}\beta^2 = \frac{1}{x_1 dy_1/dx_1}, \text{ when } x_1 = \frac{a_1}{a} \quad \dots\dots\dots\dots\dots(23)$$

This relationship has been already used for determining β_1^2.

It is obvious also that

$$\frac{\rho}{\rho_1} = \frac{x_1^3 dy_1/dx_1, \text{ when } x_1 = a_1/a}{x_1^3 dy_1/dx_1, \text{ when } x_1 = 1}$$

Therefore

$$\frac{w}{\rho} = \frac{\tfrac{1}{3}e^{y_1}}{x_1^3 dy_1/dx_1, \text{ when } x_1 = a_1/a} \quad \dots\dots\dots\dots\dots(24)$$

If w_0 be the density when $r = a$, we have

$$\frac{w_0}{\rho} = \frac{\tfrac{1}{3}e^{y_1}, \text{ when } x_1 = a_1/a}{x_1^3 dy_1/dx_1} = -\tfrac{1}{3} \cdot \frac{x_1 d^2 y_1/dx_1^2}{dy_1/dx_1}, \text{ when } x_1 = a_1/a \quad \dots(25)$$

If p_0 be the pressure when $r = a$, we have

$$p_0 = \tfrac{1}{3}v^2 w_0 = \tfrac{4}{9}\pi\mu a^2 \rho^2 . \beta^2 \frac{w_0}{\rho}$$

If, therefore, we write $P = \tfrac{4}{9}\pi\mu a^2 \rho^2$,

$$\frac{p_0}{P} = \tfrac{1}{3}\beta^2 . \frac{w_0}{\rho} = -\frac{d^2 y_1/dx_1^2}{(dy_1/dx_1)^2}, \text{ when } x_1 = a_1/a \quad \dots\dots\dots\dots(26)$$

By (26) we are able to find how the pressure on an envelope of given radius a varies with the variation of the temperature of a given mass M of gas contained in it. By means of the formulæ (23), (25), (26), we are now able to obtain from the original solution any number of other ones; for, after the changes have been effected in the notation, we may proceed to magnify or diminish all the various values of a until they are of one size, and we shall thus obtain the solution for a gas at any temperature whatever.

I shall now proceed to give a table of results when the standard radius a, which may be conveniently called the boundary, is placed successively infinitely near the centre, where $r = 0 \times a_1$, at $r = \cdot 1 \times a_1$, $r = \cdot 2 \times a_1$, and so on. The first line of entries gives the various values of $\frac{1}{3}\beta^2$ (computed from (23)), on which the elasticity of the gas depends; the second line gives w_0/ρ (computed from (25)), or the ratio of the boundary density to the mean density of all inside it; the third line gives p_0/P (computed from (26)), by which we trace the variations of pressure at the boundary.

TABLE II.

Value of a by reference to former solution $\dfrac{a}{a_1} =$	0	·1	·2	·3	·4	·5	·6	·6264	·7	·8
$\dfrac{\frac{1}{3}v^2}{\mu M/a} = \frac{1}{3}\beta^2 \left[= \dfrac{1}{x_1\, dy_1/dx_1} \right] =$	∞	1·7577	·6741	·4826	·4236	·4031	·3974	·3972	·3984	·4027
$\dfrac{w_0}{\rho} \left[= \dfrac{-\frac{1}{3}x_1\, d^2y_1/dx_1^2}{dy_1/dx_1} \right] =$	1·0000	·8841	·6838	·5333	·4383	·3791	·3410	$\frac{1}{3}$	·3158	·2986
$\dfrac{p_0}{P} \left[= \dfrac{-d^2y_1/dx_1^2}{[dy_1/dx_1]^2} \right] =$	∞	4·662	1·383	·772	·557	·458	·407	·397	·377	·361

Value of a by reference to former solution $\dfrac{a}{a_1} =$	·9	1·0	1·25	1·5	2·0	2·5	3·0	∞
$\dfrac{\frac{1}{3}v^2}{\mu M/a} = \frac{1}{3}\beta^2 \left[= \dfrac{1}{x_1\, dy_1/dx_1} \right] =$	·4085	·4152	·4325	·449	·476	·487	·497	$\frac{1}{2}$
$\dfrac{w_0}{\rho} \left[= \dfrac{-\frac{1}{3}x_1\, d^2y_1/dx_1^2}{dy_1/dx_1} \right] =$	·2867	·2785	·2676	·264	·267	·269	·273	$\frac{1}{3}$
$\dfrac{p_0}{P} \left[= \dfrac{-d^2y_1/dx_1^2}{[dy_1/dx_1]^2} \right] =$	·351	·347	·347	·356	·382	·392	·406	$\frac{1}{2}$

The minimum value of w_0/ρ occurs when $a/a_1 = 1\cdot6$ very nearly, for, when $a/a_1 = 1\cdot4$, $1\cdot5$, $1\cdot6$, I find $w_0/\rho = \cdot26521$, $\cdot26437$, $\cdot26425$ respectively[*]. When $r/a_1 = 1\cdot6$, $y_1 = -\cdot38435$ and $dy_1/dx_1 = 3\cdot5180$. The minimum value of p_0/P occurs when $a/a_1 = 1\cdot1$ very nearly, for, when $a/a_1 = 1\cdot0$, $1\cdot1$, $1\cdot2$, I find $p_0/P = \cdot3469$, $\cdot3455$, $\cdot3462$ respectively.

When w_0/ρ is a minimum, the density at the centre is 381 times that at the boundary, and, when p_0/P is a minimum, the density at the centre is 129 times that at the boundary. M. Ritter finds the pressure to be a minimum when this ratio is 258, instead of 129. As this corresponds to $a/a_1 = 1\cdot5$, this discrepancy between our solutions is not so large as might be expected from the great discrepancy between these results, and I cannot but think that my result is more accurate than his.

* Mr Hill finds that the minimum value of w_0/ρ approximates to $\frac{4}{15}$, or ·2667. The agreement between our results is satisfactory.

The minimum value of $\frac{1}{3}\beta^2$ occurs when $a/a_1 = \cdot6264$, and its value is $\cdot39723$. This value makes the surface density exactly one-third of the mean density, for $\frac{1}{3}\beta^2$ is a minimum when $x_1 dy_1/dx_1$ is a maximum, and this occurs when $x_1 d^2y_1/dx_1^2 + dy_1/dx_1 = 0$; and, when this relationship is satisfied, $w_0/\rho = \frac{1}{3}$.

It is interesting to note that in this case β^2 is very nearly equal to $\frac{6}{5}$, so that the total internal kinetic energy of agitation of the sphere of gas at minimum temperature limited by the radius a is $\frac{1}{2}(\frac{6}{5}\mu M/a)\,M = \frac{3}{5}\mu M^2 a$ *very nearly*. Now, the energy lost in the concentration of a homogeneous sphere M from a condition of infinite dispersion is *exactly* $\frac{3}{5}\mu M^2/a$. It might, therefore, be suspected that $\cdot39723$ is only an approximation to $\frac{2}{5}$, which may be the rigorous value. But my numerical calculations were carried out with so much care that I find it almost impossible to believe that there is an error as large as 3 in the third place of decimals, or, indeed, any error at all in the third figure. Moreover, it would be expected that, if this very simple relationship is rigorously correct, it would be possible to prove it rigorously, just as it is rigorously shown above that $w_0/\rho = \frac{1}{3}$; but I am unable to find any analytical relationships by which the minimum value of $\frac{1}{3}\beta^2$ can be deduced. If my arithmetical process be correctly carried out, then we ought to find that, when $r = \cdot6264$, dy_1/dx_1 should be equal to $-x_1 d^2y_1/dx_1^2$ or e^{y_1}/x_1^3. Now, I find that, when $r = \cdot6264$, $dy_1/dx_1 = 1\cdot57703$ and $e^{y_1}/x_1^3 = 1\cdot5770$, so that the two agree to four places of decimals. I conclude, therefore, that the true minimum of $\frac{1}{3}\beta^2$ is $\cdot3972$*.

It will be observed that, as a/a_1 increases to infinity, $\frac{1}{3}\beta^2$ terminates by being equal to $\frac{1}{2}$. M. Ritter has found that it rises above $\frac{1}{2}$, and oscillates about that value an indefinite number of times with diminishing amplitude, gradually settling down to $\frac{1}{2}$ as a/a_1 becomes infinite. The values in the preceding table are not, however, carried far enough to exhibit these oscillations of $\frac{1}{3}\beta^2$. A consequence of this result is that there are a number of modes of equilibrium of a gas at a given temperature, provided that the temperature lies within certain narrow limits. This very remarkable conclusion is rendered more intelligible by Mr Hill's treatment than by M. Ritter's.

This point has, however, no bearing on the present investigation.

In any one of the solutions comprised in Table II. we may complete the table of densities by the formula (24), viz.,

$$\frac{w}{\rho} = \frac{\frac{1}{3}e^{y_1}}{x_1^3 dy_1/dx_1\,(x_1 = a_1/a)}$$

---

* This is confirmed by Mr Hill. His equation $s = z$ is equivalent to $x_1^2 d^2y_1/dx_1^2 + x_1 dy_1/dx_1 = 0$, and it appears from his tables that $s = z = 2\cdot517$. Now, $s = 3/\beta^2$, and the reciprocal of $2\cdot517$ is $\cdot397$.

and I shall later proceed to do this in the one case which has interest for our present problem, namely, where the temperature is a minimum, so that a/a_1 is ·6264. The full numerical results may be more conveniently given hereafter, and it will only be now necessary to indicate how they are to be computed.

When, for example, $r = ·1 \times a_1$, $r/a = ·1/·6264 = ·1596$; thus, our equidistant values of the density and other functions will proceed by multiples of ·1596a up to ·9578a, and the limit of the isothermal sphere is where $r = a$.

When the temperature is a minimum $\frac{1}{3}\beta^2 = ·39723$, and we have $w_0 = \frac{1}{3}\rho$; therefore, $w/w_0 = w/\frac{1}{3}\rho$, and, therefore, if $y_{1,0}$ be the value of y_1, when $r = ·6264a_1$, $w/\frac{1}{3}\rho = e^{y_1 - y_{1,0}}$. Thus, for example, at the centre, $w/\frac{1}{3}\rho$ is 32·14, and when $r = ·4789a$ it is 5·7417.

The proportion of the mass M which is included in radius a/x is $\frac{1}{3}\beta^2 dy/dx = \frac{1}{3}\beta^2 a_1 dy_1/a \, dx_1 = \dfrac{·39723}{·6264} dy_1/dx_1$. Hence, the masses may be computed.

At any part of the isothermal sphere gravity g is to be found from

$$g = \tfrac{1}{3}\beta^2 \mu M \frac{dy}{dx} \cdot \frac{x^2}{a^2}$$

or, expressing g in terms of G gravity at the surface, we have, since $G = \mu M/a^2$,

$$\frac{g}{G} = \tfrac{1}{3}\beta^2 x^2 \frac{dy}{dx} \quad \dots\dots\dots\dots\dots\dots\dots\dots(27)$$

The angular velocity of a body moving in a circular orbit at any part of the nebula, and its linear velocity v are also easily to be found.

§ 5.   *On an Atmosphere in Convective Equilibrium.*

I shall now suppose that a sphere of gas of mass M at minimum temperature is bounded by an atmosphere in convective equilibrium, with continuity of temperature and density at the sphere of discontinuity of radius a. Let v_0^2 be the mean square of velocity of agitation in the isothermal sphere, and v^2 that at any other radius r. Then throughout the isothermal sphere $v^2 = v_0^2$, but in the layer outside v^2 gradually decreases to zero.

Let w_0 be the density and p_0 the pressure at radius a, and w, p the same things at radius r.

Then, if the ratio of the two specific heats be that deduced from the simple kinetic theory of gases, without any allowance for intra-molecular vibrations, we have that ratio equal to $\frac{5}{3}$.

Hence, $$p = p_0 \left(\frac{w}{w_0}\right)^{\frac{5}{3}}$$

and
$$\tfrac{1}{3}v^2 = p_0 \frac{w^{\frac{2}{3}}}{w_0^{\frac{5}{3}}} = \tfrac{1}{3}v_0^2 \cdot \left(\frac{w}{w_0}\right)^{\frac{2}{3}}$$

also
$$\frac{dp}{w} = \tfrac{5}{3} \cdot \frac{p_0}{w_0^{\frac{5}{3}}} w^{-\frac{1}{3}} dw = \tfrac{5}{2} \cdot \tfrac{1}{3}v_0^2 d\left(\frac{w}{w_0}\right)^{\frac{2}{3}}$$

Now, the equation for the hydrostatic equilibrium of the layer is

$$\frac{r^2}{w}\frac{dp}{dr} + \mu M + 4\pi\mu \int_a^r wr^2 dr = 0 \ldots\ldots\ldots\ldots\ldots(28)$$

Let
$$x = \frac{a}{r}, \qquad z = \frac{v^2}{v_0^2} = \left(\frac{w}{w_0}\right)^{\frac{2}{3}}$$

and we have
$$\frac{r^2 dp}{w\,dr} = -\tfrac{5}{2} \cdot \tfrac{1}{3}v_0^2 a \frac{dz}{dx}$$

$$\mu M = \frac{v_0^2 a}{\beta^2}$$

$$4\pi\mu a^3 = \frac{3\mu M}{\rho} = \frac{\mu M}{w_0}, \quad \text{since } w_0 = \tfrac{1}{3}\rho \text{ rigorously,}$$

$$= \frac{v_0^2 a}{\beta^2 w_0}$$

Hence, our equation is

$$\mu M\left\{-\tfrac{5}{6}\beta^2\frac{dz}{dx} + 1 + \int_x^1 \frac{z^{\frac{3}{2}}}{x^4} dx\right\} = 0 \ \ldots\ldots\ldots\ldots(29)$$

It is obvious that $\tfrac{5}{6}\beta^2 M\,dz/dx$ is the whole mass (expressed in terms of the mass of the isothermal sphere) enclosed inside radius a/x. The differential equation to be satisfied is

$$\tfrac{5}{6}\beta^2\frac{d^2 z}{dx^2} + \frac{z^{\frac{3}{2}}}{x^4} = 0 \ \ldots\ldots\ldots\ldots\ldots(30)$$

We have seen in the last section that $\tfrac{1}{3}\beta^2 = \cdot39723$, and, hence, $\tfrac{5}{6}\beta^2 = \cdot99308$.

This equation is not so easy to solve as that in the last section, and I have not succeeded in finding the general law of the coefficients in an expansion. Nevertheless it is easy to find a series which will do all that is required.

Let c be any value of x for which we know z and dz/dx, and let

$$\xi = x - c$$

Assume
$$z = z_0\{1 + A_1\xi + A_2\xi^2 + A_3\xi^3 + \ldots\}$$

Then, if the suffix 0 indicates the value of a symbol when $x = c$ and $\xi = 0$, we have

$$z_0 = z_0$$

$$\left(\frac{dz}{dx}\right)_0 = A_1 z_0$$

$$\left(\frac{d^2 z}{dx^2}\right)_0 = 2A_2 z_0$$

But
$$\left(\frac{d^2z}{dx^2}\right)_0 = -\frac{6}{5\beta^2}\cdot\frac{z_0^{\frac{3}{2}}}{c^4}$$

and
$$2A_2z_0 = -\frac{6}{5\beta^2}\cdot\frac{z_0^{\frac{3}{2}}}{c^4}, \quad \text{or} \quad A_2 = -\frac{3}{5\beta^2}\frac{z_0^{\frac{1}{2}}}{c^4}$$

so that, if z_0 and $(dz/dx)_0$ are known, A_2 is known.

The differential equation (30) which we have to satisfy is

$$\tfrac{5}{6}\beta^2(\xi+c)^4\frac{d^2z}{d\xi^2} = -z^{\frac{3}{2}}$$

or
$$\frac{1}{A_2}\left(\frac{\xi}{c}+1\right)^4\frac{d^2(z/z_0)}{d\xi^2} = 2\left(\frac{z}{z_0}\right)^{\frac{3}{2}}$$

Now, by expansion,

$$2\left(\frac{z}{z_0}\right)^{\frac{3}{2}} = 2 + 3A_1\xi + 3\left[A_2 + \tfrac{1}{4}A_1^2\right]\xi^2 + 3\left[A_3 + \tfrac{1}{2}A_1A_2 - \tfrac{1}{24}A_1^3\right]\xi^3$$

$$+ 3\left[A_4 + \tfrac{1}{2}(A_1A_3 + \tfrac{1}{2}A_2^2) - \tfrac{1}{8}A_1^2A_2 + \tfrac{1}{64}A_1^4\right]\xi^4$$

$$+ 3\left[A_5 + \tfrac{1}{2}(A_1A_4 + A_2A_3) - \tfrac{1}{8}(A_1^2A_3 + A_1A_2^2) + \tfrac{1}{16}A_1^3A_2 - \tfrac{1}{128}A_1^5\right]\xi^5$$

$$+ 3\left[A_6 + \tfrac{1}{2}(A_1A_5 + A_2A_4 + \tfrac{1}{2}A_3^2) - \tfrac{1}{8}(A_1^2A_4 + 2A_1A_2A_3 + \tfrac{1}{3}A_2^3)\right.$$
$$\left. + \tfrac{1}{16}(A_1^3A_3 + \tfrac{3}{2}A_1^2A_2^2) - \tfrac{5}{128}A_1^4A_2 + \tfrac{7}{1536}A_1^6\right]\xi^6$$

$$+ 3\left[A_7 + \tfrac{1}{2}(A_1A_6 + A_2A_5 + A_3A_4) - \tfrac{1}{8}(A_1^2A_5 + 2A_1A_2A_4 + A_1A_3^2 + A_2^2A_3)\right.$$
$$+ \tfrac{1}{16}(A_1^3A_4 + 3A_1^2A_2A_3 + A_1A_2^3) - \tfrac{5}{128}(A_1^4A_3 + 2A_1^3A_2^2)$$
$$\left. + \tfrac{1}{256}A_1^5A_2 - \tfrac{3}{1024}A_1^7\right]\xi^7 + \cdots \quad\quad\quad \cdots\cdots\cdots\cdots(31)$$

And

$$\frac{1}{A_2}\left(\frac{\xi}{c}+1\right)^4\frac{d^2}{d\xi^2}(z/z_0)$$

$$= 2 + \left(3\cdot2\frac{A_3}{A_2} + \frac{4}{c}\cdot2\cdot1\right)\xi + \left(4\cdot3\cdot\frac{A_4}{A_2} + \frac{4}{c}\cdot3\cdot2\frac{A_3}{A_2} + \frac{6}{c^2}\cdot2\cdot1\right)\xi^2$$

$$+ \left(5\cdot4\frac{A_5}{A_2} + \frac{4}{c}\cdot4\cdot3\frac{A_4}{A_2} + \frac{6}{c^2}\cdot3\cdot2\frac{A_3}{A_2} + \frac{4}{c^3}\cdot2\cdot1\right)\xi^3$$

$$+ \left(6\cdot5\frac{A_6}{A_2} + \frac{4}{c}\cdot5\cdot4\frac{A_5}{A_2} + \frac{6}{c^2}\cdot4\cdot3\frac{A_4}{A_2} + \frac{4}{c^3}\cdot3\cdot2\frac{A_3}{A_2} + \frac{1}{c^4}\cdot2\cdot1\right)\xi^4$$

$$+ \left(7\cdot6\frac{A_7}{A_2} + \frac{4}{c}\cdot6\cdot5\frac{A_6}{A_2} + \frac{6}{c^2}\cdot5\cdot4\frac{A_5}{A_2} + \frac{4}{c^3}\cdot4\cdot3\frac{A_4}{A_2} + \frac{1}{c^4}\cdot3\cdot2\frac{A_3}{A_2}\right)\xi^5$$

$$+ \cdots \quad\cdots\cdots(32)$$

By equating the coefficients in (31) and (32) we are able to determine the A's. The law of the series (32) is obvious, and sufficient of the series (31) is written down to enable us to find A_9. We can, however, obtain a good approximation to higher coefficients, because the coefficients in (31) become relatively unimportant.

We now begin the solution with

$$c = 1, \quad z_0 = 1, \quad \left(\frac{dz}{dx}\right)_0 = \frac{1}{\frac{5}{6}\beta^2} = 1\cdot0070, \quad \left(\frac{d^2z}{dx^2}\right)_0 = -\frac{1}{\frac{5}{6}\beta^2} = -1\cdot0070$$

Hence, $\qquad\qquad\qquad A_1 = 1\cdot0070, \quad A_2 = -\cdot5035$

whence I compute

$$A_3 = +\cdot41782, \quad A_4 = -\cdot30068, \quad A_5 = +\cdot16175, \quad A_6 = -\cdot01306, \quad A_7 = -\cdot1333$$

$$A_8 = +\cdot266, \quad A_9 = -\cdot378, \quad A_{10} = +\cdot48, \quad A_{11} = -\cdot6$$

With these coefficients I find

$$\left.\begin{array}{cccccc}\frac{r}{a} = & \frac{12}{11}, & \frac{12}{10}, & \frac{12}{9}, & \frac{12}{8}, & \frac{12}{7}, & \frac{12}{6} \\[4pt] z = \cdot9123 & \cdot8160 & \cdot7089 & \cdot5887 & \cdot4525 & \cdot2982\end{array}\right\}\ldots(33)$$

Then, evaluating $x^{-4}z^{\frac{3}{2}}$, and combining the several values by the rules for integration of the calculus of finite differences, I find

$$\left.\begin{array}{ccccccc}\frac{r}{a} = & \frac{12}{11}, & \frac{12}{10}, & \frac{12}{9}, & \frac{12}{8}, & \frac{12}{7}, & \frac{12}{6} \\[4pt] \tfrac{5}{6}\beta^2\frac{dz}{dx} = \ldots & & 1\cdot21 & 1\cdot35 & 1\cdot527 & 1\cdot729 & 1\cdot9513\end{array}\right\}\ldots\ldots(34)$$

When $r = 2$, we begin a new series with

$$c = \tfrac{1}{2}, \quad z_0 = \cdot2982, \quad A_1 = \left(\frac{1}{z}\frac{dz}{dx}\right)_0 = \frac{1\cdot9513}{\cdot9907 \times \cdot2982} = +6\cdot5894$$

$$A_2 = \left(\frac{1}{2z}\frac{d^2z}{dx^2}\right)_0 = \frac{-z_0^{\frac{1}{2}}}{2\left(\frac{1}{2}\right)^4} = -4\cdot3686$$

From these I compute $A_3 = -2\cdot744, \quad A_4 = +21\cdot365, \quad A_5 = -45\cdot409,$ $A_6 = +9\cdot932, \quad A_7 = +\cdot319.$

It appears that z vanishes when $x - c = -\cdot141$ or $x = \cdot359.$

It follows, therefore, that four equidistant values of x lying between $r = 2a$ and $r = a/\cdot359 = 2\cdot786a$ correspond to $x - c = 0, \quad x - c = -\cdot047,$ $x - c = -\cdot094, \quad x - c = -\cdot141.$

For the first of these, where $r = 2a$, we have $z = \cdot2982$, and for the last, where $r = 2\cdot786a$, $z = 0$; and, when $x - c = -\cdot047$, or $r = a/\cdot453 = 2\cdot208a$, I find $z = \cdot2031$; and, when $x - c = -\cdot094$, or $r = a/\cdot406 = 2\cdot463a$, I find $z = \cdot1033.$

Finding $x^{-4}z^{\frac{3}{2}}$ for these four values and combining them by the rules of integration, I find

$$\tfrac{5}{6}\beta^2\frac{dz}{dx} = 2\cdot1767, \quad \text{when } r = 2\cdot786a \ldots\ldots\ldots\ldots(35)$$

We thus see that the mass of the whole system is 2·1767 times the mass of the isothermal nucleus, and its radius is 2·786 times the radius of the nucleus.

The mass of the isothermal nucleus is thus 46 per cent. of the whole. M. Ritter, taking the ratio of the specific heats as $\frac{7}{5}$ instead of $\frac{5}{3}$, says that the proportion is about 40 per cent.

§ 6. *On a Gaseous Sphere in "Isothermal-Adiabatic" Equilibrium.*

M. Ritter calls a sphere, with isothermal nucleus and a layer in convective equilibrium above it, a case of isothermal-adiabatic equilibrium. Since the height of an atmosphere in convective equilibrium depends only on the temperature at the base, and since the isothermal nucleus in our numerical example is at minimum temperature, the thickness of the adiabatic layer is a minimum, and the isothermal nucleus a maximum.

We are now in a position to collect together all the numerical results of the last two sections in a form appropriate for our subsequent investigation. It will be convenient to refer all the densities and masses to the mean density and mass of the isothermal nucleus. Gravity may also be referred to gravity G at the limit of the isothermal nucleus, and velocity to v_0^2, the mean square of velocity of agitation in the isothermal nucleus.

TABLE III. Isothermal-Adiabatic Sphere.

Radius $\frac{r}{a}$ =	0	·1596	·3193	·4789	·6385	·7932	·9578	1	1·0909	1·2	1·3333	1·5	1·7143	2·0	2·208	2·463	2·786
Square of velocity of agitation $\frac{v^2}{v_0^2}$ =	1	1	1	1	1	1	1	1	·912	·816	·709	·589	·452	·298	·203	·103	0
Density $\frac{w}{\frac{1}{3}\rho}$ =	32·14	23·52	11·86	5·742	3·024	1·759	1·115	1	·871	·737	·597	·452	·304	·163	·092	·033	0
Mass in terms of M =	0	·0361	·1881	·3942	·5988	·7866	·9574	1	...	1·207	1·349	1·527	1·729	1·951	...	...	2·177
Gravity $\frac{g}{G}$ =	0	1·4156	1·8467	1·7188	1·4685	1·2346	1·0436	1	...	·8385	·7589	·6785	·5882	·4878	...	...	·28
Square of velocity of satellite $\frac{v^2}{v_0^2}$ =	0	·1896	·4945	·6908	·7869	·8269	·8888	·8391	...	·844	·849	·854	·846	·818	...	...	·66
$\log \frac{\frac{1}{3}\rho}{w}$ =	8·4929	8·6286	8·9260	9·2410	9·5194	9·7547	9·9529	·0000	·0598	·1325	·2241	·3452	·5166	·7882	1·0385	1·4791	∞
$\frac{1}{\log\left[w/\frac{1}{3}\rho\right][v/v_0]}$ =	8·4929	8·6286	8·9260	9·2410	9·5194	9·7547	9·9529	·0000	·0797	·1766	·2988	·4603	·6889	1·0510	1·3846	1·9722	∞
$\log F_1 = \log\left[\dfrac{3\pi}{8\beta^2}\dfrac{g/G}{[v^2/v_0^2][w/\frac{1}{3}\rho]}\right]$ =	$-\infty$	8·7745	9·1872	9·4712	9·6813	9·8413	9·9664	9·9950	...	·1393	·2487	·4019	·6256	·9970	...	...	∞
$\log F_2 = \log\left[\dfrac{1}{\pi}\dfrac{[g/G]^{\frac{1}{4}}x^{\frac{3}{4}}}{[v/v_0][w/\frac{1}{3}\rho]}\right]$ =	$-\infty$	8·6053	8·8098	9·0213	9·2031	9·3523	9·4744	9·5029	...	9·5967	9·6793	9·7909	9·9594	·2475	...	...	∞

§ 7. *On the Kinetic Energy of Agitation and its Distribution in an Isothermal-Adiabatic Sphere of Gas.*

We shall now consider what would be the distribution of kinetic energy in the nebula if each meteorite (or molecule) were to fall from infinity to the neighbourhood where we find it, and were to retain that energy afterwards. This will give the distribution of energy in a swarm of the supposed arrangement of density, if the rate of diffusion of kinetic energy were to be infinitely slow, and if there were no loss of energy through imperfect elasticity.

The square of the velocity of a satellite in a circular orbit is one-half of the square of the velocity acquired by the fall from infinity to the distance of the satellite from the centre. If the concentration has proceeded as far as radius r, and if a meteorite falls from infinity to distance r, then, if U be its velocity, and v the velocity in a circular orbit at distance r,

$$\tfrac{1}{2}U^2 = v^2 = \tfrac{1}{3}\beta^2 \frac{\mu M}{a} . x \frac{dy}{dx} = \tfrac{1}{3}v_0^2 x \frac{dy}{dx}, \text{ in the isothermal sphere}$$

$$= \tfrac{5}{6}\beta^2 \frac{\mu M}{a} . x \frac{dz}{dx} = \tfrac{5}{6}v_0^2 x \frac{dz}{dx}, \text{ in the adiabatic layer}$$

In these formulæ, by the definitions of y and z,

$$y = \log_e \left(\frac{9w}{\beta^2 \rho}\right) \text{ in the first, and } z = \left(\frac{w}{w_0}\right)^{\tfrac{2}{3}} \text{ in the second}$$

From these formulæ v^2 was computed in Table III. The value of v^2 or $\tfrac{1}{2}U^2$ gives what may be called the theoretical value of the kinetic energy, because it gives us a measure of the amount of redistribution of energy by diffusion and loss of energy by imperfect elasticity, which must take place before the whole system can assume the form of an isothermal adiabatic sphere.

We will now go on to consider the total potential energy lost in condensation.

We have seen that the potential energy lost by the fall of a single meteorite is $\tfrac{1}{3}v_0^2 x \, dy/dx$ in the isothermal part, and $\tfrac{5}{6}v_0^2 x \, dz/dx$ in the outer part.

Now, in the isothermal part a spherical element of mass is

$$-\tfrac{1}{3}M\beta^2 . \frac{d^2y}{dx^2} \, dx$$

and the energy lost by its fall is

$$-\tfrac{1}{9}M\beta^2 v_0^2 . x \frac{dy}{dx}\frac{d^2y}{dx^2} \, dx$$

Hence, the whole energy lost in the concentration of the isothermal nucleus is

$$\tfrac{1}{9}M\beta^2 v_0^2 \int_1^\infty x \frac{dy}{dx}\frac{d^2y}{dx^2}\, dx$$

But

$$-\int_1^\infty x \frac{dy}{dx}\frac{d^2y}{dx^2}\, dx = \int_1^\infty \frac{e^y}{x^3}\frac{dy}{dx}\, dx = 3\int_1^\infty \frac{e^y}{x^4}\, dx - e^{y_0}$$

$$= 3\left(\frac{dy}{dx}\right)_0 - e^{y_0}$$

$$= \frac{9}{\beta^2} - \frac{9w_0}{\beta^2\rho}$$

Hence, the energy lost is $Mv_0^2\left(1 - \dfrac{w_0}{\rho}\right)$. But in an isothermal sphere of minimum temperature $w_0 = \tfrac{1}{3}\rho$, and thus the total lost energy is $\tfrac{2}{3}Mv_0^2$.

Again, in the adiabatic layer an element of mass is

$$-\tfrac{5}{6}M\beta^2\frac{d^2z}{dx^2}\, dx = +M\frac{z^{\frac{3}{2}}}{x^4}\, dx$$

and, therefore, the energy lost by its fall from infinity is

$$\tfrac{5}{6}Mv_0^2 \cdot \frac{z^{\frac{3}{2}}}{x^3}\frac{dz}{dx}\, dx$$

and the whole loss of energy is the integral of this from $x = 1$ to $x = \cdot359$. When $x = 1$, $z = 1$, and when $x = \cdot359$, $z = 0$.   Hence

$$\int_{\cdot359}^1 \frac{z^{\frac{3}{2}}}{x^3}\frac{dz}{dx}\, dx = \tfrac{2}{5} + \tfrac{6}{5}\int_{\cdot359}^1 \frac{z^{\frac{5}{2}}}{x^4}\, dx$$

Thus, the whole energy lost in the adiabatic layer is

$$Mv_0^2\left[\tfrac{1}{3} + \int_{\cdot359}^1 \frac{z^{\frac{5}{2}}}{x^4}\, dx\right]$$

Add this to the energy found before for the isothermal part, and the whole lost energy of the system is found to be

$$Mv_0^2\left[1 + \int_{\cdot359}^1 \frac{z^{\frac{5}{2}}}{x^4}\, dx\right] \quad\dotfill\quad(36)$$

Now let us evaluate the total kinetic energy existing in the form of agitation of molecules.

In the isothermal part it is clearly $\tfrac{1}{2}Mv_0^2$. In the adiabatic part it is half the element of mass into the square of velocity of agitation integrated through the layer, that is to say, $\tfrac{1}{2} \cdot M\dfrac{z^{\frac{3}{2}}}{x^4}\, dx \times v^2$, and, since $z = \dfrac{v^2}{v_0^2}$, we have

$$\tfrac{1}{2}Mv_0^2\left[1 + \int_{\cdot359}^1 \frac{z^{\frac{5}{2}}}{x^4}\, dx\right]$$

for the total internal kinetic energy of agitation. This is rigorously one-half of the energy lost in concentration.

Hence, if a meteor swarm concentrates into this arrangement of density, one half of the original energy is occupied in vaporising and heating parts of the meteorites on impact, and the other half is retained as kinetic energy of agitation.

I find by quadrature that $\int_{\cdot359}^{1} \frac{z^{\frac{5}{2}}}{x^4} dx = \cdot643$. Hence, the potential energy lost in concentration is $Mv_0{}^2 (1\cdot643)$, and that part of it which is retained as energy of agitation is $\frac{1}{2} Mv_0{}^2 (1\cdot643)$. The whole mass of the system is $2\cdot1767\ M$, and we may, therefore, write these

$$\cdot7548\ (2\cdot1767\ M)\ v_0{}^2 \quad \text{and} \quad \tfrac{1}{2} \times \cdot7548\ (2\cdot1767\ M)\ v_0{}^2$$

It is clear then that the average mean square of velocity of agitation of the *whole* system is $\cdot7548\ v_0{}^2$*. Or, shortly, the average temperature is very nearly $\frac{3}{4}$ of the temperature of the isothermal nucleus.

It follows from this whole investigation that for any given mass of matter, arranged in an isothermal-adiabatic sphere of given dimensions, the actual velocities of agitation are determinable throughout.

§ 8.   *On the "Sphere of Action."*

When two meteorites pass near to one another, each will be deflected from its straight path by the attraction of the other. The question arises as to whether the amount of such deflection can be so great that the passage of two meteorites near to one another ought to be estimated as an encounter in the kinetic theory.

We shall now, therefore, find the deflection of two meteorites, moving with the mean relative velocity, when they pass so close as just to graze one another.

The mean square of relative velocity in the isothermal portion is $2v_0{}^2$, and this may be taken as the square of the velocity at infinity in the relative hyperbola described. The angle between the asymptotes of the hyperbola is the deflection due to this sort of encounter.

Let α, ϵ be the semi-axis and eccentricity of the hyperbola. Then, if ϵ be large, the angle between the asymptotes is $1/\epsilon$; and, if $\frac{1}{2}s$ be the radius of either meteorite, the pericentral distance (when they graze) is s. Therefore,

$$s = \alpha\ (\epsilon - 1)$$

---

* M. Ritter gives $\cdot741$ in place of $\cdot755$, but, as already remarked, he uses a different value for the ratio of the specific heats.

By the law of central orbits

$$2v_0^2 = \frac{\mu m}{\alpha}$$

Therefore,

$$\epsilon = \frac{2v_0^2 s}{\mu m} + 1$$

But, since $v_0^2 = \beta^2 \mu M/a$, we have

$$\epsilon = 2\beta^2 \frac{Ms}{ma} + 1$$

The unity on the right-hand side is negligible, and, since $180/\pi\epsilon$ is the deflection in degrees, that deflection is

$$\frac{180}{2\pi\beta^2} \frac{ma}{Ms} \text{ degrees}$$

Now, if δ be the density of the body of a meteorite, $m = \frac{1}{6}\pi\delta s^3$, and, therefore, this expression becomes

$$\frac{15°}{\beta^2} \times \frac{\delta a s^2}{M}$$

Let us find what s must be if the deflection is $10°$; we have

$$s = \beta\sqrt{\frac{2M}{3a\delta}}$$

We may, for a rough evaluation, take β as unity instead of $\sqrt{6/5}$, and suppose a to be equal to the distance of Neptune from the sun (viz., $4\cdot5 \times 10^{14}$ cm.), and, as a very high estimate of the value of δ, let us suppose the density of a meteorite is 10. Then, since the sun, $M_0 = 2 \times 10^{33}$ grammes, and M is about a half of the sun's mass, we have

$$s = \left[\frac{2 \times 10^{33}}{3 \times 4\cdot5 \times 10^{14} \times 10}\right]^{\frac{1}{2}} = (15 \times 10^{16})^{\frac{1}{2}} = 4 \times 10^8$$

Hence, $m = \frac{1}{6}\pi\delta s^3 = \frac{1}{6}\pi \times 10 \times 64 \times 10^{24} = 3 \times 10^{26}$ grammes, in round numbers.

But the earth's mass is 6×10^{27} grammes, and therefore the meteorites are one-twentieth of the mass of the earth.

It follows, therefore, that, with such small masses as those with which the present theory deals, the deflection due to gravity is insensible, and we need only estimate actual impacts as encounters.

Hence, the radius of the sphere of action of a meteorite is identical with the diameter of its body.

§ 9. *On the Criterion for the Applicability of Hydrodynamics to a Swarm of Meteorites.*

The question at issue is to determine within what limits the quasi-gas formed by a swarm of colliding meteorites may be treated as a plenum, subject to the laws of hydrodynamics. The doctrines of the nebular hypothesis depend on the stability of a rotating mass of fluid, and that stability depends on the frequencies of its gravitational oscillations. Now the works of Poincaré and others seem to show that instability, at least in a homogeneous fluid, first arises from one of the graver modes of oscillation, and the period of the gravest mode does not differ much from the period of a satellite grazing the surface of the mass of fluid. Then, in order that hydrodynamical treatment may be applicable for the discussion of such questions of stability, the mean free time between collisions must be small compared with the period of such a satellite. Another way of stating this is that the mean free path of a meteorite shall be but little curved, and that the velocity of a meteorite shall be but little changed by gravity in the interval between two collisions. This must be fulfilled not only at the limits of the swarm, but at every point of it. The condition above stated will be satisfied if the space through which a meteorite falls from rest, at any part of the swarm, in the mean interval between collisions is small compared with the mean free path. If this criterion is fulfilled, then, in most respects which we are likely to discuss, the swarm will behave like a gas, and we must at present confine the consideration of the matter to this general criterion.

It would be laborious to determine exactly the space fallen through from rest, because gravity varies as the meteorite falls, but a sufficiently close approximation may be found by taking gravity constant throughout the fall and equal to its value at the point from which the meteorite starts.

We have already denoted by g the value of gravity at any part of the swarm, and have tabulated it in Table III. in terms of G or $\mu M/a^2$.

Now the mean interval is $T = L/(v \sqrt{8/3\pi})$. Hence, if D be the distance fallen in this time,

$$D = \tfrac{1}{2}gT^2 = \tfrac{1}{2}\frac{gL^2}{v^2} \cdot \frac{3\pi}{8}$$

But $$L = l\left(\frac{M_0}{M}\right)\frac{\tfrac{1}{3}\rho}{w} \cdot \frac{m}{s^2} \quad\text{and}\quad \frac{G}{v_0^2} = \frac{1}{\beta^2 a}$$

Therefore,

$$\frac{D}{L} = \frac{l}{2a} \cdot \frac{M_0}{M} \cdot \frac{m}{s^2} \left\{\frac{3\pi}{8\beta^2} \cdot \frac{[g/G]}{[v^2/v_0^2][w/\tfrac{1}{3}\rho]}\right\} = \frac{l}{2a} \cdot \frac{M_0}{M} \cdot \frac{m}{s^2} \cdot F_1 \quad\ldots\ldots(37)$$

The factor F_1 has been tabulated above in Table III., and it increases from the centre to the outside.

This criterion may be regarded from another point of view, for, if the meteorite be describing a circular orbit about the centre of the swarm, D is the deflection from the straight path in the mean interval between two collisions. Then the criterion is that the deflection shall be small compared with the mean free path.

We may consider the criterion from again another point of view, and state that the arc of circular orbit described in the mean interval shall be a small fraction of the whole circumference.

The linear velocity v in the circular orbit is given by

$$v^2 = g\frac{a}{x} = \frac{v_0^2}{\beta^2 x} \cdot \frac{g}{G}$$

And the mean interval $T = L/[v \sqrt{8/3\pi}]$. Hence, if A be the arc described with velocity v in time T,

$$A^2 = \frac{3\pi}{8\beta^2} \cdot \frac{L^2}{v^2/v_0^2} \cdot \frac{g}{Gx} = \frac{L^2}{v^2/v_0^2} \cdot \frac{g}{Gx} \text{ nearly, since } \frac{3\pi}{8\beta^2} = \cdot 988$$

But the whole arc of circumference C is $2\pi a/x$.

Therefore,

$$\frac{A}{C} = \frac{L}{2a} \cdot \frac{1}{\pi} \cdot \frac{[g/G]^{\frac{1}{2}} x^{\frac{1}{2}}}{v/v_0}$$

$$= \frac{l}{2a} \cdot \frac{M_0}{M} \cdot \frac{m}{s^2} \cdot \left\{ \frac{1}{\pi} \cdot \frac{[g/G]^{\frac{1}{2}} x^{\frac{1}{2}}}{[v/v_0] [w/\frac{1}{3}\rho]} \right\} = \frac{l}{2a} \cdot \frac{M_0}{M} \cdot \frac{m}{s^2} \cdot F_2 \quad \ldots\ldots\ldots(38)$$

The factor F_2 has been tabulated above, in Table III.

§ 10.   On the Density of Meteorites and Numerical Application.

It is necessary to make assumptions both as to the mass and the density of the meteorites. We have a right to assume, I think, that the density δ is a little less than that of iron, say about 6, and we may put $\frac{4}{3}\pi\delta$ equal to 25. Then we have

$$m = \frac{1}{6}\pi\delta s^3 = \tfrac{25}{8}s^3, \text{ and } \frac{m}{s^2} = \tfrac{25}{8}s$$

There is but little information about the average size of meteorites; but, if we retain the symbol s, it will be easy, by merely shifting the decimal point in the final results, to obtain results for all sizes. Thus, if $s = 1$ cm., $m = 3\frac{1}{8}$ grammes; if $s = 10$ cm., $m = 3\frac{1}{8}$ kilogrammes; if $s = 100$ cm., $m = 3\frac{1}{8}$ tonnes, and if $s = 1000$ cm., $m = 3125$ tonnes. I shall, therefore, keep s in the analytical formulæ, and put it equal to unity in the numerical results.

In the first place, making no assumptions as to the density or masses of the meteorites, we have

$$M_0 = 2\text{·}1767 \times M, \quad \tfrac{1}{3}\beta^2 = \text{·}39723$$

Then, by substitution in (10) and (11), we have

$$\left.\begin{aligned}
v_0 &= u \times 10^{9\text{·}86918-10} \\[4pt]
L &= l \times 10^{0\text{·}33781}\, \frac{m/s^2}{w/\frac{1}{3}\rho} \\[4pt]
T &= \tau \times 10^{0\text{·}46863}\, \frac{m/s^2}{[v/v_0]\,[w/\frac{1}{3}\rho]} \\[4pt]
\frac{D}{L} &= \frac{l}{2a} \times 10^{0\text{·}33781} \times F_1 \times \frac{m}{s^2} \\[4pt]
\frac{A}{C} &= \frac{l}{2a} \times 10^{0\text{·}33781} \times F_2 \times \frac{m}{s^2}
\end{aligned}\right\} \quad \cdots\cdots\cdots\cdots(39)$$

We will now apply this solution to a case which will put the theory to a severe test. Suppose that the limit of the sphere of uniformly distributed energy of agitation is nearly as far as the planet Uranus, so that, say $a = 16a_0$. Then the extreme limit of the swarm is at $44\frac{1}{2}a_0$; but the orbit of the planet Neptune is at $30a_0$, so that the limit is further beyond Neptune than Saturn is from the Sun.

Now, if $a/a_0 = 16$, I find

$$\left.\begin{aligned}
u &= 10^{5\text{·}86117} \text{ cm. per sec.} \\[4pt]
&= 10^{0\text{·}18795} a_0 \text{ per annum} \\[4pt]
\tau &= 10^{4\text{·}46808} \text{ seconds} \\[4pt]
&= 10^{6\text{·}96899-10} \text{ years}
\end{aligned}\right\} \quad \cdots\cdots\cdots\cdots\cdots(40)$$

Introducing these values in (39) and putting $\frac{2.5}{8}s$ for m/s^2, I find

$$\left.\begin{aligned}
v &= 1\text{·}141a_0 \text{ per annum} = 5\text{·}374 \text{ kilom. per sec.} \\[4pt]
\frac{L}{a_0} &= 10^{7\text{·}9540-10} \times \frac{s}{[w/\frac{1}{3}\rho]} \\[4pt]
T &= 10^{7\text{·}9325-10} \times \frac{s}{[v/v_0]\,[w/\frac{1}{3}\rho]} \\[4pt]
\frac{D}{L} &= 10^{6\text{·}4489-10}\, sF_1 \\[4pt]
\frac{A}{C} &= 10^{6\text{·}4489-10}\, sF_2
\end{aligned}\right\} \quad \cdots\cdots\cdots(41)$$

We have in Table III. the logarithms of the several factors, which occur last in these formulæ (41), at various distances from the centre.

It will suffice for our purpose only to take every other value from Table III. The distances from the centre are expressed in terms of the astronomical unit distance, viz., the earth's mean distance from the sun. The mean free path is expressed both in the same unit and in kilometres; and the mean intervals between collisions in days. The criteria D/L and A/C are, of course, pure numbers. Table IV., as it stands, is applicable to meteorites weighing $3\frac{1}{8}$ grammes, but by shifting the decimal point one place to the right in the last four rows of entries it becomes kilogrammes, one more and it becomes tonnes, and another, thousands of tonnes, and so on.

IV.—TABLE OF RESULTS.

The meteorites weigh $3\frac{1}{8}$ grammes, and have the density of iron. The swarm extends to $44\frac{1}{2}a_0$, a_0 being earth's distance from sun.

		Sun	Asteroids	Saturn			Uranus		Neptune	
Distance from centre	$\dfrac{r}{a_0}=$	0	2·55	7·66	12·77	16	19·2	24	32	$44\frac{1}{2}$
Velocity of mean square in kilometres per sec.	$v=$	5·37	5·37	5·37	5·37	5·37	4·85	4·12	2·93	0
Mean free path,	$\dfrac{L}{a_0}=$	·00028	·00038	·00157	·00511	·00900	·0122	·0199	·0552	∞
	L kilom. $=$	41,600	57,000	233,000	760,000	1,340,000	1,810,000	2,960,000	8,210,000	∞
Mean free time, in days	$T=$	·097	·133	·545	1·78	3·13	4·70	9·02	35·17	∞
Criterion,	$\dfrac{D}{L}=$	...	·0000167	·0000832	·000195	·000278	·000387	·000709	·00279	∞
Criterion,	$\dfrac{A}{C}=$	...	·0000113	·0000295	·0000633	·0000895	·000111	·000174	·000497	∞

The incidence of the several planets in the scale of distance is roughly indicated by the names written above.

The criteria show that, if the meteorites weigh $3\frac{1}{8}$ kilogrammes, the collisions are frequent enough, even beyond the orbit of Neptune, to allow the kinetic theory of gases to be applicable for such problems as are in contemplation. For, when $r/a = 32$, the two criteria (with decimal point shifted one place to the right) are ·028 and ·005, both small fractions. But, if the meteorites weigh $3\frac{1}{8}$ tonnes, the criteria cease to be very small, about $r/a = 24$. If they weigh 3125 tonnes, the applicability will cease somewhat beyond where the asteroids now are.

I conclude, then, from this discussion that we are justified in applying hydrodynamical treatment to a swarm of meteorites from which the solar system originated, even in the earliest stages of the history of the swarm.

This discussion has, of course, no bearing on the fundamental hypothesis that meteorites *can* glance from one another on impact with a virtually high degree of elasticity; nor does it do anything to justify the assumption that a swarm will consist approximately of a quasi-isothermal nucleus with a quasi-adiabatic layer over it. This latter assumption I have been led to by the considerations to which we now pass.

§ 11. *On the Diffusion of Kinetic Energy and on the Viscosity.*

In order to discuss these questions, it will be well to begin with a simple case of fluid motion.

Consider two-dimensional motion, in which there are a number of streams of equal breadth moving parallel to y with velocity V, and, interpolated between them, let there be strata of quiescent fluid; suppose then that we wish to find the motion at any time after this initial state. Let the boundaries of the streams V be from $x = ml$ to $\frac{1}{2}(2m+1)l$. Then, if u be the velocity at x, parallel to y at time t, and ν the kinetic modulus of viscosity, the equation of motion is

$$\frac{du}{dt} = \nu \frac{d^2u}{dx^2}$$

The solution of this being of the form $e^{-p^2\nu t}\cos px$, the complete solution satisfying the initial condition is

$$u = \tfrac{1}{2}V + \frac{2V}{\pi}\left[e^{-\pi^2\nu t/l^2}\cos\frac{\pi x}{l} - \tfrac{1}{3}e^{-9\pi^2\nu t/l^2}\cos\frac{3\pi x}{l} + \tfrac{1}{5}e^{-25\pi^2\nu t/l^2}\cos\frac{5\pi x}{l} - \dots \right]$$

If we refer time to a period τ, where $\tau = l^2/\pi^2\nu$, then after a time $\theta\tau$, which is greater than τ, the solution is sensibly

$$u = \tfrac{1}{2}V\left[1 + \frac{4}{\pi e^\theta}\cos\frac{\pi x}{l}\right]$$

It is clear that the maximum of u occurs when $x = 0$, and the minimum when $x = l$, and that they are

$$\tfrac{1}{2}V\left[1 \pm \frac{4}{\pi e^\theta}\right]$$

Hence, the difference between the maximum and minimum is $4V/\pi e^\theta$. Therefore, the ratio of the greatest difference of velocities after time $\theta\tau$ to the initial difference of velocities is $4/\pi e^\theta$. When θ is 1, 2, 3, this ratio assumes the values 1/2·135, 1/5·804, 1/15·73 respectively. Thus, after three times the interval τ, the difference of velocities is small. The time τ may be therefore taken as a convenient measure of viscosity.

In our problem the streams must be taken of a width comparable with the linear dimensions of the solar system. I therefore take l, the width of the streams, as equal to a_0, the Earth's distance from the Sun, and we have

$$\tau = \frac{a_0^2}{\pi^2 \nu}$$

Now, according to the kinetic theory of gases, the kinetic modulus of viscosity is $1/\pi$ into the mean free path multiplied by the mean velocity. Hence,

$$\nu = \frac{1}{\pi} L \left(v \sqrt{\frac{8}{3\pi}} \right)^*$$

Hence, we have

$$\tau = \left(\frac{a_0}{L} \right) \left(\frac{a_0}{v} \right) \sqrt{\frac{3}{8\pi}} \quad \dots\dots\dots\dots\dots\dots(42)$$

If we apply this formula to the solution which has been already found in Table IV., we obtain the following results :—

$\dfrac{r}{a_0}$	0,	2·55,	7·66,	12·77,	16,	19·2,	24,	32
τ years	1082,	792,	193,	59·2,	33·7,	27·5,	19·8,	61·7

These results are applicable to meteorites weighing $3\frac{1}{8}$ grammes in a swarm extending to $44\frac{1}{2} a_0$. If the meteorites weigh $3\frac{1}{8}$ kilogrammes, the values of τ would be one-tenth of the tabulated values. If the streams were ten times as broad, the periods would be a hundred times as long.

Now the periods τ in the above table, even if multiplied by a thousand, must be considered as short in the history of a stellar system. It thus appears that the quasi-viscosity must be such that a swarm of meteors will, if revolving, move nearly without relative motion of its parts, at least in the early stages of its evolution.

But let us consider the values of τ at different epochs in the history of the same system. If a be the radius of the isothermal sphere the formulæ (9) and (10) show that L/a_0 varies as a^3, whilst v/a_0 varies as $a^{-\frac{1}{2}}$. Hence τ varies inversely as $a^{\frac{5}{2}}$. Thus, as the swarm contracts, the periods τ increase rapidly.

Thus, later in the history, the viscosity will probably fall off so much that equalisation of angular velocity may be no longer attained, and we should then have the central portion rotating more rapidly than the outside, with a gradual transition from one angular velocity to the other.

The modulus ν gives, besides the viscosity, the rate of equalisation of the kinetic energy of agitation; this corresponds in a true gas with the con-

---

* Meyer, *Kinetische Theorie der Gase*, p. 321. The $1/\pi$ is derived from a numerical quadrature which gives the value ·318, and it is apparently only accidentally equal to $1/\pi$. The $v \sqrt{(8/3\pi)}$ is the mean velocity denoted Ω by Meyer.

duction of heat.   The conclusion at which we thus arrive appears to justify the assumption that the whole of the central part of the swarm is endued with uniform kinetic energy of agitation, and that the mass of the quasi-isothermal nucleus is the greatest possible.   With regard to the assumption that the nucleus is coated with a layer in adiabatic or convective equilibrium, it may be remarked that the velocity of agitation must decrease when we get to the outskirts of the swarm, and convective equilibrium will probably satisfy the conditions of the case better than any other.   Further considerations will be adduced on this point in the Summary.

§ 12.   On the Rate of Loss of Kinetic Energy through Imperfect Elasticity, and on the Heat Generated.

In a collision between two meteorites the loss of energy is probably proportional to their relative kinetic energy before impact.   Therefore, the amount of heat generated by a single meteorite per unit time is proportional to the kinetic energy (say h) and to the frequency of collision.   By (10) the frequency, or reciprocal of T, varies as vws^2/m; but $m^{\frac{1}{2}}v$ is equal to $(2h)^{\frac{1}{2}}$, and $s^2m^{-\frac{3}{2}}$ varies as $m^{-\frac{5}{6}}$.   Hence, the frequency of collision varies as $h^{\frac{1}{2}}wm^{-\frac{5}{6}}$, and the amount of heat generated by a single meteorite per unit time varies as $h^{\frac{3}{2}}wm^{-\frac{5}{6}}$.   But, if p be the quasi-hydrostatic pressure, p varies as hwm^{-1}, and, therefore, the heat generated by a single meteorite varies as $h^{\frac{1}{2}}pm^{\frac{1}{6}}$.

Then, to find the total heat generated per unit time and volume, we have to multiply this by the number of meteorites per unit volume, that is to say, by wm^{-1}, which is equal to $3ph^{-1}$.

Thus the amount of heat generated per unit time and volume is proportional to $p^2m^{\frac{1}{6}}h^{-\frac{1}{2}}$.   With meteorites of uniform size, and with uniform kinetic energy of agitation, this becomes simply the square of the hydrostatic pressure.

The mean temperature of the gases volatilised by collisions must depend on a variety of considerations, but it would seem as if the temperature would follow, more or less closely, the variations of heat generated per unit time and volume.

§ 13.   On the Fringe of a Swarm of Meteorites.

The law of distribution of meteorites found above depends on the frequency of collisions.   But at some distance from the centre collisions must have become so rare that the statistical method is inapplicable.   There must then be a sort of fringe to the swarm, which I attempt to represent by supposing that beyond a certain radius a (not the same as the former a) collisions never occur, and each meteorite describes an orbit under gravity.

Now, at any point gravity depends on the mass of all the matter lying inside a sphere whose radius is equal to the distance of that point from the centre of the swarm. Hence, the value of gravity depends on the law of density of distribution of the meteorites, which is the thing which we are seeking to discover.

We suppose, then, that from every point of a sphere of radius a a fountain of meteorites is shot up, at all inclinations to the vertical, and with velocities grouped about a mean velocity, according to the exponential law appropriate to the case. As many meteorites are supposed to fall back on to the surface as leave it, and this inward cannonade against the boundary of the sphere exactly balances the quasi-gaseous pressure on the inside of the sphere. Thus, the ideal surface may be annihilated. Since the falling half of the orbit of a meteorite is the facsimile of the rising half, we need only trace the body from projection to apocentre, and then double the distribution of density which is deduced on the hypothesis that all the meteorites are rising. Again, since every element of the sphere shoots out a similar fountain, and since collisions are precluded by hypothesis, we need only consider the velocity along the radius vector. As far as concerns the distribution of density, it is the same as if each element shot up a vertical fountain; but, of course, in determining the vertical velocity, we must pay attention to the inclination to the vertical at which the meteorite was shot out.

The mass of the matter inside the sphere, whose attraction affords the principal part of the force under which the meteorites move, is say M, and, for the sake of simplicity of notation, we shall take $2\mu M/a$ as being unit square of velocity.

Let $\tfrac{1}{2}\phi(r)$ be the potential at the point whose radius is r, and suppose that a meteorite is shot out from a point on the sphere with a velocity u, and at an inclination ϵ to the vertical; then, if r, θ be the radius vector and longitude of the meteorite at the time t, the equations of conservation of moment of momentum, and of energy are

$$r^2 \frac{d\theta}{dt} = ua \sin \epsilon$$

$$\left(\frac{dr}{dt}\right)^2 + \left(r \frac{d\theta}{dt}\right)^2 - \phi(r) = u^2 - \phi(a)$$

If we write $f(r) = \phi(a) - \phi(r)$, and eliminate $d\theta/dt$, we get

$$r^2 \frac{dr}{dt} = r \{r^2(u^2 - f(r)) - u^2 a^2 \sin^2 \epsilon\}^{\frac{1}{2}}$$

Now, we are to regard dr/dt as the vertical velocity in a fountain squirting up from a point on the sphere. Then, since $f(a) = 0$, it follows that at the foot of the fountain $r^2 dr/dt$ is equal to $a^2 u \cos \epsilon$. If, therefore, δ be the density at the height r, and δ_0 at the foot, the equation of continuity is

$$\delta r^2 \frac{dr}{dt} = \delta_0 a^2 u \cos \epsilon$$

Therefore,

$$\frac{\delta}{\delta_0} = \frac{a^2 u \cos \epsilon}{r \left\{ r^2 \left(u^2 - f(r) \right) - u^2 a^2 \sin^2 \epsilon \right\}^{\frac{1}{2}}}$$

But now let us suppose that the meteorites are not only shot out at inclination ϵ, but at all possible inclinations from $0°$ to $90°$. It is then clear that this expression must be multiplied by $\sin \epsilon \, d\epsilon$, and integrated. Hence, if δ now denotes the integral density,

$$\delta = C \int_0^{\epsilon_0} \frac{a^2 u \cos \epsilon \sin \epsilon \, d\epsilon}{r \left\{ r^2 \left(u^2 - f(r) \right) - u^2 a^2 \sin^2 \epsilon \right\}^{\frac{1}{2}}}$$

where C is a constant which it will be unnecessary to determine, and where the limit ϵ_0 will be the subject of future consideration.

Effecting the integration, we have

$$\delta = -\frac{C}{ur} \left\{ r^2 \left(u^2 - f(r) \right) - u^2 a^2 \sin^2 \epsilon \right\}^{\frac{1}{2}}, \text{ between limits}$$

$$= \frac{C}{ur} \left[\left\{ r^2 \left(u^2 - f(r) \right) \right\}^{\frac{1}{2}} - \left\{ r^2 \left(u^2 - f(r) \right) - u^2 a^2 \sin^2 \epsilon_0 \right\}^{\frac{1}{2}} \right]$$

It is obvious that, if u^2 is greater than $r^2 f(r)/(r^2 - a^2)$, the square root involved in dr/dt does not vanish for any value of ϵ; and, hence, we must simply take $\epsilon_0 = 90°$. If, on the other hand, u^2 is less than this critical value, ϵ_0 is that value of ϵ which makes dr/dt vanish.

Thus, our formula divides into three, viz. :—

1st. u^2 greater than $\dfrac{f(r)}{1 - a^2/r^2}$;

$$\delta = \frac{C}{u} \left[\left(u^2 - f(r) \right)^{\frac{1}{2}} - \left(1 - \frac{a^2}{r^2} \right)^{\frac{1}{2}} \left(u^2 - \frac{f(r)}{1 - a^2/r^2} \right)^{\frac{1}{2}} \right]$$

2nd. u^2 less than $\dfrac{f(r)}{1 - a^2/r^2}$;

$$\delta = \frac{C}{u} \left(u^2 - f(r) \right)^{\frac{1}{2}}$$

3rd. u^2 less than $f(r)$; $\delta = 0$.

The physical meaning of this division is as follows: If we take a station near the surface of the sphere, meteorites shot out at all inclinations, even horizontally, reach the height of our station; and, when they are shot out horizontally, $\epsilon = 90°$. If we go, however, to a higher region, there is a certain inclination which just brings the meteorites at apocentre, where $dr/dt = 0$, to our height; but those shot out more nearly horizontally fail to reach us. Still higher, not even a meteorite shot up vertically can reach us, and the density vanishes.

These results only correspond to a single velocity u; but, if v^2 be the mean square of the velocity, the number of meteorites whose velocities range between u and $u + du$ is proportional to $u^2 e^{-3u^2/2v^2} du$*. Hence, we have to multiply δ by this expression, and integrate from $u = \infty$ to $u = 0$.

Now, the first term of the first form for δ is the same as the second form; and in the third form δ is zero; hence, this first term when multiplied by the exponential has to be integrated from $u^2 = \infty$ to $f(r)$. The second term of the first form of δ has to be multiplied by the exponential, and integrated from $u^2 = \infty$ to $f(r)/(1 - a^2/r^2)$.

For the first term put

$$\frac{3}{2v^2}(u^2 - f(r)) = x^2$$

therefore,

$$u(u^2 - f(r))\ du = (\tfrac{2}{3}v^2)^{\frac{3}{2}} x^2 dx$$

and the limits of x are ∞ to 0.

Hence, the first term is

$$C\,(\tfrac{2}{3}v^2)^{\frac{3}{2}}\, e^{-3f(r)/2v^2} \int_0^{} x^2 e^{-x^2} dx$$

Again, for the second term put

$$\frac{3}{2v^2}\left(u^2 - \frac{f(r)}{1 - a^2/r^2}\right) = x^2$$

and similarly introduce it into the second term, and we have

$$-C\,(\tfrac{2}{3}v^2)^{\frac{3}{2}}\left(1 - \frac{a^2}{r^2}\right)^{\frac{1}{2}} e^{3f(r)/2v^2(1-a^2/r^2)} \int_0^{\infty} x^2 e^{-x}\ dx$$

From these expressions we may omit the constant factors; and, if w be the density at height r, whilst w_0 is the density at the sphere,

$$\frac{w}{w_0} = e^{-3f(r)/2v^2} - \left(1 - \frac{a^2}{r^2}\right)^{\frac{1}{2}} e^{-3f(r)/2v^2(1-a^2/r^2)}$$

In this formula unit square of velocity is $2\mu M/a$; but we have elsewhere written $v^2 = \beta^2 \mu M/a$; hence, if the special unit of velocity be given up, we may write β^2 in place of $2v^2$, and the result becomes

$$\frac{w}{w_0} = e^{-3f(r)/\beta^2} - \left(1 - \frac{a^2}{r^2}\right)^{\frac{1}{2}} e^{-3f(r)/\beta^2(1-a^2/r^2)} \quad\ldots\ldots\ldots\ldots(43)$$

* Oskar Meyer, *Die Kinetische Theorie der Gase*, 1877, pp. 271—2.

It is interesting to observe the connection between this law of density and that which would have held if the gaseous law (due to collisions) had obtained. In that case, since $\frac{1}{2}\phi(r)$ is the potential, we should have had

$$\frac{1}{w}\frac{dp}{dr} - \frac{1}{2}\frac{d}{dr}\phi(r) = 0$$

Now, $p = \frac{1}{3}v^2 w$, and, therefore,

$$\log w - \frac{3}{2v^2}\phi(r) = \text{const.}$$

$$= \log w_0 - \frac{3}{2v^2}\phi(a)$$

Thus,

$$\log \frac{w}{w_0} = -\frac{3}{2v^2}[\phi(a) - \phi(r)] = -\frac{3}{2v^2}f(r)$$

or

$$\frac{w}{w_0} = e^{-3f(r)/2v^2} = e^{-3f(r)/\beta^2}$$

The first term of our result, then, is exactly that resulting from the gaseous law, and the second subtractive term represents the action of the diminished velocity with which the meteorites move in the higher regions, when they are liberated from the equalising effects of continual impacts.

By previous definition, $\mu M f(r)/a$ is the excess of the potential at radius a above its value at radius r; hence,

$$\frac{\mu M}{a}f(r) = \int_a^r \frac{4\pi\mu}{r^2}\int_0^r wr^2 dr \, . \, dr$$

Now, since $f(r)$ is only required for values of r greater than a, we may put w equal to its mean value ρ, between the limits 0 and a. Thus,

$$\int_0^r wr^2 dr = \int_a^r wr^2 dr + \frac{1}{3}\rho a^3$$

Hence,

$$f(r) = \frac{4\pi a}{M}\left[\int_a^r \frac{1}{r^2}\int_a^r wr^2 dr + \frac{1}{3}\rho a^3 \int_a^r \frac{dr}{r^3}\right] = \left(1 - \frac{a}{r}\right) + 3\int_1^{r/a}\frac{1}{z^2}\int_1^z \frac{w}{\rho}z^2 dz$$

If this form for $f(r)$ were substituted in (43), we should obtain a very complicated differential equation for w. We may, however, find two values of $f(r)$ within which the truth must lie.

First, if we neglect the attraction of all the matter lying outside radius a, the second term vanishes, and we have

$$f(r) = 1 - \frac{a}{r}$$

and the law of density is

$$\frac{w}{w_0} = e^{-3(1-a/r)/\beta^2} - \left(1 - \frac{a^2}{r^2}\right)^{\frac{1}{2}} e^{3/(1+a/r)\beta^2} \quad \ldots\ldots\ldots\ldots\ldots(44)$$

Secondly, we may suppose the density to go on diminishing according to the inverse square of the distance. We have seen in the preceding solution and tables that this is roughly the law of diminution for a long way outside the isothermal nucleus. According to this assumption, $w = w_0 a^2 / r^2$. Hence, in the second term of $f(r)$ we put $w = w_0 a^2 / r^2 = w_0 / z^2$.

Hence,
$$\int_1^z \frac{w}{\rho} z^2 dz = \frac{w_0}{\rho} \int_1^z dz = \frac{w_0}{\rho} (z - 1)$$

and
$$f(r) = 1 - \frac{a}{r} + \frac{w_0}{\frac{1}{3}\rho} \int_1^{r/a} \frac{z-1}{z^2} dz = \left(1 - \frac{w_0}{\frac{1}{3}\rho}\right)\left(1 - \frac{a}{r}\right) + \frac{w_0}{\frac{1}{3}\rho} \log \frac{r}{a} \dots(45)$$

The substitution of this value in (43) gives the law of density.

In order to see the kind of results to which these formulæ lead, let us suppose that, when we have reached radius 2 in the adiabatic layer, collisions have become so rare as to be negligible. Then the symbols in the formulæ of this section have numerical values; and, in order to distinguish them, let them be accented, so that, for example, we write a', β'^2, ρ', &c., in (43), (44), and (45), in place of a, β^2, ρ.

Now Table III. shows that, when $a' = 2a$, $M' = 1 \cdot 95 M = 2M$ nearly. Hence, $M'/a' = M/a$ nearly. But, at radius $2a$ in Table III., $v^2/v_0^2 = \cdot 298 = \cdot 3$, and this v^2 is what we now write v'^2 or $\beta'^2 \mu M'/a'$, whilst $v_0^2 = \beta^2 \mu M/a$.

But $\beta^2 = \frac{6}{5}$ very nearly; hence, $\beta'^2/\beta^2 = \cdot 3$, or $\beta'^2 = \cdot 36$.

Thus, $3/\beta'^2 = 8 \cdot 333$.

Then, substituting $2a$ for a', and noticing that in Table III., $w/\frac{1}{3}\rho = \cdot 163$, when $r = 2a$, the first law of density (44) becomes

$$\frac{w}{\frac{1}{3}\rho} = \cdot 163 \left[e^{-\frac{25}{3}(1 - 2a/r)} - \left(1 - 4\frac{a^2}{r^2}\right)^{\frac{1}{2}} e^{-\frac{25}{3}/(1 + 2a/r)} \right] \dots \dots(46)$$

Again, since $M' = 2M$, and $a' = 2a$, $\rho' = \frac{1}{4}\rho$, $\dfrac{w_0'}{\frac{1}{3}\rho'} = 4 \times \dfrac{w_0'}{\frac{1}{3}\rho} = 4 \times \cdot 163$ by Table III., and $\dfrac{w_0'}{\frac{1}{3}\rho'} = \cdot 65 = \frac{2}{3}$ nearly.

Thus, according to the second assumption, we have by (45)

$$3f(r) = \left(1 - \frac{2a}{r}\right) + \log\left(\frac{r}{2a}\right)^2, \quad \text{and, since} \quad \frac{1}{\beta'^2} = 2 \cdot 78$$

$$\left.\begin{array}{l} \dfrac{3f(r)}{\beta'^2} = 2 \cdot 78 \left(1 - \dfrac{2a}{r}\right) + 2 \cdot 78 \log\left(\dfrac{r}{2a}\right)^2 \\[3mm] \dfrac{3f(r)}{\beta'^2 (1 - 4a^2/r^2)} = \dfrac{2 \cdot 78}{1 + 2a/r} + \dfrac{2 \cdot 78}{1 - 4a^2/r^2} \log\left(\dfrac{r}{2a}\right)^2 \end{array}\right\}$$

and the law of density is

$$\frac{w}{\frac{1}{3}\rho} = \cdot 163 \left[e^{-3f(r)/\beta'^2} - \left(1 - \frac{4a^2}{r^2}\right)^{\frac{1}{2}} e^{-3f(r)/\beta'^2(1 - 4a^2/r^2)} \right] \dots \dots(47)$$

The values computed from these alternative formulæ (46) and (47) will be comparable with those in Table III.

In Table III. we have the value of $w/\frac{1}{3}\rho$ computed at distances $r/a = 2\cdot208$, $2\cdot463$, $2\cdot786$. The following short table gives the result extracted from Table III. for comparison with the values computed from (46) and (47):—

$$r/a = \quad 2\cdot0, \qquad 2\cdot208, \qquad 2\cdot463, \qquad 2\cdot786$$

Convective equilib. $\dfrac{w}{\frac{1}{3}\rho} = \quad \cdot163, \qquad \cdot092, \qquad \cdot033, \qquad 0$

First hypoth. (46) $\quad \dfrac{w}{\frac{1}{3}\rho} = \quad \cdot163, \qquad \cdot074, \qquad \cdot033, \qquad \cdot015$

Second hypoth. (47) $\dfrac{w}{\frac{1}{3}\rho} = \quad \cdot163, \qquad \cdot071, \qquad \cdot029, \qquad \cdot011$

It appears, therefore, that the results from the two hypotheses differ but little for some distance outside the region of collisions, and either line may be taken as near enough to the correct result. We see then that the effect of annulling collisions and allowing each body to describe an orbit is that the density at first falls off more rapidly than if the medium were in convective equilibrium, and that further away the density falls off less rapidly. At more remote distances the density would be found to tend to vary as the inverse square of this distance. Thus, the formulæ would make the mass of the system infinite. In other words, the existence of meteorites with nearly parabolic and hyperbolic orbits necessitates an infinite number, if the loss to the system is constantly made good by the supply.

The subject of this section is considered further, from a physical point of view, in the Summary at the end.

§ 14. *On the Kinetic Theory where the Meteorites are of all sizes.*

In an actual swarm of meteorites all sizes occur, for, even if this were not the case initially, inequality of size would soon arise through fractures. Hence, it becomes of interest to examine the kinetic theory on the hypothesis that the colliding bodies are of all possible sizes, grouped about some mean value according to some law of frequency.

If there be two sets of elastic spheres in such numbers that there are respectively A and B in unit volume, and if the mean squares of the velocities of the two are α^2 and β^2 respectively, and if a and b are the radii of the spheres of the two sets, then it is proved that the number of collisions between them per unit time and volume is

$$2AB(a+b)^2 \left[\tfrac{2}{3}\pi(\alpha^2 + \beta^2)\right]^{\frac{1}{2}} *$$

* *The Kinetic Theory of Gases*, by H. W. Watson, p. 11.

We shall now change the notation, and for a and b write s_1 and s_2, and for α and β write u_1 and u_2.

Then, if δ be the density of the spheres, their masses are $\frac{4}{3}\pi\delta s_1^3$ and $\frac{4}{3}\pi\delta s_2^3$.

The condition for the permanence of condition is that the spheres of all masses shall have the same mean kinetic energy. Hence, we refer the mass to a mean sphere of radius ς, and the velocity to a square of velocity V^2.

Then
$$s_1^3 u_1^2 = s_2^3 u_2^2 = \varsigma^3 V^2$$

Thus, our formula may be written
$$2AB(s_1+s_2)^2\left[\left(\frac{\varsigma}{s_1}\right)^3+\left(\frac{\varsigma}{s_2}\right)^3\right]^{\frac{1}{2}}(\tfrac{2}{3}\pi V^2)^{\frac{1}{2}}$$

But now suppose that there are spheres of all possible sizes, and that in unit volume the number whose radius lies between s and $s+ds$ is
$$\frac{4n}{\sigma^3\sqrt{\pi}}\,s^2 e^{-s^2/\sigma^2}\,ds\,*$$

Since the integral of this from ∞ to 0 is n, it follows that n is the number of spheres of all sizes in unit volume.

If ρ be the total mass in unit volume, or the density of distribution,
$$\rho = \frac{4n}{\sqrt{\pi}}\cdot\int_0^\infty \tfrac{4}{3}\pi\delta s^3\cdot\frac{s^2}{\sigma^3}\,e^{-s^2/\sigma^2}\,ds$$
$$= \frac{4n}{\sqrt{\pi}}\cdot\tfrac{4}{3}\pi\delta\sigma^3\int_0^\infty x^5 e^{-x^2}\,dx$$
$$= \frac{4n}{\sqrt{\pi}}\cdot\tfrac{4}{3}\pi\delta\sigma^3$$

If m be the mean mass, $m=\rho/n$; but $m=\tfrac{4}{3}\pi\delta\varsigma^3$; hence,
$$\varsigma^3 = \frac{4}{\sqrt{\pi}}\sigma^3$$

and
$$\left(\frac{\varsigma}{s_1}\right)^3+\left(\frac{\varsigma}{s_2}\right)^3 = \frac{4}{\sqrt{\pi}}\left[\left(\frac{\sigma}{s_1}\right)^3+\left(\frac{\sigma}{s_2}\right)^3\right]$$

If the A spheres of radii s_1 are those whose radii lie between s_1 and s_1+ds_1, and the B spheres of radii s_2 are those whose radii lie between s_2 and s_2+ds_2,
$$A = \frac{4n}{\sqrt{\pi}}\left(\frac{s_1}{\sigma}\right)^2 e^{-s_1^2/\sigma^2}\frac{ds_1}{\sigma}$$
$$B = \frac{4n}{\sqrt{\pi}}\left(\frac{s_2}{\sigma}\right)^2 e^{-s_2^2/\sigma^2}\frac{ds_2}{\sigma}$$

* If the spheres are grouped about a mean mass, instead of about a mean radius, according to a law of this kind, the subsequent integrals become very troublesome. Any law of the kind suffices for the discussion. If, however, I had foreseen the investigation of § 16, I should not have taken this law of frequency.

Hence, the formula for collisions between the A's and B's is

$$\frac{64 n^2}{\pi^{\frac{5}{4}}} \cdot (\tfrac{2}{3}\pi V^2)^{\frac{1}{2}} \cdot (s_1 + s_2)^2 \left[\left(\frac{\sigma}{s_1}\right)^3 + \left(\frac{\sigma}{s_2}\right)^3 \right]^{\frac{1}{2}} \frac{s_1^2 s_2^2}{\sigma^4} e^{-(s_1{}^2 + s_2{}^2)/\sigma^2} \frac{ds_1}{\sigma} \frac{ds_2}{\sigma}$$

or, if we write $x = s_1/\sigma$, $y = s_2/\sigma$, it is

$$\cdot \frac{64 n^2}{\pi^{\frac{5}{4}}} (\tfrac{2}{3}\pi V^2)^{\frac{1}{2}} \sigma^2 (x+y)^2 (x^3 + y^3)^{\frac{1}{2}} (xy)^{\frac{1}{2}} e^{-x^2 - y^2} dx\, dy \quad \ldots\ldots\ldots(48)$$

But $\qquad \sigma^2 = \dfrac{\pi^{\frac{1}{3}}}{2^{\frac{2}{3}}} s^2,$ and $\dfrac{64}{\pi^{\frac{5}{4}}} (\tfrac{2}{3}\pi)^{\frac{1}{2}} \dfrac{\pi^{\frac{1}{3}}}{2^{\frac{2}{3}}} = \dfrac{32}{\pi^{\frac{5}{12}}} \cdot \dfrac{2^{\frac{1}{6}}}{3^{\frac{1}{2}}}$

Hence, the number of collisions per unit time and volume between spheres whose radii range between s_1 and $s_1 + ds_1$, and others with radii between s_2 and $s_2 + ds_2$, is

$$\frac{32}{\pi^{\frac{5}{12}}} \frac{2^{\frac{1}{6}}}{3^{\frac{1}{2}}} \cdot V s^2 n^2 \cdot (x+y)^2 (x^3 + y^3)^{\frac{1}{2}} (xy)^{\frac{1}{2}} e^{-x^2 - y^2} dx\, dy$$

The number of collisions of a single sphere per unit time is $1/n$ of this, and, since $n = \rho/m$, we have for the collisions of a single sphere the factor $\dfrac{V s^2}{m/\rho}$ instead of $V s^2 n^2$.

Then the total number of collisions of all kinds in unit time, or the reciprocal of the mean free time, is the double integral of this from ∞ to 0.

For the purpose of carrying out the integration, we may conveniently, as an algebraic artifice, change from the rectangular axes x, y to the polar coordinates r, θ. Thus,

$$\int_0^\infty \int_0^\infty (x+y)^2 (x^3 + y^3)^{\frac{1}{2}} (xy)^{\frac{1}{2}} e^{-x^2 - y^2} dx\, dy$$

$$= \int_0^\infty r^{\frac{11}{2}} e^{-r^2} dr \int_0^{\frac{1}{2}\pi} (\sin\theta + \cos\theta)^{\frac{5}{2}} (1 - \sin\theta \cos\theta)^{\frac{1}{2}} (\sin\theta \cos\theta)^{\frac{1}{2}} d\theta$$

Now, if we put $r = z^2$,

$$\int_0^\infty r^{\frac{11}{2}} e^{-r^2} dr = 2 \int_0^\infty z^{12} e^{-z^4} dz = 2 \cdot \frac{9 \cdot 5 \cdot 1}{4 \cdot 4 \cdot 4} \int_0^\infty e^{-z^4} dz$$

For the transformation of the second integral, put

$$z = \cos\theta - \sin\theta$$

and we find

$$\int_0^{\frac{1}{2}\pi} (\sin\theta + \cos\theta)^{\frac{5}{2}} (1 - \sin\theta \cos\theta)^{\frac{1}{2}} (\sin\theta \cos\theta)^{\frac{1}{2}} d\theta$$

$$= \int_{-1}^{+1} \tfrac{1}{2} (2 - z^2)^{\frac{3}{4}} (1 - z^4)^{\frac{1}{2}} dz$$

$$= \int_0^{+1} (2 - z^2)^{\frac{3}{4}} (1 - z^4)^{\frac{1}{2}} dz$$

Hence, the whole integral is

$$\tfrac{4 \cdot 5}{3 \cdot 2} \int_0^\infty e^{-z^4} dz \int_0^1 (2 - z^2)^{\frac{3}{4}} (1 - z^4)^{\frac{1}{2}} dz$$

and the mean frequency of collision of a single ball per unit time is

$$\tfrac{15}{4} \cdot \frac{3^{\frac{1}{2}} \cdot 2^{\frac{1}{6}}}{\pi^{\frac{5}{12}}} \frac{V(2\varsigma)^2}{m/\rho} \int_0^\infty e^{-z^4} dz \int_0^1 (2 - z^2)^{\frac{3}{4}} (1 - z^4)^{\frac{1}{2}} dz$$

The second of these two integrals cannot, I think, be evaluated algebraically, but its value is easily found by quadratures. I find, then,

$$\int_0^1 (2 - z^2)^{\frac{3}{4}} (1 - z^4)^{\frac{1}{2}} dz = 1 \cdot 2999$$

The former of the two integrals may be evaluated as follows :—

Let

$$I = \int_0^\infty e^{-x^4} dx$$

then,

$$4 I^2 = 4 \int_0^\infty \int_0^\infty e^{-x^4 - y^4} dx \, dy$$

$$= 4 \int_0^\infty \int_0^{\frac{1}{2}\pi} e^{-r^4 (1 - \frac{1}{2} \sin^2 2\theta)} r \, dr \, d\theta$$

$$= \int_0^\infty \int_0^\pi e^{-z^2 (1 - \frac{1}{2} \sin^2 \phi)} dz \, d\phi$$

$$= 2 \int_0^{\frac{1}{2}\pi} \int_0^\infty \frac{e^{-t^2}}{(1 - \frac{1}{2} \sin^2 \phi)} dt \, d\phi$$

$$= \pi^{\frac{1}{2}} \int_0^{\frac{1}{2}\pi} \frac{d\phi}{(1 - \frac{1}{2} \sin^2 \phi)^{\frac{1}{2}}}$$

$$= \pi^{\frac{1}{2}} F(45°)$$

where F is the complete elliptic integral with modulus sin 45°.

Hence,

$$I = \tfrac{1}{2} \pi^{\frac{1}{4}} F^{\frac{1}{2}} *$$

We thus have the frequency of collision given by

$$\tfrac{15}{8} \cdot \frac{3^{\frac{1}{2}} \cdot 2^{\frac{1}{6}}}{\pi^{\frac{1}{6}}} \cdot F^{\frac{1}{2}} \cdot 1 \cdot 2999 \cdot \frac{V(2\varsigma)^2}{m/\rho}$$

Now, Legendre's Tables give

$$\log F = \cdot 2681272$$

***

* I owe this to Mr Forsyth, and the result verifies an evaluation by quadratures which I had made.

with which value we easily find for T the mean free time, or $1/T$ the frequency,

$$\frac{1}{T} = 5 \cdot 3318 \frac{V(2\varsigma)^2}{m/\rho} = \frac{16}{3} \frac{V(2\varsigma)^2}{m/\rho} \text{ nearly } \dots\dots\dots(49)$$

If $1/T_0$ be the frequency of collision when the spheres are all of the same size and mass ς and m, and are agitated with mean square of velocity V^2, we have, by the ordinary theory,

$$\frac{1}{T_0} = 4 \sqrt{\frac{\pi}{3}} \cdot \frac{V(2\varsigma)^2}{m/\rho} = 4 \cdot 0935 \frac{V(2\varsigma)^2}{m/\rho} \dots\dots\dots(50)$$

It follows, therefore, that in our case collisions are more frequent than if the balls were all of the same size in about the proportion of 4 to 3.

In order to find the mean free path, we require to find the mean velocity.

If u^2 be the mean square of the velocity for any size s, the proportion of all the spheres of that size which move with velocities lying between v and $v + dv$ is

$$\frac{4}{\sqrt{\pi}} y^2 e^{-y^2} dy$$

where $y^2 = 3v^2/2u^2$.

But the number of spheres of size between s and $s + ds$, in unit volume, is

$$\frac{4n}{\sqrt{\pi}} x^2 e^{-x^2} dx$$

where $x = s/\sigma$.

Hence, the mean velocity U is given by

$$U = \frac{16}{\pi} \int_0^\infty \int_0^\infty v x^2 y^2 e^{-x^2 - y^2} dx \, dy$$

Now,

$$v = \sqrt{\tfrac{2}{3}} \cdot uy, \quad \text{and} \quad s^3 u^2 = \varsigma^3 V^2, \quad \text{or} \quad x^3 u^2 = \left(\frac{\varsigma}{\sigma}\right)^3 V^2 = \frac{4}{\sqrt{\pi}} V^2$$

so that

$$u = \frac{2}{\pi^{\frac{1}{4}}} x^{-\frac{3}{2}} V, \quad \text{and} \quad v = \frac{2\sqrt{2}}{\pi^{\frac{1}{4}} \sqrt{3}} x^{-\frac{3}{2}} y V$$

Therefore,

$$U = \frac{32\sqrt{2}}{\pi^{\frac{5}{4}} \sqrt{3}} V \int_0^\infty \int_0^\infty x^{\frac{1}{2}} y^3 e^{-x^2 - y^2} dx \, dy$$

But

$$\int_0^\infty y^3 e^{-y^2} dy = \tfrac{1}{2}, \quad \text{and} \quad \int_0^\infty x^{\frac{1}{2}} e^{-x^2} dx = 2 \int_0^\infty z^2 e^{-z^4} dz$$

therefore,

$$U = \frac{32\sqrt{2}}{\pi^{\frac{5}{4}} \sqrt{3}} V \int_0^\infty z^2 e^{-z^4} dz$$

This integral may be evaluated as follows :—

Let
$$J = \int_0^\infty x^2 e^{-x^4} dx$$

$$4J^2 = 4 \int_0^\infty \int_0^\infty x^2 y^2 e^{-x^4 - y^4} dx\, dy$$

$$= \int_0^\infty \int_0^{\frac{1}{2}\pi} r^4 \sin^2 2\theta\, e^{-r^4(1 - \frac{1}{2}\sin^2 2\theta)}\, r\, dr\, d\theta$$

$$= \tfrac{1}{4} \int_0^\infty \int_0^\pi z^2 \sin^2 \phi\, e^{-z^2(1 - \frac{1}{2}\sin^2 \phi)}\, dz\, d\phi$$

$$= \tfrac{1}{2} \int_0^\infty \int_0^{\frac{1}{2}\pi} \frac{\sin^2 \phi}{(1 - \frac{1}{2}\sin^2 \phi)^{\frac{3}{2}}}\, t^2 e^{-t^2}\, dt\, d\phi$$

$$= \tfrac{1}{4}\pi^{\frac{1}{2}} \int_0^{\frac{1}{2}\pi} \frac{\frac{1}{2}\sin^2 \phi}{(1 - \frac{1}{2}\sin^2 \phi)^{\frac{3}{2}}}\, d\phi$$

$$= \tfrac{1}{4}\pi^{\frac{1}{2}} \left[\int_0^{\frac{1}{2}\pi} \frac{d\phi}{(1 - \frac{1}{2}\sin^2 \phi)^{\frac{3}{2}}} - F \right]$$

Now,
$$\int_0^{\frac{1}{2}\pi} \frac{d\phi}{(1 - k^2 \sin^2 \phi)^{\frac{3}{2}}} = \frac{E}{k'^2}$$

and in the present case $k^2 = k'^2 = \frac{1}{2}$.

Hence,
$$J = \tfrac{1}{4}\pi^{\frac{1}{4}} [2E - F]^{\frac{1}{2}} *$$

where E and F are the complete elliptic integrals with modulus sin 45°.

In Legendre's Tables, we find

$$E = 1\cdot350644, \quad F = 1\cdot854075, \quad \text{and} \quad 2E - F = \cdot847213$$

Then,
$$\frac{U}{V} = \frac{8}{\pi} \sqrt{\tfrac{2}{3}} \sqrt{(2E - F)} = 1\cdot91377$$

The mean free path

$$L = UT = 1\cdot9138\, VT = \frac{1\cdot9138}{5\cdot3318} \frac{m/\rho}{(2s)^2}$$

and thus
$$L = \frac{1}{2\cdot786} \frac{m/\rho}{(2s)^2} \quad \dots\dots\dots\dots\dots\dots(51)$$

If the spheres had all been of the same size, we should have had

$$L_0 = \frac{m/\rho}{\pi (2s)^2 \sqrt{2}} = \frac{1}{4\cdot44} \frac{m/\rho}{(2s)^2} \quad \dots\dots\dots\dots\dots(52)$$

Hence, finally from (49) to (52), if there be a number of spherical meteorites, of uniform density, of all sizes with radii grouped about a mean

* I owe this to Mr Forsyth.

radius according to the law of error, and if S be the *diameter* of the meteorite of mean mass m, and ρ be the density of the distribution of meteorites in space, and $\frac{1}{2}mV^2$ their mean kinetic energy of agitation, then the mean free path L, mean free time T, and mean velocity U are given by

$$L = \frac{1}{2\cdot786}\frac{m/\rho}{S^2} = \tfrac{5}{14}\frac{m/\rho}{S^2}\ \text{nearly}$$

$$T = \frac{1}{5\cdot332}\frac{m/\rho}{VS^2} = \tfrac{3}{16}\frac{m/\rho}{VS^2}\ \text{nearly}\quad \Bigg\}\ \dots\dots\dots\dots(53)$$

$$U = 1\cdot9138V = 2V\ \text{nearly}$$

Also the mean free path is about $\tfrac{7}{11}$ths, and the mean free time about $\tfrac{3}{4}$ of that which would have held if the meteorites had all been of the same size m and had had the same mean kinetic energy $\frac{1}{2}mV^2$.

§ 15. *On the Variation of Mean Frequency of Collision and Mean Free Path for the several sizes of balls.*

Each size of ball has its own mean frequency of collision, and mean free path, and it is well to trace how the total means evaluated in the last section are made up.

We have already seen in (48) that (substituting for σ its value in terms of ς) the number of collisions per unit time and volume between balls of sizes s to $s + ds$ and balls of sizes s' to $s' + ds'$ is

$$\frac{64n^2}{\pi^{\frac{5}{4}}}(\tfrac{2}{3}\pi V^2)^{\frac{1}{2}}\cdot\frac{\pi^{\frac{1}{3}}}{2^{\frac{4}{3}}}\varsigma^2(x+y)^2(x^3+y^3)^{\frac{1}{2}}(xy)^{\frac{1}{2}}e^{-x^2-y^2}\,dx\,dy$$

where $x = s/\sigma,\ y = s'/\sigma$.

But the number of balls of size s to $s + ds$ in unit volume is

$$\frac{4n}{\sqrt{\pi}}x^2 e^{-x}\,dx$$

Hence, the mean frequency of collision for a ball of size s with all others is

$$\frac{\pi^{\frac{1}{2}}}{4n}\cdot\frac{64n^2}{\pi^{\frac{5}{4}}}(\tfrac{2}{3}\pi V^2)^{\frac{1}{2}}\frac{\pi^{\frac{1}{3}}}{2^{\frac{4}{3}}}\varsigma^2\int_0^\infty(x+y)^2(x^3+y^3)^{\frac{1}{2}}x^{-\frac{3}{2}}y^{\frac{1}{2}}e^{-y^2}\,dy$$

Now,

$$x^{-\frac{3}{2}} = \left(\frac{\sigma}{s}\right)^{\frac{3}{2}} = \tfrac{1}{2}\pi^{\frac{1}{4}}\cdot\left(\frac{\varsigma}{s}\right)^{\frac{3}{2}}$$

Therefore, if we write $1/\tau$ for the frequency of collision of a ball of size s with all others, we have

$$\frac{1}{\tau} = \frac{2^{\frac{1}{6}}\cdot\pi^{\frac{1}{3}}}{3^{\frac{1}{2}}}\cdot\frac{V(2\varsigma)^2}{m/\rho}\left(\frac{\varsigma}{s}\right)^{\frac{3}{2}}\int_0^\infty(x+y)^2(x^3+y^3)^{\frac{1}{2}}y^{\frac{1}{2}}e^{-y^2}\,dy$$

Now, the mean frequency for all sizes is given by

$$\frac{1}{T} = 5\cdot3318 \cdot \frac{V(2\varsigma)^2}{m/\rho}$$

Hence, $$\frac{T}{\tau} = \frac{1}{5\cdot3318} \cdot \frac{2^{\frac{1}{6}} \cdot \pi^{\frac{1}{3}}}{3^{\frac{1}{2}}} \cdot \left(\frac{\varsigma}{s}\right)^{\frac{3}{2}} \int_0^\infty (x+y)^2 (x^3+y^3)^{\frac{1}{2}} y^{\frac{1}{2}} e^{-y^2} dy$$

$$= \cdot1780 \cdot \left(\frac{\varsigma}{s}\right)^{\frac{3}{2}} \int_0^\infty (x+y)^2 (x^3+y^3)^{\frac{1}{2}} y^{\frac{1}{2}} e^{-y^2} dy \quad\ldots\ldots\ldots\ldots(54)$$

The integral involved here cannot in general be determined algebraically; but, if x be very small, or very great, we can find an approximate value for it.

If x be very small, the integral becomes

$$\int_0^\infty y^4 e^{-y^2} dy = \tfrac{3}{8}\sqrt{\pi}, \quad\text{and}\quad \frac{T}{\tau} = \cdot118 \left(\frac{\varsigma}{s}\right)^{\frac{3}{2}}$$

If x be very large, the integral becomes

$$x^{\frac{7}{2}} \int_0^\infty y^{\frac{1}{2}} e^{-y^2} dy = 2x^{\frac{7}{2}} \int_0^\infty z^2 e^{-z^4} dz$$

Now, $$x^{\frac{7}{2}} = \left(\frac{s}{\sigma}\right)^{\frac{7}{4}} = \frac{2^{\frac{7}{6}}}{\pi^{\frac{7}{12}}} \left(\frac{s}{\varsigma}\right)^{\frac{7}{2}} \quad\text{and}\quad \int_0^\infty z^2 e^{-z^4} dz = \tfrac{1}{4}\pi^{\frac{1}{4}}(2E-F)^{\frac{1}{2}}$$

Therefore, the integral becomes

$$\frac{2^{\frac{4}{3}}}{\pi^{\frac{1}{3}}} (2E-F)^{\frac{1}{2}} \left(\frac{s}{\varsigma}\right)^{\frac{7}{2}}$$

and with the known values of E and F this gives us

$$\frac{T}{\tau} = \cdot282 \left(\frac{s}{\varsigma}\right)^2$$

For intermediate values of s recourse must be had to quadratures for evaluating the integral.   I have therefore determined, by a rough numerical process, sufficient values of the integral to render possible the drawing of a curve for the values of T/τ for all values of s.   The following table gives the results for the integral

$$\int_0^\infty (x+y)^2 (x^3+y^3)^{\frac{1}{2}} y^{\frac{1}{2}} e^{-y^2} dy$$

which may be denoted by K :—

	K
$s = \tfrac{1}{2}\varsigma$	1·71
$s = \tfrac{3}{4}\varsigma$	2·90
$s = \varsigma$	4·94
$s = \tfrac{3}{2}\varsigma$	12·97
$s = 2\varsigma$	28·75

If these values be introduced in the formula

$$\frac{T}{\tau} = \cdot 1780 \, K \left(\frac{\varsigma}{s}\right)^{\frac{3}{2}}$$

we obtain

	T/τ
$s = \frac{1}{2}\varsigma$	$\cdot 86$
$s = \frac{3}{4}\varsigma$	$\cdot 80$
$s = \varsigma$	$\cdot 88$
$s = \frac{3}{2}\varsigma$	$1\cdot 26$
$s = 2\varsigma$	$1\cdot 81$

These values are used for forming the curve, entitled "frequency of collision," in Fig. 1 below, and they are supplemented by the values found above for T/τ, in the case where s/ς is either very small or very large.

The frequency becomes infinite when the balls are infinitely small, because of the infinite velocity with which they move, and again infinite for infinitely large balls, because of their infinite size. But it must be remembered that there are infinitely few balls of these two limiting sizes.

We have now to consider the mean free path, say λ, for the several sizes.

If u^2 be the mean square of velocity for the size s, the mean velocity for that size is $u\sqrt{(8/3\pi)}$, by the ordinary kinetic theory.

From the constancy of mean kinetic energy for all sizes, we have

$$s^3 u^2 = \varsigma^3 V^2$$

so that the mean velocity for size s is

$$V(\varsigma/s)^{\frac{3}{2}}\sqrt{(8/3\pi)}$$

But, if U be the mean velocity, and L the mean free path, and T the mean free time for all sizes together, we have

$$V = \frac{U}{1\cdot 9138} = \frac{1}{1\cdot 9138}\frac{L}{T}$$

Therefore, the mean velocity for size s is

$$\frac{\sqrt{(8/3\pi)}}{1\cdot 9138}\left(\frac{\varsigma}{s}\right)^{\frac{3}{2}}\frac{L}{T} = \cdot 4815\left(\frac{\varsigma}{s}\right)^{\frac{3}{2}}\frac{L}{T}$$

But the mean velocity for size s is λ/τ; hence,

$$\frac{\lambda}{L} = \cdot 4815\left(\frac{\varsigma}{s}\right)^{\frac{3}{2}}\frac{\tau}{T} = \frac{4815}{1780}\cdot\frac{1}{K}$$

$$= \frac{2\cdot 705}{K}$$

When s is very small, we find $\lambda/L = 4$, and, when s is very large, $\lambda/L = 1\cdot7\,(s/s)^{\frac{7}{2}}$. Thus, for small values of s, the mean free path reaches a constant limit 4, and for large values it becomes infinitely small.

The intermediate values, sufficient for drawing a curve, are given in the following short table:—

	λ/L
$s = \tfrac{1}{2}s$	1·58
$s = \tfrac{3}{4}s$	·93
$s = s$	·55
$s = \tfrac{3}{2}s$	·21
$s = 2s$	·09

These values are set out in the annexed figure in the curve marked "free path," and are supplemented by the values found above for small and large values of s. The constant limit 4 falls outside the figure. The horizontal portion of the curve is asymptotic to the s-axis.

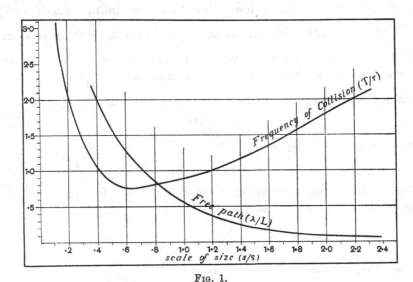

Fig. 1.

No immediate use is made of these conclusions, but it was proper to examine this point in the theory.

§ 16. *On the Sorting of Meteorites according to size and its Results.*

It is a well-known result of the kinetic theory of gases, that if a number of different gases co-exist, each gas has the same density as though it alone existed, and was subject to the resultant forces of the system; also the mean kinetic energy of agitation of each gas is the same. From this it follows that the elasticity of each gas is inversely proportional to the mass of its molecule.

Carrying on this conclusion to meteorites, we see that the elasticity of the gas formed by large meteorites is less than that for small; and, hence, there is a greater concentration of large meteorites towards the centre, and there will be a sorting according to size. The object of this section is to investigate this point.

In §§ 14 and 15, the laws of a kinetic theory were investigated when the gas consisted of molecules of all masses, grouped, according to a law of frequency, about a certain mean radius, and molecules of infinite mass were considered to be admissible, with, of course, infinite rarity. Now, if we were to continue to use that law of frequency of masses in the present investigation, we should find, as an analytical result, that the mean mass in the centre of the swarm becomes infinite. The existence of very large meteorites in sufficient numbers to give statistical constancy in a volume which is not a considerable fraction of the volume of the whole swarm is physically improbable. We shall, therefore, treat the case best by absolutely excluding very large masses. When such masses occur, they must not be treated statistically; this is a question which I hope to consider in a future paper. Had I foreseen this conclusion when the investigations of the last two sections were carried out, a different law of frequency of mass would have been assumed. But the results of those sections are amply sufficient to indicate the conclusions which would have been reached with another law of frequency, and, therefore, it does not seem worth while to recompute the results by means of a fresh series of laborious quadratures.

Any law of frequency would suffice for our purpose which excludes masses greater than a certain limit and rises to a maximum for a certain mean mass. For the present, I do not specify that law precisely, but merely assume that at some radius, which may conveniently be taken as that of the isothermal sphere, where $r = a$, the number of meteorites whose masses lie between x and $x + \delta x$ is $f(x)\,\delta x$; it is also assumed that x may range from M* to zero.

The meteorites whose masses range from x to $x + \delta x$ may be deemed to constitute a gas. Suppose that at radius r the number of its molecules per

---

* This M is not to be confused with M, the mass of the isothermal sphere.

unit volume is δn, its density δw, its pressure δp, and let the same symbols, with suffix 0, denote the same things at radius a.   Since all the partial gases are in the permanent state, they all have the same mean kinetic energy of agitation, equal to $\frac{1}{2}h$, suppose.   Throughout the isothermal sphere, this h is constant, and equal, say, to h_0, but varies with the radius in the adiabatic layer over it.   It follows, therefore, that the mean square of the velocity of the particular partial gas x to $x + \delta x$ is equal to h/x, and the relation between δp and δw is

$$\delta p = \tfrac{1}{3}\frac{h}{x}\,\delta w$$

Let $-\chi$ be the excess of the gravitation potential of the *whole* swarm at radius r above its value at radius a.

Then, since each partial gas behaves as though it existed by itself, the equation of hydrostatic equilibrium of the partial gas x to $x + \delta x$ is

$$\frac{1}{\delta w}\frac{d\delta p}{dr} + \frac{d\chi}{dr} = 0$$

The investigation must now divide into two, according as whether we are considering the isothermal sphere or the adiabatic layer.

The Isothermal Sphere.

Here we have h a constant and equal to h_0, and δp varies as δw, so that

$$\tfrac{1}{3}\frac{h_0}{x}\log\frac{\delta w}{\delta w_0} = -\chi$$

or

$$\frac{\delta w}{\delta w_0} = e^{-3\chi x/h_0}$$

Now it is obvious that $\delta n/\delta n_0 = \delta w/\delta w_0$; and, therefore,

$$\delta n = e^{-3\chi x/h_0}\,\delta n_0$$

But, by the definition of $f(x)$,

$$\delta n_0 = f(x)\,\delta x$$

hence,

$$\delta n = e^{-3\chi x/h_0}\,f(x)\,\delta x \quad\ldots\ldots\ldots\ldots\ldots\ldots\ldots(55)$$

This is the law of frequency of mass x to $x + \delta x$ at radius r.

Now, if m, m_0 be the mean masses at radii r and a respectively,

$$m = \frac{\displaystyle\int_0^M x e^{-3\chi x/h_0} f(x)\,dx}{\displaystyle\int_0^M e^{-3\chi x/h_0} f(x)\,dx} \quad\ldots\ldots\ldots\ldots\ldots\ldots\ldots(56)$$

and, if we put $\chi = 0$, we obtain m_0 from the same formula.

It is also clear that, if w be the total density of the swarm at radius r,

$$w = \int x\,dn = \int_0^M x e^{-3\chi x/h_0} f(x)\,dx \quad\ldots\ldots\ldots\ldots\ldots(57)$$

By the definition of χ, and in consequence of the supposed spherical arrangement of matter, we have

$$\chi = \int_a^r \frac{1}{r^2} \left(\int_0^r 4\pi\mu wr^2 dr \right) dr$$

If this value were substituted in (57), we should obtain a very complicated differential equation to determine w, the solution of which is hopelessly difficult. We may, however, assume without much error that the w in the integral expressing χ is the density of meteorites, all of which are of the same size m', and which are agitated with mean kinetic energy $\frac{1}{2}h_0$. If this density be written w, we then clearly have

$$\chi = -\frac{h_0}{3m'} \log \frac{w}{w_0}$$

The values of w and w_0 may be extracted from Table III. of solutions in § 6.

Then we have $\qquad -\frac{3\chi x}{h_0} = \frac{x}{m'} \log \frac{w}{w_0} = qx$, suppose

where q is rigorously equal to $-3\chi/h_0$; but for computing the approximate value $(1/m') \log (w/w_0)$ is to be employed.

In order to proceed to the evaluation of the mean mass at various distances, we must assume some form for $f(x)$.

I assume, then, that $\qquad f(x) = \frac{6n_0}{M^3} x(M-x)$

It is easy to show that

$$\int_0^M f(x)\, dx = n_0, \quad \text{and} \quad \frac{1}{n_0} \int_0^M x f(x)\, dx = \frac{1}{2}M$$

Hence, the mean mass $m_0 = \frac{1}{2}M$, and the maximum frequency is for masses equal to m_0.

Then, by (56), we have for the mean mass at radius r

$$m = \frac{\displaystyle\int_0^M x^2 (M-x) e^{qx}\, dx}{\displaystyle\int_0^M x (M-x) e^{qx}\, dx}$$

But

$$\left.\begin{aligned}
\int_0^M x^2 (M-x) e^{qx}\, dx &= \frac{1}{q^4} \left[e^{Mq} (M^2 q^2 - 4Mq + 6) - 2(Mq+3) \right] \\
\int_0^M x (M-x) e^{qx}\, dx &= \frac{1}{q^3} \left[e^{Mq} (Mq-2) + (Mq+2) \right]
\end{aligned}\right\} \dots(58)$$

It may be remarked that, if Mq be treated as small, we have the first of these integrals equal to $\frac{1}{12}M^4(1+\frac{3}{5}Mq)$, and the second equal to $\frac{1}{6}M^3(1+\frac{1}{2}Mq)$, and the ratio of the first to the second is $\frac{1}{2}M(1+\frac{1}{10}Mq)$.

In order to evaluate m, we proceed to introduce the approximate value for q.

Now $$q = \frac{1}{m'} \log \frac{w}{w_0}, \quad \text{and} \quad e^{Mq} = \left(\frac{w}{w_0}\right)^{M/m'}$$

then, writing for brevity, $$P = \log \left(\frac{w}{w_0}\right)^{M/m'}$$

we have $$\frac{m}{\frac{1}{2}M} = \frac{2}{P} \cdot \frac{\left(\frac{w}{w_0}\right)^{M/m'} (P^2 - 4P + 6) - 2(P+3)}{\left(\frac{w}{w_0}\right)^{M/m'} (P-2) + (P+2)} \quad \dots\dots\dots\dots(59)$$

Also, if P be small, the approximate result is

$$\frac{m}{\frac{1}{2}M} = 1 + \tfrac{1}{10}P$$

Before proceeding to give numerical values for the fall of mean mass as we proceed outwards from the centre of the isothermal sphere, we must consider

The Adiabatic Layer.

In this case we assume, as before, that the ratio of the two specific heats is $1\frac{2}{3}$, and we therefore have for the relationship between δp and δw at radius r,

$$\frac{\delta p}{\delta p_0} = \left(\frac{\delta w}{\delta w_0}\right)^{\frac{5}{3}}$$

Hence, $$\frac{1}{\delta w}\frac{d\,\delta p}{dr} = \tfrac{5}{6}\left[\frac{3\,\delta p_0}{\delta w_0}\right]\frac{d}{dr}\left(\frac{\delta w}{\delta w_0}\right)^{\frac{2}{3}}$$

But, since δp_0, δw_0 apply to the radius a where $h = h_0$, a constant,

$$\frac{3\,\delta p_0}{\delta w_0} = \frac{h_0}{x}$$

Thus, in the adiabatic layer the equation of hydrostatic equilibrium is

$$\tfrac{5}{6}\frac{h_0}{x}\frac{d}{dr}\left(\frac{\delta w}{\delta w_0}\right)^{\frac{2}{3}} + \frac{d\chi}{dr} = 0$$

whence, $$\chi = \tfrac{5}{6}\frac{h_0}{x}\left(1 - \left(\frac{\delta w}{\delta w_0}\right)^{\frac{2}{3}}\right) \dots\dots\dots\dots\dots(60)$$

or $$\delta w = \delta w_0 \left[1 - \frac{6\chi x}{5h_0}\right]^{\frac{3}{2}}$$

The investigation now follows a line parallel to that taken before.

We have $\delta n/\delta n_0 = \delta w/\delta w_0$, and $\delta n_0 = f(x)\,\delta x$, so that

$$\delta n = \left(1 - \frac{6\chi x}{5h_0}\right)^{\frac{3}{2}} f(x)\,\delta x$$

This is the law of frequency of masses lying between x and $x + \delta x$ at radius r.

As δn can never be negative, we see that there can be no mass greater than $\frac{5}{6}h_0/\chi$; and, if M be the greatest positive value of the expression $f(x)$, there can be no mass greater than the smaller of $\frac{5}{6}h_0/\chi$ or M.

Thus, if m be the mean mass at radius r,

$$m = \frac{\int_0^a x \left(1 - \frac{6\chi x}{5h_0}\right)^{\frac{3}{2}} f(x)\, dx}{\int_0^a \left(1 - \frac{6\chi x}{5h_0}\right)^{\frac{3}{2}} f(x)\, dx} \quad\quad\dots\dots\dots\dots\dots(61)$$

where a is the smaller of $\frac{5}{6}h_0/\chi$ and M.

If we put $\chi = 0$ in (61), we obtain m_0, the mean mass at radius a.

It is clear also that, if w be the total density of the swarm at radius r,

$$w = \int x\, dn = \int_0^a x \left(1 - \frac{6\chi x}{5h_0}\right)^{\frac{3}{2}} f(x)\, dx \dots\dots\dots\dots\dots(62)$$

By definition of χ, and in consequence of the supposed spherical arrangement of matter, we have

$$\chi = \int_a^r \frac{1}{r^2} \left(\int_0^r 4\pi \mu w r^2\, dr\right) dr$$

If this value were substituted in (62), we might obtain a complicated differential equation for w. It is clear, however, that an adequate approximation may be obtained by assuming that the w in the integral expressing χ is the density of meteorites, all of which are of the same size m', arranged in a layer in convective equilibrium, and with kinetic energy of agitation at the limit $r = a$ equal to $\frac{1}{2}h_0$.

If this density be written w, and if v^2 be the mean square of velocity of agitation at radius r, we have, by (60), and in consequence of the relationship $(w/w_0)^{\frac{2}{3}} = (v/v_0)^2$,

$$\chi = \frac{5}{6}\frac{h_0}{m'}\left(1 - \left(\frac{w}{w_0}\right)^{\frac{2}{3}}\right) = \frac{5}{6}v_0^2\left(1 - \frac{v^2}{v_0^2}\right)$$

and

$$\frac{6}{5}\frac{\chi x}{h_0} = \frac{x}{m'}\left(1 - \frac{v^2}{v_0^2}\right)$$

Let

$$\frac{1}{\beta} = \frac{1}{m'}\left(1 - \frac{v^2}{v_0^2}\right)$$

for brevity; then, adopting the law of frequency $f(x) = \dfrac{6n_0}{M^3} x (M - x)$, as before, we have for the mean mass at radius r

$$m = \frac{\displaystyle\int_0^a x^2 (M - x)\left(1 - \frac{x}{\beta}\right)^{\frac{3}{2}} dx}{\displaystyle\int_0^a x (M - x)\left(1 - \frac{x}{\beta}\right)^{\frac{3}{2}} dx} \qquad\qquad \dots\dots\dots\dots\dots(63)$$

where a is equal to the smaller of M and β.

The solution now becomes different according as M or β is the smaller.

First, suppose M is the smaller. Then the limits of integration are M and 0.

If we put
$$z = 1 - \frac{x}{\beta}$$

$$x^n \left(1 - \frac{x}{\beta}\right)^{\frac{3}{2}} dx = - \beta^{n+1} z^{\frac{3}{2}} \left(1 - nz + \frac{n \cdot n - 1}{1 \cdot 2} z^2 - \dots\right) dz$$

so that the numerator and denominator of m are easily integrable.

If now we write
$$Q = 1 - \frac{M}{\beta}$$

$$\int_0^M x^2 (M - x)\left(1 - \frac{x}{\beta}\right)^{\frac{3}{2}} dx$$

$$= 2\beta^4 \left[\tfrac{1}{5}\left(\frac{M}{\beta} - 1\right)(1 - Q^{\frac{5}{2}}) - \tfrac{1}{7}\left(2\frac{M}{\beta} - 3\right)(1 - Q^{\frac{7}{2}})\right.$$

$$\left. + \tfrac{1}{9}\left(\frac{M}{\beta} - 3\right)(1 - Q^{\frac{9}{2}}) + \tfrac{1}{11}(1 - Q^{\frac{11}{2}})\right]$$

$$= 2\beta^4 \left[\frac{8}{7 \cdot 9 \cdot 11} - \frac{8}{5 \cdot 7 \cdot 9} Q + \frac{2}{5 \cdot 7} Q^{\frac{7}{2}} - \frac{4}{7 \cdot 9} Q^{\frac{9}{2}} + \frac{2}{9 \cdot 11} Q^{\frac{11}{2}}\right]$$

$$\int_0^M x (M - x)\left(1 - \frac{x}{\beta}\right)^{\frac{3}{2}} dx$$

$$= 2\beta^3 \left[\tfrac{1}{5}\left(\frac{M}{\beta} - 1\right)(1 - Q^{\frac{5}{2}}) - \tfrac{1}{7}\left(\frac{M}{\beta} - 2\right)(1 - Q^{\frac{7}{2}}) - \tfrac{1}{9}(1 - Q^{\frac{9}{2}})\right]$$

$$= 2\beta^3 \left[\frac{2}{7 \cdot 9} - \frac{2}{5 \cdot 7} Q + \frac{2}{5 \cdot 7} Q^{\frac{7}{2}} - \frac{2}{7 \cdot 9} Q^{\frac{9}{2}}\right]$$

Then, since $\beta = (1 - Q)/M$, we have

$$\frac{m}{M} = \frac{\tfrac{4}{11} - \tfrac{4}{5}Q + \tfrac{9}{5}Q^{\frac{7}{2}} - 2Q^{\frac{9}{2}} + \tfrac{7}{11}Q^{\frac{11}{2}}}{(1 - Q)(1 - \tfrac{9}{5}Q + \tfrac{9}{5}Q^{\frac{7}{2}} - Q^{\frac{9}{2}})} \qquad \dots\dots\dots\dots(64)$$

This expression has a high order of indeterminateness when $Q = 1$, but I find that when Q is nearly equal to unity

$$\frac{m}{M} = \tfrac{1}{2}[1 - \tfrac{3}{10}(1 - Q^{\frac{1}{2}})] \text{ nearly} \dots\dots\dots\dots\dots(65)$$

Thus, the mean mass is $\frac{1}{2}M$ where $r = a$, which we know to be correct.

Secondly, suppose that β is smaller than M. Then effecting the integrations in the same manner as before, we have

$$\int_0^\beta x^2 (M - x) \left(1 - \frac{x}{\beta}\right)^{\frac{3}{2}} dx = 2\beta^4 \left[\frac{1}{5}\left(\frac{M}{\beta} - 1\right) - \frac{1}{7}\left(\frac{2M}{\beta} - 3\right) + \frac{1}{9}\left(\frac{M}{\beta} - 3\right) + \frac{1}{11}\right]$$

$$= \frac{2 \cdot 8}{5 \cdot 7 \cdot 9} \beta^4 \left(\frac{M}{\beta} - \frac{6}{11}\right)$$

$$\int_0^\beta x (M - x) \left(1 - \frac{x}{\beta}\right)^{\frac{3}{2}} dx = 2\beta^3 \left[\frac{1}{5}\left(\frac{M}{\beta} - 1\right) - \frac{1}{7}\left(\frac{M}{\beta} - 2\right) - \frac{1}{9}\right]$$

$$= \frac{2 \cdot 2}{5 \cdot 7} \beta^3 \left[\frac{M}{\beta} - \frac{4}{9}\right]$$

Therefore
$$m = \frac{4}{9}\beta \cdot \frac{\dfrac{M}{\beta} - \frac{6}{11}}{\dfrac{M}{\beta} - \frac{4}{9}}$$

or
$$\frac{m}{M} = \frac{4}{9} \cdot \frac{\dfrac{m'}{M}}{1 - \dfrac{v^2}{v_0^2}} : \frac{\dfrac{M}{m'}\left(1 - \dfrac{v^2}{v_0^2}\right) - \frac{6}{11}}{\dfrac{M}{m'}\left(1 - \dfrac{v^2}{v_0^2}\right) - \frac{4}{9}} \quad \ldots\ldots\ldots\ldots(66)$$

In order to compute from the formulæ, (59), (64), (66), it is necessary to make an assumption as to the value of m' the mass of the meteorites of uniform size whose arrangement of density is supposed to be the same as that of the heterogeneous meteorites.

We have supposed that the law of frequency of masses is known at radius a, and that the mean mass is there equal to $\frac{1}{2}M$. Now, inside that radius the larger masses are more frequent, and outside it the smaller masses. I suppose, then, that throughout the isothermal sphere m' lies half-way between m_0 or $\frac{1}{2}M$ and the maximum mass M, and in the adiabatic layer that it lies half-way between m_0 or $\frac{1}{2}M$ and the minimum mass 0.

Thus, inside I take $m' = \frac{3}{4}M$, and outside $m' = \frac{1}{4}M$.

As we only want to consider the general nature of the sorting process, these assumptions will suffice. It may also be remarked that a large variation of m' is required to make any considerable difference in the numerical results.

We now have—

In the isothermal sphere (where $w_0 = \frac{1}{3}\rho$),

$$\frac{M}{m'} = \frac{4}{3}, \qquad P = \log_e \left(\frac{w}{\frac{1}{3}\rho}\right)^{\frac{4}{3}}, \qquad \frac{1}{2}M = m_0$$

In the adiabatic layer,

$$\frac{M}{m'} = 4, \quad Q = 1 - \frac{M}{m'}\left(1 - \frac{v^2}{v_0^2}\right) = 4\frac{v^2}{v_0^2} - 3$$

Thus, our formulæ are:—

In the isothermal sphere, from (59),

$$\frac{m}{m_0} = \frac{2}{P} \cdot \frac{\left(\frac{w}{\frac{1}{3}\rho}\right)^{\frac{4}{3}}(P^2 - 4P + 6) - 2(P + 3)}{\left(\frac{w}{\frac{1}{3}\rho}\right)^{\frac{4}{3}}(P - 2) + (P + 2)} \quad \dots\dots\dots\dots(67)$$

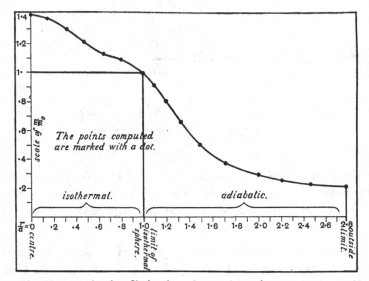

The points computed are marked with a dot.

FIG. 2.   Diagram showing diminution of mean mass from centre to outside.

In the adiabatic layer,

when $\dfrac{v^2}{v_0^2} > \frac{3}{4}$, from (64),

$$\frac{m}{m_0} = \frac{\frac{4}{11} - \frac{4}{5}Q + \frac{9}{5}Q^{\frac{7}{5}} - 2Q^{\frac{9}{5}} + \frac{7}{11}Q^{\frac{11}{5}}}{\frac{1}{2}(1 - Q)(1 - \frac{9}{5}Q + \frac{9}{5}Q^{\frac{7}{5}} - Q^{\frac{9}{5}})} \quad \dots\dots\dots\dots(68)$$

when $\dfrac{v^2}{v_0^2} < \frac{3}{4}$, from (66),

$$\frac{m}{m_0} = \frac{2}{9(1 - v^2/v_0^2)} \cdot \frac{\frac{19}{22} - v^2/v_0^2}{\frac{8}{9} - v^2/v_0^2} \quad \dots\dots\dots\dots(69)$$

The values of $w/\frac{1}{3}\rho$ and of v^2/v_0^2 are tabulated in Table III., and from these I compute—

	isothermal				$\frac{v^2}{v_0^2} > \frac{3}{4}$		$\frac{v^2}{v_0^2} < \frac{3}{4}$						
$\frac{r}{a}=0$	·16	·48	·80	1·0	1·09	1·2	1·33	1·50	1·71	2·00	2·21	2·46	2·79
$\frac{m}{m_0}=1·41$	1·38	1·22	1·11	1·0	·92	·83	·66	·49	·38	·30	·27	·24	·22

These values (together with two others in the isothermal part) are set out in Fig. 2, and show the law of diminution of mean mass from centre to outside.

The evaluation of mean mass in the fringe (see § 13), where collisions are supposed to be non-existent, is not very difficult, although it involves some troublesome algebra. I do not give the investigation, merely remarking that it leads to almost exactly the same kind of law of diminution of mean mass as we have found in the adiabatic layer.

§ 17. *Summary.*

The first and second sections only involved arguments of a general character in which mathematical analysis was unnecessary. The reader who does not wish to concern himself with details may therefore be supposed to have passed from §§ 1 and 2 to this Summary.

In order to submit the theory to an adequate test, it is necessary to discuss some definite case of the aggregation of a swarm of meteorites, and it is obvious that the only system of which we possess any knowledge is our own. It is accordingly supposed that a number of meteorites have fallen together from a condition of wide dispersion, and have ultimately coalesced so as to leave the sun and planets as their progeny. The object of this paper is to consider the mechanical condition of the system after the cessation of any considerable supply of meteorites from outside, and before the coalescence of the swarm into a star with attendant planets.

For the sake of simplicity, the meteorites are considered to be spherical, and are treated, at least in the first instance, as being of uniform size.

It is assumed provisionally that the kinetic theory of gases may be applied for the determination of the distribution of the meteorites in space. No account being taken of the rotation of the system, the meteorites will be arranged in concentric spherical layers of equal density of distribution, and the quasi-gas, whose molecules are meteorites, being compressible, the density will be greater towards the centre of the swarm.

The elasticity of a gas depends on the kinetic energy of agitation of its molecules; and, therefore, in order to determine the law of density in the swarm, we must know the distribution of kinetic energy of agitation. It is assumed that, when the swarm comes under our notice, uniformity of distribution of energy has been attained throughout a central sphere, which is surrounded by a layer of meteorites with that distribution of kinetic energy which in a gas corresponds to convective equilibrium.   In other words, we have a quasi-isothermal sphere surrounded by what may be called an atmosphere in convective equilibrium, and with continuity of density and velocity of agitation at the sphere of separation.   Since in a gas in convective equilibrium the law connecting pressure and density is that which holds when the gas is contained in a vessel impermeable to heat, such an arrangement of gas has been called by M. Ritter "an isothermal-adiabatic sphere," and the same term is adopted here as applicable to a swarm of meteorites. The justifiability of these assumptions will be considered later.

The first problem which presents itself, then, is the equilibrium of an isothermal sphere of gas under its own gravitation.   The law of density is determined in § 4 ; but it will here suffice to remark that, if a given mass be enclosed in an envelope of given radius, there is a minimum temperature (or energy of agitation) at which isothermal equilibrium is possible.   The minimum energy of agitation is found to be such that the mean square of velocity of the meteorites is almost exactly $\frac{6}{5}$ (viz. 1·1917) of the square of the velocity of a satellite grazing the surface of the sphere in a circular orbit.

As indicated above, it is supposed that in the meteor-swarm the rigid envelope bounding the isothermal sphere is replaced by a layer of meteorites in convective equilibrium.   The law of density in the adiabatic layer is determined in § 5, and it appears that, when the isothermal sphere has minimum temperature, the mass of the adiabatic atmosphere is a minimum relatively to that of the isothermal sphere.   Numerical calculation shows, in fact, that the isothermal sphere cannot amount in mass to more than 46 per cent. of the mass of the whole isothermal-adiabatic sphere, and that the limit of the adiabatic atmosphere is at a distance equal to 2·786 times the radius of the isothermal sphere*.   A table of various quantities in such a system, at various distances from the centre, is given in Table III., § 6.

It is next proved, in § 7, that the total energy, existing in the form of energy of agitation in an isothermal-adiabatic sphere, is exactly one-half of the potential energy lost in the concentration of the matter from a condition of infinite dispersion.   This result is brought about by a continual transfer of energy from a molar to a molecular form, for a portion of the kinetic energy of a meteorite is constantly being transferred into the form of thermal energy in the volatilised gases generated on collision.   The thermal energy is then lost by radiation.

* These results had been previously discovered by M. Ritter.

It is impossible as yet to sum up all the considerations which go to justify the assumption of the isothermal-adiabatic arrangement; but it is clear that uniformity of kinetic energy of agitation in the isothermal sphere must be principally brought about by a process of diffusion. It is, therefore, interesting to consider what amount of inequality in the kinetic energy would have to be smoothed away.

The arrangement of density in the isothermal-adiabatic sphere being given, it is easy to compute what the kinetic energy would be at any part of the swarm, if each meteorite fell from infinity to the neighbourhood where we find it, and there retained all the velocity due to such fall. The variation of the square of this velocity gives an indication of the amount of inequality of kinetic energy which has to be degraded by conversion into heat and redistributed by diffusion in the attainment of uniformity. This may be called "the theoretical value of the kinetic energy"; it is tabulated in Table III., on the line called "square of velocity of satellite." It rises from zero at the centre of the sphere to a maximum, which is attained nearly half-way through the adiabatic layer, and then falls again. If the radius of the isothermal sphere be unity, then from $\frac{1}{2}$ to 2 the variations of this theoretical value of the kinetic energy are small. Since this "theoretical value of the kinetic energy" is zero at the centre, there must have been diffusion of energy from without inwards, and considerations of the same kind show that when a planet consolidates there must be a cooling of the middle strata both outwards and inwards.

We must now consider the nature of the criterion which determines whether the hydrodynamical treatment of a swarm of meteorites is permissible.

The hydrodynamical treatment of an ideal plenum of gas leads to the same result as the kinetic theory with regard to any phenomenon involving purely a mass, when that mass is a large multiple of the mass of a molecule; to any phenomenon involving purely a length, when the cube of that length contains a large number of molecules; and to any phenomenon involving purely a time, when that time is a large multiple of the mean interval between collisions. Again, any velocity to be justly deduced from hydrodynamical principles must be expressible as the edge of a cube containing many molecules passed over in a time containing many collisions of a single molecule; and a similar statement must hold of any other function of mass, length, and time.

Beyond these limits, we must go back to the kinetic theory itself, and in using it care must be taken that enough molecules are considered at once to impart statistical constancy to their properties.

There are limits, then, to the hydrodynamical treatment of gases, and the like must hold of the parallel treatment of meteorites.

The principal question involved in the nebular hypothesis seems to be the stability of a rotating mass of gas; but, unfortunately, this has remained up to now an untouched field of mathematical research. We can only judge of probable results from the investigations which have been made concerning the stability of a rotating mass of liquid. Now, it appears that the instability of a rotating mass of liquid first enters through the graver modes of gravitational oscillation. In the case of a rotating spheroid of revolution the gravest mode of oscillation is an elliptic deformation, and its period does not differ much from that of a satellite which revolves round the spheroid so as to graze its surface. Hence, assuming for the moment that a kinetic theory of liquids had been formulated, we should not be justified in applying the hydrodynamical method to this discussion of stability unless the periodic time of such a satellite were a large multiple of the analogue of the mean free time of a molecule of liquid*.

Carrying, then, this conclusion on to the kinetic theory of meteorites, it seems probable that hydrodynamical treatment must be inapplicable for the discussion of such a theory as the meteoric-nebular hypothesis, unless a similar relation holds good.

These considerations, although of a very general character, will afford a criterion of the applicability of hydrodynamics to the discussion of the mechanical conditions of a swarm of meteorites in the kind of problem suggested by the nebular hypothesis.

In § 9 two criteria, suggested by this line of thought, are found. They measure, roughly speaking, the degree of curvature of the average path pursued by a meteorite between two collisions. These two criteria, denoted D/L and A/C, will afford a measure of the applicability of hydrodynamics in the sense above indicated.

After these preliminary investigations, we have to consider what kind of meeting of two meteorites will amount to an "encounter" within the meaning of the kinetic theory. Is it possible, in fact, that two meteorites can considerably bend their paths under the influence of gravitation when they pass near one another? This question is answered in § 8, where a formula is found for the deflection of the path of each of a pair of meteorites, when, moving with their mean relative velocity, they graze past one another without striking. It appears from the formula that, unless they have the dimensions of small planets, the mutual gravitational influence is practically insensible. Hence, nothing short of absolute impact is to be considered an encounter in the kinetic theory; and what is called the radius of "the sphere of action" is simply the distance between the centres of a pair when they graze, and is,

---

* If the molecules of liquid describe orbits about one another, the analogue would probably be the mean periodic time of one molecule about another.

therefore, the sum of their radii, or, if of uniform size, the diameter of one of them.

The next point to consider is the mass and size which must be attributed to the meteorites.

The few samples which have been found on the earth prove that no great error can be committed if the average density of a meteorite be taken as a little less than that of iron, and I accordingly suppose their density to be six times that of water.

Undoubtedly, in a swarm of meteorites all sizes exist (a supposition considered hereafter); for, even if originally of one uniform size, they would, by subsequent fracture, be rendered diverse. But in the first consideration of the problem they have been treated as of uniform size, and, as actual average sizes are nearly unknown, results are given in the numerical table for meteorites weighing $3\frac{1}{8}$ grammes. By merely shifting the decimal point one, two, or three places to the right the results become applicable to meteorites weighing $3\frac{1}{8}$ kilogrammes, $3\frac{1}{8}$ tonnes, 3125 tonnes, and so on.

It is known that meteorites are actually of irregular shapes, but certainly no material error can be incurred when we treat them as being spheres.

The object of all these investigations is to apply the formulæ to a concrete example. The mass of the system is therefore taken as equal to that of the sun, and the limit of the swarm at any arbitrary distance from the present sun's centre. The theory is, of course, most severely tested the wider the dispersion of the swarm; and, accordingly, in the numerical example the outside limit of the solar swarm is taken at $44\frac{1}{2}$ times the earth's distance from the sun, or further beyond the planet Neptune than Saturn is from the sun. This assumption makes the limit of the isothermal sphere at distance 16, about half-way between Saturn and Uranus.

The results, applicable to meteorites of $3\frac{1}{8}$ grammes, are exhibited in Table IV., § 10.

The velocity of mean square in the isothermal sphere is $\sqrt{(6/5)}$ of the linear velocity of a planet at distance 16, revolving about a central body with a mass equal to 46 per cent. of that of the sun, viz., $5\frac{1}{4}$ kilometres per second; and in the adiabatic layer it diminishes down to zero at distance $44\frac{1}{2}$. This velocity is independent of the size of the meteorites.

The mean free path between collisions ranges from 42,000 kilometres at the centre, to 1,300,000 kilometres at radius 16, and to infinity at radius $44\frac{1}{2}$. The mean interval between collisions ranges from a tenth of a day at the centre, to three days at radius 16, and to infinity at radius $44\frac{1}{2}$. The criterion D/L ranges from $\frac{1}{60000}$ at the distance of the asteroids, to $\frac{1}{3600}$ at radius 16, and to infinity at radius $44\frac{1}{2}$. The criterion A/C is somewhat smaller than

D/L. All these quantities are ten times as great for meteorites of $3\frac{1}{8}$ kilogrammes, and a hundred times as great for meteorites of $3\frac{1}{8}$ tonnes.

From a consideration of the table it appears that, with meteorites of $3\frac{1}{8}$ kilogrammes, the collisions are sufficiently frequent, even beyond the orbit of Neptune, to allow the kinetic theory to be applicable in the sense explained. But, if the meteorites weigh $3\frac{1}{8}$ tonnes, the criteria cease to be very small about distance 24; and, if they weigh 3125 tonnes, they cease to be very small at about the orbit of Jupiter.

It may be concluded, then, that, as far as frequency of collision is concerned, the hydrodynamical treatment of a swarm of meteorites is justifiable.

Although these numerical results are necessarily affected by the conjectural values of the mass and density of the meteorites, yet it was impossible to arrive at any conclusion whatever as to the validity of the theory without numerical values, and such a discussion as the above was therefore necessary. If the particular values used are not such as to commend themselves to the judgment of the reader, it is easy to substitute others in the formulæ, and so submit the theory to another test.

I now pass on to consider some results of this view of a swarm of meteorites, and to consider the justifiability of the assumption of an isothermal-adiabatic arrangement of density.

With regard to the uniformity of distribution of kinetic energy in the isothermal sphere, it is important to ask whether or not sufficient time can have elapsed in the history of the system to allow of the equalisation by diffusion.

In § 11 the rate of diffusion of the kinetic energy of agitation is considered, and it is shown that, in the case of our numerical example, primitive inequalities of distribution would, in a few thousand years, be sensibly equalised over a distance some ten times as great as our distance from the sun. This result, then, goes to show that we are justified in assuming an isothermal sphere as the centre of the swarm. As, however, the swarm contracts, the rate of diffusion diminishes as the inverse $\frac{5}{2}$ power of its linear dimensions, whilst the rate of generation of inequalities of distribution of kinetic energy, through the imperfect elasticity of the meteorites, increases. Hence, in a late stage of the swarm inequalities of kinetic energy would be set up; thus, there would be a tendency to the production of convective currents, and the whole swarm would probably settle down to the condition of convective equilibrium throughout.

It may be conjectured, then, that the best hypothesis in the early stages of the swarm is the isothermal-adiabatic arrangement, and later an adiabatic sphere. It has not seemed to me worth while to discuss the latter hypothesis in detail at present.

The investigation of § 11 also gives the coefficient of viscosity of the quasi-gas, and shows that it is so great that the meteor-swarm must, if rotating, revolve nearly without relative motion of its parts, other than the motion of agitation. But, as the viscosity diminishes when the swarm contracts, this would probably not be true in the later stages of its history, and the central portion would probably rotate more rapidly than the outside. It forms, however, no part of the scope of this paper to consider the rotation of the system.

In § 12 the rate of loss of kinetic energy through imperfect elasticity is considered, and it appears that the rate estimated per unit time and volume must vary directly as the square of the quasi-pressure and inversely as the mean velocity of agitation. Since the kinetic energy lost is taken up in volatilising solid matter, it follows that the heat generated must follow the same law. The mean temperature of the gases generated in any part of the swarm depends on a great variety of circumstances, but it seems probable that its variation would be according to some law of the same kind. Thus, if the spectroscope enables us to form an idea of the temperature in various parts of a nebula, we shall at the same time obtain some idea of the distribution of density.

It has been assumed that the outer portion of the swarm is in convective equilibrium, and therefore there is a definite limit beyond which it cannot extend. Now, a medium can only be said to be in convective equilibrium when it obeys the laws of gases, and the applicability of those laws depends on the frequency of collisions. But at the boundary of the adiabatic layer the velocity of agitation vanishes, and collisions become infinitely rare. These two propositions are mutually destructive of one another, and it is impossible to push the conception of convective equilibrium to its logical conclusion. There must, in fact, be some degree of rarity of density, and of collisions, at which the statistical treatment of the medium breaks down.

I have sought to obtain some representation of the state of things by supposing that collisions never occur beyond a certain distance from the centre of the swarm. Then, from every point of the surface of the sphere, which limits the regions of collisions, a fountain of meteorites is shot out, in all azimuths and inclinations to the vertical, and with velocities grouped about a mean according to the law of error. These meteorites ascend to various heights without collision, and, in falling back on to the limiting sphere, cannonade its surface, so as to counterbalance the hydrostatic pressure at the limiting sphere.

The distribution of meteorites, thus shot out, is investigated in § 13, and it is found that near the limiting sphere the decrease in density is somewhat more rapid than the decrease corresponding to convective equilibrium. But at more remote distances the decrease is less rapid, and the density ultimately tends to vary inversely as the square of the distance from the centre.

It is clear, then, that, according to this hypothesis, the mass of the system is infinite in a mathematical sense, for the existence of meteorites with nearly parabolic and hyperbolic orbits necessitates an infinite number, if the loss of the system shall be made good by the supply*.

But, if we consider the subject from a physical point of view, this conclusion appears unobjectionable. The ejection of molecules with exceptionally high velocities from the surface of a liquid is called evaporation, and the absorption of others is called condensation. The general history of a swarm, as stated in § 2, may then be put in different words, for we may say that at first a swarm gains by condensation, that condensation and evaporation balance, and, finally, that evaporation gains the day.

If the hypothesis of convective equilibrium be pushed to its logical conclusion, we reach a definite limit to the swarm, whereas, if collisions be entirely annulled, the density goes on decreasing inversely as the square of the distance. The truth must clearly lie between these two hypotheses. It is thus certain that even the very small amount of evaporation shown by the formulæ derived from the hypothesis of no collisions must be in excess of the truth; and it may be that there are enough waifs and strays in space, ejected from other systems, to make up for the loss. Whether or not the compensation is perfect, a swarm of meteorites would pursue its evolution without being sensibly affected by a slow evaporation.

Up to this point the meteorites have been considered as of uniform size, but it is well to examine the more truthful hypothesis, that they are of all sizes, grouped about a mean according to a law of error.

It appears, from the investigation in § 14, that the larger stones move slower, the smaller ones faster; and the law is that the mean kinetic energy is the same for all sizes.

It is proved that the mean path between collisions is shorter in the proportion of 7 to 11, and the mean frequency greater in the proportion of 4 to 3, than if the meteorites were of uniform mass, equal to their mean. Hence, the previous numerical results for uniform size are applicable to non-uniform meteorites of mean mass about a third greater than the uniform mass; for example, the results for uniform meteorites of $3\frac{1}{4}$ tonnes apply to non-uniform ones of mean mass, a little over 4 tonnes.

The means here spoken of refer to all sizes grouped together, but there are a separate mean free path and a mean frequency appropriate to each size.

---

* It must also be borne in mind that the very high velocities, which occur occasionally in a medium with perfectly elastic molecules, must happen with great rarity amongst meteorites. An impact of such violence that it *ought* to generate a hyperbolic velocity will probably merely cause fracture.

These are investigated in § 15, and their various values are illustrated in Fig. 1. The horizontal scale in that figure gives the ratio of the radius of each size to the radius of the meteorite of mean mass. The vertical scales are the ratio of the mean free path of any size to that of all sizes together, and the ratio of the mean frequency for any size to that of all sizes together. The figure shows that collisions become infinitely frequent for the infinitely small ones, because of their infinite velocity; and again infinitely frequent for the infinitely large ones, because of their infinite size. There is a minimum frequency of collision for a certain size, a little less in radius than the mean, and considerably less in mass than the mean mass.

For infinitely small meteorites, the mean free path reaches a finite limit, equal to about four times the grand mean free path; but this could not be shown in the figure without a considerable extension of it upwards. For infinitely large ones, the mean free path becomes infinitely short. It must be borne in mind that there are infinitely few of the infinitely large and small meteorites.

Variety of size does not, then, so far, materially affect the results.

But a difference arises when we come to consider the different parts of the swarm. The larger meteorites, moving with smaller velocities, form a quasi-gas of less elasticity than do the smaller ones. Hence, the larger meteorites are more condensed towards the centre than are the smaller ones, or the large ones have a tendency to sink down, whilst the small ones have a tendency to rise. Accordingly, the various kinds are to some extent sorted according to size.

In § 16, an investigation is made of the mean mass of the meteorites at various distances from the centre, both inside and outside the isothermal sphere, and Fig. 2 is drawn to illustrate the law of diminution of mean mass.

It is also clear that the loss of the system through evaporation must fall more heavily on the small meteorites than on the large ones.

After the foregoing summary, it will be well to briefly recapitulate the principal conclusions which seem to be legitimately deducible from the whole investigation; and, in this recapitulation, qualifications must necessarily be omitted, or stated with great brevity.

When two meteorites are in collision, they are virtually highly elastic, although ordinary elasticity must be nearly inoperative.

A swarm of meteorites is analogous with a gas, and the laws governing gases may be applied to the discussion of its mechanical properties. This is true of the swarm from which the solar system was formed, when it extended beyond the orbit of the planet Neptune.

When the swarm was very widely dispersed, the arrangement of density and of velocity of agitation of the meteorites was that of an isothermal-adiabatic sphere.   Later in its history, when the swarm had contracted, it was probably throughout in convective equilibrium.

The actual mean velocity of the meteorites is determinable in a swarm of given mass, when expanded to a given extent.

The total energy of agitation in an isothermal-adiabatic sphere is half the potential energy lost in the concentration from a condition of infinite dispersion.

The half of the potential energy lost, which does not reappear as kinetic energy of agitation, is expended in volatilising solid matter and heating the gases produced on the impact of meteorites.   The heat so generated is gradually lost by radiation.

The amount of heat generated per unit time and volume varies as the square of the quasi-hydrostatic pressure, and inversely as the mean velocity of agitation.   The temperature of the gases volatilised probably varies by some law of the same nature.

The path of the meteorites is approximately straight, except when abruptly deflected by a collision with another.   This ceases to be true at the outskirts of the swarm, where the collisions have become rare.   The meteorites here describe orbits, under gravity, which are approximately elliptic, parabolic, and hyperbolic.

In this fringe to the swarm the distribution of density ceases to be that of a gas under gravity, and, as we recede from the centre, the density at first decreases more rapidly, and afterwards less rapidly, than if the medium were a gas.

Throughout all stages of the history of a swarm there is a sort of evaporation, by which the swarm very slowly loses in mass, but this loss is more or less counterbalanced by condensation.   In the early stages, the gain by condensation outbalances the loss by evaporation; they then equilibrate; and, finally, the evaporation may be greater than the condensation.

Throughout the swarm the meteorites are partially sorted, according to size.   As we recede from the centre, the number of small ones preponderates more and more, and, thus, the mean mass continually diminishes with increasing distance.   The loss to the system by evaporation falls principally on the smaller meteorites.

A meteor-swarm is subject to gaseous viscosity, which is greater the more widely diffused is the swarm.   In consequence of this, a widely extended swarm, if in rotation, will revolve like a rigid body, without relative movement of its parts.   Later in its history, the viscosity will, probably, not suffice to secure uniformity of rotation, and the central portion will revolve more rapidly than the outside.

The kinetic theory of meteorites may be held to present a fair approximation to the truth in the earlier stages of the evolution of the system. But ultimately the majority of the meteors must have been absorbed by the central sun and its attendant planets, and amongst the meteors which remain free the relative motion of agitation must have been largely diminished. These free meteorites—the dust and refuse of the system—probably move in clouds, but with so little remaining motion of agitation that (except, perhaps, near the perihelion of very eccentric orbits) it would scarcely be permissible to treat the cloud as in any respects possessing the mechanical properties of a gas.

The value of this whole investigation will appear very different to different minds. To some it will stand condemned, as altogether too speculative; others may think that it is better to risk error on the chance of winning truth. To me, at least, it appears that the line of thought flows in a true channel; that it may help to give a meaning to the observations of the spectroscopist; and that many interesting problems, here barely alluded to, may, perhaps, be solved with sufficient completeness to throw light on the evolution of nebulæ and of planetary systems.

19.

ON THE PERTURBATION OF A COMET IN THE NEIGHBOUR-HOOD OF A PLANET.

[Proceedings of the Cambridge Philosophical Society, Vol. VII. (1892), pp. 314—319.]

In Chapter II. of Book IX. of the *Mécanique Céleste*, Laplace considers the transformation of the orbit of a comet when it passes a large planet. His object is to show that the action of Jupiter suffices to account for the disappearance of Lexell's comet after 1779.

He remarks that if a comet passes very near to Jupiter, it will throughout a small portion of its orbit move round the planet almost as though it were unperturbed by the sun, and that both before its approach to and after its recession from the planet it will move round the sun almost as though it were unperturbed by the planet. The nature of the orbit of the comet will usually be much transformed by its encounter with the planet. It is clear then that there must be some surface surrounding the planet which separates the region, inside which the comet moves nearly round the planet, from the region in which it moves nearly round the sun. Such a surface is to be found by the comparison of the ratio of the perturbing force to the central force in the motion round the sun with its value in the motion round the planet. There is a certain surface at which this ratio will be the same in the two cases, and this is the surface required for the proposed approximate treatment of the problem.

Now it does not appear to me that Laplace makes any attempt to show that such a surface is even approximately spherical, but he assumes that what has been called "the sphere of Jupiter's activity" is a true sphere, and determines its radius by the consideration of a special case.

The object of the present note is then to treat this problem more fully than does Laplace, and to investigate the nature of the surface in question.

It will appear that whilst Laplace's result is accurate enough for the purpose for which it is intended, yet a slightly different value for the radius of the sphere of activity would be more nearly correct.

Let R, r be the radii vectores of Jupiter and of the comet, and let ρ be the distance of the comet from Jupiter.

Let ω be the angle between R and r, and θ the angle between R produced and ρ.

Let S, M, m be the masses of the sun, Jupiter and the comet.

Let P, T be the disturbing forces along and perpendicular to ρ, which act on the comet in its motion round the sun; let F be the resultant of P, T; and let C be the central force acting on the comet.

Let $\mathfrak{P}, \mathfrak{T}, \mathfrak{F}, \mathfrak{C}$ be the similar things, also with reference to ρ, in the motion of the comet round Jupiter.

Now we want to find a surface with reference to Jupiter such that outside it the comet moves approximately in a conic section round the sun, and inside it in a conic section round Jupiter.

If we consider a surface such that

$$\frac{F}{C} = \frac{\mathfrak{F}}{\mathfrak{C}}$$

we shall have what is required.

By the ordinary theory the disturbing function for the motion of the comet round the sun as perturbed by Jupiter, is

$$M\left\{\frac{1}{\rho} - \frac{r}{R^2}\cos\omega\right\}$$

But $r\cos\omega = \rho\cos\theta + R$, hence the disturbing function is

$$M\left\{\frac{1}{\rho} - \frac{1}{R} - \frac{\rho}{R^2}\cos\theta\right\}$$

Differentiating with respect to ρ and θ, we have

$$P = -M\left\{\frac{1}{\rho^2} + \frac{1}{R^2}\cos\theta\right\}$$

$$T = \frac{M}{R^2}\sin\theta$$

Hence
$$F^2 = M^2\left\{\frac{1}{\rho^4} + \frac{1}{R^4} + \frac{2}{\rho^2 R^2}\cos\theta\right\}$$

$$= \frac{M^2}{\rho^4}\left\{1 + 2\frac{\rho^2}{R^2}\cos\theta + \frac{\rho^4}{R^4}\right\}$$

But
$$C = \frac{S+m}{r^2}$$

and thus
$$\frac{F^2}{C^2} = \frac{M^2}{(S+m)^2} \frac{r^4}{\rho^4} \left\{ 1 + 2\frac{\rho^2}{R^2}\cos\theta + \frac{\rho^4}{R^4} \right\}$$

Again the disturbing function for the motion of the comet round Jupiter as perturbed by the sun is

$$S \left\{ \frac{1}{r} + \frac{\rho}{R^2}\cos\theta \right\}$$

It will be seen that the sign of the second term is here +, because the angle between R and ρ is $\pi - \theta$.

In this formula we have

$$r^2 = \rho^2 + R^2 + 2\rho R \cos\theta$$

Hence differentiating with respect to ρ and θ, we have

$$\mathfrak{P} = -S\left\{ \frac{\rho}{r^3} + R\left(\frac{1}{r^3} - \frac{1}{R^3}\right)\cos\theta \right\}$$

$$\mathfrak{T} = SR\left(\frac{1}{r^3} - \frac{1}{R^3}\right)\sin\theta$$

Now we might proceed to square these two and add them together to find $\mathfrak{F}^2$, and so go on to find the rigorous expression for $\mathfrak{F}/\mathfrak{C}$, which equated to F/C would give the rigorous equation to the required surface; but the result would be so complex as to be of little value because not easily intelligible. I therefore at once proceed to approximation.

S being very large compared with M and m, ρ will be small compared with r.

Then since
$$r^2 = R^2 + \rho^2 + 2R\rho\cos\theta$$

$$\frac{R^3}{r^3} = 1 - 3\frac{\rho}{R}\cos\theta - \frac{3}{2}\frac{\rho^2}{R^2} + \frac{15}{2}\frac{\rho^2}{R^2}\cos^2\theta$$

Therefore approximately

$$\mathfrak{P} = -\frac{S}{R^2}\left\{ \frac{\rho}{R}\left(1 - \frac{3\rho}{R}\cos\theta\right) + \left(-\frac{3\rho}{R}\cos\theta - \frac{3}{2}\frac{\rho^2}{R^2} + \frac{15}{2}\frac{\rho^2}{R^2}\cos^2\theta\right)\cos\theta \right\}$$

$$= -\frac{S\rho}{R^3}\left\{ 1 - 3\cos^2\theta - \frac{9}{2}\frac{\rho}{R}\cos\theta + \frac{15}{2}\frac{\rho}{R}\cos^3\theta \right\}$$

and

$$\mathfrak{P}^2 = \frac{S^2\rho^2}{R^6}\left\{ (1 - 3\cos^2\theta)^2 + 3\cos\theta\,(5\cos^2\theta - 3)(1 - 3\cos^2\theta)\frac{\rho}{R} \right\}$$

Again

$$\mathfrak{T} = -\frac{S\rho}{R^3}\left\{ 3\cos\theta + \frac{3}{2}\frac{\rho}{R} - \frac{15}{2}\frac{\rho}{R}\cos^2\theta \right\}\sin\theta$$

and

$$\mathbb{C}'^2 = \frac{S^2\rho^2}{R^6}\left\{9\cos^2\theta\,(1-\cos^2\theta) - 9\cos\theta\,(5\cos^2\theta - 1)\,(1-\cos^2\theta)\,\frac{\rho}{R}\right\}$$

whence

$$\mathbb{F}^2 = \frac{S^2\rho^2}{R^6}\left\{1 + 3\cos^2\theta - 12\cos^3\theta\,\frac{\rho}{R}\right\}$$

Now $\qquad \mathbb{C} = \dfrac{M+m}{\rho^2}$

And $\qquad \dfrac{\mathbb{F}^2}{\mathbb{C}^2} = \left(\dfrac{S}{M+m}\right)^2 \dfrac{\rho^6}{R^6}\left\{1 + 3\cos^2\theta - 12\cos^3\theta\,\dfrac{\rho}{R}\right\}$

We now have to introduce a similar approximation in the value of F^2/C^2.

We have $\qquad\qquad \dfrac{r^4}{R^4} = 1 + 4\,\dfrac{\rho}{R}\cos\theta$

and therefore $\qquad \dfrac{F^2}{C^2} = \left(\dfrac{M}{S+m}\right)^2 \dfrac{R^4}{\rho^4}\left(1 + 4\,\dfrac{\rho}{R}\cos\theta\right)$

Equating F^2/C^2 to $\mathbb{F}^2/\mathbb{C}^2$, we get

$$\left(\frac{R}{\rho}\right)^{10} = \left(\frac{S}{M}\right)^2 \left(\frac{S+m}{M+m}\right)^2 \left\{1 + 3\cos^2\theta - 4\cos\theta\,(1 + 6\cos^2\theta)\,\frac{\rho}{R}\right\}$$

Therefore

$$\frac{R}{\rho} = \left(\frac{S}{M}\right)^{\frac{1}{5}} \left(\frac{S+m}{M+m}\right)^{\frac{1}{5}} (1 + 3\cos^2\theta)^{\frac{1}{10}}\left\{1 - \frac{2}{5}\,\frac{\cos\theta\,(1 + 6\cos^2\theta)}{1 + 3\cos^2\theta}\,\frac{\rho}{R}\right\}$$

Thus the equation to the surface is approximately

$$\frac{R}{\rho} = \left(\frac{S}{M}\right)^{\frac{1}{5}} \left(\frac{S+m}{M+m}\right)^{\frac{1}{5}} (1 + 3\cos^2\theta)^{\frac{1}{10}}\left\{1 - \frac{2}{5}\left(\frac{M}{S}\right)^{\frac{1}{5}}\left(\frac{M+m}{S+m}\right)^{\frac{1}{5}}\frac{\cos\theta\,(1 + 6\cos^2\theta)}{(1 + 3\cos^2\theta)^{\frac{11}{10}}}\right\}$$

It is usually the case that m is negligible compared with M, and that M is also small compared with S, and in this case we may write the equation with sufficient accuracy

$$\frac{R}{\rho} = \left(\frac{S}{M}\right)^{\frac{2}{5}} (1 + 3\cos^2\theta)^{\frac{1}{10}}$$

Laplace gives a formula for the radius of the sphere of activity which is virtually derivable from the above investigation on the special hypothesis that the three bodies lie in a straight line. Thus he puts θ equal to zero or $180°$ and finds,

$$\frac{R}{\rho} = 4^{\frac{1}{10}}\left(\frac{S}{M}\right)^{\frac{2}{5}}$$

But to find the true mean value of $(1 + 3 \cos^2 \theta)^{\frac{1}{10}}$, we must estimate it all over the sphere.

Now $$\frac{1}{4\pi} \iint (1 + 3 \cos^2 \theta)^{\frac{1}{10}} \sin \theta \, d\theta \, d\phi = \int_0^1 (1 + 3x^2)^{\frac{1}{10}} \, dx$$

This integral, evaluated by quadratures, is found to be equal to 1·063.

Thus the true mean gives

$$\frac{R}{\rho} = 1·063 \left(\frac{S}{M}\right)^{\frac{2}{5}}$$

Laplace makes it

$$\frac{R}{\rho} = 4^{\frac{1}{10}} \left(\frac{S}{M}\right)^{\frac{2}{5}} = 1·149 \left(\frac{S}{M}\right)^{\frac{2}{5}}$$

The ratio of the least to the greatest value of ρ in the formula suggested in this note is 1·149, and Laplace takes the minimum value of ρ as the radius of his sphere.

In the case of Jupiter, Laplace's formula gives $\rho = ·054 \, R$, and my formula gives $\rho = ·058 \, R$.

It follows that Laplace's conclusion is sufficiently accurate for the purpose for which it is intended.

20.

THE EULERIAN NUTATION OF THE EARTH'S AXIS*.

[*Bulletin de l'Académie Royale de Belgique (Sciences)*, 1903, pp. 147—161.]

THE latitude of any place is the mean of the altitudes of a star at its two transits above and below the pole, and the meridian is the great circle through the zenith bisecting the small circle described by a circumpolar star. Thus both the latitude and the meridian depend on the instantaneous axis of the earth's rotation, the observed latitude being the altitude of the instantaneous axis, and the meridian a great circle passing through that axis and the zenith of the place of observation.

But the meridian may also be taken to mean a great circle fixed in the earth passing through the geographical pole and the place. We may call these two circles the astronomical and geographical meridians. The relationship between the two meridians may be illustrated by a figure which I leave to the reader to draw for himself. If C be the geographical pole, I the instantaneous axis and P the place of observation, then IP is the astronomical meridian and CP the geographical meridian. The Eulerian nutation of an absolutely rigid earth would be such that I would describe a circle round C in a period of 306 days; if we allow the earth to yield elastically to centrifugal force the period of the circular motion is augmented and observation shows that the period of 306 days becomes one of about 430 days; lastly it appears that in actuality the simple circular motion is perturbed by other inequalities. In the present discussion it will suffice to consider only the nutation of the ideal unyielding earth, although I shall draw attention in passing to the modifications to be introduced in order to allow for the elastic yielding of the planet.

When P, the place of observation, lies near the circle described by I the instantaneous axis, the oscillations of the astronomical about the geo-

---

* A similar investigation by M. Ch. Lagrange had been published previously in *Bull. Acad. Roy. Belg.*, Vol. LXV., 3rd Series (1895), p. 257.

graphical meridian will be large; and when it is inside the circle the astronomical meridian will describe a complete circuit relatively to the geographical meridian, but with variable angular velocity.

The linear dimensions of the circle or curve described by the instantaneous axis are actually excessively small compared with the earth's radius, and it is not likely that observations will be taken within a very short distance of the pole, yet when we are endeavouring to secure the highest possible accuracy in astronomical observations the oscillation of the astronomical meridian ought to be examined.

It is clear that in discussing this subject the precession and forced nutations, due to the attractions of the sun and moon, may be omitted and that we need only consider the free Eulerian nutation. We have then to determine the position at any time of a system of rectangular axes fixed in the earth relatively to a system of rectangular axes fixed in space. The axes fixed in the earth will be C the geographical north pole and a pair of mutually perpendicular axes A, B in the geographical equator. Since *ex hypothesi* the system moves free from the action of external force, we may take the axis of resultant moment of momentum as axis of Z fixed in space, and X, Y perpendicular to one another in the principal plane of the system, also fixed in space.

The position of ABC relatively to XYZ may be defined by the Eulerian angles θ_0, ϕ_0, ψ_0 as shown in the annexed figure. If p, q, r are the components of angular velocity of the earth about the axes A, B, C; and if A, A, C are the moments of inertia about the same axes, the equations of motion are

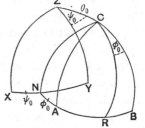

$$A\frac{dp}{dt} = (A - C)\, qr, \quad A\frac{dq}{dt} = (C - A)\, pr, \quad C\frac{dr}{dt} = 0$$

These equations have to be modified if we take account of the elastic yielding of the earth to centrifugal force. Supposing that a spherical planet of the same mass, law of internal density, and elasticity as the actual earth were made to rotate with the same angular velocity as the actual earth, it would become strained under the influence of centrifugal force. This spherical planet being ideal, we may further suppose that the materials of which it is composed are such that the strain does not exceed the limits of perfect elasticity. Let C_1, A_1 denote the principal moments of inertia of the strained spherical planet; C, A the moments of inertia of the actual earth; and $A' = A + (C_1 - A_1)$. Then it may be proved that the equations of motion of the elastic earth have the same form as before, except that A' replaces A. Hence the solution will be the same in form as that for the unyielding earth, and we may henceforth revert to our original hypothesis. I may however remark that if the Eulerian nutation has its period augmented from 306 to 430 days, we must have $A' = 1\cdot00094\,A$.

From the third of the differential equations, we have $r = n$, a constant; and from the first two, with appropriate choice of the epoch,

$$p = \mathrm{F} \sin \frac{C - A}{A} nt, \qquad q = -\mathrm{F} \cos \frac{C - A}{A} nt$$

where F is a constant.

If k is the resultant moment of momentum of the system, the direction cosines of Z referred to ABC are Ap/k, Aq/k, Cr/k; hence

$$\frac{Ap}{k} = -\sin \theta_0 \sin \phi_0, \qquad \frac{Aq}{k} = -\sin \theta_0 \cos \phi_0, \qquad \frac{Cr}{k} = \cos \theta_0$$

Since $r = n$, the last of these shows that θ_0 is a constant, say γ.

Therefore $\quad p = -\dfrac{Cn}{A} \tan \gamma \sin \phi_0, \qquad q = -\dfrac{Cn}{A} \tan \gamma \cos \phi_0$

Comparison with the solution of the differential equation shows that

$$\phi_0 = -\frac{C - A}{A} nt$$

One of the geometrical equations derivable from the figure is

$$\sin \theta_0 \frac{d\psi_0}{dt} = -p \sin \phi_0 - q \cos \phi_0$$

Whence $\qquad\qquad \dfrac{d\psi_0}{dt} = \dfrac{Cn}{A} \sec \gamma$

Hence when time is measured from the instant when the right ascension and east longitude of the instantaneous axis are both 270°, the solution is

$$\theta_0 = \gamma, \qquad \phi_0 = -\frac{C - A}{A} nt, \qquad \psi_0 = \frac{Cnt}{A} \sec \gamma$$

$$p = \frac{Cn}{A} \tan \gamma \sin \frac{C - A}{A} nt, \qquad q = -\frac{Cn}{A} \tan \gamma \cos \frac{C - A}{A} nt, \qquad r = n$$

It is interesting to remark that $d\psi_0/dt$ is an angular velocity a little more rapid than n, and $d\phi_0/dt$ is a slow negative angular velocity, the two being so adjusted as to maintain a constant angular velocity n about C.

If ω be the resultant angular velocity

$$\omega^2 = p^2 + q^2 + r^2 = \left(\frac{Cn}{A \cos \gamma}\right)^2 \left(1 - \frac{C^2 - A^2}{C^2} \cos^2 \gamma\right)$$

Therefore the direction cosines of the instantaneous axis are

$$\frac{\sin\gamma\sin\dfrac{C-A}{A}nt}{\sqrt{\left(1-\dfrac{C^2-A^2}{C^2}\cos^2\gamma\right)}},\qquad \frac{-\sin\gamma\cos\dfrac{C-A}{A}nt}{\sqrt{\left(1-\dfrac{C^2-A^2}{C^2}\cos^2\gamma\right)}},\qquad \frac{\dfrac{A}{C}\cos\gamma}{\sqrt{\left(1-\dfrac{C^2-A^2}{C^2}\cos^2\gamma\right)}}$$

If α denotes the angle between the instantaneous axis I and C the geographical pole, these direction cosines are also $-\sin\alpha\sin\phi_0$, $-\sin\alpha\cos\phi_0$, $\cos\alpha$.

Hence
$$\tan\gamma=\frac{A}{C}\tan\alpha$$

so that $p=n\tan\alpha\sin\dfrac{C-A}{A}nt,\qquad q=-n\tan\alpha\cos\dfrac{C-A}{A}nt$

Since γ denotes the angle CZ and α the angle CI, and since C is greater than A, the instantaneous axis I lies on CZ produced beyond Z in the figure.

We also have $\tan IZ=\dfrac{C-A}{A}\dfrac{\tan\alpha}{1+\dfrac{A}{C}\tan^2\alpha}$

Since $\dfrac{A}{C}$ is very nearly unity, we have to a very close degree of approximation

$$\tan IZ=\frac{C-A}{C}\sin\alpha\cos\alpha$$

In the case of the earth α is known to be about $0''\cdot15$ and $(C-A)/A$ is about $\frac{1}{306}$ (or $\frac{1}{430}$, when A' replaces A). Hence IZ is about $0''\cdot0005$ (or $0''\cdot00034$). Thus for observational purposes the instantaneous axis is practically identical with the axis of resultant moment of momentum. Hence the free nutation does not sensibly affect the obliquity of the ecliptic as derived from observation, and there is no observable nutation of this kind in longitude.

However in order to examine the subject fully I continue to maintain the distinction between the instantaneous axis and that of resultant moment of momentum.

I now desire to find the relationship between the astronomical and geographical poles and equators. Hitherto the investigation has been rigorous, but I shall now proceed approximately.

If λ be the geographical latitude of the place of observation P lying on the circle CA, IP being the astronomical colatitude may be put equal to $\frac{1}{2}\pi-\lambda-\delta\lambda$, so that $\delta\lambda$ is the variation of latitude.

Let us replace I by C' as designating the instantaneous axis, then if the great circle IP or C'P be produced until PA' is equal to $\lambda+\delta\lambda$, C' and A' are the extremities of two axes at right angles to one another. Adding to these a third axis B' at right angles both to A' and C', we have a new set of rectangular axes such that C' is the astronomical pole and A'B' the astronomical equator.

The problem is to determine the position of A', B', C' with respect to axes X_0, Y_0, Z_0 fixed in space, where X_0Y_0 is the ecliptic and X_0 is the descending node of the principal plane of the earth, namely our former XY, on the ecliptic our present X_0Y_0.

For the moment, I have no need to refer to $X_0Y_0Z_0$ but fix my attention on the systems ABC and A'B'C'.

If A'B'C' is derived from ABC by three small rotations θ_1, θ_2, θ_3 about A, B, C respectively, and if x, y, z are the coordinates of any point fixed with reference to ABC, whilst $x + \delta x$, $y + \delta y$, $z + \delta z$ are the coordinates of the same point when referred to A'B'C', we have

$$\delta x = \theta_3 y - \theta_2 z$$
$$\delta y = \theta_1 z - \theta_3 x$$
$$\delta z = \theta_2 x - \theta_1 y$$

First suppose that the fixed point lies at C' or I on a sphere of unit radius. We found above the rigorous expressions for the direction cosines of I the instantaneous axis; on reducing these to an approximate form, we have to the first order of small quantities

$$x = \sin \alpha \sin \frac{C-A}{A} nt, \qquad y = -\sin \alpha \cos \frac{C-A}{A} nt, \qquad z = 1$$

But its coordinates referred to A'B'C' are 0, 0, 1.

Therefore

$$\delta x = -\sin \alpha \sin \frac{C-A}{A} nt, \qquad \delta y = \sin \alpha \cos \frac{C-A}{A} nt, \qquad \delta z = 0$$

In this case y and x are both small quantities of the first order, and θ_3 is also small; therefore $\theta_3 y$ in the first equation and $-\theta_3 x$ in the second are of the second order and negligible.

Hence the first two equations become

$$\delta x = -\theta_2, \qquad \delta y = \theta_1$$

these give $\qquad \theta_2 = \sin \alpha \sin \frac{C-A}{A} nt, \qquad \theta_1 = \sin \alpha \cos \frac{C-A}{A} nt$

The value of θ_3 is not to be determined from this, save that it must be a small quantity of the first order.

Secondly let the fixed point be P lying on the circle CA on a sphere of unit radius. Its coordinates referred to A, B, C are

$$x = \cos \lambda, \qquad y = 0, \qquad z = \sin \lambda$$

Referred to A′, B′, C′, its coordinates are

$$x + \delta x = \cos(\lambda + \delta\lambda), \qquad y + \delta y = 0, \qquad z + \delta z = \sin(\lambda + \delta\lambda)$$

Therefore $\delta x = -\delta\lambda \sin \lambda, \qquad \delta y = 0, \qquad \delta z = \delta\lambda \cos \lambda$

Hence $-\delta\lambda \sin \lambda = -\theta_2 \sin \lambda, \qquad 0 = \theta_1 \sin \lambda - \theta_3 \cos \lambda, \qquad \delta\lambda \cos \lambda = \theta_2 \cos \lambda$

Therefore $\theta_2 = \delta\lambda, \qquad \theta_3 = \theta_1 \tan \lambda$

The angular displacements whereby A, B, C is brought to A′, B′, C′ are accordingly

$$\theta_1 = \sin \alpha \cos \frac{C - A}{A} nt$$

$$\theta_2 = \sin \alpha \sin \frac{C - A}{A} nt$$

$$\theta_3 = \tan \lambda \sin \alpha \cos \frac{C - A}{A} nt$$

Further the variation of latitude is

$$\delta\lambda = \sin \alpha \sin \frac{C - A}{A} nt$$

Since in our figure the angle ACR is $\frac{1}{2}\pi - \phi_0$ or $\frac{1}{2}\pi + \dfrac{C - A}{A} nt$ it follows from the approximate treatment of the spherical triangle C′CP or ICP that the angle $\delta\chi$ between the astronomical and geographical meridians at the point P is given by

$$\delta\chi = \frac{\sin \alpha}{\cos \lambda} \cos \frac{C - A}{A} nt *$$

Having obtained these results, we are in a position to introduce the axes X_0, Y_0, Z_0.

Let us suppose that the position of A, B, C is defined with reference to the ecliptic by the Eulerian angles θ, ϕ, ψ. The same figure will serve again with θ, ϕ, ψ occupying the positions of θ_0, ϕ_0, ψ_0 respectively. $X_0 Y_0$, which replaces XY, will be the ecliptic, and X_0 is the first point of Aries. The angle θ being the obliquity of the ecliptic will no longer be small as was θ_0.

If $\theta + \delta\theta$, $\phi + \delta\phi$, $\psi + \delta\psi$ are the Eulerian angles defining A′, B′, C′ with respect to X_0, Y_0, Z_0, we require to find $\delta\theta$, $\delta\phi$, $\delta\psi$.

They may be determined at once by exactly the same method as that by which the geometrical equations are obtained connecting the rates of change

---

* The following are the rigorous formulæ for $\delta\lambda$ and $\delta\chi$:

$$\sin(\lambda + \delta\lambda) = \sin \lambda \cos \alpha - \cos \lambda \sin \alpha \sin \phi_0$$

$$\tan \delta\chi = \frac{\tan \alpha \cos \phi_0}{\cos \lambda + \tan \alpha \sin \lambda \sin \phi_0}$$

of the Eulerian angles with the angular velocities p, q, r. We have merely to replace velocities by displacements and thus obtain

$$\delta\theta = -\theta_1 \cos\phi + \theta_2 \sin\phi$$

$$\sin\theta\,\delta\psi = -\theta_1 \sin\phi - \theta_2 \cos\phi$$

$$\delta\phi + \cos\theta\,\delta\psi = \theta_3$$

We have also the geometrical equations

$$\frac{d\theta}{dt} = -p\cos\phi + q\sin\phi$$

$$\sin\theta\,\frac{d\psi}{dt} = -p\sin\phi - q\cos\phi$$

$$\frac{d\phi}{dt} + \cos\theta\,\frac{d\psi}{dt} = r$$

Then substituting for θ_1, θ_2, θ_3 and p, q, r their values, with $\sin\alpha$ for $\tan\alpha$ as is permissible, we have

$$\delta\theta = -\sin\alpha\cos\left(\phi + \frac{C-A}{A}nt\right)$$

$$\sin\theta\,\delta\psi = -\sin\alpha\sin\left(\phi + \frac{C-A}{A}nt\right)$$

$$\delta\phi + \cos\theta\,\delta\psi = \tan\lambda\sin\alpha\cos\frac{C-A}{A}nt$$

$$\frac{d\theta}{dt} = -n\sin\alpha\sin\left(\phi + \frac{C-A}{A}nt\right)$$

$$\sin\theta\,\frac{d\psi}{dt} = -n\sin\alpha\cos\left(\phi + \frac{C-A}{A}nt\right)$$

$$\frac{d\phi}{dt} + \cos\theta\,\frac{d\psi}{dt} = n$$

We now differentiate the first three of these equations, and only retaining terms of the first order add them respectively to the second three. We thus obtain

$$\frac{d}{dt}(\theta + \delta\theta) = -n\sin\alpha\left(1 - \frac{d\phi}{ndt} - \frac{C-A}{A}\right)\sin\left(\phi + \frac{C-A}{A}nt\right)$$

$$\sin(\theta + \delta\theta)\frac{d}{dt}(\psi + \delta\psi)$$

$$= n\sin\alpha\left(1 - \frac{d\phi}{ndt} - \frac{C-A}{A}\right)\cos\left(\phi + \frac{C-A}{A}nt\right)$$

$$\frac{d}{dt}(\phi + \delta\phi) + \cos(\theta + \delta\theta)\frac{d}{dt}(\psi + \delta\psi)$$

$$= n - \frac{C-A}{A}n\tan\lambda\sin\alpha\sin\frac{C-A}{A}nt$$

To the order zero θ is constant and $d\psi/dt$ is a quantity of the first order of small quantities, and the same is true of $\cos\theta\, d\psi/dt$. Therefore to the order zero $d\phi/dt = n$, and to the same order

$$1 - \frac{d\phi}{ndt} - \frac{C-A}{A} = -\frac{C-A}{A}$$

Therefore to the first order of small quantities

$$\frac{d}{dt}(\theta + \delta\theta) = \frac{C-A}{A}\, n \sin\alpha \sin\left(\phi + \frac{C-A}{A}\, nt\right)$$

$$\sin(\theta + \delta\theta)\frac{d}{dt}(\psi + \delta\psi) = -\frac{C-A}{A}\, n \sin\alpha \cos\left(\phi + \frac{C-A}{A}\, nt\right)$$

These give the Eulerian diurnal nutations in obliquity and longitude, and as was foreseen they are excessively small.

The third equation gives the correction to the sidereal time. If h denotes the true sidereal hour-angle determined by the motion of the axes A, B, C,

$$\frac{dh}{dt} = \frac{d\phi}{dt} + \cos\theta\, \frac{d\psi}{dt} = n$$

Thus h increases uniformly.

If $h + \delta h$ denotes the observed sidereal hour-angle

$$\frac{d}{dt}(h + \delta h) = n - \frac{C-A}{A}\, n \tan\lambda \sin\alpha \sin\frac{C-A}{A}\, nt$$

The correction to be applied to the observed hour-angle to find the true is $-\delta h$.

Hence by integration

$$-\delta h = -\tan\lambda \sin\alpha \cos\frac{C-A}{A}\, nt$$

It is easy to verify geometrically that if the great circle IP be produced to meet the geographical equator in A'' the expression just found with its sign changed is AA''. It is pretty clear that AA'' is the excess of the observed above the true hour-angle, and therefore the correction to pass from the observed to the true hour-angle is the opposite of AA''. I have however thought it more satisfactory to give a more complete investigation.

If we attribute to α its observed value of $0''\cdot15$ and reduce it to time, it becomes $0^s\cdot01$. Therefore the correction to the observed sidereal hour-angle is

$$-0^s\cdot01 \tan\lambda \cos\frac{C-A}{A}\, nt \quad.$$

In latitude $89° 26'$ the coefficient of the correction is 1^s, and in latitude $84° 20'$ it is $0^s\cdot1$. It is clear that even in the higher of these latitudes the change of length in any given sidereal day is altogether insensible.

The whole range of the amplitude of the oscillation of the astronomical about the geographical meridian is $2 \sin \alpha \sec \lambda$. Since sec 89° 25' is 100, sec 84° 15' is 10, sec 60° is 2, we find, with α still equal to 0″·15, the range to be 30″ in latitude 89° 25'; it is 3″ in latitude 84° 20'; 0″·6 in latitude 60° (that of Helsingfors), and 0″·5 in that of Cambridge.

It would seem that the difficulty of determining the meridian with accuracy must increase with the latitude, and I believe that the meridian can never be found with the same accuracy as the latitude. I presume, therefore, that for all latitudes the variations of latitude afford the better subject for analysis. But ought not astronomers to take the wandering of the pole into account when correcting the meridian of their instruments?

21.

THE ANALOGY BETWEEN LESAGE'S THEORY OF GRAVITATION AND THE REPULSION OF LIGHT.

[Proceedings of the Royal Society, Vol. 76 (1905), A, pp. 387—410.]

I AM not aware that anyone has taken the trouble to work out Lesage's theory, except in the case where the particles of gross matter, subjected to the bombardment of ultramundane corpuscles, are at a distance apart which is a large multiple of the linear dimensions of either of them. Some years ago I had the curiosity to investigate the case where the particles are near together, and having been reminded of my work by reading Professor Poynting's paper on the pressure of radiation*, I have thought it might be worth while to publish my solution, together with some recent additions thereto.

If a corpuscle of mass m moving with velocity v impinges on a plane surface, so that the inclination of its direction of motion before impact to the normal to the surface is ϑ, it communicates to the surface normal momentum $kmv \cos \vartheta$, and tangential momentum $k'mv \sin \vartheta$; where k is 1 for complete inelasticity, and 2 for perfect elasticity, and k' is 0 for perfect smoothness and 1 for perfect roughness.

In the following paper the effects are investigated of the bombardment by Lesagian corpuscles of two spheres, which are taken to be types of the atoms or molecules of gross matter. The effects of the normal and tangential components of the momentum communicated by each blow from a corpuscle will be treated separately.

---

* *Phil. Trans.*, A, Vol. 202, pp. 525—552. I have had the advantage of corresponding with him on the present subject.

§ 1. *The Normal Component of the Impacts.*

Suppose that there are in space n corpuscles per unit volume, moving indiscriminately in all directions with velocity v; let the mass of each corpuscle be m, and ρ the density of the medium, so that $\rho = mn$.

Suppose, further, that the corpuscles are so small that collisions between them are rare enough to be negligible. The system thus described is that of Lesage's mechanism for explaining gravitation, provided that the corpuscles are not perfectly elastic.

As stated in the introductory remarks, when any corpuscle strikes a plane surface at such an angle that ϑ is the inclination of its motion before impact to the normal to the plane, it communicates to the plane normal momentum $kmv \cos \vartheta$, where k is a factor, lying between 1 and 2, being 1 for perfectly inelastic corpuscles, and 2 for perfectly elastic ones.

If two bodies, A and B, are exposed to the bombardment of corpuscles, the Lesagian attraction between them will vanish if the corpuscles are perfectly elastic, because the corpuscles reflected between A and B will on the whole cause such a repulsion between the bodies as exactly to counterbalance the attraction.

But in the present investigation we shall only consider those corpuscles which come from infinity, and pay no attention to their behaviour after impact on the surfaces of A and B. This can, I suppose, be only completely justified when the corpuscles are perfectly inelastic.

If the velocity of each of the n corpuscles in unit volume be represented by a vector drawn from a centre, the ends of the vectors will lie equally spaced on the surface of a sphere of radius v. If from the centre of the sphere of velocities a cone of small solid angle $\delta\omega$ be drawn, the number of corpuscles per unit volume, whose directions of motion fall within the cone is $n\delta\omega/4\pi$.

Let us consider the pressure on a surface exposed to bombardment. On the surface draw a unit area, which may clearly be treated as plane, and let ϑ be the zenith distance of the axis of the cone $\delta\omega$. On the unit area draw an oblique prism, with edges of length v, parallel to the axis of the cone $\delta\omega$; the volume of this prism is $v \cos \vartheta$. Hence the number of corpuscles, whose directions of motion fall within the cone $\delta\omega$, lying inside the oblique prism at any moment is $nv \cos \vartheta \, \delta\omega/4\pi$. At the end of unit time elapsing after the moment under consideration all these corpuscles will have struck the plane, and each will have communicated momentum towards the nadir equal to $kmv \cos \vartheta$. Thus the momentum towards the nadir communicated by this class of corpuscles is $kmnv^2 \cos^2 \vartheta \, \delta\omega/4\pi$.

Since $mn = \rho$, the pressure (or momentum towards the nadir communicated per unit time and area) is

$$\frac{k\rho v^2}{4\pi} \Sigma \cos^2 \vartheta \, \delta\omega$$

where the summation is effected for all directions whence corpuscles may come.

If the particles may come from all over the sky, the summation is for half of angular space, and $\Sigma \cos^2 \vartheta \, \delta\omega = \frac{2}{3}\pi$. Therefore the pressure p on the surface is given by

$$p = \tfrac{1}{6}k\rho v^2$$

If we put $k = 2$, this is the well-known result of the kinetic theory of gases.

If, however, a certain cone of space is screened off, so that no corpuscle can come from thence, the pressure is given by

$$p = \tfrac{1}{6}k\rho v^2 - \frac{k\rho v^2}{4\pi} \Sigma \cos^2 \vartheta \, \delta\omega$$

where the summation is taken over the screening area.

In Fig. 1 let the origin be at the unit area, and let P be the intersection with the celestial sphere of the normal to the area; let the screen be circular round the Z axis, and let AA′ be one quadrant of the screen. Let the radius of the screen be α, and the zenith distance (for an observer standing on the unit area) of the centre of the screen be β. Then

$$ZA = ZA' = \alpha, \quad ZP = \beta$$

Let θ, ϕ be the polar co-ordinates of an element $\sin\theta \, d\theta \, d\phi$ of the screen at Q, so that the $\delta\omega$ of the formula for the pressure is $\sin\theta \, d\theta \, d\phi$. Also, since the ϑ of the formula is the angle PQ, we have

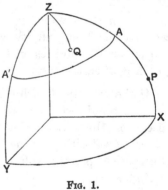

Fig. 1.

$$\cos\vartheta = \cos\theta\cos\beta + \sin\theta\sin\beta\cos\phi$$

Hence, taking the summation over the area of the screen,

$$\frac{1}{2\pi}\Sigma \cos^2\vartheta\,\delta\omega = \frac{1}{2\pi}\int_0^{2\pi}\int_0^{a}(\cos\theta\cos\beta + \sin\theta\sin\beta\cos\phi)^2 \sin\theta\,d\theta\,d\phi$$

$$= \int_0^{a}(\cos^2\theta\cos^2\beta + \tfrac{1}{2}\sin^2\theta\sin^2\beta)\sin\theta\,d\theta$$

$$= \int_0^{a}[(\tfrac{3}{2}\cos^2\beta - \tfrac{1}{2})\cos^2\theta + \tfrac{1}{2}\sin^2\beta]\sin\theta\,d\theta$$

$$= \tfrac{1}{2}(\cos^2\beta\cos\alpha\sin^2\alpha + \tfrac{2}{3} + \tfrac{1}{3}\cos^3\alpha - \cos\alpha) \quad\ldots\ldots\ldots\ldots(1)$$

Thus $\quad p = \frac{1}{6}k\rho v^2 - \frac{1}{4}k\rho v^2 (\cos^2 \beta \cos \alpha \sin^2 \alpha + \frac{2}{3} + \frac{1}{3}\cos^3 \alpha - \cos \alpha)$

Now let there be two spheres exposed to bombardment, a sphere A with radius a, and a sphere B with radius b; and let R be the distance between their centres. As stated in the introduction, A and B are meant to typify atoms or molecules of gross matter.

Let an element at P of the sphere A be determined by colatitude θ from the axis BA, and by longitude ϕ measured from some fixed plane passing through BA, and let the element be expressed by $a^2 \sin \theta\, d\theta\, d\phi$.

This element is screened by the sphere B, and therefore the screen is circular. The semi-angle of the screening cone was denoted above by α, and the observer's zenith distance of the axis of the screening cone was β.

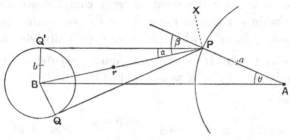

Fig. 2.

If then we denote the distance PB from the element to the centre of the sphere B by r, we have from Fig. 2,

$$\sin \alpha = \frac{b}{r}, \quad \sin \beta = \frac{R}{r}\sin \theta, \quad \cos \beta = \frac{R\cos \theta - a}{r}$$

$$r^2 = R^2 + a^2 - 2Ra \cos \theta$$

It is to be noted that θ has not the same meaning as that used in evaluating (1).

If the angle QPA is obtuse, the whole of the sphere B is visible to the observer at P. This case will be considered first, and afterwards those regions on the sphere A from which B is partially below the horizon of the observer.

We want to evaluate the total resultant force on the sphere A, due to the pressure on all its elements, in the direction A to B. As far as concerns the element under consideration, the resultant force in the required direction is $-p \cos \theta . a^2 \sin \theta\, d\theta\, d\phi$. In order to determine the whole of that portion of the resultant which is due to pressure on the portions of A from which B is completely visible, this expression will have to be integrated from $\phi = 2\pi$ to $\phi = 0$, and from $\cos \theta = (a + b)/R$ to $\cos \theta = 1$—that is to say, from the colatitude in which B just begins to set until B is in the zenith.

The constant part of the expression for the pressure, namely, $\frac{1}{8}k\rho v^2$, is the same all round the sphere, and it cannot contribute anything to the resultant force; hence it may be omitted.

Since the expression for the pressure does not involve ϕ, the integration with respect to ϕ simply introduces the factor 2π. Thus the resultant force estimated towards B, in as far as it is due to the elements from which B is wholly visible, is

$$\tfrac{1}{2}\pi k\rho a^2 v^2 \int [\cos^2\beta \cos\alpha \sin^2\alpha + \tfrac{2}{3} + \tfrac{1}{3}\cos^3\alpha - \cos\alpha]\cos\theta \sin\theta\, d\theta$$

where the limits of θ are $\cos^{-1}\left(\dfrac{a+b}{R}\right)$ to 0.

This expression may be integrated rigorously, but it would not be expedient to do so, because the result is very complicated, and because I have failed to obtain a rigorous integral for the remaining portion of the complete expression, namely, where B is only partially visible.

It will now save trouble if we omit the factor $\frac{1}{2}\pi k\rho a^2 v^2$ and reintroduce it later, and if besides we choose R as our unit of length.

Hence we have

$$\sin\alpha = \frac{b}{r}, \quad \sin\beta = \frac{1}{r}\sin\theta, \quad \cos\beta = \frac{\cos\theta - a}{r}, \quad r^2 = 1 + a^2 - 2a\cos\theta$$

and the limits of θ are $\cos^{-1}(a+b)$ to 0.

Cos α can be expanded convergently in powers of b/r, so long as the two spheres do not actually touch, and we easily obtain

$$\tfrac{2}{3} - \tfrac{1}{3}\cos^3\alpha - \cos\alpha = \tfrac{1}{4}\frac{b^4}{r^4} + \tfrac{1}{12}\frac{b^6}{r^6} + \tfrac{3}{64}\frac{b^8}{r^8} + \cdots$$

$$\sin^2\alpha \cos\alpha = \frac{b^2}{r^2} - \tfrac{1}{2}\frac{b^4}{r^4} - \tfrac{1}{8}\frac{b^6}{r^6} - \tfrac{1}{16}\frac{b^8}{r^8} - \cdots$$

also

$$\cos^2\beta = \frac{1}{r^2}(\cos^2\theta - 2a\cos\theta + a^2)$$

Hence the integral becomes

$$\int\left(\frac{b^2}{r^4} - \tfrac{1}{2}\frac{b^4}{r^6} - \tfrac{1}{8}\frac{b^6}{r^8} \cdots\right)(\cos^2\theta - 2a\cos\theta + a^2)\cos\theta\sin\theta\, d\theta$$

$$+ \tfrac{1}{4}b^2\int\left(\frac{b^2}{r^4} + \tfrac{1}{3}\frac{b^4}{r^6} + \tfrac{3}{16}\frac{b^6}{r^8} \cdots\right)\cos\theta\sin\theta\, d\theta$$

Now

$$r^2 = (1+a^2)\left(1 - \frac{2a}{1+a^2}\cos\theta\right)$$

If, then, we write

$$A_1 = \frac{2a}{1+a^2}, \quad B_{2n} = \frac{b^{2n}}{(1+a^2)^{n+1}}$$

we have

$$r^2 = (1+a^2)(1 - A_1\cos\theta)$$

Whence

$$\frac{b^{2n}}{r^{2n+2}} = B_{2n} \left\{ 1 + (n+1) A_1 \cos\theta + \frac{(n+1)(n+2)}{1.2} A_1^2 \cos^2\theta + \dots \right\}$$

Integrating between the appropriate limits, we find

$$\int \frac{b^{2n}}{r^{2n+2}} \cos^m\theta \sin\theta \, d\theta = B_{2n} \left\{ \frac{1}{m+1} + \frac{n+1}{m+2} A_1 + \frac{(n+1)(n+2)}{1.2(m+3)} A_1^2 + \dots \right\}$$

$$- B_{2n} \left\{ \frac{(a+b)^{m+1}}{m+1} + \frac{n+1}{m+2} A_1 (a+b)^{m+2} \right.$$

$$\left. + \frac{(n+1)(n+2)}{1.2(m+3)} A_1^2 (a+b)^{m+3} + \dots \right\}$$

It will be observed that the suffixes of the B's and of A_1 indicate their orders in the powers of $1/R$.

This formula may be applied to each term of the preceding integrals, so that the result may be obtained in terms of A_1 and B_{2n}.

The several terms must then be expanded in powers of a^2, and terms rearranged in their several orders.

After some tedious analysis I find, on reintroducing R, which was treated as being the unit of length, the following result as far as the ninth order:—

$$\tfrac{1}{8}\pi k\rho \, \frac{v^2 a^2 b^2}{R^2} \left\{ 1 + \frac{8}{3.5} \frac{a}{R} + \frac{4a(4a^2+7b^2)}{3.5.7R^3} + \frac{a(8a^4+36a^2b^2+21b^4)}{3.5.7R^5} \right.$$

$$+ \frac{a}{R^7} \left(\tfrac{32}{693} a^6 + \tfrac{8}{21} a^4 b^2 + \tfrac{4}{7} a^2 b^4 + \tfrac{1}{6} b^6 \right)$$

$$\left. - \frac{7b^3}{R^4} \left(\tfrac{1}{3} a + \tfrac{5}{24} b \right) - \frac{2b^3}{R^6} \left(\tfrac{7}{3} a^3 + \tfrac{35}{8} a^2 b + \tfrac{21}{10} a b^2 - \tfrac{35}{192} b^3 \right) \right\} \quad \dots\dots(2)$$

This formula gives the force on the sphere A towards B in so far as it is produced by the normal component of the bombardment on those parts of A from which B is completely visible. It remains to investigate the regions whence B is only partially visible.

In Fig. 3, as in Fig. 1, let the origin be at a unit area on sphere A, and let P be the intersection of the normal to the area with the celestial sphere. Then if PC is a right-angle, CY will be the horizon which bounds the screen. Let Z be the axis of the screening cone, and DBE the cone itself. As in the previous case, the semi-angle of the cone is denoted by α, and PZ is β.

The half of the screen which is shown in the figure is CBEZ.

If Q be any point in the screen, the pressure on the unit area on the sphere A is

$$\tfrac{1}{6} k\rho v^2 - \frac{k\rho v^2}{4\pi} \Sigma \cos^2 PQ \, \delta\omega$$

the summation being carried out over the whole screen. As before, we may neglect the constant part of the expression for the pressure.

In this figure more than half the sphere B is visible, and the case when less than half is visible will be treated subsequently.

If θ, ϕ are the polar co-ordinates of Q,

$$\frac{1}{2\pi} \Sigma \cos^2 PQ \, \delta\omega = \frac{1}{2\pi} \iint (\cos\theta \cos\beta + \sin\theta \sin\beta \cos\phi)^2 \sin\theta \, d\theta \, d\phi$$

where the integral is taken over the area CBEZ and the other half not shown in the figure.

The area of integration may conveniently be divided into two parts—viz., the part inside the dotted circle and the part between CF and BE.

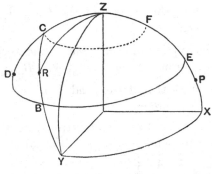

FIG. 3.

The part of the result corresponding to the dotted circle may be obtained at once from our previous result (1), for CZ the radius of the cone is equal to $\frac{1}{2}\pi - \beta$, and therefore the result is

$$\frac{1}{2}\left\{\cos^4\beta \sin\beta + \frac{2}{3} + \frac{1}{3}\sin^3\beta - \sin\beta\right\}$$

For the remaining portion, take the limits of integration so as only to include CBEF; then doubling the preceding formula, so as to permit us to halve the range of ϕ, we write the integral in the form

$$\frac{1}{\pi} \iint (\cos^2\theta \cos^2\beta + \tfrac{1}{2}\sin^2\theta \sin^2\beta + \tfrac{1}{2}\sin^2\theta \sin^2\beta \cos 2\phi$$
$$+ 2\sin\theta \cos\theta \sin\beta \cos\beta \cos\phi) \sin\theta \, d\theta \, d\phi$$

If R is any point on the boundary CB in colatitude θ, its longitude ϕ is equal to $\pi - CZR$; also $CZ = \frac{1}{2}\pi - \beta$. Then from the right-angled spherical triangle CZR we have

$$\cos\phi = -\cot\beta \cot\theta$$

Thus the limits of ϕ are $\cos^{-1}[-\cot\beta \cot\theta]$ to 0; and the limits of θ are α to $\frac{1}{2}\pi - \beta$.

At the upper limit of ϕ we have

$$\sin\phi = \sqrt{(1 - \cot^2\beta \cot^2\theta)} = \frac{\sqrt{(\sin^2\theta - \cos^2\beta)}}{\sin\beta \sin\theta}$$

and $$\sin 2\phi = -2\,\frac{\cos \beta \cos \theta}{\sin^2 \beta \sin^2 \theta}\,\sqrt{(\sin^2 \theta - \cos^2 \beta)}$$

Integrating with respect to ϕ, we obtain

$$\frac{1}{\pi}\int(\cos^2 \theta \cos^2 \beta + \tfrac{1}{2}\sin^2 \theta \sin^2 \beta)\cos^{-1}(-\cot \beta \cot \theta)\,d\theta$$

$$-\frac{1}{2\pi}\int\cos \beta \cos \theta\,\sqrt{(\sin^2 \theta - \cos^2 \beta)}\sin \theta\,d\theta$$

$$+\frac{2}{\pi}\int\cos \beta \cos \theta\,\sqrt{(\sin^2 \theta - \cos^2 \beta)}\sin \theta\,d\theta$$

The last two terms involve the same integral and only differ in the coefficients; they therefore fuse together with a coefficient $3/2\pi$.

The whole may now be written

$$-\frac{1}{\pi}\int\cos^{-1}(-\cot \beta \cot \theta)\frac{d}{d\theta}\left[\tfrac{1}{3}\cos^3 \theta\,(\tfrac{3}{2}\cos^2 \beta - \tfrac{1}{2}) + \tfrac{1}{2}\sin^2 \beta \cos \theta\right]d\theta$$

$$+\frac{3}{2\pi}\int\sin \theta \cos \theta \cos \beta\,\sqrt{(\sin^2 \theta - \cos^2 \beta)}\,d\theta$$

The last term is a perfect integral, and the first may be integrated by parts. Effecting this, I find that the indefinite integral of the whole is

$$-\frac{1}{2\pi}\cos \theta\left[\cos^2 \theta\,(\cos^2 \beta - \tfrac{1}{3}) + \sin^2 \beta\right]\cos^{-1}(-\cot \beta \cot \theta)$$

$$+\frac{1}{3\pi}\sin^{-1}\frac{\cos \beta}{\sin \theta} + \frac{\cos \beta}{2\pi}(\sin^2 \theta - \tfrac{1}{3})\,\sqrt{(\sin^2 \theta - \cos^2 \beta)}$$

Then taking the expression, between limits, and adding the part corresponding to the dotted circle, we have

$$\tfrac{1}{2}\sin \beta\left[\sin^2 \beta\,(\cos^2 \beta - \tfrac{1}{3}) + \sin^2 \beta\right]$$

$$-\frac{1}{2\pi}\cos \alpha\left[\cos^2 \alpha\,(\cos^2 \beta - \tfrac{1}{3}) + \sin^2 \beta\right]\cos^{-1}(-\cot \beta \cot \alpha)$$

$$+\frac{1}{3\pi}\sin^{-1}\frac{\cos \beta}{\sin \alpha} - \frac{1}{6}$$

$$+\frac{\cos \beta}{2\pi}(\sin^2 \alpha - \tfrac{1}{3})\,\sqrt{(\sin^2 \alpha - \cos^2 \beta)}$$

$$+\tfrac{1}{2}\left[\cos^4 \beta \sin \beta + \tfrac{2}{3} + \tfrac{1}{3}\sin^3 \beta - \sin \beta\right]$$

This expression admits of reduction, and it gives

$$\frac{1}{2\pi}\Sigma \cos^2 \mathrm{PQ}\,\delta\omega = \frac{1}{6} - \frac{1}{2\pi}\cos \alpha\left[\tfrac{2}{3} + \sin^2 \alpha\,(\tfrac{1}{3} - \cos^2 \beta)\right]\cos^{-1}(-\cot \beta \cot \alpha)$$

$$+\frac{1}{2\pi}\cos \beta\,(\sin^2 \alpha - \tfrac{1}{3})\,\sqrt{(\sin^2 \alpha - \cos^2 \beta)} + \frac{1}{3\pi}\sin^{-1}\frac{\cos \beta}{\sin \alpha} \quad\ldots\ldots(3)$$

The pressure is the same quantity multiplied by $\tfrac{1}{2}k\rho v^2$.

In the case when less than half of the sphere B is visible, the dotted circle vanishes. It appears that at the upper limit $\theta = \alpha$, at the lower $\theta = \beta - \frac{1}{2}\pi$; the angle $\cos^{-1}(-\cot\beta\cot\theta)$ is the angle ϕ at the point C, and therefore vanishes; and at the same point $\sin^{-1}(\cos\beta/\sin\theta)$ is equal to $-\frac{1}{2}\pi$. It is, then, easily verified that the formula (3) holds good, and we may therefore apply it both when more and when less than half of B is visible.

If p is the pressure as given by this formula, and $a^2 \sin\theta\, d\theta\, d\phi$ is the element of area of the sphere A, the component of pressure in the direction AB is $-p\cos\theta.\, a^2\sin\theta\, d\theta\, d\phi$. This has to be integrated over the "sunset" region of the sphere A. The integration with respect to ϕ is effected at once, and we are left with

$$-2\pi a^2 \int p \sin\theta \cos\theta\, d\theta$$

integrated from $\theta = \cos^{-1}(a-b)/R$ to $\theta = \cos^{-1}(a+b)/R$, or from $\cos^{-1}(a-b)$ to $\cos^{-1}(a+b)$, when R is taken as the unit of length.

We have accordingly to multiply (3) by $\pi k\rho a^2 v^2 \sin\theta \cos\theta\, d\theta$ and integrate it from $\theta = \cos^{-1}(a-b)$ to $\cos^{-1}(a+b)$.

I have not been able to effect the integration rigorously, and shall not give the full details of the tedious analysis involved in the approximate integration.

It is convenient, in the first place, to change the independent variable. Putting then $\sin\psi = (a-\cos\theta)/b$, so that the limits become $\pm\frac{1}{2}\pi$, we have

$$\sin\theta\cos\theta\, d\theta = (ab\cos\psi - b^2\sin\psi\cos\psi)\, d\psi$$

$$r^2 = 1 - a^2 + 2ab\sin\psi, \quad \cos\beta = -\frac{b}{r}\sin\psi = -\sin\alpha\sin\psi$$

so that $\qquad \psi = -\sin^{-1}\dfrac{\cos\beta}{\sin\alpha}, \quad \sqrt{(\sin^2\alpha - \cos^2\beta)} = \dfrac{\cot\psi}{\cos\alpha}$

If we drop the factor $\pi k\rho a^2 v^2$ temporarily, it will be found that the subject of integration may be written in the form

$$\tfrac{1}{4}b^4(1+2a^2)(\tfrac{1}{4}+\sin^2\psi) - \tfrac{1}{8}b^6(\sin^2\psi - \tfrac{1}{6}) + \frac{2}{\pi}ab^5\psi\sin\psi(\tfrac{1}{4}+\sin^2\psi)$$

$$+ \frac{1}{6\pi}ab^5\sin^2\psi\cos\psi(2\sin^2\psi + 13)$$

$$- ab^5\sin\psi(\tfrac{1}{4}+\sin^2\psi) - \frac{1}{2\pi}\psi\{b^4(1+2a^2)(\tfrac{1}{4}+\sin^2\psi) - \tfrac{1}{2}b^6(\sin^2\psi - \tfrac{1}{6})\}$$

$$- \frac{1}{24\pi}b^4(1+2a^2)\sin\psi\cos\psi(2\sin^2\psi + 13)$$

$$+ \frac{b^6}{3\pi}\sin\psi\cos\psi(\tfrac{1}{8}+\tfrac{2}{3}\sin^2\psi - \tfrac{1}{6}\sin^4\psi)$$

multiplied by $\qquad (ab\cos\psi - b^2\sin\psi\cos\psi)\, d\psi$

The first two lines of the first factor are even functions of ψ, and the last three lines are odd functions; the first term of the second factor is even, and the second term is odd. Then since the limits of ψ are $\pm \frac{1}{2}\pi$, it follows that we need only multiply the first two lines of the first factor by $ab \cos \psi \, d\psi$ and the last three lines by $-b^2 \sin \psi \cos \psi \, d\psi$. All the integrals involved are then known, viz.:—

$$\int_{-\frac{1}{2}\pi}^{\frac{1}{2}\pi} \sin^{2n} \psi \cos \psi \, d\psi \;=\; \frac{2}{2n+1}$$

$$\int_{-\frac{1}{2}\pi}^{\frac{1}{2}\pi} \psi \sin \psi \cos \psi \, d\psi \;=\; \frac{1}{4}\pi$$

$$\int_{-\frac{1}{2}\pi}^{\frac{1}{2}\pi} \psi \sin^3 \psi \cos \psi \, d\psi \;=\; \frac{5}{32}\pi$$

$$\int_{-\frac{1}{2}\pi}^{\frac{1}{2}\pi} \sin^{2n} \psi \cos^{2m} \psi \, d\psi \;=\; \frac{(2n-1)(2n-3)\ldots 3.1}{(2m+2n)(2m+2n-2)\ldots 2}\pi$$

By means of these I find that the component force on the sphere A from A towards B due to this pressure is

$$\tfrac{1}{8}\pi k \rho \frac{v^2 a^2 b^2}{R^2} \left\{ \frac{7b^3}{R^4} \left(\tfrac{1}{3}a + \tfrac{5}{24}b \right) + \frac{2b^3}{R^6} \left(\tfrac{7}{3}a^3 + \tfrac{35}{8}a^2 b + \tfrac{21}{10}ab^2 - \tfrac{35}{192}b^3 \right) \right\}$$

Now on comparing this with the expression for the first part of the force in (2) we see that it exactly annuls the terms of the sixth and eighth orders; and we may feel confident that the term of the tenth order would be similarly annulled. The result is of so simple a character that it must surely be possible to prove it in some shorter way. However this may be, our final result for the resultant force on A towards B, which I may call F_A, is given by

$$F_A = \tfrac{1}{8}\pi k\rho \frac{v^2 a^2 b^2}{R^2} \left\{ 1 + \frac{8a}{15R} + \frac{4a(4a^2 + 7b^2)}{105R^3} + \frac{a(8a^4 + 36a^2 b^2 + 21b^4)}{105R^5} \right.$$

$$\left. + \frac{a}{R^7} \left(\tfrac{32}{693}a^6 + \tfrac{8}{21}a^4 b^2 + \tfrac{4}{7}a^2 b^4 + \tfrac{1}{8}b^6 \right) \right\} \quad \ldots\ldots (4)$$

We may observe that, as a rough approximation,

$$F_A = \tfrac{1}{8}\pi k\rho \frac{v^2 a^2 b^2}{(R - \frac{4}{15}a)^2}$$

By symmetry the full expression for the force on B towards A is

$$F_B = \tfrac{1}{8}\pi k\rho \frac{v^2 a^2 b^2}{R^2} \left\{ 1 + \frac{8b}{15R} + \frac{4b(4b^2 + 7a^2)}{105R^3} + \frac{b(8b^4 + 36a^2 b^2 + 21a^4)}{105R^5} + \ldots \right\}$$

In considering the excess of F_A above F_B, it will suffice if we drop the terms of the seventh and ninth orders, since the reader will easily be able to extend the result if he desires to do so.

Then $\quad F_A - F_B = \frac{1}{8}\pi k\rho \frac{v^2 a^2 b^2}{R^2} \cdot \frac{8}{15}\left(\frac{a-b}{R}\right)\left[1 + \frac{4a^2 - 3ab + 4b^2}{14R^2}\right]$

It follows, therefore, that, as the result of the normal component of the bombardment, the larger sphere is more strongly urged towards the smaller than the smaller towards the larger.

§ 2. *The Tangential Component of the Impacts; Effect of both Components together.*

When a corpuscle m strikes a plane surface so that ϑ is the inclination of its motion before impact to the normal to the plane, the tangential component of its momentum is $mv\sin\vartheta$, and it communicates to the plane tangential momentum equal to $k'mv\sin\vartheta$, where k' lies between 0 and 1, being 1 for perfect roughness and 0 for perfect smoothness of the surface.

Following a procedure exactly similar to that adopted previously, we see that the corpuscles whose directions of motion before impact lie in a small cone of solid angle $\delta\omega$ communicate a tangential force to the plane equal to

$$\frac{k'\rho v^2}{4\pi}\sin\vartheta\cos\vartheta\,\delta\omega$$

Let l, m, n be the direction cosines of the normal to the plane bombarded, and l', m', n' the direction cosines of the axis of the cone $\delta\omega$. Then the direction cosines of the projection of the axis of the cone on the plane, which is identical with but opposite in direction to that of the tangential force communicated to the plane, are

$$\frac{l'-l\cos\vartheta}{\sin\vartheta}, \quad \frac{m'-m\cos\vartheta}{\sin\vartheta}, \quad \frac{n'-n\cos\vartheta}{\sin\vartheta}$$

Hence if X, Y, Z are the components of the force thus imparted to the plane by the bombardment,

$$X = -\frac{1}{4\pi}k'\rho v^2 \Sigma \cos\vartheta\,(l'-l\cos\vartheta)\,\delta\omega$$

$$Y = -\frac{1}{4\pi}k'\rho v^2 \Sigma \cos\vartheta\,(m'-m\cos\vartheta)\,\delta\omega$$

$$Z = -\frac{1}{4\pi}k'\rho v^2 \Sigma \cos\vartheta\,(n'-n\cos\vartheta)\,\delta\omega$$

where the summations are carried out over all the directions from whence corpuscles arrive at the plane.

If corpuscles come from all over the sky, $X = Y = Z = 0$; hence the forces are

$$X = \frac{1}{4\pi}k'\rho v^2 \Sigma \cos\vartheta\,(l'-l\cos\vartheta)\,\delta\omega, \text{ and two others}$$

where the summations are carried out over the screening area.

In Fig. 1 the direction cosines of the normal to the bombarded surface are

$$l = \sin \beta, \quad m = 0, \quad n = \cos \beta$$

and
$$l' = \sin \theta \cos \phi, \quad m' = \sin \theta \sin \phi, \quad n' = \cos \theta$$

$$\cos \vartheta = \cos \theta \cos \beta + \sin \theta \sin \beta \cos \phi$$

The limits of integration are

$$\phi = 2\pi \text{ to } 0, \quad \theta = \alpha \text{ to } 0, \quad \text{and } \delta\omega = \sin \theta \, d\theta \, d\phi$$

Therefore

$$X = \frac{1}{4\pi} k'\rho v^2 \int_0^{2\pi} \int_0^\alpha [\sin \theta \cos \phi \, (\cos \theta \cos \beta + \sin \theta \sin \beta \cos \phi)$$
$$- \sin \beta \, (\cos \theta \cos \beta + \sin \theta \sin \beta \cos \phi)^2] \sin \theta \, d\theta \, d\phi$$

$$Y = \frac{1}{4\pi} k'\rho v^2 \int_0^{2\pi} \int_0^\alpha \sin \theta \sin \phi \, (\cos \theta \cos \beta + \sin \theta \sin \beta \cos \phi) \sin \theta \, d\theta \, d\phi$$

$$Z = \frac{1}{4\pi} k'\rho v^2 \int_0^{2\pi} \int_0^\alpha [\cos \theta \, (\cos \theta \cos \beta + \sin \theta \sin \beta \cos \phi)$$
$$- \cos \beta \, (\cos \theta \cos \beta + \sin \theta \sin \beta \cos \phi)^2] \sin \theta \, d\theta \, d\phi$$

Integrating with respect to ϕ, we find

$$X = \tfrac{1}{2} k'\rho v^2 \sin \beta \cos^2 \beta \int_0^\alpha (\tfrac{1}{2} - \tfrac{3}{2} \cos^2 \theta) \sin \theta \, d\theta$$

$$= - \tfrac{1}{4} k'\rho v^2 \sin \beta \cos^2 \beta \sin^2 \alpha \cos \alpha$$

$$Y = 0$$

$$Z = \tfrac{1}{2} k'\rho v^2 \sin^2 \beta \cos \beta \int_0^\alpha (\tfrac{3}{2} \cos^2 \theta - \tfrac{1}{2}) \sin \theta \, d\theta$$

$$= \tfrac{1}{4} k'\rho v^2 \sin^2 \beta \cos \beta \sin^2 \alpha \cos \alpha$$

In Fig. 2 P is at the bombarded element of surface, and is the origin of Fig. 1; then we take PB as the axis of Z, and the dotted line PX as the axis of X. Accordingly the component force in the direction A to B is

$$Z \cos (\beta - \theta) + X \sin (\beta - \theta) = \tfrac{1}{4} k'\rho v^2 \sin \beta \cos \beta \sin^2 \alpha \cos \alpha \sin \theta \ldots(5)$$

The element of surface at P is $a^2 \sin \theta \, d\theta \, d\phi$, and it is to be noted that θ is used in a different sense from that employed in evaluating X, Y, Z.

Hence the total force on the sphere A from A to B, due to the tangential component arising from the bombardment, for all those parts of A from which B is wholly visible, is

$$\tfrac{1}{4} k'\rho v^2 a^2 \iint \sin \beta \cos \beta \sin^2 \alpha \cos \alpha \sin \theta \cdot \sin \theta \, d\theta \, d\phi$$

the limits of ϕ being 2π to 0, and of θ from $\cos^{-1}(a + b)/R$ to 0.

The integration for ϕ involves only multiplication by 2π.

As before, I take temporarily R as unit of length, so that the limits of θ are $\cos^{-1}(a + b)$ to 0.

Then

$$\sin^2 \alpha \cos \alpha = \frac{b^2}{r^2} - \frac{1}{2}\frac{b^4}{r^4} - \frac{1}{8}\frac{b^6}{r^6} \dots, \quad \sin \beta \cos \beta = \frac{1}{r^2}\sin \theta\,(\cos \theta - a)$$

Hence the integral becomes

$$\tfrac{1}{2}\pi k'\rho v^2 a^2 \int\left(\frac{b^2}{r^4} - \frac{1}{2}\frac{b^4}{r^6} - \frac{1}{8}\frac{b^6}{r^8}\dots\right)\sin^3\theta\,(\cos\theta - a)\,d\theta$$

If we use the same abbreviations as before in the development in inverse powers of r, and note that

$$\int \sin^3\theta \cos^{m+1}\theta\,d\theta = \frac{2}{(m+2)(m+4)} - \frac{(a+b)^{m+2}}{m+2} + \frac{(a+b)^{m+4}}{m+4}$$

we find

$$\int \frac{b^{2n}}{r^{2n+2}}\sin^3\theta\cos\theta\,d\theta = B_{2n}\left\{\frac{2}{2.4} + \frac{(n+1)}{1!}\frac{2}{3.5}A_1 + \frac{(n+1)(n+2)}{2!}\frac{2}{4.6}A_1^2\right.$$

$$+ \frac{(n+1)(n+2)(n+3)}{3!}\frac{2}{5.7}A_1^3 + \dots$$

$$- \frac{(a+b)^2}{2} - \frac{(n+1)}{1!}\frac{(a+b)^3}{3}A_1 - \frac{(n+1)(n+2)}{2!}\frac{(a+b)^4}{4}A_1^2 - \dots$$

$$\left. + \frac{(a+b)^4}{4} + \frac{(n+1)}{1!}\frac{(a+b)^5}{5}A_1 + \frac{(n+1)(n+2)}{2!}\frac{(a+b)^6}{6}A_1^2 + \dots\right\}$$

$$\int \frac{b^{2n}}{r^{2n+2}}\sin^3\theta\,d\theta = B_{2n}\left\{\frac{2}{1.3} + \frac{(n+1)}{1!}\frac{2}{2.4}A_1 + \frac{(n+1)(n+2)}{2!}\frac{2}{3.5}A_1^2 + \dots\right.$$

$$- \frac{(a+b)}{1} - \frac{(n+1)}{1!}\frac{(a+b)^2}{2}A_1 - \frac{(n+1)(n+2)}{2!}\frac{(a+b)^3}{3}A_1^2 - \dots$$

$$\left. + \frac{(a+b)^3}{3} + \frac{(n+1)}{1!}\frac{(a+b)^4}{4}A_1 + \frac{(n+1)(n+2)}{2!}\frac{(a+b)^5}{5}A_1^2 + \dots\right\}$$

These formulæ may be applied to each term of the integral, and the result is thus obtained in terms of A_1 and B_{2n}. We next expand these in powers of a^2, and rearrange the result in their several orders. Finally on reintroducing R I find the following result:—

$$\tfrac{1}{8}\pi k'\rho\,\frac{v^2 a^2 b^2}{R^2}\left\{1 - \frac{8}{3.5}\frac{a}{R} - \frac{4a(4a+7b)}{3.5.7R^3} - \frac{a(8a^4 + 36a^2b^2 + 21b^4)}{3.5.7R^5}\right.$$

$$- \frac{a}{R^7}\left(\tfrac{32}{693}a^6 + \tfrac{8}{21}a^4b^2 + \tfrac{4}{7}a^2b^4 + \tfrac{1}{6}b^6\right)$$

$$\left. - \frac{5}{2}\frac{b^2}{R^2} - \frac{b^2}{R^4}\left(\tfrac{5}{2}a^2 + \tfrac{8}{3}ab - \tfrac{15}{8}b^2\right) - \frac{b^2}{R^6}\left(\tfrac{5}{2}a^4 + \tfrac{16}{3}a^3b + \tfrac{5}{12}a^2b^2 - \tfrac{88}{15}ab^3 + \tfrac{5}{16}b^4\right)\right\}\quad (6)$$

This gives the force on the sphere A towards B as arising from the tangential component of the impacts, in so far as it is due to those parts of A from which B is completely visible. It remains to consider the regions whence B is only partially visible.

By proceeding in the same way as before I find, for the portion of the result which does not involve that part of the screen which is represented by the dotted circle in Fig. 3, after the integration with respect to ϕ,

$$X = \frac{1}{2\pi} k'\rho v^2 \int_{\frac{1}{2}\pi - \beta}^{a} [\sin\beta\cos^2\beta\,(\tfrac{1}{2} - \tfrac{3}{2}\cos^2\theta)\cos^{-1}(-\cot\beta\cot\theta)$$
$$+ \cot\beta\,(\tfrac{1}{2} - \tfrac{3}{2}\sin^2\beta)\cos\theta\,\sqrt{(\sin^2\theta - \cos^2\beta)}]\sin\theta\,d\theta$$

$$Z = \frac{1}{2\pi} k'\rho v^2 \int_{\frac{1}{2}\pi - \beta}^{a} [\sin^2\beta\cos\beta\,(\tfrac{3}{2}\cos^2\theta - \tfrac{1}{2})\cos^{-1}(-\cot\beta\cot\theta)$$
$$+ (1 - \tfrac{3}{2}\cos^2\beta)\cos\theta\,\sqrt{(\sin^2\theta - \cos^2\beta)}]\sin\theta\,d\theta$$

The last terms in each of these expressions are integrable as they stand, and for the first terms of each

$$\int (\tfrac{3}{2}\cos^2\theta - \tfrac{1}{2})\cos^{-1}(-\cot\beta\cot\theta)\,d\theta$$
$$= \tfrac{1}{2}\sin^2\theta\cos\theta\cos^{-1}(\cot\beta\cot\theta) + \tfrac{1}{2}\cos\beta\,\sqrt{(\sin^2\theta - \cos^2\beta)}$$

In proceeding to the limits, it is to be noted that $\cos^{-1}(-\cot\beta\cot\theta)$ is π, when $\theta = \tfrac{1}{2}\pi - \beta$. Thus after integration we shall have, for that portion which depends on the term involving $\cos^{-1}(-\cot\beta\cot\theta)$, when $\theta = \tfrac{1}{2}\pi - \beta$,

$$X = +\frac{1}{2\pi}k'\rho v^2 . \frac{\pi}{2}\sin^2\beta\cos^4\beta, \quad Z = -\frac{1}{2\pi}k'\rho v^2 . \frac{\pi}{2}\sin^3\beta\cos^3\beta$$

Now, since the resultant force is given by $Z\cos(\beta - \theta) + X\sin(\beta - \theta)$, this portion of the resultant is equal to $-\tfrac{1}{4}k'\rho v^2\sin^2\beta\cos^3\beta\sin\theta$, where θ has the meaning indicated in Fig. 3.

The force due to the portion of the screen represented by the dotted circle is given by the result (5), namely $\tfrac{1}{4}k'\rho v^2\sin\beta\cos\beta\sin^2\alpha\cos\alpha\sin\theta$, when α is put equal to ZC, which is equal to $\tfrac{1}{2}\pi - \beta$; it is, therefore, equal to $\tfrac{1}{4}k'\rho v^2\sin^2\beta\cos^3\beta\sin\theta$, and exactly annuls the term referred to above as resulting from the term in $\cos^{-1}(-\cot\beta\cot\theta)$ with $\theta = \tfrac{1}{2}\pi - \beta$. Thus I find that the complete values for X and Z, inclusive of the dotted circle, are given by

$$X = \frac{1}{4\pi}k'\rho v^2\,\{-\sin\beta\cos^2\beta\sin^2\alpha\cos\alpha\cos^{-1}(-\cot\beta\cot\alpha)$$
$$+ \tfrac{1}{3}\cot\beta\,[\sin^2\alpha - \cos^2\beta - 3\sin^2\alpha\sin^2\beta]\,\sqrt{(\sin^2\alpha - \cos^2\beta)}\}$$

$$Z = \frac{1}{4\pi}k'\rho v^2\,\{+\sin^2\beta\cos\beta\sin^2\alpha\cos\alpha\cos^{-1}(-\cot\beta\cot\alpha)$$
$$- \tfrac{1}{3}[\sin^2\alpha - \cos^2\beta - 3\sin^2\alpha\sin^2\beta]\,\sqrt{(\sin^2\alpha - \cos^2\beta)}\}$$

These formulæ may be shown to be equally true of the case where less than half of B is visible, and they are therefore applicable throughout. The component force of A towards B is equal to $Z \cos (\beta - \theta) + X \sin (\beta - \theta)$, and becomes

$$\frac{1}{4\pi} k' \rho v^2 \left\{ \sin \beta \cos \beta \sin^2 \alpha \cos \alpha \sin \theta \cos^{-1} (- \cot \beta \cot \alpha) \right.$$

$$\left. - \frac{\sin \theta}{3 \sin \beta} [\sin^2 \alpha - \cos^2 \beta - 3 \sin^2 \alpha \sin^2 \beta] \sqrt{(\sin^2 \alpha - \cos^2 \beta)} \right\}$$

In this formula θ is used in the sense indicated by Fig. 3.

This expression has to be multiplied by $a^2 \sin \theta \, d\theta \, d\phi$, and integrated from $\phi = 2\pi$ to 0, and from $\theta = \cos^{-1} (a - b)/R$ to $\cos^{-1} (a + b)/R$. The integration with respect to ϕ merely involves multiplication by 2π. For the integration with respect to θ, I change the variable to ψ, and, as before, develop the expression and effect the various integrations.

The final outcome of some tedious analysis is that the result is the same as the last three terms of (6) with the sign changed. Hence, when we add this contribution to the force to (6), those three terms simply disappear.

Thus the tangential component of the impacts gives as a resultant, say F'_A, acting on the sphere A towards B,

$$F'_A = \tfrac{1}{8} \pi k' \rho \, \frac{v^2 a^2 b^2}{R^2} \left\{ 1 - \frac{8}{3.5} \frac{a}{R} - \frac{4a (4a + 7b)}{3.5.7 R^3} - \frac{a (8a^4 + 36a^2 b^2 + 21 b^4)}{3.5.7 R^5} \right.$$

$$\left. - \frac{a}{R^7} \left(\tfrac{32}{693} a^6 + \tfrac{8}{21} a^4 b^2 + \tfrac{4}{7} a^2 b^4 + \tfrac{1}{6} b^6 \right) \right\} \quad \ldots (7)$$

As a rough approximation we have

$$F'_A = \tfrac{1}{8} \pi k' \rho \, \frac{v^2 a^2 b^2}{(R + \tfrac{4}{15} a)^2}$$

If we form F'_B, or the force acting on the other sphere, it is clear that $F'_A - F'_B$ has a form similar to that found previously for $F_A - F_B$, but it has the opposite sign. Hence, as the result of the tangential component of the bombardment, the larger sphere is less strongly urged towards the smaller one, than the smaller towards the larger.

On comparing (7) with the result for F_A in (4), we see that they only differ in the signs of all the terms after the first, and in the fact that k' replaces k. The result is of so simple a character that it is probable that it may be derived by some elementary considerations which escape me.

Thus, including both the tangential and normal components, we have

$$F_A + F'_A = \tfrac{1}{8} \pi \rho \, \frac{v^2 a^2 b^2}{R^2} \left\{ (k + k') + \frac{8}{3.5} \frac{a}{R} (k - k') + \frac{4a (4a + 7b)}{3.5.7 R^3} (k - k') \right.$$

$$\left. + \frac{a (8a^4 + 36a^2 b^2 + 21 b^4)}{3.5.7 R^5} (k - k') + \frac{a}{R^7} \left(\tfrac{32}{693} a^6 + \tfrac{8}{21} a^4 b^2 + \tfrac{4}{7} a^2 b^4 + \tfrac{1}{6} b^6 \right) (k - k') \right\}$$

$$\ldots \ldots (8)$$

In the case where the momentum of the impinging corpuscles is completely absorbed by the surface struck, we have $k = k' = 1$, and the force is

$$\tfrac{1}{4}\pi\rho\,\frac{v^2 a^2 b^2}{R^2}$$

Thus, in this case the force varies rigorously as the inverse square of the distance. No doubt this simple result may be proved much more shortly in the case of the complete absorption of the momentum of impacts*. In any other case the result can only be regarded as approximate, because we have neglected reflected particles. But in general it seems certain that the interaction between the two spheres will not be equal and opposite.

§ 3. *Repulsion of Radiation.*

When the Lesagian corpuscles deliver their whole momentum on impact, we have the exact converse of the case of radiation, for in the one case we consider all the particles which converge on to a given element of surface, and in the other case they all diverge. A corpuscular theory of light would give the same result as the electro-magnetic theory as regards repulsion; hence we see that two radiating and perfectly absorbing spheres at the same temperature will repel one another rigorously as the inverse square of the distance†. The case of a perfectly reflecting sphere which receives radiation

---

* See a note to the following section.

† Professor Larmor has given me a direct proof of the above result. This I paraphrase in my own words as follows :—

A sphere radiating from its surface in the manner of a perfect radiator may be replaced by a uniform distribution of radiating spherules inside it, if we suppose the radiations from the several spherules not to interfere with one another. This follows from the fact that the radiation issuing towards any zenith distance is proportional to the depth of the crowd of spherules beneath it in the given direction. Now for a sphere that depth is a chord of the sphere, and is therefore proportional to the cosine of the zenith distance. Hence the law of radiation of the crowd of spherules is the same as the natural law of radiation from the surface itself.

Suppose that one of the spherules P emits n' corpuscles of mass m with velocity v per unit time, and let it be distant r from the centre A of an absorbing sphere of radius a. With origin at the spherule measure colatitude θ from PA, and longitude from some fixed plane passing through PA.

The number of particles emitted per unit time through solid angle $\sin\theta\,d\theta\,d\phi$ is $n'\sin\theta\,d\theta\,d\phi/4\pi$. Each of them carries momentum mv, and therefore the component of momentum along PA absorbed per unit time by the sphere A is

$$\frac{n'mv}{4\pi}\sin\theta\cos\theta\,d\theta\,d\phi$$

To find the whole thrust on the sphere A we must integrate this from $\phi = 2\pi$ to 0, and from $\theta = \sin^{-1}\dfrac{a}{r}$ to 0.

Hence the resultant repulsion exercised by the spherule on the sphere is $\tfrac{1}{4}n'mv\,a^2/r^2$.

is analogous to the case treated in § 1, where we consider only the effect of the normal component of the impacts. If two perfectly absorbing spheres have different temperatures, the action and reaction between them will not be equal and opposite. The analogue of this case in the Lesagian hypothesis would be that the velocities of the particles which strike one of the spheres, should be different from those which strike the other.

If one side of a body of any shape be at a higher temperature than the other side, the body will be subject to a force tending to propel the cooler side forward, and to drag the warmer side after it. This follows from the fact that the recoil of the emission on the warmer side is greater than that on the cooler side. The result that a hot sphere will pursue a cold one is a special instance of this more general conclusion.

In the Lesagian hypothesis the pressure has a definite relationship to the amount of energy received per unit area and per unit time, and the law is the same as that which governs the relationship between radiation and the recoil of light. The result has so great a physical interest that it seems worth while to investigate this matter more closely. We have already seen in § 1 that the number of Lesagian corpuscles, which strike unit area in unit time is $\Sigma \dfrac{nv \cos \theta}{4\pi} \delta\omega$, and each corpuscle before impact carries energy $\frac{1}{2}mv^2$.

Now suppose that there are N spherules arranged uniformly in a sphere B, of radius b, whose centre is distant R from the centre of the sphere A.

Since each spherule repels the sphere A inversely as the square of the distance from its centre, it follows, as in the theory of attractions, that the aggregate of them repels inversely as the square of the distance between the centres of the spheres A and B.

Hence the total repulsion of the crowd of spherules must be equal to

$$\tfrac{1}{4}Nn'mv\,\frac{a^2}{R^2}$$

It remains to find the value of Nn', in terms of the equivalent radiation from the surface of the sphere B.

Nn' is the total number of corpuscles emitted per unit time by the whole crowd, and this must be equal to the total number emitted from the surface of the sphere B.

Now we have taken in the text above n to represent the number of corpuscles emitted from unit area of the surface of the sphere. Hence we have

$$Nn' = 4\pi b^2 n$$

It follows that the repulsion between the spheres is

$$\pi nmv\,\frac{a^2b^2}{R^2}$$

But it appears from the latter portion of § 3 that

$$nm = \tfrac{1}{2}\sigma v = \tfrac{1}{4}\rho v$$

Thus the repulsion is $\frac{1}{4}\pi\rho\,\dfrac{v^2a^2b^2}{R^2}$, and this is identical with the result obtained at the end of § 2 above.

Hence if the surface completely absorbs all the energy on impact, we can at once find I, the total absorption of energy. It is given by

$$I = \iint \tfrac{1}{2}mv^2 \cdot \frac{nv\cos\theta}{4\pi} \sin\theta\, d\theta\, d\phi = \tfrac{1}{8}nmv^3$$

But nm is the mass of the Lesagian medium per unit volume, and has been denoted by ρ, and therefore

$$I = \tfrac{1}{8}\rho v^3$$

Complete absorption of energy corresponds to the case $k = 1$, and therefore the Lesagian pressure is $p = \tfrac{1}{6}\rho v^2$.

Hence

$$p = \frac{4I}{3v}$$

This result is the converse of the case of a corpuscular theory of radiation, and I is then the radiation of the surface, whilst p is the pressure of radiation.

We shall now see how this same result may be obtained, when the subject is considered from the point of view of radiation.

Suppose that $n\cos\theta\,\delta\omega/\pi$ be the number of corpuscles of mass m emitted per unit time with velocity v from unit area of a surface through an elementary cone $\delta\omega$ in zenith distance θ. The total number of corpuscles emitted per unit time is n, because

$$\int_0^{2\pi}\int_0^{\frac{1}{2}\pi} \frac{n}{\pi}\cos\theta\sin\theta\, d\theta\, d\phi = n$$

Each corpuscle carries $\tfrac{1}{2}mv^2$ energy, and the radiation is given by

$$I = \tfrac{1}{2}mnv^2$$

But mn has not the same meaning that it had in § 1, and therefore we must consider what it is.

The number of particles radiated per unit area and time through the elementary cone $\delta\omega$ towards zenith distance θ is $n\cos\theta\,\delta\omega/\pi$, and therefore for a small element of area δs it is $n\cos\theta\,\delta\omega\,\delta s/\pi$.

We may then concentrate the radiation from δs at its centre, and consider the distribution of corpuscles emitted. Since the corpuscles have mass m and move with velocity v, the total mass of the particles at any moment in an element of volume $r^2\delta\omega$ distant r from the centre of δs is $\dfrac{mn}{\pi v}\cos\theta\,\delta\omega\,\delta s$.

Hence the density of corpuscles radiated from δs at a point distant r from it, in zenith distance θ, is $\dfrac{mn\cos\theta}{\pi v r^2}\delta s$.

We may now find the density at a point distant z from an infinite plane radiating surface. Take origin in the plane vertically under the point at which the density is to be found, and let ρ, ϕ be the distance and azimuth of an element of radiating surface. For such an element

$$\delta s = \rho\,\delta\rho\,\delta\phi, \quad \cos\theta = \frac{z}{\sqrt{(\rho^2+z^2)}}, \quad r^2 = \rho^2 + z^2$$

Then the density at the point under consideration is the sum of the contributions of all elements of the plane, and if we denote that density by σ, we have

$$\sigma = \int_0^{2\pi}\int_0^{\infty} \frac{mn}{\pi v}\frac{z}{(\rho^2+z^2)^{\frac{3}{2}}}\rho\,d\rho\,d\phi = \frac{2mn}{v}$$

Hence
$$I = \tfrac{1}{4}\sigma v^3$$

The same result will be true infinitely near a curved radiating surface.

We found above in considering Lesagian bombardment that $I = \tfrac{1}{8}\rho v^3$. Hence it appears that $\sigma = \tfrac{1}{2}\rho$. This result might have been foreseen, because near the plane half the corpuscles are screened off, and so the density must be half that in free space.

The normal recoil of the radiation is clearly equal to the normal component of the momentum radiated per unit time and area, and therefore

$$p = \int_0^{2\pi}\int_0^{\frac{1}{2}\pi} \frac{n\cos\theta}{\pi}\,.\,mv\cos\theta\,.\,\sin\theta\,d\theta\,d\phi$$

Completing the integration and substituting for mn its value, we have

$$p = \tfrac{1}{3}\sigma v^2 = \frac{4I}{3v}$$

The density of energy in space is clearly $\tfrac{1}{2}\sigma v^2$, and this is equal to $\tfrac{3}{2}p$ or $2I/v$.

In the electro-magnetic theory of light the density of energy in space is also $2I/v$, but I understand that the pressure is only half that computed from the corpuscular theory and is equal to $\tfrac{2}{3}I/v$; thus in that theory the density of energy in space is three times the pressure.

§ 4. The Resistance to a Sphere moving with Uniform Velocity.

We may find the force on such a sphere by imparting to the Lesagian corpuscles a uniform drift in superposition on their common velocities v.

Let us first find the pressure on a plane unit area. Take the normal to the area as axis of Z, and suppose that the direction cosines of the axis of an infinitesimal cone $\delta\omega$ are $\sin\theta\cos\phi$, $\sin\theta\sin\phi$, $\cos\theta$. Then, if U, V, W, are the components of the uniform drift of the corpuscles, the velocity of the corpuscles whose direction of motion is parallel to the axis of the cone is

$$v + U \sin \theta \cos \phi + V \sin \theta \sin \phi + W \cos \theta$$

The momentum towards the nadir communicated by this class of corpuscles per unit time is therefore

$$\frac{k\rho}{4\pi} (v + U \sin \theta \cos \phi + V \sin \theta \sin \phi + W \cos \theta)^2 \cos^2 \theta \, \delta\omega$$

We must then take $\delta\omega = \sin \theta \, d\theta \, d\phi$, and integrate through half of angular space. The integration with respect to ϕ may be effected at once, and the expression becomes

$$\tfrac{1}{2} k\rho \int_0^{\frac{1}{2}\pi} [v^2 + \tfrac{1}{2}(U^2 + V^2) \sin^2 \theta + W^2 \cos^2 \theta + 2v W \cos \theta] \cos^2 \theta \, d\theta$$

$$= \tfrac{1}{2} k\rho \int_0^1 [v^2 + \tfrac{1}{2}(U^2 + V^2)(1 - \mu^2) + W^2 \mu^2 + 2v W \mu] \mu^2 \, d\mu$$

$$= \tfrac{1}{6} k\rho [v^2 + \tfrac{1}{5}(U^2 + V^2 + W^2) + \tfrac{2}{5} W^2 + \tfrac{3}{2} v W]$$

Now, suppose a sphere of radius 'a to be moving with velocity u in the direction of the axis of Z, and that $a^2 \sin \theta \, d\theta \, d\phi$ is any element of its surface.

When we reduce the sphere to rest, we have the case just considered, and W of the preceding investigation is equal to $u \cos \theta$. The portion of the pressure corresponding to the terms $v^2 + \tfrac{1}{5}(U^2 + V^2 + W^2)$ is the same all round the sphere, and will produce no effect; it may, therefore, be omitted.

Then the component of the pressure on the element in the direction of the new axis of Z is

$$\tfrac{1}{6} k\rho \left[\tfrac{2}{5} u^2 \cos^2 \theta + \tfrac{3}{2} vu \cos \theta \right] \cos \theta \, . \, a^2 \sin \theta \, d\theta \, d\phi$$

This must be integrated all over the sphere. After integration with respect to ϕ, which merely involves multiplication by 2π, we have

$$\tfrac{1}{3} \pi k\rho a^2 \int_0^\pi \left(\tfrac{2}{5} u^2 \cos^2 \theta + \tfrac{3}{2} vu \cos \theta \right) \cos \theta \sin \theta \, d\theta$$

$$= \tfrac{1}{3} \pi k\rho a^2 \int_{-1}^1 \tfrac{3}{2} vu \mu^2 \, d\mu = \tfrac{1}{3} \pi k\rho a^2 vu$$

If the whole momentum is absorbed, $k = 1$.

Next consider the tangential component of the momentum. In the investigation of § 2, put $l = 0$, $m = 0$, $n = 1$; $l' = \sin \theta \cos \phi$, $m' = \sin \theta \sin \phi$, $n' = \cos \theta$. Then the components of force are

$$X = -\frac{1}{4\pi} k' \rho \Sigma \sin \theta \cos \theta \cos \phi \, [v + U \sin \theta \cos \phi + V \sin \theta \sin \phi + W \cos \theta]^2 \, \delta\omega$$

$$Y = \dots\dots\dots\dots\dots \sin \phi \dots\dots\dots\dots\dots\dots\dots\dots\dots\dots\dots\dots$$

$$Z = 0$$

These must be integrated through half of angular space. Effecting the integration with respect to ϕ, we have

$$X = -\tfrac{1}{4}k'\rho \int_0^{\frac{1}{2}\pi} \sin\theta \cos\theta \, (vU\sin\theta + UW\sin\theta\cos\theta)\sin\theta\, d\theta$$

$$Y = -\tfrac{1}{4}k'\rho \int_0^{\frac{1}{2}\pi} \sin\theta \cos\theta \, (vV\sin\theta + VW\sin\theta\cos\theta)\sin\theta\, d\theta$$

From this we easily find

$$X = -\tfrac{1}{8}k'\rho\,(vU + \tfrac{8}{15}UW), \qquad Y = -\tfrac{1}{8}k'\rho\,(vV + \tfrac{8}{15}VW)$$

In applying this to the case of the sphere, we have, as before, $W = u\cos\theta$, and if we take the meridian as the X axis of the preceding investigation, $U = -u\sin\theta$, and $V = 0$. Hence

$$X = \tfrac{1}{8}k'\rho\,(vu\sin\theta + \tfrac{8}{15}u^2\sin\theta\cos\theta)$$

The component force on the sphere, parallel to the motion u, is $-X\sin\theta$.

Hence, the whole force on the sphere, due to the tangential component, is

$$\tfrac{1}{8}k'\rho a^2 \iint (vu\sin\theta + \tfrac{8}{15}u^2\sin\theta\cos\theta)\sin^2\theta\, d\theta\, d\phi$$

$$= \tfrac{1}{4}\pi k'\rho a^2 \int_0^{\frac{1}{2}\pi} vu\sin^3\theta\, d\theta = \tfrac{1}{3}\pi k'\rho a^2 vu$$

For complete absorption of momentum $k' = 1$. Hence, in this case the normal and tangential components contribute an equal amount to the resistance, which becomes

$$\tfrac{2}{3}\pi\rho a^2 vu$$

The uniform normal pressure, say p, on the sphere when at rest is $\tfrac{1}{6}\rho v^2$. Hence, the resistance is

$$4\pi a^2 p\,\frac{u}{v}$$

I take it that a similar formula gives the resistance to a radiating sphere, because the recoil corresponding to the emission of a particle is exactly equal to the momentum communicated by one impinging.

§ 5.   Summary.

Various hypotheses may be adopted as to the form and constitution of the elementary portions of matter and of Lesage's ultramundane corpuscles. In this paper I consider the elementary portion of matter to be a sphere, and I suppose the sphere to be either smooth or rough, and the corpuscles to be either perfectly elastic or inelastic in their collisions.

With perfectly smooth spheres and perfectly elastic corpuscles, it is clear that the total energy of the system remains unchanged when two spheres, immersed in the medium, are made to approach or recede from one another. As no work is done by such movements, there can be no force on either sphere. It would be excessively difficult to prove the vanishing of the force between the two spheres by a detailed examination of the impacts of the corpuscles, because it would be necessary to take into account the corpuscles which are reflected from either sphere so as to strike the other. It is, however, certain that in this case there would be no force, and therefore imperfect elasticity or roughness in the spheres are necessary conditions for the applicability of Lesage's theory.

In the case of partial elasticity and roughness, it would be even more difficult than in the former case to trace the effect of reflected corpuscles which strike the other sphere. But in proportion as the inelasticity and roughness increase, so will a solution, which only takes into account first impacts, increase in accuracy. We may fairly conjecture that a very moderate degree of inelasticity would suffice to make such a solution fairly correct. However this may be, no attempt is made in this paper to consider these repeated impacts.

A fundamental objection to the physical truth of Lesage's hypothesis lies in the fact that it demands a continual creation of energy at infinity to supply the gravific machinery. But Lord Kelvin has suggested a manner in which this physical absurdity might be avoided*. He supposes that the corpuscles are capable of absorbing energy in the form of internal agitation. On each impact some of the energy of translation is converted into energy of agitation, and a repartition of the energies of translation and agitation is effected by the mutual collisions of corpuscles according to Clausius's law. If, however, the work of this paper is correct, this suggestion will not serve to remove all the defects of Lesage's hypothesis.

I here suppose two spheres to be subjected to bombardment, and evaluate the effects of the normal and tangential components of the several impacts separately. It is thus possible to make various hypotheses as to the degrees of elasticity and roughness. It appears that neither the normal nor the tangential components of the impacts give rise to forces of attraction between the spheres which vary rigorously as the inverse square of the distance between their centres. In fact the resultant force acting on one of the spheres, due to the normal component, varies (approximately) as the inverse square of a distance equal to the distance between the two centres diminished by $\frac{4}{15}$ of the radius of the sphere in question; and the

---

* "On the Ultramundane Corpuscles of Lesage," *Phil. Mag.*, May, 1873, Vol. XLV., p. 321, fourth series.

resultant force, due to the tangential component, similarly demands the augmentation of the distance by $\frac{4}{15}$ of the radius of the sphere.

If the two spheres are unequal in size, then, as far as concerns the normal component, the diminution of the distance, so as to maintain the law of inverse square, is greater for the larger sphere than for the smaller; and the like is true for the augmentation in the case of the effect of the tangential component.

It follows that if the normal component is only effective, the larger sphere is more strongly impelled towards the smaller than the smaller towards the larger, and the converse is true for the separate action of the tangential component. In general, the sum of the two effects will not insure equality of action and reaction, nor the rigorous truth of the law of the inverse square. If these be necessary conditions for the truth of any theory of gravitation, then Lesage's hypothesis and Lord Kelvin's modification stand condemned. It is true that the inequality of action and interaction may be avoided by supposing that all elementary portions of matter are rigorously of the same size, but this still leaves the law of inverse square imperfectly fulfilled.

There is, however, one limiting case in which these particular imperfections in the theory are avoided. If the inelasticity of the corpuscles is complete and the roughness of the spheres such as absolutely to annul the tangential velocity of a corpuscle on impact—in other words, if the absorption of momentum on impact is total—the law of inverse square becomes rigorous, and action and reaction become equal. This supposition leaves the necessity for the creation of energy at infinity in its acutest form.

The case of the total absorption of energy on impact is strictly analogous to the repulsion of light, for the emission of light may be regarded as the exact converse of Lesage's mechanism. Thus the preceding investigation proves that two radiating and completely absorbing spheres at the same temperature repel one another rigorously as the inverse square of the distance between their centres.

If they are not at the same temperature they will tend to move (as indicated by Poynting) so that the cooler sphere leads. This appears to be a special case of a more general law, namely, that a body with one end hot and the other cold will tend to move with the cold end leading, because the recoil of the emission of radiation from the hot end will be greater than that from the cold end.

Another effect of the recoil of the emission of light may, perhaps, be of importance in solar physics. Poynting shows that the impulse of solar radiation at the earth is 5.8×10^{-5} dyne per square centimetre. Now the earth is distant 214.4 solar radii from the sun; hence the recoil of light at the sun's surface must be $5\cdot8 \times 10^{-5} \times (214\cdot4)^2 = 2\frac{2}{3}$ dynes per square centimetre. An "atmosphere" is about 10^6 dynes per square centimetre, and although the pressure just computed is very small compared with a terrestrial standard atmosphere, yet its effect may be worthy of consideration.

In the last section the resistance is evaluated which an isolated sphere suffers when moving with uniform velocity in a Lesagian gravific medium. It appears that the resistance is equal to the area of surface of the sphere multiplied by the pressure per unit area on a surface at rest, and by the ratio of the velocity of the sphere to the velocity of the gravific corpuscles. I take it that the same result will give the resistance to motion of a radiating sphere.

PART IV

PAPERS ON TIDES

SUPPLEMENTARY TO VOLUME I

22.

THE TIDAL OBSERVATIONS OF THE BRITISH
ANTARCTIC EXPEDITION, 1907.

[*Proceedings of the Royal Society*, Vol. 84 (1910), A, pp. 403—422.]

THE present investigation was undertaken at the request of Sir Ernest Shackleton; the expense of the reduction was defrayed by him, and this paper was communicated to the Royal Society by his permission. It will ultimately be republished as a contribution to the volume of the physical results of the expedition.

The first section, describing the method of observing, is by Mr James Murray. The second section explains the reduction of the observations and gives a comparison between the new results and those obtained by the *Discovery* in 1902–3. The third section is devoted to the discussion of certain remarkable oscillations of mean sea-level and to speculations as to their cause and meaning.

§ 1. *On the Method of Observing the Tides.*

Early in June, 1908, preparations were begun for the erection of a tide-gauge, the most important feature of which was to be a recording apparatus made from a modified barograph. Owing to various delays and mishaps it was not before the middle of July that the gauge was completed in its final form, and the continuous record begun, which was carried on for more than three months, subject only to the loss of half an hour weekly, while the paper was being changed.

Dr Mackay undertook the erection of the instrument, the apparatus was devised by the joint suggestions of Messrs David, Mackay, Mawson, and Murray, while Mr Day did the more delicate part of the work, namely, the alteration of the barograph.

The diagram (Fig. 1) shows the chief parts of the apparatus and their relations to one another. The ice is shown in section, with the tripod and recording apparatus erected on it. A weight A, consisting of a box filled with stones, rests on the sea bottom. A piece of iron tubing B is let through the ice vertically and fastened. It is filled with paraffin oil, the object of which is to prevent the wire being frozen in, an idea used with success by the officers of the *Discovery*. A wire C is taken from the weight on the sea bottom, passed through the oil-filled iron tube B, over the pulley D, and fixed to the end of the bamboo lever E, where it is kept

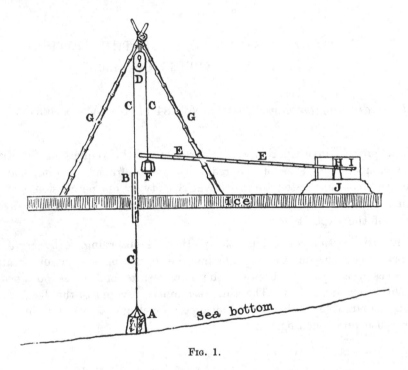

Fig. 1.

taut by the smaller weight F. The pulley D is suspended from a tripod of bamboo poles, of which two legs G are shown. The long lever E works on a spindle at H, and its short end I is connected by a cross-piece with the pen of the barograph. The details of this part are too small to be shown in this diagram, and will be illustrated in another figure. From this diagram there are omitted several parts, such as the guides which prevent the long lever from swinging during a blizzard, which are not essential to the under- standing of the instrument. The barograph was of necessity covered by a box to keep out the snow. The lever entered through a slit in the end of the box, and an arrangement of canvas kept the snow out. The box containing the barograph was raised on a little mound of snow J, in order to give the lever equal play above and below, or in other words, to allow of the mean sea-level being recorded about half-way up the drum. Of course the mean level had to be ascertained by a little observation.

The second diagram (Fig. 2) is a plan on a larger scale of the recording part of the apparatus. The circle A is the drum of the barograph; B is the pen making the tracing on the drum; C is the axle on which the lever bearing the pen works. This lever is continued beyond the axle to a distance rather greater than that of the part bearing the pen. This end of the lever D is made much heavier than the other. The long bamboo lever E, of which only a small part is shown, is borne on an axle F, which is in line with the axle C in the barograph, though of course quite unconnected with it, being outside the glass box of the barograph. Attached to the end of the bamboo is a stout wire G, bent round so that it passes under the end D of the pen lever, which rests upon it by its own weight, and rises and falls with it, but, being quite free from it, is not affected by any vibration of the bamboo under the influence of the wind. The barograph pen has, of course, been uncoupled from the aneroid capsules, which are not indicated in the plan.

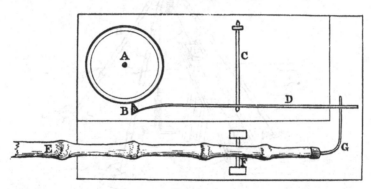

FIG. 2.

Dr Mackay, with much assistance from Prof. David, had the tide-gauge set up, all but the recording part, by June 22. In order to utilise the facilities we now had for noting the changes of level, while waiting for the recording instrument to be finished, Dr Mackay devised a very simple arrangement for ascertaining the amount of the tide. It was simply an inclined plane on which a paper marked with lines an inch apart was pinned. On this there slid a heavy block of wood which was attached to the end of the wire coming over the pulley. A lead pencil was inserted through a hole in the block of wood, which was kept in position by two guides.

This arrangement is shown in Fig. 3, which is drawn in perspective. The pulley A is suspended from the tripod B, B, B. The wire C is attached to the wood block D, which slides on the inclined board E between the guides F, F. The pencil G is fixed so as to project a little below the block of wood; H is the line traced by the pencil.

This simple device was not intended to give anything but a straight line but it was hoped by frequent inspections to ascertain the turn of the tide, and Dr Mackay kept vigil one night, and visited the gauge at intervals of about an hour.

Owing to a general slackness of the parts of the instrument the line traced was not a straight line, but a zigzag one, which proved of much greater interest. The pencil in descending did not follow the same course as it made going up, but swerved a little, and thus we got the first indications of what we believe to be seiches, at any rate of regular oscillations of much shorter intervals than those of the tides. The tracing obtained thus accidentally was too vague to enable us to count with certainty the number of periods per hour, but at any rate it demonstrated oscillations of a period of a few minutes and an amplitude of a few inches.

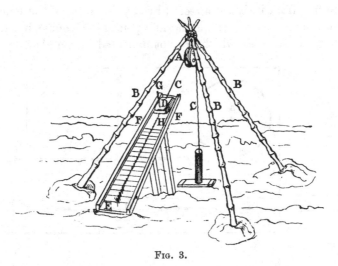

Fig. 3.

A number of records were taken in this rough manner till July 3, when the wire was found to be broken. A new situation was then selected for the tide-gauge, nearer the house, about 100 yards from shore, where the depth was 13 fathoms. The ice being now of considerable thickness, it was with no little labour that Dr Mackay, with the help of Mr Marston, got a new hole made to put the weight down.

By July 8 the apparatus was completed at the new place, this time with the recording attachment. A preliminary record was got from July 8 to 11, and the instrument was then stopped for readjustment. On July 14 it was finally started, and ran without mishap till nearly the end of October.

Prof. David usually changed the papers weekly. It was impossible to do this in the field, as it required bare hands. The barograph was therefore disconnected and carried to the house, where a new paper was put on and fresh ink put in the pen. The whole changing did not take more than half an hour.

The scale on which the curve was traced was about one-nineteenth of actuality—the long end of the lever being 11 feet and the short end 7 inches. The value of the factor of reduction of amplitude has therefore been taken to be 7/132 or 1/18·857.

One of the first records for a complete week was analysed. The curve appeared a simple one with one maximum daily, but a slight flattening of the minima indicated that other elements were present. The analysis showed that there was a smaller tide, having two maxima daily. The whole range of the tide at its highest was about 3 feet. The greatness of it surprised us, as the tide cracks usually showed a difference of level of not more than from 1 foot to 1½ feet. This may have been because the free edge of the crack had not room to sink to the full extent of the tide, but came on bottom, and the ice then sagged away to the level part.

Towards the end of October the tripod was blown down during a blizzard, and the wire was snapped. The ice was by this time so thick that it was found impracticable to cut another hole to put a weight down, and so the observations were discontinued.

The curve traced on the drum gave very frequent indications of seiches in the form of festooning, but the scale was so small and the clock-motion so slow that these indications were blurred and useless for study. It was intended, and would have been easy, to substitute a clock of about 10 times the speed belonging to a Callendar recorder which we had with us, but it was late in the season before we could try it, and the breaking of the wire put a stop to the attempt.

§ 2. *The Reduction of the Tidal Observations.*

The motion of the ice carrying the tide-gauge relatively to the sea-bed was transmitted to the drum by means of a lever, as explained above. Accordingly, it is very nearly exactly the chord of the arc turned through by the lever which ought to have been measured. Yet it is the arc itself which is recorded on the curvilinear scale on the drum. The angle turned through by the lever is, however, sufficiently small to permit us safely to neglect the correction in strictness required for the conversion of arcs to chords, and the arcs have been accepted as giving the changes of water level with sufficient accuracy.

The tidal record extended from July 14 to October 25, 1908, but the sheet which bore the record from October 12 to 18 is missing, and the record actually treated ends with October 11.

It was possible by means of a few interpolations to obtain an unbroken record from 0 h. astronomical time of July 14 to 23 h. of October 11. About an hour was generally lost once a week, while the paper was being changed, but it was always easy to complete the curve over this short interval by a pencil line, and this was regarded as equivalent to the actual curve.

On September 13 the pen failed to mark, but the curves on the 12th and 14th were unusually regular in character, so that a good interpolation for the 13th was easily obtained. The following is a list of the interpolated readings, and the hours are given inclusively in astronomical time :—

July 14, 0 h. to 3 h. (extrapolated); July 19, 19 h. to 23 h.;

September 12, 17 h. to 23 h.; September 13, 0 h. to 23 h.;

October 11, 22 h. and 23 h.

The errors of the clock do not seem to have been great enough to demand attention, and in fact, they are not always noted on the diagrams.

The clock was kept to apparent time, and was reset as the equation of time changed sensibly. But the scheme of reduction assumes that mean time has been used. This error may be taken into account with sufficient accuracy by certain changes in the true longitude of the place of observation, which was 166° 12′ E.

The observations were broken into three groups of a month each, for which the epochs were:

(1) 0 h., July 14; (2) 0 h., August 13; (3) 0 h., September 12, 1908.

To allow for the equation of time the longitude for the first month was taken as 6 m. of time or 1° 30′ further east than in reality; in the second month the longitude was regarded as correct, and in the third it was shifted 10 m. or 2° 30′ to the west. The correction for the last month is less satisfactory than for the other two, because at that time of year the equation of time is changing rapidly, and differs considerably at the beginning and end of the month.

The unit adopted in tabulating the height was 1/10 of an inch of the scale on the drum. Since 1 inch on the drum corresponds to 18·857 inches of water, the heights as derived from the harmonic analysis of the drum readings were converted to inches on multiplication by 1·8857.

For the tides M and O the observations were also treated as appertaining to a single period of three months, without regard to the equation of time; and a similar treatment was also extended to the tides S, K_2, K_1, and P_2 as will be explained more fully hereafter.

The reductions were made, under my supervision, by Mr F. Finch with my apparatus[*], and in the first instance the three months were discussed independently. The semi-diurnal tides were derived from months of 30 days, and the diurnal tides from months of 27 days. In this treatment it is

---

[*] *Roy. Soc. Proc.*, 1892, Vol. LII., p. 345, or *Scientific Papers*, Vol. I., Paper 6.

necessary to assume that the phase of the tide K_2 is the same as that of S_2, and that the amplitude of K_2 is 3/11ths of that of S_2. Similarly, we must assume identity of phases for K_1 and P, and that the amplitude of P is 1/3rd of that of K_1.

The following are the results* :—

	1 { July 14— Aug. 11	2 Aug. 12— Sept. 11	3 Sept. 12— Oct. 11
M_2 H = 2·55 in. κ = 358°		2·64 in. 12°	2·11 in. 6°
S_2 H = 1·08 in. κ = 293°		1·21 in. 282°	1·20 in. 267°
K_2 H = 0·29 in. κ = Same	as for S_2	0·33 in.	0·33 in.
K_1 H = 7·66 in. κ = 6°		8·57 in. 9°	10·06 in. 10°
P H = 2·56 in. κ = Same	as for K_1	2·86 in.	3·35 in.
O H = 6·95 in. κ = 357°		7·94 in. 6°	8·68 in. 359°

In these results there appears to be some evidence of a progressive change as the season advances, such as was noted in the case of the observations made by the *Discovery* in 1902–3†, and I shall return later to this subject. But in the case of the tides S_2, K_2, K_1, P, this might easily arise from an erroneous assumption as to the heights and phases of K_2 and P relatively to those of S_2 and K_1 respectively. It is therefore advisable to discuss these tides without making the assumptions which are necessary when each month is treated quite independently of the others.

In explaining my procedure, I adopt the notation of my paper " On an Apparatus for Facilitating the Reduction of Tidal Observations‡."

The heights and phases of the tides S_2, K_2, K_1, P are denoted respectively by H_s, κ_s; H'', κ''; H', κ'; H_p, κ_p.

* [A mistake of principle was made in the reductions (explained in Paper 23 below), and this has now been corrected by the augmentation of the phases of M_2 and O, as originally evaluated, by 1°. The same mistake also affected the reduction of the *Discovery* observations (see Vol. I., p. 372), but it was quite unimportant. It is however now corrected as far as concerns this present paper.]

† *National Antarctic Expedition*, 1901—4, *Physical Observations* (1908), p. 3; or my *Scientific Papers*, Vol. I. (1907), Paper 12, p. 372.

‡ *Roy. Soc. Proc.*, 1892, Vol. LII., pp. 345—389; or *Scientific Papers*, Vol. I. (1907), Paper 6.

The pair of harmonic constituents for diurnal tides, when 27 consecutive days are analysed, are denoted by $\mathfrak{A}_1$, $\mathfrak{B}_1$, and the theory shows that

$$\left.\begin{matrix}\mathfrak{A}_1\\\mathfrak{B}_1\end{matrix}\right\} = \frac{f'\,H'}{\mathfrak{F}_1}\,\frac{\cos}{\sin}\,(\kappa'-V'-13°\!\cdot\!29) - \frac{H_p}{\mathfrak{F}_1}\,\frac{\cos}{\sin}\,(\kappa'-V'-13°\!\cdot\!29+2h_0-\nu'$$
$$+26°\!\cdot\!58+\kappa_p-\kappa')$$

Similarly, when 30 consecutive days are analysed, and when P denotes the mean value for the month of the ratio of the cube of the sun's parallax to his mean parallax, the pair of semi-diurnal constituents are given by

$$\left.\begin{matrix}\mathfrak{A}_2\\\mathfrak{B}_2\end{matrix}\right\} = PH_s\,\frac{\cos}{\sin}\,\kappa_s + \frac{f''\,H''}{\mathfrak{F}_2}\,\frac{\cos}{\sin}\,(\kappa_s-2h_0+2\nu''-29°\!\cdot\!53+\kappa''-\kappa_s)$$

In treating each single month independently, we assumed $\kappa_p=\kappa'$, $\kappa_s=\kappa''$, $\frac{H_p}{H'}=\frac{1}{3}$, $\frac{H'}{H_s}=\frac{3}{11}$, but we now no longer make that supposition.

If we put

$$\left.\begin{matrix}a'\\b'\end{matrix}\right\} = \frac{f'}{\mathfrak{F}_1}\,\frac{\cos}{\sin}\,(V'+13°\!\cdot\!29);\qquad \left.\begin{matrix}a_p\\b_p\end{matrix}\right\} = \frac{1}{\mathfrak{F}_1}\,\frac{\cos}{\sin}\,(V_p-13°\!\cdot\!29)$$

$$\left.\begin{matrix}A'\\B'\end{matrix}\right\} = H'\,\frac{\cos}{\sin}\,\kappa';\qquad\qquad \left.\begin{matrix}A_p\\B_p\end{matrix}\right\} = H_p\,\frac{\cos}{\sin}\,\kappa_p$$

a', b', a_p, b_p, are known functions, and each month gives the pair of equations

$$\mathfrak{A}_1 = \quad a'A'+b'B'+a_pA_p+b_pB_p$$
$$\mathfrak{B}_1 = -\,b'A'+a'B'-b_pA_p+a_pB_p$$

Thus the three months afford six equations for the determination of A', B', A_p, B_p, from which the heights and phases of K_1 and P are easily found.

Again, if we put

$$\left.\begin{matrix}a_s=P\\b_s=0\end{matrix}\right\};\qquad \left.\begin{matrix}a''\\b''\end{matrix}\right\} = \frac{f''}{\mathfrak{F}_2}\,\frac{\cos}{\sin}\,(V''+29°\!\cdot\!53)$$

$$\left.\begin{matrix}A_s\\B_s\end{matrix}\right\} = H_s\,\frac{\cos}{\sin}\,\kappa_s;\qquad \left.\begin{matrix}A''\\B''\end{matrix}\right\} = H''\,\frac{\cos}{\sin}\,\kappa''$$

each month gives for the semi-diurnal tides the pair of equations

$$\mathfrak{A}_2 = \quad a_sA_s+b_sB_s+a''A''+b''B''$$
$$\mathfrak{B}_2 = -\,b_sA_s+a_sB_s-b''A''+a''B''$$

and the three months give six equations for determining A_s, B_s, A'', B'', from which the heights and phases of S_2 and K_2 are easily found.

On solving the diurnal group of equations by least squares, I find $H'=8\cdot311$ inches, $\kappa'=11°\,50'$, $H_p=1\cdot795$ inch, $\kappa_p=12°\,11'$. The ratio of H' to H_p is $4\cdot63$, instead of the 3 assumed from theoretical considerations in the separate treatment of the months, but the phases are virtually identical. The similar treatment of the semi-diurnal group gives

$$H_s = 0.938 \text{ inch}, \quad \kappa_s = 273° 25'; \quad H'' = 0.584 \text{ inch}, \quad \kappa'' = 257° 35'$$

The ratio of H_s to H'' is 1.605, instead of $3\frac{2}{3}$, as assumed from theory.

It thus appears that the theoretical hypotheses were considerably in error, and results probably more in accordance with the truth will be obtained from the several months if we assume

$$H' = 4.63 H_p, \quad \kappa_p - \kappa' = 0° 22'; \quad H_s = 1.605 H'', \quad \kappa_s - \kappa'' = 15° 50'$$

With these assumptions the three months now give :—

		1	2	3
K_1	$H =$ $\kappa =$	8.20 in. $10°$	8.31 in. $16°$	8.58 in. $10°$
P	$H =$ $\kappa =$	1.77 in. $11°$	1.79 in. $17°$	1.85 in. $11°$
S_2	$H =$ $\kappa =$	0.96 in. $272°$	0.93 in. $276°$	0.94 in. $272°$
K_2	$H =$ $\kappa =$	0.60 in. $256°$	0.58 in. $260°$	0.59 in. $256°$

The appearance of progressive seasonal change in this group of tides has now almost disappeared, although the middle month is slightly discordant from the other two.

It is interesting to note that, in the result of the treatment by least squares, κ' (for K_1) is practically identical with κ_p (for P), but that there is a considerable divergence between κ_s (for S_2) and κ'' (for K_2).

The difference between the phase of M_2 (which we may take as given by $\kappa_m = 6°$) and that of S_2 given by $\kappa_s = 273°$ is very large, although their difference of speeds is not great. Hence we should expect that a small difference of speed in a semi-diurnal tide would make a sensible difference in phase.

If phase varies simply as difference of speed, we shall have the following results :—

$$\text{Speed of } S_2 - \text{speed of } M_2 = 1°.016 \text{ per hour}$$

$$\kappa_s - \kappa_m = 273° - 366° = -93°$$

$$\text{Speed of } K_2 - \text{speed of } S_2 = 0°.082 \text{ per hour}$$

Hence we ought to find

$$\frac{\kappa'' - \kappa_s}{\kappa_s - \kappa_m} = \frac{0.082}{1.016}, \quad \text{or} \quad \kappa'' = \kappa_s - \frac{82}{1016} \times 93° = \kappa_s - 7\frac{1}{2}°$$

As a fact, we find $\kappa'' = \kappa_s - 16°$, and thus the direction of the difference of phases is such as was to be expected, although the amount is not quite satisfactory. With tides of such small amplitude, however, and with only three months on which to rely, the amount of agreement is all that is to be expected.

The results of the analysis for the tides M_2 and O, when the months are taken independently, are given above. If, however, we neglect the equation of time, the whole period of three months may be treated as a single group of observations. In this way I obtain for M_2

$$H_m = 2·4233 \text{ inches}$$

$$\kappa_m = 6° 37'$$

If we take the three values of each of the quantities $H_m \cos \kappa_m$, $H_m \sin \kappa_m$, from our previous results, and form means of these functions, we obtain

$$H_m = 2·4195 \text{ inches}$$

$$\kappa_m = 5° 39'$$

The latter method has the advantage that it takes the equation of time into account; the former is somewhat more likely to eliminate casual inequalities. We may safely take $H_m = 2·42$ inches, $\kappa_m = 6°$, as being very near the truth.

Similarly the whole series when treated for the O tide gives

$$H_o = 8·1645 \text{ inches}$$

$$\kappa_o = 2° 8'$$

But the means of the three values of $H_o \cos \kappa_o$, $H_o \sin \kappa_o$, give the somewhat discordant result

$$H_o = 7·84 \text{ inches}$$

$$\kappa_o = 0° 54'$$

I should have expected the two evaluations to be closer together, as was the case with M_2, and I think we must accept

$$H_o = 8·0 \text{ inches}$$

$$\kappa_o = 1°$$

as being as nearly accurate as is possible from our data.

In the reduction of the *Discovery* observations it was known that there had frequently been a small change of the zero point in consequence of the shift in the ship, and I did not think it was worth while to attempt to combine the several months by least squares so as to separate the tides S_2 from K_2, and K_1 from P. I now think that it was a pity that the attempt was not made to separate them, and therefore I have gone back to the old work and discussed the numbers by least squares with the results given below. In the course of this revision it appeared that there had been a small mistake in the value assigned to P for each month, which, however, made little change in the values assigned to the tide S_2, and did nothing to remove the considerable discrepancies between the results from each of the 12 months.

FINAL TABLE OF RESULTS FOR *NIMROD*, TOGETHER WITH COMPARISON WITH *DISCOVERY*.

Nimrod, 1908	*Discovery*, 1902–3	*Discovery*, new reduction
M_2 $H = 2\cdot42$ in. $= 0\cdot202$ ft. $\kappa = \quad 6°$	$1\cdot966$ in. $= 0\cdot164$ ft. $11°$	
S_2 $H = 0\cdot94$ in. $= 0\cdot078$ ft. $\kappa = \quad 273°$	$1\cdot142$ in. $= 0\cdot095$ ft. $272°$	$1\cdot129$ in. $= 0\cdot094$ ft. $272°$
K_2 $H = 0\cdot584$ in. $= 0\cdot049$ ft. $\kappa = \quad 258°$	$0\cdot311$ in. $= 0\cdot024$ ft. $272°$	$0\cdot396$ in. $= 0\cdot033$ ft. $294°$
K_1 $H = 8\cdot31$ in. $= 0\cdot693$ ft. $\kappa = \quad 12°$	$9\cdot245$ in. $= 0\cdot770$ ft. $14°$	$10\cdot177$ in. $= 0\cdot848$ ft. $14°$
P $H = 1\cdot795$ in. $= 0\cdot150$ ft. $\kappa = \quad 12°$	$3\cdot082$ in. $= 0\cdot257$ ft. $14°$	$3\cdot228$ in. $= 0\cdot269$ ft. $3°$
O $H = 8\cdot0$ in. $= 0\cdot67$ ft. $\kappa = \quad 1°$	$9\cdot264$ in. $= 0\cdot772$ ft. $1°$	

The agreement between these two sets of constants, deduced from observations taken at places some 25 miles apart, seems to be very good. The later observations were taken further north than the earlier ones, and the greater value of M_2 in the more northerly series is probably a reality. The two days of observation made by Dr Wilson in 1904 close to the *Nimrod* station agree with our present results in indicating a slightly increased value of the semi-diurnal tide.

In discussing the *Discovery* tides, I was led to suspect that there were semi-diurnal nodal lines to the northward, but that the node for S_2 was nearer than that for M_2. The fall in the amplitude of S_2 agrees with this, and possibly the amplitude of M_2 has begun to increase as we go northward previously to its subsequent decrease to the zero value at the node.

The ratio of M_2 to S_2 for *Nimrod* is $2\cdot57$, and for *Discovery* $1\cdot74$; the former value is more nearly normal than the latter.

The sums of the heights of M_2, S_2, K_2, are respectively $3\cdot94$ inches for *Nimrod* and $3\cdot49$ for *Discovery*.

The sums for K_1, P, O are $18\cdot1$ inches for *Nimrod* and $22\cdot7$ inches for *Discovery*. Thus for *Nimrod* the greatest diurnal tides are $4\cdot6$ times as great as the greatest semi-diurnal tides, while for *Discovery* the greatest diurnal tides are 6·5 times as great as the greatest semi-diurnal tides. This again emphasises the diminishing importance of the semi-diurnal tides as we penetrate to the south.

It should be remarked that the difference of phase of K_2 from that of S_2 in the new reduction for *Discovery* is not in accordance with the theoretical considerations adduced in support of the corresponding difference for *Nimrod*. However, too much stress should not be placed on results derived from these very small tidal oscillations.

On the whole, I conclude that we now know the tidal constants at this part of the Antarctic Ocean with as much accuracy as is desirable, and I refer the reader to the discussion of the *Discovery* observations for the conclusions which may be drawn from the values found.

In discussing the *Discovery* observations, I saw reason to suspect a remarkable seasonal change in the amplitude and phase of the tide M_2; it is therefore interesting to see whether these new observations tend to confirm that conclusion. The results in my previous paper were discussed by means of curves, but I will now merely examine the matter numerically.

The results for each month which has been reduced, viz. 12 for *Discovery* and 3 for *Nimrod*, may be held to appertain to the middle of the month under consideration, that is to say 15 days after the corresponding epoch.

The following table exhibits the values of H_m and κ_m for each of the 15 months, together with the dates to which they may be held to apply. The new results are marked with an asterisk.

	Date	H_m	κ_m
		inches	°
	Apr. 21, 1903............	1·91	−11
	May 24, 1903............	2·20	− 4
	„ 27, 1902............	2·27	2
	June 20, 1902............	2·29	3
	„ 30, 1903............	2·33	5
	July 29, 1903............	2·41	10
*	„ 29, 1908............	2·55 (2·08)	− 4
	Aug. 8, 1902............	2·18	15
*	„ 28, 1908............	2·64 (2·15)	12
	„ 29, 1903............	2·18	16
	Sept. 7, 1902............	1·93	23
*	„ 27, 1908............	2·11 (1·72)	6
	Oct. 8, 1902............	1·74	27
	„ 28, 1902............	1·56	33
	Nov. 28, 1902............	1·21	32

In order to judge of the progression in the heights, we should note that the mean H_m for the northern place is 2·42, and for the southern is 1·97.

Hence we ought, perhaps, to reduce the three heights for *Nimrod*, viz., 2·55, 2·64, 2·11, in the proportion of 197 to 242. The corresponding numbers as so reduced are written in parentheses after the actual numbers. The progression in the heights appears to be fairly consistent, but that of the phases is not nearly so clear. The phase of the first of the *Nimrod* months is some 15° away from what we should expect if the progressive change is an actuality. If we convert this into time, it means that the high water should be changed by about half an hour to fit into the supposed progression. The middle month fits into its place fairly well, but the high water for the third month should be shifted some 40 minutes. Such changes are not, however, large, when we consider that the range from high to low water is only about 5 inches. On the whole, I should say that the new results do not tend to confirm the truth of the progressive change in any marked degree, but they can hardly be held to invalidate it.

If we examine the results of the three months for the O tides, we find some traces of a seasonal progression, for the heights are 6·95, 7·94, 8·64; but the progression of phases is again not clearly marked, for they are $-4°, +6°, -1°$. I was not able to detect any evidence of progression in the case of the O tide as observed by the *Discovery*.

§ 3. *On Sea-seiches in the Antarctic Ocean.*

In the course of the reduction of the tidal observations the mean daily heights of the water were computed, so as to furnish a cross verification of the summations necessary in the harmonic analysis. In view of the arduous conditions under which the observations were made it also seemed well to test the series of means, so as to detect any accidental shift in the zero of the gauge which might have occurred. Unfortunately, no such systematic examination of the *Discovery* observations had been carried out, because it was well known that there had been frequent small changes of zero due to the shift of the ship. It did not occur to me that a graphical illustration of mean sea-levels, known to be subject to somewhat frequent changes of zero, might give indications of anything worthy of notice.

A cursory examination of a table of the daily mean sea-levels of the present series at once revealed considerable inequalities. The paper on the drum was changed once a week, yet there was no sign of any weekly discontinuity, and the observers did not think there was any reason to suspect a change between each paper and the next. A zigzag of daily mean sea-levels was accordingly plotted, as shown on a reduced scale in the firm line of Fig. 4. I was surprised to see a somewhat regular rise and fall of the water with a period of about three days, for nearly five weeks on end. Although the rise and fall was then interrupted, this seemed to be a fact worth looking into.

A line drawn so as to bisect the zigzags clearly undergoes changes of considerable amount, for which it is only possible to guess the causes. Distant barometric changes and distant gales may be responsible for most of the effect. There are also probably annual and semi-annual meteorological tides, fortnightly and monthly astronomical tides, and some small apparent inequality with a period of a fortnight due to the residual effects of the tides of short period. But these causes obviously could not produce the shorter zigzags, so that we may consider these as being embroidered, to use M. Forel's phrase, on a slowly variable curve.

Local barometric changes must affect the mean sea-level, and pressure above the mean will correspond with depressed sea-level, at the rate of about $13\frac{1}{2}$ inches of water to one of mercury, and *vice versâ*. Mr James Murray has given me the mean barometric heights both in a tabular and in a graphical form. The means of pressure are given in civil time, while those of sea-level were computed according to astronomical time. I therefore made a rough estimate from the curve of the mean pressure according to the latter time. The mean pressure for the 90 days of observation was then found, and a correction was applied to the sea-levels at the rate of 14 inches of water to one of mercury above or below the estimated mean. A rather high value for the correction is taken, because it seemed desirable to give the barometric changes every possible chance of annulling the sea changes; and further, because, by the use of the factor 19 (instead of 18·857), the zigzags had been very slightly exaggerated. In any case, the correction is quite exact enough for such a rough allowance for barometric pressure as is possible.

The corrected mean sea-levels are shown in the dotted curve of Fig. 4. It will be seen that the zigzags are sensibly diminished, but not annulled, and that in one or two places a new maximum or minimum has been introduced. We may conjecture that distant barometric changes and distant gales may have annulled some maxima and minima which would otherwise have been visible.

For observations of this uncertain kind mathematical treatment for the detection of partially veiled periodicity seems inappropriate. I have therefore only examined the zigzag for maxima and minima and have noted their incidences in the following table. In five cases a mark of *quære* is added because another observer might deny the existence of a maximum or minimum which I conceived to be there, but partially masked by the general rise or fall of an ideal line bisecting the zigzags of the dotted line. In the second column of each half of the table I give the differences between the dates in days, and these numbers will give the period of the suspected inequality.

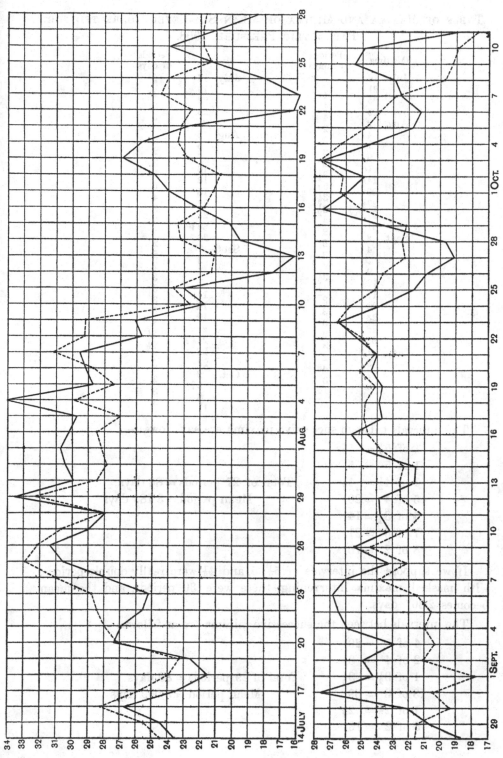

Fig. 4. Daily means of sea-level, and the same corrected for barometric pressure, referred to an arbitrary zero and expressed in inches.

TABLE OF MAXIMA AND MINIMA OF MEAN SEA-LEVEL CORRECTED FOR
BAROMETRIC PRESSURE, 1908.

Maxima	Periods in days	Minima	Periods in days
July 16	5	July 14	5
21	4	19	4
25	4	23	5
29	4	28	3
Aug. 2	2	31	3
4	3	Aug. 3	2
7	2	5	2
9	2	7	3
11	3$\frac{1}{2}$	10	3
14$\frac{1}{2}$	5$\frac{1}{2}$	13	5
20	3$\frac{1}{2}$	18	4
23$\frac{1}{2}$	3$\frac{1}{2}$	22	3
27	4	25	3
31	2	28 ?	2
Sept. 2	2	30	2
4	3	Sept. 1	2
7	2	3	2
9	3$\frac{1}{2}$	5	3
12$\frac{1}{2}$	4$\frac{1}{2}$	8	3
17	3	11	3
20	3	14	3
23	3	19	5
26 ?	2	21	2
28	3	24$\frac{1}{2}$?	3$\frac{1}{2}$
Oct. 1	2	27	2$\frac{1}{2}$
3	4	29	2
7 ?	3	Oct. 2	3
10		5 ?	3
		8	3
		11	3

The intervals between successive maxima are as follows :—

1 of 5$\frac{1}{2}$ = 5$\frac{1}{2}$

1 of 5 = 5

1 of 4$\frac{1}{2}$ = 4$\frac{1}{2}$ Total of 27 periods = 86 days

5 of 4 = 20   Mean period = 3·185 days

4 of 3$\frac{1}{2}$ = 14

7 of 3 = 21

8 of 2 = 16

If we suppose the intervals of 5$\frac{1}{2}$, 5, and 4$\frac{1}{2}$ were really double periods, with masked maxima intervening, there were 30 periods, and the mean becomes 2·867 days.

The intervals between the successive minima are as follows :—

4 of 5 = 20

2 of 4 = 8

1 of 3$\frac{1}{2}$ = 3$\frac{1}{2}$ Total of 29 periods = 89 days

13 of 3 = 39   Mean period = 3·069 days

1 of 2$\frac{1}{2}$ = 2$\frac{1}{2}$

8 of 2 = 16

If the intervals of five days were really double periods with masked minima intervening, there were 33 periods and the mean period becomes 2·697 days.

Taking both estimates as of equal weight we get a mean interval of 3·127 days, or allowing for possible masked maxima or minima, as explained above, of 2·782 days. I think then that there is some evidence of the existence of an oscillation with a period of about three days.

In a paper in the *Philosophical Magazine* (Jan. 1908, p. 88), Messrs Honda, Terada, and Isitani discuss "Secondary Undulations of Oceanic Tides" or sea-seiches. These seiches occur in bays, and they find that the period depends on the size and depth of the bay. In some bays the period is fairly constant, but in others it changes "continuously and through certain ranges." They show that for a bay of length l and depth h the main period T is given by the formula $4l \div \sqrt{(gh)}$, where g is gravity. The period as so computed is subject to a correction due to the opening into the sea, but as we only now want a very rough estimate of the period, the correction may be neglected. The formula is the same as that for the period of the uninodal seiche in a lake of length $2l$. The authors in fact regard the end of the bay as resembling the end of a lake, while the seaward opening is equivalent to the middle of the lake. Accordingly the second half of the lake, which would stretch out into the sea, is suppressed. The formula gives results in accordance with the seiches observed in many Japanese bays, and they remark that bays are also sometimes disturbed by seiches of shorter period, which they regard as transverse seiches from side to side of the bay, just as if it were an enclosed basin.

In none of the examples given by these authors has the seiche a period at all comparable with that of which we have reason to suspect the existence in the Antarctic Sea, but that affords no reason for refraining to apply the theory to such prolonged oscillations. In most inland lakes the seiches have periods of 10 minutes to one or two hours, yet in Lake Erie the seiche is found to have a period of 13 hours, while in the Lakes of Michigan and Huron conjointly a seiche of 45 hours is suspected[*]. Thus we have justification for the application of the theory to oscillations of very long period.

In the case of the Antarctic Sea, if there is a great bay running far back into the Antarctic continent behind the ice barrier, its length and depth are quite unknown. Hence there are elements of great uncertainty in the application of the theory.

It seems likely, at any rate, that the bay extends for a considerable

---

[*] Dr Anton Endrös, *Petermann's Geograph. Mitteilungen*, Heft ɪɪ., 1908.

distance, and speculations have even been made as to whether there may not be an arm of the sea stretching through to the Weddell Sea almost diametrically across what was supposed to be a continent.

It might, perhaps, be thought that the thick ice of the barrier would serve to damp out oscillations of sea-level; but, unless, indeed, the sea is solid to the bottom, I conceive that the ice would behave like an elastic skin, and would hardly exercise any damping effect on oscillations with a period of more than an hour or two.

It seems almost impossible that the remarkable changes of sea-level which are observed should arise from errors of observation, and if they exist at Backdoor Bay, the neighbouring sea along the barrier must necessarily also partake of the motion.   If the sea rises and falls, the barrier itself must move with it; and it may be suspected that it is subject to a true tidal rise and fall.

If we accept the existence of a sea-seiche with a period of three days, the formula gives some indication as to the length and depth of the bay behind the barrier. We cannot assume the sea to be very shallow, because if it were so it would inevitably be frozen solid to the bottom.   Moreover, a shallow sea would certainly be broken up by shoals, so that it could not oscillate as a single system.   A little consideration shows that to produce a seiche of three days the bay must be of enormous length, and for the reasons assigned it would be useless to assume it to be very shallow.   The few soundings near the barrier give depths of between 200 and 300 fathoms, and perhaps a somewhat smaller depth might suffice to allow of the required seiche.   I propose to guess the length of the bay and to find what depth of sea is required to produce a seiche of three-day period.

I guess then that the bay behind the barrier stretches past the South Pole and a little to the east of it as far as latitude 80°. Such an inlet would have a length of 25° to 30° of latitude.   It seems likely that if it is really an arm of the sea through to Weddell's Sea, with a constriction about the place where we place the end of the bay, the seiche would be much the same.

The length of our supposed bay in centimetres will be 25 or 30 times $60 \times 1.852 \times 10^5$ cm., and these I take as two assumed values of l.   On completing the multiplications I find that $4l$ will be $1\frac{1}{9} \times 10^9$ cm. or $1\frac{1}{3} \times 10^9$ cm.

The period of oscillation is three days, or 2.592×10^5 sec.; also g is 981. Thus, numbering our two alternatives as (1) and (2), we get:—

$$(1) \quad 2.592 \times 10^5 = \frac{1\frac{1}{9} \times 10^9}{\sqrt{981h}}; \qquad (2) \quad 2.592 \times 10^5 = \frac{1\frac{1}{3} \times 10^9}{\sqrt{981h}}$$

Whence

$$(1) \quad h = \frac{1}{981} \left(\frac{1\frac{1}{9} \times 10^4}{2\cdot592} \right)^2 \qquad\qquad (2) \quad h = \frac{1}{981} \left(\frac{1\frac{1}{3} \times 10^4}{2\cdot592} \right)^2$$

$$= 18{,}732 \text{ cm.} \qquad\qquad\qquad = 26{,}975 \text{ cm.}$$

$$= 102\cdot4 \text{ fathoms} \qquad\qquad\quad = 147\cdot5 \text{ fathoms}$$

Thus a sea of from 100 to 150 fathoms in such an immense bay as has been conjectured would oscillate with a period of three days, and the observed results are seen to be consistent with the existence of a deep inlet, almost or quite cutting the Antarctic continent in two.

Such a conclusion is interesting, but it would not be right to attribute to it a high degree of probability, because there are elements of uncertainty on every side.

In view of the interest of our result it has seemed well to revert to the observations made by Captain Scott's expedition, notwithstanding the known uncertainty in the zero of the gauge. I have therefore examined 175 days of the *Discovery's* record, viz., 113 days of 1902 and 62 days of 1903. No correction has been applied for barometric pressure, and thus periodic inequalities have doubtless sometimes been masked by contemporaneous changes of pressure, and perhaps by the shift of zero. Thus I should expect to find rather a larger proportion of long intervals between consecutive maxima and minima than in the *Nimrod* results as reduced for pressure. I found, in fact, on analysing the zigzag in the way already explained that, amongst the periods as deduced from maxima, there were :

2 of 7 d.,   1 of $6\frac{1}{2}$ d.,   10 of 5 d.,   1 of $4\frac{1}{2}$ d.

and amongst the periods, as deduced from minima, there were :

1 of 9 d.,   1 of $6\frac{1}{2}$ d.,   4 of 6 d.,   3 of 5 d.,   1 of $4\frac{1}{2}$ d.

Taking maxima and minima together there were 83 periods amounting to 323 days, thus giving a mean period of 3·9 days.

But if we postulate that periods from 7 days to $4\frac{1}{2}$ days were really double periods with masked maxima or minima intervening, and that the 9-day period is really triple, we get 105 periods for 323 days, with a mean of 3·1 days.

These results are generally confirmatory of the preceding ones, but seem to indicate a slightly longer period.

In this rough examination there is undoubtedly a danger of finding a false periodicity under the influence of unconscious bias. I thought it advisable therefore to examine other tidal records, for it might be possible to perceive periodicity even in cases where there was but small likelihood of its real existence. Colonel Burrard then kindly sent me tables of daily mean sea-levels for the year 1880 from May 1 to June 30 and from October 1 to November 30 for Aden, Karachi, Madras and Port Blair, Andaman Islands. These old observations were chosen because it had been usual at that time to have each daily mean "cleared" of the residual effects of the tides of short period, and thus one slight source of error was obviated. I also proceeded in the case of Aden to deduct the tides of long period, but as this correction clearly made no difference in the kind of inequality I was looking for, I did not carry out that laborious task in the other cases.

The tabulated numbers were then plotted out in a number of curves.

An imaginative investigator might possibly fancy he could detect signs of periodicity with a period of two or three days at Aden and at Port Blair, but as the range from crest to hollow was not more than half an inch, it seems safer to say that no periodicity could be traced.

In the curve for Madras there are considerable irregularities, but it seemed impossible even to imagine any periodicity. At Karachi there does seem to be an inequality with a period of two to three days and a range of two or three inches. A succession of waves with three to five crests one after the other is observable at several parts of the curve. It seems quite likely that sea-seiches may exist in the Indian Ocean, and Karachi would be well placed for observing them.

These Indian results were not corrected for barometric pressure, and it may be worth while hereafter to submit them to a more systematic examination. For the present, however, I am satisfied with the conclusion that periodicity is not to be seen in all cases, and that the oscillations of mean sea-level in the Antarctic Sea are many times as great as those in the Indian Ocean and Bay of Bengal. Thus it seems unlikely that imagination is responsible for the existence of the Antarctic sea-seiches, and we may hope that the investigations of Captain Scott's second expedition will throw some further light on the subject, and possibly also on the existence of a deep bay behind the barrier.

23.

ON A MISTAKE IN THE INSTRUCTIONS FOR THE USE OF
A CERTAIN APPARATUS IN TIDAL REDUCTIONS.

[*Proceedings of the Royal Society*, Vol. 84 (1910), A, pp. 423—425.]

THE apparatus is described in the *Proceedings of the Royal Society*, 1892, Vol. 52, p. 345 (or my *Collected Scientific Papers*, Vol. I. p. 216). I first correct an obvious misprint in Fig. 3, where 24β is inserted in place of 24ϵ.

In § 6 the incidence is determined of the exact mean solar hours of a given day amongst the hours of a special time scale, when the 12 h. of solar time is assumed to fall within half a special hour of an exact hour of special time. Starting from this 12 h. of solar time we proceed upwards by subtracting 1, 2...12 hours of solar time, but expressed in special time, adding 24 h. to the integral special hours when necessary so as to make that integral number lie between 0 and 23 inclusive. Similarly we proceed downwards by adding 1, 2...11 hours of solar time expressed in special time, and similarly reduce the integral number of special hours so that it shall lie between 0 and 23. There are thus 24 lines in the schedule, 12 lying above the middle and 11 below.

If t denotes an exact solar hour, τ an exact special hour, and δ an error expressed in special time, each line of the schedule is of the form

$$t = \tau + \delta$$

where both t and τ are whole numbers lying between 0 and 23 inclusive.

The frequency of the error δ is investigated, and is shown to conform to a certain law which is identically the same for each special hour from 0 to 23 inclusive.

Each harmonic tide goes through its period n times (with n equal to 1, 2, 3, 4, or 6) in its appropriate special day. The height of the water is supposed to be observed at the exact solar hours, and the problem is to determine the error in the result when the observations are deemed to appertain to the corresponding exact hours of special time.

M. M. H. van Beresteyn wrote to me from the Hague expressing his opinion that my procedure was erroneous, and I regret to find that his suspicion is well founded. Fortunately the error thus introduced is insignificant as regards practical tidal work.

I will not explain how I came to go wrong, but will consider the correct procedure and show how the rules of computation must be amended.

Let ω denote the speed of any one of the harmonic tides expressed in degrees per mean solar hour. Then since we may without loss of generality take the amplitude of the tide as unity and the phase as zero, the 24 observations at the mean solar hours will be 24 values of $\cos n\omega t$ corresponding to t equal to 0, 1, 2...23, and as before n is one of the numbers 1, 2, 3, 4 or 6. We might equally well have proceeded from $\sin n\omega t$ and it is on this account that the phase is immaterial for our discussion.

We have to translate $\cos n\omega t$ into special time. By the definition of a special day, ω when translated becomes 15° per special hour, and by the schedule of incidence t becomes $\tau + \delta$, thus we have

$$\cos n\omega t = \cos 15°n (\tau + \delta)$$

In my paper I virtually wrote 15°τ as θ, and 15°δ as x, so that x was error reduced to angular measure at the rate of 15° per special hour.

Hence the function to be considered is $\cos n (\theta + x)$, where x is subject to a certain known law of frequency, say $f(x)$.

It was at this point that I made the mistake, for I erroneously considered the function $\cos n (\theta - x)$.

The required mean value of $\cos n (\theta + x)$ is clearly to be determined from the fraction

$$\int_{-\infty}^{+\infty} f(x) \cos n (\theta + x) \, dx \div \int_{-\infty}^{+\infty} f(x) \, dx$$

The integral in the denominator was found correctly, but that in the numerator was wrong because of the wrong sign of x under the cosine.

The factor in the result which was given as $\cos n [\theta - \frac{1}{2}(a - b)]$ must be corrected so as to stand as $\cos n [\theta + \frac{1}{2}(a - b)]$.

It is proved that $a - b$ is equal to ϵ, where

$$\epsilon = 15° \left[1 - \frac{\text{m. s. day}}{\text{special day}} \right]$$

The correct final result is

$$\frac{1}{\mathcal{F}_n} \cos n (\theta + \tfrac{1}{2}\epsilon)$$

The $\mathfrak{H}_n$ was correct, but the term in ϵ was given in the paper with the wrong sign.

Accordingly the paper may be corrected by changing the sign of every term involving ϵ and in the computation forms the corrections in ϵ have been applied with the wrong sign although the numerical values remain correct.

Copies of the computation forms have been sold by the Cambridge Scientific Instrument Company for use with the apparatus, and I shall try to reach the purchasers by circular pointing out that all the corrections on p. 12 of the forms which involve ϵ have been systematically applied with the wrong signs. I notice on p. 12 under the heading M_6 a misprint of $\frac{5}{2}\epsilon$ instead of 3ϵ, but the number attached, viz., $1°\cdot57$ is correct, except of course as to its sign.

The correction of the phases κ derived from the erroneous instructions may be at once effected without recurring to the original computations by adding to each κ twice the value tabulated in the paper for the corresponding $\frac{1}{2}n\epsilon$*.

If this note should be seen by anyone concerned who may not receive a circular I beg him to notice the correction.

It has already been remarked that the error is practically insignificant. The only tide in which it could possibly be appreciable is M_2, and since in this case the correction ϵ is equivalent to one minute of time in high water, the mistake caused by the erroneous instructions has been two minutes of time or $1°$ in κ. Even for M_2 the discrepancies in κ from year to year are often as great or greater than $1°$, and in the smaller tides they are frequently far greater; moreover the solar group of tides has been unaffected by the mistake.

Although practically the error is of little importance, it is clear that it ought to be corrected.

* See *Roy. Soc. Proc., loc. cit.*, p. 372, or *Scientific Papers, loc. cit.*, p. 241.

PART V

ADDRESSES TO SOCIETIES

24.

GEOLOGICAL TIME.

[Address to the Mathematical Section of the British Association in 1886. *Reports of the British Association (Birmingham)*, 1886, pp. 511—518.]

A MERE catalogue of facts, however well arranged, has never led to any important scientific generalisation. For in any subject the facts are so numerous and many-sided that they only lead us to a conclusion when they are marshalled by the light of some leading idea. A theory is then a necessity for the advance of science, and we may regard it as the branch of a living tree, of which facts are the nourishment. In the struggle between competing branches to reach the light some perish, and others form vigorous limbs. And as in a tree the shape of the young shoot can give us but little idea of the ultimate form of the branch, so theories become largely transformed in the course of their existence, and afford in their turn the parent stem for others.

The success of a theory may be measured by the extent to which it is capable of assimilating facts, and by the smallness of the change which it must undergo in the process. Every theory which is based on a true perception of facts is to some extent fertile in affording a nucleus for the aggregation of new observations. And a theory, apparently abandoned, has often ultimately appeared to contain an element of truth, which receives acknowledgment by the light of later views.

It will, I think, be useful to avail myself of the present occasion to direct your attention to a certain group of theories, which are still in an undeveloped and somewhat discordant condition, but which must form the nucleus round which many observations have yet to be collected before these theories and their descendants can make a definitely accepted body of truth. If I am disposed to criticise some of them in their actual form, I shall not be understood as denying the great service which has been rendered to science by their formulation.

Great as have been the advances of geology during the present century, we have no precise knowledge of one of its fundamental units. The scale of time on which we must suppose geological history to be drawn is important not only for geology itself, but it has an intimate relation with some of the profoundest questions of biology, physics, and cosmogony.

We can hardly hope to obtain an accurate measure of time from pure geology, for the extent to which the events chronicled in strata were contemporaneous is not written in the strata themselves, and there are long intervals of time of which no record has been preserved.

An important step has been taken by Alfred Tylor, Croll, and others, towards the determination of the rate of action of geological agents[*]. From estimates of the amount of sediment carried down by rivers, it appears that it takes from 1000 to 6000 years to remove one foot of rock from the general surface of a river basin.

From a consideration of the denuding power of rivers, and a measurement of the thickness of stratified rock, Phillips has made an estimate of the period of time comprised in geological history, and finds that, from stratigraphical evidence alone, we may regard the antiquity of life on the earth as being possibly between 38 and 96 millions of years[†].

Now while we should perhaps be wrong to pay much attention to these figures, yet at least we gain some insight into the order of magnitude of the periods with which we have to deal, and we may feel confident that a million years is not an infinitesimal fraction of the whole of geological time.

It is hardly to be hoped, however, that we shall ever attain to any very accurate knowledge of the geological time scale from this kind of argument.

But there is another theory which is precise in its estimate, and which, if acceptable from other points of view, will furnish exactly what is requisite. Mr Croll claims to prove that great changes of climate must be brought about by astronomical events of which the dates are known or ascertainable[‡]. The perturbation of the planets causes a secular variability in the eccentricity of the earth's orbit, and we are able confidently to compute the eccentricity for many thousands of years forward and backward from to-day, although it appears that, in the opinion of Newcomb and Adams, no great reliance can be placed on the values deduced from the formulæ at dates so remote as those of which Mr Croll speaks. According to Mr Croll, when the eccentricity of the earth's orbit is at its maximum, that hemisphere which has its winter in aphelion would undergo a glacial period. Now, as

* Geikie, *Textbook of Geology*, 1882, p. 442.
† Phillips, *Life on the Earth*. Rede Lecture, 1860, p. 119.
‡ *Climate and Time*.

the date of great eccentricity is ascertainable, this would explain the great ice-age and give us its date.

The theory has met with a cordial acceptance on many sides, probably to a great extent from the charm of the complete answer it affords to one of the great riddles of geology.

Adequate criticism of Mr Croll's views is a matter of great difficulty on account of the diversity of causes which are said to co-operate in the glaciation. In the case of an effect arising from a number of causes, each of which contributes its share, it is obvious that if the amount of each cause and of each effect is largely conjectural the uncertainty of the total result is by no means to be measured by the uncertainty of each item, but is enormously augmented. Without going far into details it may be said that these various concurrent causes result in one fundamental proposition with regard to climate, which must be regarded as the keystone of the whole argument. That proposition amounts to this—that climate is unstable.

Mr Croll holds that the various causes of change of climate operate *inter se* in such a way as to augment their several efficiencies. Thus the trade-winds are driven by the difference of temperature between the frigid and torrid zones, and if from the astronomical cause the northern hemisphere becomes cooler the trade-winds on that hemisphere encroach on those of the other, and the part of the warm oceanic current, which formerly flowed into the cold north zone, will be diverted into the southern hemisphere. Thus the cold of the northern hemisphere is augmented, and this in its turn displaces the trade-winds further, and this again acts on the ocean-currents, and so on; and this is neither more nor less than instability.

But if climate be unstable, and if from some of those temporary causes, for which no reasons can as yet be assigned, there occurs a short period of cold, then surely some even infinitesimal portion of the second link in the chain of causation must exist; and this should proceed as in the first case to augment the departure from the original condition, and the climate must change.

In a matter so complex as the weather, it is at least possible that there should be instability when the cause of disturbance is astronomical, whilst there is stability in an ordinary sense. If this is so, it might be explained by the necessity for a prolonged alteration in the direction of prevailing winds in order to affect oceanic currents*.

However this may be, so remarkable a doctrine as the instability of climate must certainly be regarded with great suspicion, and we should require abundant proof before accepting it. Now there is one result of

---

* Zöppritz, *Phil. Mag.*, 1878.

Mr Croll's theory which should afford almost a crucial test of its accept-ability. In consequence of the precession of the equinoxes the conditions producing glaciation in one hemisphere must be transferred to the other every 10,000 years. If there is good geological evidence that this has actually been the case, we should allow very great weight to the astronomical theory, notwithstanding the difficulties in its way. Mr Croll has urged that there is such evidence, and this view has been recently strongly supported by M. Blytt*. Other geologists do not, however, seem convinced of the conclusiveness of the evidence.

Thus Mr Wallace†, whilst admitting that there was some amelioration of climate from time to time during the last glacial period, cannot agree in the regular alternations of cold and warm demanded by Mr Croll's theory. To meet this difficulty he proposes a modification. According to his view large eccentricity in the earth's orbit will only produce glaciation when accompanied by favourable geographical conditions. And when extreme glaciation has once been established in the hemisphere which has its winter in aphelion, the glaciation will persist, with some diminution of intensity, when precession has brought round the perihelion to the winter. In this case, according to Wallace, glaciation will be simultaneous on both hemi-spheres.

Again he contends that, if the geographical conditions are not favourable, astronomical causes alone are not competent to produce glaciation.

There is agreement between the two theories in admitting instability of climate at first, when glaciation is about to begin under the influence of great eccentricity of the orbit, but afterwards Wallace demands great stability of climate. Thus he maintains that there is great stability in extreme climates, either warm or cold, whilst there is instability in moderate climates. I cannot perceive that we have much reason from physical con-siderations for accepting these remarkable propositions, and the acceptance or rejection of them demands an accurate knowledge of the most nicely balanced actions, of which we have as yet barely an outline.

Ocean currents play a most important part in these theories, but at this moment our knowledge of the principal oceanic circulation, and of its annual variability, is very meagre. In the course of a few years we may expect a considerable accession to our knowledge, when the Meteorological Office shall have completed a work but just begun—viz., the analysis of ships' logs for some sixty years, for the purpose of laying down in charts the oceanic currents.

With regard to the great atmospheric currents even the general scheme is not yet known. Nearly thirty years ago Professor James Thomson gave

* *Nature*, July 8 and 15, 1886.　　　　　　　　† *Island Life*.

before this Association at Dublin an important suggestion on this point. As it has been passed over in complete silence ever since, the present seems to be a good opportunity of redirecting attention to it.

According to Halley's theory of atmospheric circulation, the hot air rises at the equator and floats north and south in two grand upper currents, and it then acquires a westward motion relatively to the earth's surface, in consequence of the earth's rotation. Also the cold air at the pole sinks and spreads out over the earth's surface in a southerly current, at first with a westerly tendency, because the air comes from the higher regions of the atmosphere, and afterwards due south, and then easterly, when it is left behind by the earth in its rotation.

Now Professor Thomson remarks that this theory disagrees with fact in as far as that in our latitudes, the winds, though westerly, have a poleward tendency, instead of the reverse.

In the face of this discrepancy he maintains that "the great circulation already described does actually occur, but occurs subject to this modification, that a thin stratum of air on the surface of the earth in the latitudes higher than 30°—a stratum in which the inhabitants of those latitudes have their existence, and of which the movements constitute the observed winds of those latitudes—being by friction and impulses on the surface of the earth, retarded with reference to the rapid whirl or vortex motion from west to east of the great mass of air above it, tends to flow towards the pole, and actually does so flow to supply the partial void in the central parts of that vortex, due to the centrifugal force of its revolution. Thus it appears that in the temperate latitudes there are three currents at different heights— that the uppermost moves towards the pole, and is part of a grand primary circulation between equatorial and polar regions; that the lowermost moves also towards the pole, but is only a thin stratum forming part of a secondary circulation; that the middle current moves from the pole, and constitutes the return current for both the preceding; and that all these three currents have a prevailing motion from west to east[*]."

Such, then, appears to be our present state of ignorance of those great terrestrial actions, and any speculations as to the precise effect of changes in the annual distribution of the sun's heat must be very hazardous until we know more precisely the nature of the thing changed.

When looking at the astronomical theory of geological climate as a whole, one cannot but admire the symmetry and beauty of the scheme, and nourish a hope that it may be true; but the mental satisfaction derived from our survey must not blind us to the doubts and difficulties with which it is surrounded.

* *Brit. Assoc. Report*, Dublin, 1857, pp. 38—9.

And now let us turn to some other theories bearing on this important point of geological time.

Amongst the many transcendent services rendered to science by Sir William Thomson, it is not the least that he has turned the searching light of the theory of energy on to the science of geology. Geologists have thus been taught that the truth must lie between the cataclysms of the old geologists and the uniformitarianism of forty years ago. It is now generally believed that we must look for a greater intensity of geologic action in the remote past, and that the duration of the geologic ages, however little we may be able mentally to grasp their greatness, must bear about the same relation to the numbers which were written down in the older treatises on geology, as the life of an ordinary man does to the age of Methuselah.

The arguments which Sir William Thomson has adduced in limitation of geological time are of three kinds. I shall refer first to that which has been called the argument from tidal friction; but before stating the argument itself it will be convenient to speak of the data on which the numerical results are based.

Since water is not frictionless, tidal oscillations must be subject to friction, and this is evidenced by the delay of twenty-four to thirty-six hours, which is found to occur between full and change of moon and spring-tide. An inevitable result of this friction is that the diurnal rotation of the earth must be slowly retarded, and that we who accept the earth as our timekeeper must accuse the moon of a secular acceleration of her motion round the earth, which cannot be otherwise explained. It is generally admitted by astronomers that there actually is such an unexplained secular acceleration of the moon's mean motion.

No passage in Thomson and Tait's *Natural Philosophy* has excited more general interest than that in which Adams is quoted as showing that, *with a certain value for the secular acceleration*, the earth must in a century fall behind a perfect chronometer, set and rated at the beginning of the century, by twenty-two seconds. Unfortunately this passage in the first edition gave an erroneous complexion to Adams' opinion, and being quoted, without a statement of the premises, has been used in popular astronomy as an authority for establishing the statement that the earth is actually a false timekeeper to the precise amount specified.

In the second edition (in the editing of which I took part) this passage has been rewritten, and it is shown that Newcomb's estimate of the secular acceleration only gives about one-third of the retardation of the earth's rotation, which resulted from Adams' value. The last sentence of the paragraph here runs as follows:—" It is proper to add that Adams lays but little stress on the actual numerical values which have been used in this computation, and is of opinion that the amount of tidal retardation of the earth's rotation is quite uncertain." Thus, in the opinion of our great physical astronomer, a datum is still wanting for the determination of a limit to geological time, according to Thomson's argument.

However, subject to this uncertainty, with the values used by Adams in his computation, and with the assumption that the rate of tidal friction has remained constant, then a thousand million years ago the earth was rotating twice as fast as at present. In the last edition of the *Natural Philosophy* the argument from these data runs thus:—

"If the consolidation of the earth took place then or earlier, the ellipticity of the upper layers (of the earth's mass) must have been $\frac{1}{230}$ instead of about $\frac{1}{300}$, as it is at present. It must necessarily remain uncertain whether the earth would from time to time adjust itself completely to a figure of equilibrium adapted to the rotation. But it is clear that a want of complete adjustment would leave traces in a preponderance of land in equatorial regions. The existence of large continents and the great effective rigidity of the earth's mass render it improbable that the adjustment, if any, to the appropriate figure of equilibrium would be complete. The fact, then, that the continents are arranged along meridians, rather than in an equatorial belt, affords some degree of proof that the consolidation of the earth took place at a time when the diurnal rotation differed but little from its present value. It is probable, therefore, that the date of consolidation is considerably more recent than a thousand million years ago."

I trust it may not be presumptuous in me to criticise the views of my great master, at whose intuitive perception of truth in physical questions I have often marvelled, but this passage does not even yet seem to me to allow a sufficiently large margin of uncertainty.

It will be observed that the argument reposes on our certainty that the earth possesses rigidity of such a kind as to prevent its accommodation to the figure and arrangement of density appropriate to its rotation. In an interesting discussion on subaerial denudation, Croll has concluded that nearly one mile may have been worn off the equator during the past 12,000,000 years, if the rate of denudation all along the equator be equal to that of the basin of the Ganges*. Now, since the equatorial protuberance of the earth when the ellipticity is $\frac{1}{230}$ is fourteen miles greater than when it is $\frac{1}{300}$, it follows that 170,000,000 years would suffice to wear down the surface to the equilibrium figure. Now let these numbers be halved or largely reduced, and the conclusion remains that denudation would suffice to obliterate external evidence of some early excess of ellipticity.

* Croll, *Climate and Time*, 1885, p. 336.

If such external evidence be gone*, we must rely on the incompatibility of the known value of the precessional constant with an ellipticity of internal strata of equal density greater than that appropriate to the actual ellipticity of the surface. Might there not be a considerable excess of internal ellipticity without our being cognisant of the fact astronomically?

And, further, have we any right to feel so confident of the internal structure of the earth as to be able to allege that the earth would not through its whole mass adjust itself almost completely to the equilibrium figure?

Tresca has shown in his admirable memoirs on the flow of solids that when the stresses rise above a certain value the solid becomes plastic, and is brought into what he calls the state of fluidity. I do not know, however, that he determined at what stage the flow ceases when the stresses are gradually diminished. It seems probable, at least, that flow will continue with smaller stresses than were initially necessary to start it. But if this is so, then, when the earth has come to depart both internally and externally from the equilibrium condition a flow of solid will set in, and will continue until a near approach to the equilibrium condition is attained.

When we consider the abundant geological evidence of the plasticity of rock, and of the repeated elevation and subsidence of large areas on the earth's surface, this view appears to me more probable than Sir William Thomson's.

On the whole, then, I can neither feel the cogency of the argument from tidal friction itself, nor, accepting it, can I place any reliance on the limits which it assigns to geological history.

The second argument concerning geological time is derived from the secular cooling of the earth.

We know in round numbers the rate of increase of temperature, or temperature gradient, in borings and mines, and the conductivity of rock. These data enable us to compute how long ago the surface must have had the temperature of melting rock, and when it must have been too hot for vegetable and animal life.

Sir William Thomson, in his celebrated essay on this subject†, concludes from this argument that " for the last 96,000,000 years the rate of increase of temperature underground has gradually diminished from about $\frac{1}{10}$th to

---

* I find by a rough calculation that $\frac{55}{100}$ths of the land in the N. hemisphere is in the equatorial half of that hemisphere, viz. between 0° and 30° N. lat.; and that $\frac{48}{50}$ths of the land in the S. hemisphere is in the equatorial half of that hemisphere, viz. between 0° and 30° S. lat. For the whole earth, $\frac{101}{200}$ths of all the land lies in the equatorial half of its surface, between 30° N. and S. lat. In this computation the Mediterranean, Caspian, and Black Seas are treated as land.

† Republished in Thomson and Tait's *Natural Philosophy*, Appendix D.

about $\frac{1}{50}$th of a degree Fahrenheit per foot....Is not this, on the whole, in harmony with geological evidence, rightly interpreted? Do not the vast masses of basalt, the general appearance of mountain ranges, the violent distortions and fractures of strata, *the great prevalence of metamorphic action* (which must have taken place at depths of not many miles, if so much), all agree in demonstrating that the rate of increase of temperature downwards must have been much more rapid, and in rendering it probable that volcanic energy, earthquake shocks, and every kind of so-called plutonic action, have been, on the whole, more abundantly and violently operative in geological antiquity than in the present age?"

Now, while I entirely agree with the general conclusion of Sir William Thomson, it is not unimportant to indicate a possible flaw in the argument. This flaw will only be acknowledged as possible by those who agree with the previous criticism on the argument from tidal friction.

The present argument as to the date of the consolidation of the earth reposes on the hypothesis that the earth is simply a cooling globe, and there are reasons why this may not be the case. The solidification of the earth probably began from the middle and spread to the surface. Now is it not possible, if not probable, that after a firm crust had been formed, the upper portion still retained some degree of viscosity? If the interior be viscous, some tidal oscillations must take place in it, and, these being subject to friction, heat must be generated in the viscous portion; moreover the diurnal rotation of the earth must be retarded. Some years ago, in a paper on the tides of a spheroid, viscous throughout the whole mass*, I estimated the amount and distribution of the heat generated, whilst the planet's rotation is being retarded and the satellite's distance is being increased. It then appeared that on that hypothesis the distribution of the heat must be such that it would only be possible to attribute a very small part of the observed temperature gradient to such a cause. Now, with a more probable internal constitution for the earth in early times, the result might be very different. Suppose, in fact, that it is only those strata which are within some hundreds of miles of the surface which are viscous, whilst the central portion is rigid. Then, when tidal friction does its work the same amount of heat is generated as on the hypothesis of the viscosity of the whole planet, but instead of being distributed throughout the whole mass, and principally towards the middle, it is now to be found in the more superficial layers.

In my paper it is shown that with Thomson's data for the conductivity of rock and the temperature gradient, the annual loss of heat by the earth is one 260 millionth part of the earth's kinetic energy of rotation.

Also, if by tidal friction the day is reduced from D_0 hours to D hours,

---

* *Phil. Trans.*, Part II., 1879. [Vol. II., Paper 4, p. 155.]

and the moon's distance augmented from Π_0 to Π earth's radii, the energy which has been converted into heat in the process is

$$\left(\frac{D}{D_0}\right)^2 - 1 - 8\cdot 84 \left(\frac{1}{\Pi} - \frac{1}{\Pi_0}\right) \text{ times the earth's kinetic energy of rotation.}$$

From these data it results that the heat generated in the lengthening of the day from twenty-three to twenty-four hours is equal to the amount of heat lost by the earth, at its present rate of loss, in 23,000,000 years.

Now if this amount of heat, or any sensible fraction of it, was actually generated within a few hundred miles of the earth's surface, the temperature gradient in the earth must be largely due to it, instead of to the primitive heat of the mass.

Such an hypothesis precludes the assumption that the earth is simply a cooling mass, and would greatly prolong the possible extension of geological time. It must be observed that this view is not acceptable unless we admit that the earth can adjust itself to the equilibrium figure adapted to its rotation*.

It seems also worthy of suggestion that our data for the average gradient of temperature may be somewhat fallacious. Recent observations† show that the lower stratum of the ocean is occupied by water at near the freezing temperature, whilst the mean annual temperature of the earth's surface, where the borings have been made, must be at least 30° higher. It does not then seem impossible that the mean temperature gradient for the whole earth should differ sensibly from the mean gradient in the borings already made.

The foregoing remarks have not been made with a view of showing Sir William Thomson's argument from the cooling of the earth to be erroneous, but rather to maintain the scientific justice of assigning limits of uncertainty at the very least as wide as those given by him. Professor Tait‡ cuts the limit down to 10,000,000 years; he may be right, but the uncertainties of the case are far too great to justify us in accepting such a narrowing of the conclusion.

The third line of argument by which a superior limit is sought for the age of the solar system appears by far the strongest. This argument depends on the amount of radiant energy which can have been given out by the sun.

The amount of work done in the concentration of the sun from a condition of infinite dispersion may be computed with some accuracy, and we have at least a rough idea of the rate of the sun's radiation. From these data Sir William Thomson concludes§:—

---

* Since the meeting of the Association Sir William Thomson has expressed to me his absolute conviction that, with any reasonable hypothesis as to the degree of viscosity of the more superficial layers, and as to the activity of tidal friction, the disturbance of temperature gradient through internal generation of heat must be quite infinitesimal.

† *Challenger* Expedition.

‡ *Recent Advances in Physical Science* (1885).

§ Thomson and Tait, *Natural Philosophy*, Appendix E.

"It seems, therefore, on the whole most probable that the sun has not illuminated the earth for 100,000,000 years, and almost certain that he has not done so for 500,000,000 years. As for the future, we may say, with equal certainty, that inhabitants of the earth cannot continue to enjoy the light and heat essential to their life for many million years longer unless sources now unknown to us are prepared in the great storehouse of creation."

This result is based on the value assigned by Pouillet and Herschel to the sun's radiation. Langley has recently made a fresh determination, which exceeds Pouillet's in the proportion of eight to five*. With Langley's value Thomson's estimate of time would have to be reduced by the factor five-eighths.

In considering these three arguments I have adduced some reasons against the validity of the first argument, and have endeavoured to show that there are elements of uncertainty surrounding the second; nevertheless they undoubtedly constitute a contribution of the first importance to physical geology. Whilst then we may protest against the precision with which Professor Tait seeks to deduce results from them, we are fully justified in following Sir William Thomson, who says that "the existing state of things on the earth, life on the earth, all geological history showing continuity of life, must be limited within some such period of past time as 100,000,000 years."

If I have carried you with me in this survey of theories bearing on geological time, you will agree that something has been acquired to our knowledge of the past, but that much more remains still to be determined.

Although speculations as to the future course of science are usually of little avail, yet it seems as likely that meteorology and geology will pass the word of command to cosmical physics as the converse.

At present our knowledge of a definite limit to geological time has so little precision that we should do wrong summarily to reject any theories which appear to demand longer periods of time than those which now appear allowable.

In each branch of science hypothesis forms the nucleus for the aggregation of observation, and so long as facts are assimilated and co-ordinated we ought to follow our theory. Thus even if there be some inconsistencies with a neighbouring science we may be justified in still holding to a theory, in the hope that further knowledge may enable us to remove the difficulties. There is no criterion as to what degree of inconsistency should compel us to give up a theory, and it should be borne in mind that many views have been utterly condemned when later knowledge has only shown us that we were in them only seeing the truth from another side.

* Langley (*Ann. Rep. R. A. S.*, 1885) estimates that 3 calories per minute are received by a square centimetre at distance unity. This gives for the total annual radiation of the sun 4.38×10^{33} calories. Thomson gives as Pouillet's estimate 6×10^{30} times the heat required to raise 1 lb. of water 1° Cels., or 2.72×10^{33} calories.

25.

PRESENTATION OF THE MEDAL OF THE ROYAL ASTRONO-MICAL SOCIETY TO M. HENRI POINCARÉ.

[*Monthly Notices of the Royal Astronomical Society*, Vol. 60 (1900), pp. 406—415.]

THE medal of the Royal Astronomical Society is this year awarded to M. Henri Poincaré, member of the Academy of Sciences of Paris. As your President, the agreeable duty of presenting the medal to him devolves upon me, but before I do so I must endeavour to lay before you the grounds upon which the Council has made this award.

M. Poincaré's researches have been so diverse in character, and have been carried out with such a wealth of knowledge, that I feel but little confidence in my fitness to perform this arduous task; yet I cannot but rejoice that my tenure of this chair should have furnished me with the opportunity of paying the homage which is due to him for his great achievements in the field of mathematics.

A large part of his work is concerned with the development of the science of pure mathematics, and of this side of his activity I am quite incompetent to speak. But in awarding our medal we naturally think of the value of the contributions by the proposed medallist to our science. Now, although many of M. Poincaré's investigations have perhaps already found, or will at some future time find, their application in the problems of dynamical astronomy, yet it is not necessary to search for cases of possible applicability to astronomy to justify our award. I propose, then, to draw your attention to only three of his lines of research, and these have a directly astronomical scope. My choice is governed not only by the intrinsic interest of the results, but also by the fact that the subjects treated possess a special interest for me. I shall speak, then, of his researches on the dynamical theory of the tides, on figures of equilibrium of rotating masses of liquid, and on the theory of the motions of planets and satellites.

The first of these subjects is treated in two memoirs on the equilibrium and movement of the ocean*. The problem is surrounded by conditions of such intricacy that it seemed to the author advisable to consider the several difficulties separately, as a preliminary to the treatment of the question as a whole. He begins, then, by the equilibrium theory of the tides, but he proposes to take into account not only the effect of the obstructing continents, but also that of the attractions of the sea on itself. This problem was solved long ago by Bernoulli for the case when there is no land, and some thirty years ago by Lord Kelvin when there is no mutual attraction of the water on itself†. Lord Kelvin's correction to Bernoulli's theory may be taken as meaning that there are on a certain complete meridian four points at which the semi-diurnal tide vanishes; two of these points are on the same side of the Earth in equal northern and southern latitudes, and the others are antipodal to the first two. There are also four other points on another complete meridian where the diurnal tide vanishes; two of them are in one quadrant in complementary latitudes, and the other two are antipodal to the first two. Lastly, there are two parallels of equal northern and southern latitude at which the tides of long period are evanescent. There are, further, four other points of doubled semi-diurnal tide, four of doubled diurnal tide, and two parallels (sometimes, however, imaginary) of doubled tide of long period.

The positions of these points and parallels are dependent on the distribution of land and sea. The numerical quadratures necessary for their determination have been carried out, and it appears that in every case the points or lines of evanescence or of doubling lie very close to the places where the tide in question absolutely vanishes in Bernoulli's theory‡. Since it can make very little difference whether a tide of excessively small range is doubled or annulled, it follows that the correction for land is of little importance, at least when the attraction of the water is neglected.

The introduction of the effects of the mutual attraction obviously presents a problem of great difficulty. Although it is probable that the result would possess little practical importance, yet the question is undoubtedly an interesting one from a mathematical point of view. It is at this point that M. Poincaré takes up the problem, and he shows that it is possible, at least theoretically, to determine a series of harmonic functions by which the mutual gravitation of the ocean may be computed. These functions

* *Liouville's Journal*, 1896, pp. 59—102, and pp. 217—62.

† Thomson and Tait, *Natural Philosophy*, § 808.

‡ G. H. Darwin and H. H. Turner, *Proc. R. S.*, 1886, Vol. XL., pp. 303—15 [or Vol. I., Paper 8, p. 328].

degrade into ordinary spherical harmonics when the land is all submerged. Lord Kelvin had expressed the opinion that mutual attraction would only change the results to an insignificant degree, but M. Poincaré, while not contesting the justice of this view in general, adduces considerations which seem to show that a distribution of land is imaginable, which might perhaps make the correction of material importance. But the verification of this conjecture would necessitate the complete calculation, which would present an absolutely inextricable complexity on account of the capricious forms of our continents.

Although it does not seem likely that these new harmonic functions will ever be applied to terrestrial oceans, yet they may perhaps furnish a clue to the discussion of the tides of seas of simple forms, such as the quadrant or octant of the whole globe.

The author's next step is the discussion of the small free oscillations of a system about a configuration of equilibrium, and we are here concerned with principles of the widest generality. He proves that the problem of small oscillations resolves itself into the determination of the maxima and minima of the ratio of the kinetic to the potential energy; and the transition is easy from the case of free oscillations to that of forced oscillations under the action of any perturbing forces. This generalisation forms the basis of the author's whole investigation, and he remarks that he found the first suggestion of it in the work of Lord Rayleigh.

Passing from the case of a finite number to that of an infinite number of degrees of freedom, the author then applies his general principle to the oscillations of liquid standing in a vessel of small dimensions. The vessel is then enlarged so that the free surface at rest cannot any longer be treated as plane, but must be considered as a portion of a sphere concentric with the Earth's centre. It appears that the mathematical analysis is the same as that involved in finding the oscillations of a stretched membrane of unequal density and thickness in its different parts. This problem had been previously treated by the author in another memoir.

This brings us to the end of the first paper, and the second begins with the discussion of the process called by Lord Kelvin the ignoration of coordinates. The so-called gyroscopic terms now appear in the equations of motion; in the tidal problem they arise through the fact of the Earth's rotation, and they form a very essential feature of the whole.

The general principles of the motion in such cases are then applied to the consideration of the oscillations of liquid standing in a small rotating vessel. This problem had been treated previously by Lord Kelvin[*]; and he found, amongst other results, that waves propagated along a rotating canal would behave differently on the right and left banks. In this he saw a probable

---

[*] *Proc. R. S. Edinburgh*, 1879, March 17, or *Phil. Mag.*, 1880, August, pp. 109—116.

explanation of the different behaviour of the tides on the south coast of England and on the north coast of France. Well, the tides on the coasts of the two countries do not differ so much as the treatment which the problem presented by them has received from Lord Kelvin here and from M. Poincaré in France. As might be expected, the latter obtains his solution from the maxima and minima of a certain ratio, which corresponds to that between the kinetic and potential energies in the simpler problems treated before.

The problem of the rotating vessel only differs from that of the tides in all its generality in the linear scale on which it is supposed to take place; and the effects of the Earth's curvature may be introduced in the same way as in the case of the non-rotating vessel.

The object of these memoirs was not to attain a definite solution in any ideal concrete case, but was to show how the fundamental difficulties might be surmounted by mathematical analysis. Here, as elsewhere, M. Poincaré carries us far beyond the particular instance in view, and it may well be that the principles enounced will meet their actual application sooner in other fields than in the tidal problem.

Important as is the work of which I have just spoken, the memoir on the figures of equilibrium of rotating liquid* seems to me to stand on a much higher level, for it marks an epoch not only in the study of the subject itself, but also in that of many others. It may be that some of the generalisations to be found in it were floating more or less distinctly in the minds of his predecessors, but the theory of the stability of systems in equilibrium or in steady motion has undoubtedly been crystallised and rendered transparent by his efforts. So fundamental are the new conceptions introduced that a new phraseology has become necessary, and it has already been adopted in other investigations, which possess no superficial resemblance to that of which I am now speaking.

Let us imagine a number of mechanical systems which resemble one another in all respects save one, such as size or density or any other measurable quantity. Suppose, further, that the systems are all in equilibrium, and that they are arranged in the order of the magnitude of that one measurable quantity, then we may describe the whole array as a family of systems in equilibrium. Now mechanical systems are often susceptible of equilibrium in more than one configuration, and we may therefore conceive the existence of a second family, which differs from the first in that the equilibrium involves a different configuration of the parts. If we were to examine the two families, we should find that in each the arrangement of the parts changed as we passed along the series. Now it is possible that there would occur in one family a certain member which would resemble the corresponding member of the other family in all respects. If this were the case, this particular member of either family would be described as a form of bifurcation, because it would belong to both, and the two families might be regarded as branching out from it. Now, M. Poincaré proves that if we follow each family towards the form of bifurcation, the equilibrium in one of the families would be stable, while that in the other would be unstable; the same would also hold good after the passage through the form of bifurcation, but the family which was stable before would be unstable afterwards, and vice versâ. There is accordingly exchange of stabilities between the two families. Moreover, the passage of a system from stable to unstable equilibrium necessarily implies that at the moment of change we are passing through a form of bifurcation. Hence the study of the stability of a system may afford notice of the existence of hitherto unsuspected arrangements in which equilibrium is possible. These ideas possess the widest generality, but this is not the place to discuss their fertile ramifications.

If a system in steady motion or one at rest be slightly disturbed, it is said to be stable if the oscillations continue to be small throughout all time. But the existence of even the smallest amount of frictional resistance may betray the fact that stability really means two very different things. In one case even infinitesimal frictional resistance may cause the oscillations to increase to such an extent as to completely change the whole configuration; in another the oscillations may gradually die out and leave the system in the same condition as that in which it was at first. Both systems are stable when there is no friction, but the latter is said to possess secular stability. In nature only systems possessing secular stability are permanent, although it may require a very long time for a system possessing ordinary stability to break down. M. Poincaré here pays a well-merited tribute to Lord Kelvin, who appears to have been the first to indicate clearly this important distinction between these two kinds of stability.

Thus far the discussion is applicable to mechanical systems of every kind, but it finds its special application in the determination of the figures of rotating fluid. It may be well to mention, by the way, that in the course of the paper M. Poincaré justifies a number of statements as to possible forms of equilibrium and their stability, which had been made by Lord Kelvin without proof.

We now come to the main object of the investigation. A planet formed of homogeneous fluid has the form of an oblate spheroid, and its equilibrium is stable. If its angular velocity of rotation be increased, its ellipticity will increase also, but the stability will diminish. When the ellipticity has increased to a certain definite extent, the stability ceases, and for more rapid rotation the figure becomes unstable. At the critical moment of change we are passing through a form of bifurcation, and we know that there must be another series of figures which also possesses that form. This other series consists of the ellipsoids of Jacobi, which have their three axes of unequal magnitudes.   But there is one single member of Jacobi's series which is a figure of revolution, and this is identical with the form of bifurcation found by following the stability of the oblate figures.   It is true that this Jacobian is also a limiting form, since the series ends there; but we need not stop to consider that point further.   It follows from the principle of exchange of stabilities that for rotation slower than the critical value the Jacobian was stable.   All this was known before, but M. Poincaré's work has placed it in a new and clearer light.

Having followed the stable series of oblate ellipsoids of revolution up to the junction at the form of bifurcation, M. Poincaré shunts his train on into the stable branch line furnished by the Jacobian ellipsoids.   He follows this branch until he finds the Jacobian to become unstable, and announces that there is a new form of bifurcation, and that a new branch line has been reached.   And here the line is nearly blocked by mathematical obstructions, and he is only able to proceed just far enough to perceive that the new figure is pear-shaped, with its larger portion more or less spherical, and with an equatorial protuberance which we may liken to the stalk end of the pear.

This apparently abstract result elucidates the evolution of planetary systems in a very interesting way.   Let us consider a rotating liquid mass slowly cooling.   If the cooling is slow enough internal friction will cause the whole to revolve throughout with the same angular velocity.   At first, when the density is small, the figure will be an ellipsoid of revolution, but slightly flattened; and as it cools the flattening will increase until at a certain stage the figure of revolution ceases to be a figure of equilibrium, and the ellipsoid commences to have an equatorial protuberance; it passes, in fact, into one of Jacobi's ellipsoids.   The ellipsoid then lengthens, until at a certain stage it begins to acquire an unsymmetrical furrow in a plane parallel to the axis of revolution, and it becomes pear-shaped, with the axis of revolution at right angles to the core of the pear.   " The larger part of the matter tends to approach the spherical form, whilst the smaller part projects from the ellipsoid at one of the extremities of the longer axis, as though it were trying to detach itself from the larger part of the mass.   It is difficult to state with certainty what will happen then if the cooling continues, but one may suppose that the mass will go on deepening its furrow more and more, and then it will at last divide itself into two separate bodies by the throttling of the middle part."   It is clear that a process of this kind may have played its part in the evolution of celestial systems, and the speculation seems to meet with confirmation from the forms observed in many nebulæ.

M. Poincaré's paper came as a revelation to me, because, just at the time when it was published, I had attempted to attack the question from the other end, and to trace the coalescence of two detached bodies into a single one—but, alas! I have to admit that my work contained no far-reaching general principles—no light on the stability of the systems I tried to draw— nothing of all that which renders Poincaré's memoir one that will always mark an important epoch in the history not only of evolutionary astronomy, but of the wider fields of general dynamics*.

I now come to the third contribution of our medallist to astronomy, namely the work on celestial mechanics. The first of the three volumes† of which the book consists deals with the general principles of dynamics as applicable to the problem of the three bodies, the second is on the work of previous investigators, and in the third the author analyses the results at which he has arrived by the previous discussions. It is probable that for half a century to come it will be the mine from which humbler investigators will excavate their materials. The range of matter is so vast and the num- ber of new ideas so great, that I find myself in considerable difficulty as to how to speak of this book. It would clearly be impossible within the limits of such an address as this to give even an outline of the methods and conclusions.

Under these circumstances, then, I shall limit myself to only one portion of the whole, and shall speak of that at no great length. The subject to which I refer is that of the so-called periodic orbits; and my choice is natural, since I have been myself a hewer of wood and drawer of water in that field.

It was Mr G. W. Hill who first drew effective attention to this class of solutions of the problem of the three bodies, when he initiated his new method of treating the lunar theory‡. I am proud to say that our Fellow, Mr Ernest Brown, is now carrying on Hill's grand conception to its laborious end. Former mathematicians have, almost without exception, adopted the moon's elliptic orbit as their starting-point. But the first and roughest approximation to the moon's path is a circle round the earth, and the ellipse is an attempt to improve on the circle. Now Hill saw that the early intro- duction of the ellipse brought in its train a whole series of difficulties, which would be avoided if, continuing to neglect the eccentricity, he improved on the circle by introducing at once the effects of solar perturbation. The dis- torted circular orbit possesses the great advantage that it always presents the same features towards the sun and the earth. It is accordingly possible to draw, with any desired accuracy, a certain definite curve on a plane carried round with the earth in its orbital motion. This curve exhibits the leading features of the effects of solar perturbation for all time; it resembles an ellipse drawn about the earth as centre, with the major axis in quadratures,

* [See Vol. III. Paper 9, p. 135.]

† *Les Méthodes Nouvelles de la Mécanique Céleste.* Paris, Gauthier-Villars. Vol. I., 1892; Vol. II., 1893; Vol. III., 1899.

‡ *Researches in the Lunar Theory,* Cambridge, U.S.A., 1877.

and with the minor axis in syzygies. The principal solar inequality is the variation, and hence Hill describes this orbit as the variational curve, and he uses it as an orbit of reference for the moon's actual motion. If the moon had been started with a motion differing but little from actuality she would have followed the variational curve for all time*. The curve may then be appropriately termed a periodic orbit.

Hill afterwards introduced the eccentricity of the lunar orbit by reference to the variational curve, and he treated the subject by a fertile method of the highest originality†. He habitually restrains himself from every tendency towards exuberance of expression in his writing, and so I am perhaps deceived when I fancy that even he hardly realised the fundamental importance of what he had done. Speaking of Hill's papers, M. Poincaré writes: "Dans cette œuvre il est permis d'apercevoir le germe de la plupart des progrès que la science a faits depuis‡." Although he always pays warm tribute to Gyldèn and to Lindstedt, I conjecture that it may have been Hill who stimulated him to his attack on the profound questions awaiting solution in celestial mechanics. But these are almost personal questions which have little bearing on the advance of science in itself.

The variational curve obviously does not stand isolated, but is merely a single member of a series of periodic solutions of the problem of the three bodies. These solutions, according to M. Poincaré, furnish the only breach by which we may hope to penetrate the fortress of a problem hitherto deemed impregnable. Their importance becomes manifest when I say that he has proved that a periodic solution may always be found which shall differ by as small a quantity as we please from any given motion of the perturbed body, however complicated that motion may be. It is true that the required periodic orbit may need to go through a very large number of circuits before its periodicity is completed.

Orbits of this kind are divisible into three classes: in the first, the inclinations and eccentricities are zero; in the second, only the inclinations are zero; and in the third, there are no limitations in these respects. In the cases of the variational curve and of the orbits which I have traced, the orbit of the perturbing body is circular, the whole motion takes place in the plane of the circle, and the periodicity of the perturbed body is completed after a single circuit; they belong to the first group. As in the investigation of figures of equilibrium, so here also we meet with orbits of bifurcation, limit-

---

* Hill's simple variational curve was drawn on the hypothesis that the squares and higher powers of the sun's parallax are negligible. But these further approximations may be introduced, and the above statement then becomes correct. The simple curve is symmetrical both as to the lines of syzygies and of quadratures, but the corrected curve is only symmetrical as to syzygies.

† "On Part of the Motion of the Lunar Perigee, &c.,"*Acta Mathematica*, Vol. VIII., pp. 1—36.

‡ *Méc. Céleste*, Vol. I., Introduction.

ing forms and exchanges of stability; and this illustrates the wide generality of the ideas which I attempted to sketch earlier in my address.

Mr Hill had traced a certain cusped orbit in which the moon might have moved, and thinking that he had found a limiting form, described it as the moon of greatest lunation. But M. Poincaré showed that Hill had been misled, and that the cusp is succeeded by looped orbits. Lord Kelvin chose one of them as an example to illustrate a method of graphical construction by which a curve may be traced by means of its curvature*. The successive transformations of these orbits are elucidated by the series of figures which I have drawn, and I now see that they will terminate their career, at least as simply periodic orbits, in a form which has been called by M. Burrau an orbit of ejection†. As this latter name is hardly sufficiently explanatory in itself, I may say that the perturbed body is supposed to be ejected from one of the two principal bodies along the line of syzygies, or it may fall inward tangentially to that line. These orbits furnish an interesting and peculiar form of periodicity, but the limitation of my time prevents me from referring to them in greater detail.

Another class of orbits, as pointed out by Poincaré, is said to be asymptotic; they are closely akin to the periodic orbits. In this case the body draws asymptotically near to a certain curve, and after performing an indefinite number of circuits, may gradually depart therefrom again. Certain figure-of-eight orbits which I have drawn are intimately connected with these asymptotic orbits.

M. Poincaré has made a searching examination of my figures by means of his analysis, and has put his finger on a weak spot. I had carelessly treated two sets of curves as being continuous with one another, but he shows that my scheme cannot be maintained. I had already arrived at the same conclusion from an entirely different point of view in consequence of the criticism of Mr S. S. Hough, but it was too late to correct the oversight before the paper was published. Mr Hough has now a paper in the press for the *Acta Mathematica*, wherein the true sequence of the orbits will be exhibited‡. I have already by me a large part of the computations required for the actual drawing of a new family of orbits, whose existence I did not suspect.

There can, I think, be little doubt that the investigations of M. Poincaré and of his followers will ultimately afford some sort of explanation of the empirical law of Bode as to the distances of the planets from the Sun; and such an explanation will almost of necessity render intelligible the sequence of processes by which the solar system has been evolved. In the case of the satellites revolving about my ideal planet *Jove* I found that there was a tract

---

* *Phil. Mag.*, Vol. xxxiv., 1892, pp. 443—8.

† *Ast. Nachr.*, Nos. 3230 and 3251.

‡ [Paper 2, p. 114.]

near the planet which might be occupied by stable orbits, that this was surrounded by a belt, within which no stable orbit was possible, and that this was again succeeded by a belt of stability. These results are, of course, only true of simply periodic orbits; and I was perhaps rash in saying that I saw in them some indications of a law analogous to that of Bode. Nevertheless, I notice also in M. Poincaré's book a suggestion which may perhaps tend in the same direction. The inferior planets circling round the sun in my ideal problem were stable up to a certain distance, and then became unstable, and he conjectures that the further pursuit of this family of orbits will lead us back again to a belt of stability. I hope to test this interesting suggestion. But even this meagre sketch has already occupied too much time, and I will say but few words more.

The leading characteristic of M. Poincaré's work appears to me to be the immense wideness of the generalisations, so that the abundance of possible illustrations is sometimes almost bewildering. This power of grasping abstract principles is the mark of the intellect of the true mathematician; but to one accustomed rather to deal with the concrete the difficulty of completely mastering the argument is sometimes great. To the latter class of mind the easier process is the consideration of some simple concrete case, and the subsequent ascent to the more general aspect of the problem. I fancy that M. Poincaré's mind must work in another groove than this, and that he finds it easier to consider first the wider issues, from whence to descend to the more special instances. It is rare to possess this faculty in any high degree, and we cannot wonder that the possessor of it should have compiled a noble heritage for the men of science of future generations.

In handing this medal to you, M. Poincaré, I desire to say on behalf of the Society that in seeking to pay honour to you we feel ourselves honoured.

26.

COSMICAL EVOLUTION.

[Presidential Address, *Reports of the British Association*
(*South Africa*), 1905, pp. 3—32.]

BARTHOLOMEU DIAZ, the discoverer of the Cape of Storms, spent sixteen months on his voyage, and the little flotilla of Vasco da Gama, sailing from Lisbon on July 8, 1497, only reached the Cape in the middle of November. These bold men, sailing in their puny fishing smacks to unknown lands, met the perils of the sea and the attacks of savages with equal courage. How great was the danger of such a voyage may be gathered from the fact that less than half the men who sailed with da Gama lived to return to Lisbon. Four hundred and eight years have passed since that voyage, and a ship of 13,000 tons has just brought us here, in safety and luxury, in but little more than a fortnight.

How striking are the contrasts presented by these events! On the one hand compare the courage, the endurance, and the persistence of the early navigators with the little that has been demanded of us; on the other hand consider how much man's power over the forces of Nature has been augmented during the past four centuries. The capacity for heroism is probably undiminished, but certainly the occasions are now rarer when it is demanded of us. If we are heroes, at least but few of us ever find it out, and, when we read stories of ancient feats of courage, it is hard to prevent an uneasy thought that, notwithstanding our boasted mechanical inventions, we are perhaps degenerate descendants of our great predecessors.

Yet the thought that to-day is less romantic and less heroic than yesterday has its consolation, for it means that the lot of man is easier than it was. Mankind, indeed, may be justly proud that this improvement has been due to the successive efforts of each generation to add to the heritage of knowledge handed down to it by its predecessors, whereby we have been born to the accumulated endowment of centuries of genius and labour.

I am told that in the United States the phrase " I want to know " has lost the simple meaning implied by the words, and has become a mere exclamation of surprise.   Such a conventional expression could hardly have gained currency except amongst a people who aspire to knowledge.   The dominance of the European race in America, Australasia, and South Africa has no doubt arisen from many causes, but amongst these perhaps the chief one is that not only do " we want to know," but also that we are determined to find out.   And now within the last quarter of a century we have welcomed into the ranks of those who " want to know " an oriental race, which has already proved itself strong in the peaceful arts of knowledge.

I take it, then, that you have invited us because you want to know what is worth knowing; and we are here because we want to know you, to learn what you have to tell us, and to see that South Africa of which we have heard so much.

The hospitality which you are offering us is so lavish, and the journeys which you have organised are so extensive that the cynical observer might be tempted to describe our meeting as the largest picnic on record.   Although we intend to enjoy our picnic with all our hearts, yet I should like to tell the cynic, if he is here, that perhaps the most important object of these conferences is the opportunity they afford for personal intercourse between men of like minds who live at the remotest corners of the earth.

We shall pass through your land with the speed and the voracity of a flight of locusts: but, unlike the locust, we shall, I hope, leave behind us permanent fertilisation in the form of stimulated scientific and educational activity.   And this result will ensue whether or not we who have come from Europe are able worthily to sustain the lofty part of prophets of science.   We shall try our best to play to your satisfaction on the great stage upon which you call on us to act, and if when we are gone you shall, amongst yourselves, pronounce the performance a poor one, yet the fact will remain, that this meeting has embodied in a material form the desire that the progress of this great continent shall not be merely material; and such an aspiration secures its own fulfilment.   However small may be the tangible results of our meeting, we shall always be proud to have been associated with you in your efforts for the advancement of science.

We do not know whether the last hundred years will be regarded for ever as the *sœculum mirabile* of discovery, or whether it is but the prelude to yet more marvellous centuries.   To us living men, who scarcely pass a year of our lives without witnessing some new marvel of discovery or invention, the rate at which the development of knowledge proceeds is truly astonishing; but from a wider point of view the scale of time is relatively unimportant, for the universe is leisurely in its procedure.   Whether the changes which we witness be fast or slow, they form a part of a long sequence of events which begin in some past of immeasurable remoteness and tend to some end which we cannot foresee. It must always be profoundly interesting to the mind of man to trace successive cause and effect in the chain of events which make up the history of the earth and all that lives on it, and to speculate on the origin and future fate of animals, and of planets, suns, and stars. I shall try, then, to set forth in my address some of the attempts which have been made to formulate evolutionary speculation. This choice of a subject has moreover been almost forced on me by the scope of my own scientific work, and it is, I think, justified by the name which I bear. It will be my fault and your misfortune if I fail to convey to you some part of the interest which is naturally inherent in such researches.

The man who propounds a theory of evolution is attempting to reconstruct the history of the past by means of the circumstantial evidence afforded by the present. The historian of man, on the other hand, has the advantage over the evolutionist in that he has the written records of the past on which to rely. The discrimination of the truth from amongst discordant records is frequently a work demanding the highest qualities of judgment; yet when this end is attained it remains for the historian to convert the arid skeleton of facts into a living whole by clothing it with the flesh of human motives and impulses. For this part of his task he needs much of that power of entering into the spirit of other men's lives which goes to the making of a poet. Thus the historian should possess not only the patience of the man of science in the analysis of facts, but also the imagination of the poet to grasp what the facts have meant. Such a combination is rarely to be found in equal perfection on both sides, and it would not be hard to analyse the works of great historians so as to see which quality was predominant in each of them.

The evolutionist is spared the surpassing difficulty of the human element, yet he also needs imagination, although of a different character from that of the historian. In its lowest form his imagination is that of the detective who reconstructs the story of a crime; in its highest it demands the power of breaking loose from all the trammels of convention and education, and of imagining something which has never occurred to the mind of man before. In every case the evolutionist must form a theory for the facts before him, and the great theorist is only to be distinguished from the fantastic fool by the sobriety of his judgment—a distinction, however, sufficient to make one rare and the other only too common.

The test of a scientific theory lies in the number of facts which it groups into a connected whole; it ought besides to be fruitful in pointing the way to the discovery and co-ordination of new and previously unsuspected facts. Thus a good theory is in effect a cyclopædia of knowledge, susceptible of indefinite extension by the addition of supplementary volumes.

Hardly any theory is all true, and many are not all false. A theory may be essentially at fault and yet point the way to truth, and so justify its temporary existence. We should not, therefore, totally reject one or other of two rival theories on the ground that they seem, with our present knowledge, mutually inconsistent, for it is likely that both may contain important elements of truth. The theories of which I shall have to speak hereafter may often appear discordant with one another according to our present lights. Yet we must not scruple to pursue the several divergent lines of thought to their logical conclusions, relying on future discovery to eliminate the false and to reconcile together the truths which form part of each of them.

In the mouths of the unscientific evolution is often spoken of as almost synonymous with the evolution of the various species of animals on the earth, and this again is sometimes thought to be practically the same thing as the theory of Natural Selection. Of course those who are conversant with the history of scientific ideas are aware that a belief in the gradual and orderly transformation of Nature, both animate and inanimate, is of great antiquity.

We may liken the facts on which theories of evolution are based to a confused heap of beads, from which a keen-sighted searcher after truth picks out and strings together a few which happen to catch his eye, as possessing certain resemblances. Until recently, theories of evolution in both realms of Nature were partial and discontinuous, and the chains of facts were correspondingly short and disconnected. At length the theory of Natural Selection, by formulating the cause of the divergence of forms in the organic world from the parental stock, furnished the naturalist with a clue by which he examined the disordered mass of facts before him, and he was thus enabled to go far in deducing order where chaos had ruled before, but the problem of reducing the heap to perfect order will probably baffle the ingenuity of the investigator for ever.

So illuminating has been this new idea that, as the whole of Nature has gradually been re-examined by its aid, thousands of new facts have been brought to light, and have been strung in due order on the necklace of knowledge. Indeed the transformation resulting from the new point of view has been so far-reaching as almost to justify the misapprehension of the unscientific as to the date when the doctrines of evolution first originated in the mind of man.

It is not my object, nor indeed am I competent, to examine the extent to which the Theory of Natural Selection has needed modification since it was first formulated by my father and Wallace. But I am surely justified in maintaining that the general principle holds its place firmly as a permanent acquisition to modes of thought.

Evolutionary doctrines concerning inanimate nature, although of much older date than those which concern life, have been profoundly affected by the great impulse of which I have spoken. It has thus come about that the origin and history of the chemical elements and of stellar systems now occupy a far larger space in the scientific mind than was formerly the case. The subject which I shall discuss to-night is the extent to which ideas, parallel to those which have done so much towards elucidating the problems of life, hold good also in the world of matter; and I believe that it will be possible to show that in this respect there exists a resemblance between the two realms of nature, which is not merely fanciful. It is proper to add that as long ago as 1873 Baron Karl du Prel discussed the same subject, from a similar point of view, in a book entitled *The Struggle for Life in the Heavens**.

Although inanimate matter moves under the action of forces which are incomparably simpler than those governing living beings, yet the problems of the physicist and the astronomer are scarcely less complex than those which present themselves to the biologist. The mystery of life remains as impenetrable as ever, and in his evolutionary speculations the biologist does not attempt to explain life itself, but, adopting as his unit the animal as a whole, discusses its relationships to other animals and to the surrounding conditions. The physicist, on the other hand, is irresistibly impelled to form theories as to the intimate constitution of the ultimate parts of matter, and he desires further to piece together the past histories and the future fates of planets, stars, and nebulæ. If then the speculations of the physicist seem in some respects less advanced than those of the biologist, it is chiefly because he is more ambitious in his aims. Physicists and astronomers have not yet found their Johannesburg or Kimberley; but although we are still mere prospectors, I am proposing to show you some of the dust and diamonds which we have already extracted from our surface mines.

The fundamental idea in the theory of Natural Selection is the persistence of those types of life which are adapted to their surrounding conditions, and the elimination by extermination of ill-adapted types. The struggle for life amongst forms possessing a greater or less degree of adaptation to slowly varying conditions is held to explain the gradual transmutation of species. Although a different phraseology is used when we speak of the physical world, yet the idea is essentially the same.

The point of view from which I wish you to consider the phenomena of the world of matter may be best explained if, in the first instance, I refer to political institutions, because we all understand, or fancy we understand, something of politics, whilst the problems of physics are commonly far less familiar to us. This illustration will have a further advantage in that it will

* *Der Kampf um's Dasein am Himmel* (zweite Auflage), Denicke, Berlin, 1876.

not be a mere parable, but will involve the fundamental conception of the nature of evolution.

The complex interactions of man with man in a community are usually described by such comprehensive terms as the State, the Commonwealth, or the Government. Various states differ widely in their constitution and in the degree of the complexity of their organisation, and we classify them by various general terms, such as autocracy, aristocracy, or democracy, which express somewhat loosely their leading characteristics. But, for the purpose of showing the analogy with physics, we need terms of wider import than those habitually used in politics. All forms of the State imply inter-relationship in the actions of men, and action implies movement. Thus the State may be described as a configuration or arrangement of a community of men ; or we may say that it implies a definite mode of motion of man—that is to say an organised scheme of action of man on man. Political history gives an account of the gradual changes in such configurations or modes of motion of men as have possessed the quality of persistence or of stability to resist the disintegrating influence of surrounding circumstances.

In the world of life the naturalist describes those forms which persist as species ; similarly the physicist speaks of stable configurations or modes of motion of matter ; and the politician speaks of States. The idea at the base of all these conceptions is that of stability, or the power of resisting disintegration. In other words, the degree of persistence or permanence of a species, of a configuration of matter, or of a State depends on the perfection of its adaptation to its surrounding conditions.

If we trace the history of a State we find the degree of its stability gradually changing, slowly rising to a maximum, and then slowly declining. When it falls to nothing a revolution ensues, and a new form of government is established. The new mode of motion or government has at first but slight stability, but it gradually acquires strength and permanence, until in its turn the slow decay of stability leads on to a new revolution.

Such crises in political history may give rise to a condition in which the State is incapable of perpetuation by transformation. This occurs when a savage tribe nearly exterminates another tribe and leads the few survivors into slavery ; the previous form of government then becomes extinct.

The physicist, like the biologist and the historian, watches the effect of slowly varying external conditions ; he sees the quality of persistence or stability gradually decaying until it vanishes, when there ensues what is called, in politics, a revolution.

These considerations lead me to express a doubt whether biologists have been correct in looking for uniform transformation of species. Judging by analogy we should rather expect to find slight continuous changes occurring during a long period of time, followed by a somewhat sudden transformation into a new species, or by rapid extinction*. However this may be, when the stability of a mode of motion vanishes, the physicist either finds that it is replaced by a new persistent type of motion adapted to the changed conditions, or perhaps that no such transformation is possible and that the mode of motion has become extinct. The evanescent type of animal life has often been preserved for us, fossilised in geological strata; the evanescent form of government is preserved in written records or in the customs of savage tribes; but the physicist has to pursue his investigations without such useful hints as to the past.

The time-scale in the transmutation of species of animals is furnished by the geological record, although it is not possible to translate that record into years. As we shall see hereafter, the time needed for a change of type in atoms or molecules may be measured by millionths of a second, while in the history of the stars continuous changes may occupy millions of years. Notwithstanding this gigantic contrast in speed, yet the process involved seems to be essentially the same.

It is hardly too much to assert that, if the conditions which determine stability of motion could be accurately formulated throughout the universe, the past history of the cosmos and its future fate would be unfolded. How indefinitely far we stand removed from such a state of knowledge will become abundantly clear from the remainder of my address.

The study of stability and instability then furnishes the problems which the physicist and biologist alike attempt to solve. The two classes of problems differ principally in the fact that the conditions of the world of life are so incomparably more intricate than those of the world of matter that the biologist is compelled to abandon the attempt to determine the absolute amount of the influence of the various causes which have affected the existence of species. His conclusions are merely qualitative and general, and he is almost universally compelled to refrain from asserting even in general terms what are the reasons which have rendered one form of animal life stable and persistent, and another unstable and evanescent.

On the other hand, the physicist, as a general rule, does not rest satisfied unless he obtains a quantitative estimate of various causes and effects on the systems of matter which he discusses. Yet there are some problems of physical evolution in which the conditions are so complex that the physicist is driven, as is the biologist, to rest satisfied with qualitative rather than

---

* If we may illustrate this graphically, I suggest that the process of transformation may be represented by long lines of gentle slope, followed by shorter lines of steeper slope. The alternative is a continuous uniform slope of change. If the former view is correct, it would explain why it should not be easy to detect specific change in actual operation. Some of my critics have erroneously thought that I advocate specific change *per saltum*.

quantitative conclusions.   But he is not content with such crude conclusions except in the last resort, and he generally prefers to proceed by a different method.

The mathematician mentally constructs an ideal mechanical system or model, which is intended to represent in its leading features the system he wants to examine.   It is often a task of the utmost difficulty to devise such a model, and the investigator may perchance unconsciously drop out as unimportant something which is really essential to represent actuality.   He next examines the conditions of his ideal system, and determines, if he can, all the possible stable and unstable configurations, together with the circumstances which will cause transitions from one to the other.   Even when the working model has been successfully imagined, this latter task may often overtax the powers of the mathematician.   Finally it remains for him to apply his results to actual matter, and to form a judgment of the extent to which it is justifiable to interpret Nature by means of his results.

The remainder of my address will be occupied by an account of various investigations which will illustrate the principles and methods which I have now explained in general terms.

The fascinating idea that matter of all kinds has a common substratum is of remote antiquity.   In the Middle Ages the alchemists, inspired by this idea, conceived the possibility of transforming the baser metals into gold. The sole difficulty seemed to them the discovery of an appropriate series of chemical operations.   We now know that they were always indefinitely far from the goal of their search, yet we must accord to them the honour of having been the pioneers of modern chemistry.

The object of alchemy, as stated in modern language, was to break up or dissociate the atoms of one chemical element into its component parts, and afterwards to reunite them into atoms of gold.   Although even the dissociative stage of the alchemistic problem still lies far beyond the power of the chemist, yet modern researches seem to furnish a sufficiently clear idea of the structure of atoms to enable us to see what would have to be done to effect a transformation of elements.   Indeed, in the complex changes which are found to occur spontaneously in uranium, radium, and the allied metals we are probably watching a spontaneous dissociation and transmutation of elements.

Natural Selection may seem, at first sight, as remote as the poles asunder from the ideas of the alchemist, yet dissociation and transmutation depend on the instability and regained stability of the atom, and the survival of the stable atom depends on the principle of Natural Selection.

Until some ten years ago the essential diversity of the chemical elements was accepted by the chemist as an ultimate fact, and indeed the very name of atom, or that which cannot be cut, was given to what was supposed to be the final indivisible portion of matter. The chemist thus proceeded in much the same way as the biologist who, in discussing evolution, accepts the species as his working unit. Accordingly, until recently the chemist discussed working models of matter of atomic structure, and the vast edifice of modern chemistry has been built with atomic bricks.

But within the last few years the electrical researches of Lenard, Röntgen, Becquerel, the Curies, of my colleagues Larmor and Thomson, and of a host of others, have shown that the atom is not indivisible, and a flood of light has been thrown thereby on the ultimate constitution of matter. Amongst all these fertile investigators it seems to me that Thomson stands pre-eminent, because it is principally through him that we are to-day in a better position for picturing the structure of an atom than was ever the case before.

Even if I had the knowledge requisite for a complete exposition of these investigations, the limits of time would compel me to confine myself to those parts of the subject which bear on the constitution and origin of the elements.

It has been shown, then, that the atom, previously supposed to be indivisible, really consists of a large number of component parts. By various convergent lines of experiment it has been proved that the simplest of all atoms, namely that of hydrogen, consists of about 800 separate parts; while the number of parts in the atom of the denser metals must be counted by tens of thousands. These separate parts of the atom have been called corpuscles or electrons, and may be described as particles of negative electricity. It is paradoxical, yet true, that the physicist knows more about these ultra-atomic corpuscles and can more easily count them than is the case with the atoms of which they form the parts.

The corpuscles, being negatively electrified, repel one another just as the hairs on a person's head mutually repel one another when combed with a vulcanite comb. The mechanism is as yet obscure whereby the mutual repulsion of the negative corpuscles is restrained from breaking up the atom, but a positive electrical charge, or something equivalent thereto, must exist in the atom, so as to prevent disruption. The existence in the atom of this community of negative corpuscles is certain, and we know further that they are moving with speeds which may in some cases be comparable to the velocity of light, namely, 200,000 miles a second. But the mechanism whereby they are held together in a group is hypothetical.

It is only just a year ago that Thomson suggested, as representing the atom, a mechanical or electrical model whose properties could be accurately examined by mathematical methods. He would be the first to admit that his model is at most merely a crude representation of actuality, yet he has been able to show that such an atom must possess mechanical and electrical properties which simulate, with what Whetham describes as "almost Satanic exactness," some of the most obscure and yet most fundamental properties of the chemical elements. "Se non è vero, è ben trovato," and we are surely justified in believing that we have the clue which the alchemists sought in vain.

Thomson's atom consists of a globe homogeneously charged with positive electricity, inside which there are one or more thousands of corpuscles of negative electricity, revolving in regular orbits with great velocities. Since two electrical charges repel each other if they are of the same kind, and attract each other if they are of opposite kinds, the corpuscles mutually repel one another, but all are attracted by the positive electricity distributed throughout the globe. The forces called into play by these electrical inter-actions are clearly very complicated, and you will not be surprised to learn that Thomson found himself compelled to limit his detailed examination of the model atom to one containing about seventy corpuscles. It is indeed a triumph of mathematical power to have determined the mechanical conditions of such a miniature planetary system as I have described.

It appears that in general there are definite arrangements of the orbits in which the corpuscles must revolve, if they are to be persistent or stable in their motions. But the number of corpuscles in such a community is not absolutely fixed. It is easy to see that we might add a minor planet, or indeed half a dozen minor planets, to the solar system without any material derangement of the whole; but it would not be possible to add a hundred planets with an aggregate mass equal to that of Jupiter without disorganisation of the solar system. So also we might add or subtract from an atom three or four corpuscles from a system containing a thousand corpuscles moving in regular orbits without any profound derangement. As each arrangement of orbits corresponds to the atom of a distinct element, we may say that the addition or subtraction of a few corpuscles to the atom will not effect a transmutation of elements. An atom which has a deficiency of its full complement of corpuscles, which it will be remembered are negative, will be positively electrified, while one with an excess of corpuscles will be negatively electrified. I have referred to the possibility of a deficiency or excess of corpuscles because it is important in Thomson's theory; but, as it is not involved in the point of view which I wish to take, I will henceforth only refer to the normal or average number in any arrangement of corpuscles. Accordingly we may state that definite numbers of corpuscles are capable of association in stable communities of definite types.

An infinite number of communities are possible, possessing greater or lesser degrees of stability. Thus the corpuscles in one such community might make thousands of revolutions in their orbits before instability declared itself; such an atom might perhaps last for a long time as estimated in millionths of seconds, but it must finally break up and the corpuscles must disperse or rearrange themselves after the ejection of some of their number. We are thus led to conjecture that the several chemical elements represent those different kinds of communities of corpuscles which have proved by their stability to be successful in the struggle for life. If this is so, it is almost impossible to believe that the successful species have existed for all time, and we must hold that they originated under conditions about which I must forbear to follow Sir Norman Lockyer in speculating*.

But if the elements were not eternal in the past, we must ask whether there is reason to believe that they will be eternal in the future. Now, although the conception of the decay of an element and its spontaneous transmutation into another element would have seemed absolutely repugnant to the chemist until recently, yet analogy with other moving systems seems to suggest that the elements are not eternal.

At any rate it is of interest to pursue to its end the history of the model atom which has proved to be so successful in imitating the properties of matter. The laws which govern electricity in motion indicate that such an atom must be radiating or losing energy, and therefore a time must come when it will run down, as a clock does. When this time comes it will spontaneously transmute itself into an element which needs less energy than was required in the former state. Thomson conceives that an atom might be constructed after his model so that its decay should be very slow. It might, he thinks, be made to run for a million years or more, but it would not be eternal.

Such a conclusion is in absolute contradiction to all that was known of the elements until recently, for no symptoms of decay are perceived, and the elements existing in the solar system must already have lasted for millions of years. Nevertheless, there is good reason to believe that in radium, and in other elements possessing very complex atoms, we do actually observe that break-up and spontaneous rearrangement which constitute a transmutation of elements.

It is impossible as yet to say how science will solve this difficulty, but future discovery in this field must surely prove deeply interesting†. It may well be that the train of thought which I have sketched will ultimately profoundly affect the material side of human life, however remote it may now seem from our experiences of daily life.

I have not as yet made any attempt to represent the excessive minuteness of the corpuscles, of whose existence we are now so confident; but, as an introduction to what I have to speak of next, it is necessary to do so. To obtain any adequate conception of their size we must betake ourselves to a

---

* *Inorganic Evolution*, Macmillan, 1900.

† The view that the elements are not absolutely permanent seems to be gaining ground. See correspondence in *Nature*: D. Murray, December 7; Soddy and Campbell, December 14, 1905.

scheme of threefold magnification.   Lord Kelvin has shown that, if a drop of water were magnified to the size of the earth, the molecules of water would be of a size intermediate between that of a cricket-ball and of a marble.   Now each molecule contains three atoms, two being of hydrogen and one of oxygen. The molecular system probably presents some sort of analogy with that of a triple star; the three atoms, replacing the stars, revolving about one another in some sort of dance which cannot be exactly described.   I doubt whether it is possible to say how large a part of the space occupied by the whole molecule is occupied by the atoms; but perhaps the atoms bear to the molecule some such relationship as the molecule to the drop of water referred to.   Finally, the corpuscles may stand to the atom in a similar scale of magnitude.   Accordingly a threefold magnification would be needed to bring these ultimate parts of the atom within the range of our ordinary scales of measurement.

I have already considered what would be observed under the triply powerful microscope, and must now return to the intermediate stage of magnification, in which we consider those communities of atoms which form molecules.   This is the field of research of the chemist.   Although prudence would tell me that it would be wiser not to speak of a subject of which I know so little, yet I cannot refrain from saying a few words.

The community of atoms in water has been compared with a triple star, but there are others known to the chemist in which the atoms are to be counted by fifties and hundreds, so that they resemble constellations.

I conceive that here again we meet with conditions similar to those which we have supposed to exist in the atom.   Communities of atoms are called chemical combinations, and we know that they possess every degree of stability.   The existence of some is so precarious that the chemist in his laboratory can barely retain them for a moment; others are so stubborn that he can barely break them up.   In this case dissociation and reunion into new forms of communities are in incessant and spontaneous progress throughout the world.   The more persistent or more stable combinations succeed in their struggle for life, and are found in vast quantities, as in the cases of common salt and of the combinations of silicon.   But no one has ever found a mine of gun-cotton, because it has so slight a power of resistance.   If, through some accidental collocation of elements, a single molecule of gun-cotton were formed, it would have but a short life.

Stability is, further, a property of relationship to surrounding conditions; it denotes adaptation to environment.   Thus salt is adapted to the struggle for existence on the earth, but it cannot withstand the severer conditions which exist in the sun.

Thus far we have been concerned with the almost inconceivably minute, and I now propose to show that similar conditions prevail on a larger scale.

Many geological problems might well be discussed from my present point of view, yet I shall pass them by, and shall proceed at once to Astronomy, beginning with the smallest cosmical scale of magnitude, and considering afterwards the larger celestial phenomena.

The problems of cosmical evolution are so complicated that it is well to conduct the attack in various ways at the same time. Although the several theories may seem to some extent discordant with one another, yet, as I have already said, we ought not to scruple to carry each to its logical conclusion. We may be confident that in time the false will be eliminated from each theory, and when the true alone remains the reconciliation of apparent disagreements will have become obvious.

The German astronomer Bode long ago propounded a simple empirical law concerning the distances at which the several planets move about the sun. It is true that the planet Neptune, discovered subsequently, was found to be considerably out of the place which would be assigned to it by Bode's law, yet his formula embraces so large a number of cases with accuracy that we are compelled to believe that it arises in some manner from the primitive conditions of the planetary system.

The explanation of the causes which have led to this simple law as to the planetary distances presents an interesting problem, and, although it is still unsolved, we may obtain some insight into its meaning by considering what I have called a working model of ideal simplicity.

Imagine then a sun round which there moves in a circle a single large planet. I will call this planet Jove, because it may be taken as a representative of our largest planet, Jupiter. Suppose next that a meteoric stone or small planet is projected in any perfectly arbitrary manner in the same plane in which Jove is moving; then we ask how this third body will move. The conditions imposed may seem simple, yet the problem has so far overtaxed the powers of the mathematician that nothing approaching a general answer to our question has yet been given. We know, however, that under the combined attractions of the sun and Jove the meteoric stone will in general describe an orbit of extraordinary complexity, at one time moving slowly at a great distance from both the sun and Jove, at other times rushing close past one or other of them. As it grazes past Jove or the sun it may often but just escape a catastrophe, but a time will come at length when it runs its chances too fine and comes into actual collision. The individual career of the stone is then ended by absorption, and of course by far the greater chance is that it will find its Nirvana by absorption in the sun.

Next let us suppose that instead of one wandering meteoric stone or minor planet there are hundreds of them, moving initially in all conceivable directions. Since they are all supposed to be very small, their mutual attractions will be insignificant, and they will each move almost as though they were influenced only by the sun and Jove. Most of these stones will be absorbed by the sun, and the minority will collide with Jove.

When we inquire how long the career of a stone may be, we find that it depends on the direction and speed with which it is started, and that by proper adjustment the delay of the final catastrophe may be made as long as we please. Thus by making the delay indefinitely long we reach the conception of a meteoric stone which moves so as never to come into collision with either body.

There are, therefore, certain perpetual orbits in which a meteoric stone or minor planet may move for ever without collision. But when such an immortal career has been discovered for our minor planet, it still remains to discover whether the slightest possible departure from the prescribed orbit will become greater and greater and ultimately lead to a collision with the sun or Jove, or whether the body will travel so as to cross and recross the exact perpetual orbit, always remaining close to it. If the slightest departure inevitably increases as time goes on, the orbit is unstable; if, on the other hand, it only leads to a slight waviness in the path described, it is stable.

We thus arrive at another distinction: there are perpetual orbits, but some, and indeed most, are unstable, and these do not offer an immortal career for a meteoric stone; and there are other perpetual orbits which are stable or persistent. The unstable ones are those which succumb in the struggle for life, and the stable ones are the species adapted to their environment.

If, then, we are given a system of a sun and large planet, together with a swarm of small bodies moving in all sorts of ways, the sun and planet will grow by accretion, gradually sweeping up the dust and rubbish of the system, and there will survive a number of small planets and satellites moving in certain definite paths. The final outcome will be an orderly planetary system in which the various orbits are arranged according to some definite law.

But the problem presented even by a system of such ideal simplicity is still far from having received a complete solution. No general plan for determining perpetual orbits has yet been discovered, and the task of discriminating the stable from the unstable is arduous. But a beginning has been made in the determination of some of the zones surrounding the sun and Jove in which stable orbits are possible, and others in which they are impossible. There is hardly room for doubt that if a complete solution for our solar system were attainable, we should find that the orbits of the existing planets and satellites are numbered amongst the stable perpetual orbits, and should thus obtain a rigorous mechanical explanation of Bode's law concerning the planetary distances.

It is impossible not to be struck by the general similarity between the problem presented by the corpuscles moving in orbits in the atom, and that of the planets and satellites moving in a planetary system. It may not, perhaps, be fanciful to imagine that some general mathematical method devised for solving a problem of cosmical evolution may find another application to miniature atomic systems, and may thus lead onward to vast developments of industrial mechanics. Science, however diverse its aims, is a whole, and men of science do well to impress on the captains of industry that they should not look askance on those branches of investigation which may seem for the moment far beyond any possibility of practical utility.

You will remember that I discussed the question as to whether the atomic communities of corpuscles could be regarded as absolutely eternal, and that I said that the analogy of other moving systems pointed to their ultimate mortality. Now the chief analogy which I had in my mind was that of a planetary system.

The orbits of which I have spoken are only perpetual when the bodies are infinitesimal in mass, and meet with no resistance as they move. Now the infinitesimal body does not exist, and both Lord Kelvin and Poincaré concur in holding that disturbance will ultimately creep in to any system of bodies moving even in so-called stable orbits; and this is so even apart from the resistance offered to the moving bodies by any residual gas there may be scattered through space. The stability is therefore only relative, and a planetary system contains the seeds of its own destruction. But this ultimate fate need not disturb us either practically or theoretically, for the solar system contains in itself other seeds of decay which will probably bear fruit long before the occurrence of any serious disturbance of the kind of which I speak.

Before passing on to a new topic I wish to pay a tribute to the men to whom we owe the recent great advances in theoretical dynamical astronomy. As treated by the master-hands of Lagrange and Laplace and their successors, this branch of science hardly seemed to afford scope for any great new departure. But that there is always room for discovery, even in the most frequented paths of knowledge, was illustrated when, nearly thirty years ago, Hill of Washington proposed a new method of treating the theory of the moon's motion in a series of papers which have become classical. I have not time to speak of the enormous labour and great skill involved in the completion of Hill's Lunar Theory, by Ernest Brown, who I am glad to number amongst my pupils and friends; for I must confine myself to other aspects of Hill's work.

The title of Hill's most fundamental paper, namely, "On Part of the Motion of the Lunar Perigee," is almost comic in its modesty, for who would suspect that it contains the essential points involved in the determination of perpetual orbits and their stability? Probably Hill himself did not fully realise at the time the full importance of what he had done. Fortunately he was followed by Poincaré, who not only saw its full meaning but devoted his incomparable mathematical powers to the full theoretical development of the point of view I have been laying before you.

Other mathematicians have also made contributions to this line of investigation, amongst whom I may number my friend Mr Hough, chief assistant at the Royal Observatory of Cape Town, and myself. But without the work of our two great forerunners we should still be in utter darkness, and it would have been impossible to give even this slight sketch of a great subject.

The theory which I have now explained points to the origin of the sun and planets from gradual accretions of meteoric stones, and it makes no claim to carry the story back behind the time when there was already a central condensation or sun about which there circled another condensation or planet. But more than a century ago an attempt had already been made to reconstruct the history back to a yet remoter past, and, as we shall see, this attempt was based upon quite a different supposition as to the constitution of the primitive solar system. I myself believe that the theory I have just explained, as well as that to which I am coming, contains essential elements of truth, and that the apparent discordances will some day be reconciled. The theory of which I speak is the celebrated Nebular Hypothesis, first suggested by the German philosopher Kant, and later restated independently and in better form by the French mathematician Laplace.

Laplace traced the origin of the solar system to a nebula or cloud of rarefied gas congregated round a central condensation which was ultimately to form the sun. The whole was slowly rotating about an axis through its centre, and, under the combined influences of rotation and of the mutual attraction of the gas, it assumed a globular form, slightly flattened at the poles. The justifiability of this supposition is confirmed by the observations of astronomers, for they find in the heavens many nebulæ, while the spectroscope proves that their light at any rate is derived from gas. The primeval globular nebula is undoubtedly a stable or persistent figure, and thus Laplace's hypothesis conforms to the general laws which I have attempted to lay down.

The nebula must have gradually cooled by radiation into space, and as it did so the gas must necessarily have lost some of its spring or elasticity. This loss of power of resistance then permitted the gas to crowd more closely towards the central condensation, so that the nebula contracted. The contraction led to two results, both inevitable according to the laws of mechanics: first, the central condensation became hotter; and, secondly, the speed of its rotation became faster. The accelerated rotation led to an increase in the amount of polar flattening, and the nebula at length assumed the form of a lens, or of a disk thicker in the middle than at the edges. Assuming the existence of the primitive nebula, the hypothesis may be accepted thus far as practically certain.

From this point, however, doubt and difficulty enter into the argument. It is supposed that the nebula became so much flattened that it could not subsist as a continuous aggregation of gas, and a ring of matter detached itself from the equatorial regions. The central portions of the nebula, when relieved of the excrescence, resumed the more rounded shape formerly possessed by the whole. As the cooling continued the central portion in its turn became excessively flattened through the influence of its increased rotation; another equatorial ring then detached itself, and the whole process was repeated as before. In this way the whole nebula was fissured into a number of rings surrounding the central condensation, whose temperature must by then have reached incandescence.

Each ring then aggregated itself round some nucleus which happened to exist in its circumference, and so formed a subordinate nebula. Passing through a series of transformations, like its parent, this nebula was finally replaced by a planet with attendant satellites.

The whole process forms a majestic picture of the history of our system. But the mechanical conditions of a rotating nebula are too complex to admit, as yet, of complete mathematical treatment; and thus, in discussing this theory, the physicist is compelled in great measure to adopt the qualitative methods of the biologist, rather than the quantitative ones which he would prefer.

The telescope seems to confirm the general correctness of Laplace's hypothesis. Thus, for example, the great nebula in Andromeda presents a grand illustration of what we may take to be a planetary system in course of formation. In it we see the central condensation surrounded by a more or less ring-like nebulosity, and in one of the rings there appears to be a subordinate condensation.

Nevertheless it is hardly too much to say that every stage in the supposed process presents to us some difficulty or impossibility. Thus we ask whether a mass of gas of almost inconceivable tenuity can really rotate all in one piece, and whether it is not more probable that there would be a central whirlpool surrounded by more slowly moving parts. Again, is there any sufficient reason to suppose that a series of intermittent efforts would lead to the detachment of distinct rings, and is not a continuous outflow of gas from the equator more probable?

The ring of Saturn seems to have suggested the theory to Laplace; but to take it as a model leads us straight to a quite fundamental difficulty. If a ring of matter ever concentrates under the influence of its mutual attraction, it can only do so round the centre of gravity of the whole ring. Therefore the matter forming an approximately uniform ring, if it concentrates at all, can only fall in on the parent planet and be re-absorbed. Some external force other than the mutual attraction of the matter forming the ring, and therefore not provided by the theory, seems necessary to effect the supposed concentration. The only way of avoiding this difficulty is to suppose the ring to be ill-balanced or lop-sided; in this case, provided the want of balance is pronounced enough, concentration will take place round a point inside the ring but outside the planet. Many writers assume that the present distances of the planets preserve the dimensions of the primitive rings; but the argument that a ring can only aggregate about its centre of gravity, which I do not recollect to have seen before, shows that such cannot be the case.

The concentration of an ill-balanced or broken ring on an interior point would necessarily generate a planet with direct rotation—that is to say, rotating in the same direction as the earth. But several writers, and notably Faye, endeavour to show—erroneously as I think—that a retrograde rotation should be normal, and they are therefore driven to make various complicated suppositions to explain the observed facts. But I do not claim to have removed the difficulty, only to have shifted it; for the satellites of Neptune, and presumably the planet itself, have retrograde rotations; and, lastly, the astonishing discovery has just been made by William Pickering of a ninth retrograde satellite of Saturn, while the rotations of the eight other satellites, of the ring and of the planet itself, are direct. Finally, I express a doubt as to whether the telescope does really exactly confirm the hypothesis of Laplace, for I imagine that what we see indicates a spiral rather than a ring-like division of nebulæ*.

This is not the time to pursue these considerations further, but enough has been said to show that the Nebular Hypothesis cannot be considered as a connected intelligible whole, however much of truth it may contain.

In the first theory which I sketched as to the origin of the sun and planets, we supposed them to grow by the accretions of meteoric wanderers in space, and this hypothesis is apparently in fundamental disagreement with the conception of Laplace, who watches the transformations of a continuous gaseous nebula. Some years ago a method occurred to me by which these two discordant schemes of origin might perhaps be reconciled. A gas is not really continuous, but it consists of a vast number of molecules moving in all directions with great speed and frequently coming into

* Professor Chamberlin, of Chicago, has recently proposed a modified form of the Nebular Hypothesis, in which he contends that the spiral form is normal. See *Year Book*, No. 3, for 1904, of the Carnegie Institution of Washington, pp. 195—258.

collision with one another.   Now I have ventured to suggest that a swarm of meteorites would, by frequent collisions, form a medium endowed with so much of the mechanical properties of a gas as would satisfy Laplace's conditions.   If this is so, a nebula may be regarded as a quasi-gas, whose molecules are meteorites*.   The gaseous luminosity which undoubtedly is sent out by nebulæ would then be due only to incandescent gas generated by the clash of meteorites, while the dark bodies themselves would remain invisible.   Sir Norman Lockyer finds spectroscopic evidence which led him long ago to some such view as this, and it is certainly of interest to find in his views a possible means of reconciling two apparently totally discordant theories†.   However, I do not desire to lay much stress on my suggestion, for without doubt a swarm of meteors could only maintain the mechanical properties of a gas for a limited time, and, as pointed out by Professor Chamberlin, it is difficult to understand how a swarm of meteorites moving indiscriminately in every direction could ever have come into existence. But my paper may have served to some extent to suggest to Chamberlin his recent modification of the Nebular Hypothesis, in which he seeks to reconcile Laplace's view with a meteoritic origin of the planetary system‡.

We have seen that, in order to explain the genesis of planets according to Laplace's theory, the rings must be ill-balanced or even broken.   If the ring were so far from being complete as only to cover a small segment of the whole circumference, the true features of the occurrences in the births of planets and satellites might be better represented by conceiving the detached portion of matter to have been more or less globular from the first, rather than ring-shaped.   Now this idea introduces us to a group of researches whereby mathematicians have sought to explain the birth of planets and satellites in a way which might appear, at first sight, to be fundamentally different from that of Laplace.

The solution of the problem of evolution involves the search for those persistent or stable forms which biologists would call species.   The species of which I am now going to speak may be grouped in a family, which comprises all those various forms which a mass of rotating liquid is capable of assuming under the conjoint influences of gravitation and rotation.   If the earth were formed throughout of a liquid of the same density, it would be one of the species of this family; and indeed these researches date back to the time of Newton, who was the first to explain the figures of planets.

The ideal liquid planets we are to consider must be regarded as working models of actuality, and inasmuch as the liquid is supposed to be in-

---

* [Paper 18, p. 362.]

† Newcomb considers the objections to Lockyer's theory insuperable.   See p. 190 of *The Stars*, John Murray, London, 1904.

‡ See preceding reference to Chamberlin's Paper.

compressible, the conditions depart somewhat widely from those of reality. Hence, when the problem has been solved, much uncertainty remains as to the extent to which our conclusions will be applicable to actual celestial bodies.

We begin, then, with a rotating liquid planet like the earth, which is the first stable species of our family. We next impart in imagination more rotation to this planet, and find by mathematical calculation that its power of resistance to any sort of disturbance is less than it was. In other words, its stability declines with increased rotation, and at length we reach a stage at which the stability just vanishes. At this point the shape is a transitional one, for it is the beginning of a new species with different characteristics from the first, and with a very feeble degree of stability or power of persistence. As a still further amount of rotation is imparted, the stability of the new species increases to a maximum and then declines until a new transitional shape is reached and a new species comes into existence. In this way we pass from species to species with an ever-increasing amount of rotation.

The first or planetary species has a circular equator like the earth; the second species has an oval equator, so that it is something like an egg spinning on its side on a table; in the third species we find that one of the two ends of the egg begins to swell, and that the swelling gradually becomes a well-marked protrusion or filament*. Finally the filamentous protrusion becomes bulbous at its end, and is only joined to the main mass of liquid by a gradually thinning neck. The neck at length breaks, and we are left with two separated masses which may be called planet and satellite. It is fair to state that the actual rupture into two bodies is to some extent speculative, since mathematicians have hitherto failed to follow the whole process to the end†.

In this ideal problem the successive transmutations of species are brought about by gradual additions to the amount of rotation with which the mass of liquid is endowed. It might seem as if this continuous addition to the amount of rotation were purely arbitrary and could have no counterpart in nature. But real bodies cool and contract in cooling, and, since the scale of magnitude on which our planet is built is immaterial, contraction will produce exactly the same effect on shape as augmented rotation. I must ask you, then, to believe that the effects of an apparently arbitrary increase of rotation may be produced by cooling.

The figures which I succeeded in drawing, by means of rigorous calculation, of the later stages of this course of evolution, are so curious as to

---

* M. Liapounoff contends that the "pear-shaped" figure is always unstable ("Sur un problème de Tchebychef," *Acad. Imp. des Sciences de St-Pétersbourg*, 1905), but I cannot agree with this view—at least for the present. [See Vol. III., Papers 11 and 12.]

† See a paper by myself "On the Figure and Stability of a liquid Satellite" communicated to the Royal Society, January 1906. [Vol. III., Paper 15, p. 436.]

remind one of some such phenomenon as the protrusion of a filament of protoplasm from a mass of living matter, and I suggest that we may see in this almost life-like process the counterpart of at least one form of the birth of double stars, planets, and satellites.

As I have already said, Newton determined the first of these figures; Jacobi found the second and Poincaré indicated the existence of the third, in a paper which is universally regarded as one of the masterpieces of applied mathematics; finally I myself succeeded in determining the exact form of Poincaré's figure, and [I thought I had proved it to be] a true stable shape.

My Cambridge colleague Jeans has also made an interesting contribution to the subject by discussing a closely analogous problem, and he has besides attacked the far more difficult case where the rotating fluid is a compressible gas. In this case also he finds a family of types, but the conception of compressibility introduced a new set of considerations in the transitions from species to species. The problem is, however, of such difficulty that he had to rest content with results which were rather qualitative than strictly quantitative.

This group of investigations brings before us the process of the birth of satellites in a more convincing form than was possible by means of the general considerations adduced by Laplace. It cannot be doubted that the supposed Laplacian sequence of events possesses a considerable element of truth, yet these latter schemes of transformation can be followed in closer detail. It seems, then, probable that both processes furnish us with crude models of reality, and that in some cases the first and in others the second is the better representative.

The moon's mass is one-eightieth of that of the earth, whereas the mass of Titan, the largest satellite in the solar system, is $\frac{1}{4600}$ of that of Saturn. On the ground of this great difference between the relative magnitudes of all other satellites and of the moon, it is not unreasonable to suppose that the mode of separation of the moon from the earth may also have been widely different. The theory of which I shall have next to speak claims to trace the gradual departure of the moon from an original position not far removed from the present surface of the earth. If this view is correct, we may suppose that the detachment of the moon from the earth occurred as a single portion of matter, and not as a concentration of a Laplacian ring.

If a planet is covered with oceans of water and air, or if it is formed of plastic molten rock, tidal oscillations must be generated in its mobile parts by the attractions of its satellites and of the sun. Such movements must be subject to frictional resistance, and the planet's rotation will be slowly retarded by tidal friction in much the same way that a fly-wheel is gradually stopped by any external cause of friction. Since action and reaction are equal and opposite, the action of the satellites on the planet, which causes the tidal friction of which I speak, must correspond to a reaction of the planet on the motion of the satellites.

At any moment of time we may regard the system composed of the rotating planet with its attendant satellite as a stable species of motion, but the friction of the tides introduces forces which produce a continuous, although slow, transformation in the configuration. It is, then, clearly of interest to trace backwards in time the changes produced by such a continuously acting cause, and to determine the initial condition from which the system of planet and satellite must have been slowly degrading. We may also look forward, and discover whither the transformation tends.

Let us consider, then, the motion of the earth and moon revolving in company round the sun, on the supposition that the friction of the tides in the earth is the only effective cause of change. We are, in fact, to discuss a working model of the system, analogous to those of which I have so often spoken before.

This is not the time to attempt a complete exposition of the manner in which tidal friction gives rise to the action and reaction between planet and satellite, nor shall I discuss in detail the effects of various kinds which are produced by this cause. It must suffice to set forth the results in their main outlines, and, as in connection with the topic of evolution retrospect is perhaps of greater interest than prophecy, I shall begin with the consideration of the past.

At the present time the moon, moving at a distance of 240,000 miles from the earth, completes her circuit in twenty-seven days. Since a day is the time of one rotation of the earth on its axis, the angular motion of the earth is twenty-seven times as rapid as that of the moon.

Tidal friction acts as a brake on the earth, and therefore we look back in retrospect to times when the day was successively twenty-three, twenty-two, twenty-one of our present hours in length, and so on backward to still shorter days. But during all this time the reaction on the moon was at work, and it appears that its effect must have been such that the moon also revolved round the earth in a shorter period than it does now; thus the month also was shorter in absolute time than it now is. These conclusions are absolutely certain, although the effects on the motions of the earth and of the moon are so gradual that they can only doubtfully be detected by the most refined astronomical measurements.

We take the "day," regarding it as a period of variable length, to mean the time occupied by a single rotation of the earth on its axis; and the "month," likewise variable in absolute length, to mean the time occupied by the moon in a single revolution round the earth. Then, although there are now twenty-seven days in a month, and although both day and month were shorter in the past, yet there is, so far, nothing to tell us whether there were more or less days in the month in the past. For if the day is now being prolonged more rapidly than the month, the number of days in the month was greater in the past than it now is; and if the converse were true, the number of days in the month was less.

Now it appears from mathematical calculation that the day must now be suffering a greater degree of prolongation than the month, and accordingly in retrospect we look back to a time when there were more days in the month than at present. That number was once twenty-nine, in place of the present twenty-seven; but the epoch of twenty-nine days in the month is a sort of crisis in the history of moon and earth, for yet earlier the day was shortening less rapidly than the month. Hence, earlier than the time when there were twenty-nine days in the month, there was a time when there was a reversion to the present smaller number of days.

We thus arrive at the curious conclusion that there is a certain number of days to the month, namely twenty-nine, which can never have been exceeded, and we find that this crisis was passed through by the earth and moon recently; but, of course, a recent event in such a long history may be one which happened some millions of years ago.

Continuing our retrospect beyond this crisis, both day and month are found continuously shortening, and the number of days in the month continues to fall. No change in conditions which we need pause to consider now supervenes, and we may ask at once, what is the initial stage to which the gradual transformation points? I say, then, that on following the argument to its end the system may be traced back to a time when the day and month were identical in length, and were both only about four or five of our present hours. The identity of day and month means that the moon was always opposite to the same side of the earth; thus at the beginning the earth always presented the same face to the moon, just as the moon now always shows the same face to us. Moreover, when the month was only some four or five of our present hours in length the moon must have been only a few thousand miles from the earth's surface—a great contrast with the present distance of 240,000 miles.

It might well be argued from this conclusion alone that the moon separated from the earth more or less as a single portion of matter at a time immediately antecedent to the initial stage to which she has been traced. But there exists a yet more weighty argument favourable to this view, for it appears that the initial stage is one in which the stability of the species of motion is tottering, so that the system presents the characteristic of a transitional form, which we have seen to denote a change of type or species in a previous case.

In discussing the transformations of a liquid planet we saw the tendency of the single mass to divide into two portions, although we failed to extend the rigorous argument back to the actual moment of separation; and now we seem to reach a similar crisis from the opposite end, when in retrospect we trace back the system to two masses of unequal size in close proximity with one another. The argument almost carries conviction with it, but I have necessarily been compelled to pass over various doubtful points.

Time is wanting to consider other subjects worthy of notice which arise out of this problem, yet I wish to point out that the earth's axis must once have been less tilted over with reference to the sun than it is now, so that the obliquity of the ecliptic receives at least a partial explanation. Again, the inclination of the moon's orbit may be in great measure explained; and, lastly, the moon must once have moved in a nearly circular path. The fact that tidal friction is competent to explain the eccentricity of an orbit has been applied in a manner to which I shall have occasion to return hereafter.

In my paper on this subject I summed up the discussion in the following words, which I still see no reason to retract :—

"The argument reposes on the imperfect rigidity of solids, and on the internal friction of semi-solids and fluids; these are *veræ causæ*. Thus changes of the kind here discussed must be going on, and must have gone on in the past. And for this history of the earth and moon to be true throughout it is only necessary to postulate a sufficient lapse of time, and that there is not enough matter diffused through space materially to resist the motions of the moon and earth in perhaps several hundred million years.

"It hardly seems too much to say that granting these two postulates and the existence of a primeval planet, such as that above described, then a system would necessarily be developed which would bear a strong resemblance to our own.

"A theory, reposing on *veræ causæ*, which brings into quantitative correlation the lengths of the present day and month, the obliquity of the ecliptic, and the inclination and eccentricity of the lunar orbit, must, I think, have strong claims to acceptance *."

We have pursued the changes into the past, and I will refer but shortly to the future. The day and month are both now lengthening, but the day changes more quickly than the month. Thus the two periods tend again to become equal to one another, and it appears that when that goal is reached both day and month will be as long as fifty-five of our present days. The earth will then always show the same face to the moon, just as it did in the remotest past. But there is a great contrast between the ultimate and initial conditions, for the ultimate stage, with

---

* *Phil. Trans.*, Part II., 1880, p. 883.

day and month both equal to fifty-five of our present days, is one of great stability in contradistinction to the vanishing stability which we found in the initial stage.

Since the relationship between the moon and earth is a mutual one, the earth may be regarded as a satellite of the moon, and if the moon rotated rapidly on her axis, as was probably once the case, the earth must at that time have produced tides in the moon. The mass of the moon is relatively small, and the tides produced by the earth would be large; accordingly the moon would pass through the several stages of her history much more rapidly than the earth. Hence it is that the moon has already advanced to that condition which we foresee as the future fate of the earth, and now always shows to us the same face.

If the earth and moon were the only bodies in existence, this ultimate stage when the day and month were again identical in length, would be one of absolute stability, and therefore eternal; but the presence of the sun introduces a cause for yet further changes. I do not, however, propose to pursue the history to this yet remoter futurity, because our system must contain other seeds of decay which will probably bear fruit before these further transformations could take effect.

If, as has been argued, tidal friction has played so important a part in the history of the earth and moon, it might be expected that the like should be true of the other planets and satellites, and of the planets themselves in their relationship to the sun. But numerical examination of the several cases proves conclusively that this cannot have been the case. The relationship of the moon to the earth is in fact quite exceptional in the solar system, and we have still to rely on such theories as that of Laplace for the explanation of the main outlines of the solar system.

I have as yet only barely mentioned the time occupied by the sequence of events sketched out in the various schemes of cosmogony, and the question of cosmical time is a thorny and controversial one.

Our ideas are absolutely blank as to the time requisite for the evolution according to Laplace's nebular hypothesis. And again, if we adopt the meteoritic theory, no estimate can be formed of the time required even for an ideal sun, with its attendant planet Jove, to sweep up the wanderers in space. We do know, indeed, that there is a continuous gradation from stable to unstable orbits, so that some meteoric stones may make thousands or millions of revolutions before meeting their fate by collision. Accordingly, not only would a complete absorption of all the wanderers occupy an infinite time, but also the amount of the refuse of the solar system still remaining scattered in planetary space is unknown. And, indeed, it is certain that the process of clearance is still going on, for the earth is constantly meeting meteoric stones, which, penetrating the atmosphere, become luminous through the effects of the frictional resistance with which they meet.

All we can assert of such theories is that they demand enormous intervals of time as estimated in years.

The theory of tidal friction stands alone amongst these evolutionary speculations in that we can establish an exact but merely relative time-scale for every stage of the process. It is true that the value in years of the unit of time remains unknown, and it may be conjectured that the unit has varied to some extent as the physical condition of the earth has gradually changed.

It is, however, possible to determine a period in years which must be shorter than that in which the whole history is comprised. If at every moment since the birth of the moon tidal friction had always been at work in such a way as to produce the greatest possible effect, then we should find that sixty million years would be consumed in this portion of evolutionary history. The true period must be much greater, and it does not seem extravagant to suppose that 500 to 1,000 million years may have elapsed since the birth of the moon.

Such an estimate would not seem extravagant to geologists who have, in various ways, made exceedingly rough determinations of geological periods. One such determination is derived from measures of the thickness of deposited strata, and the rate of the denudation of continents by rain and rivers. I will not attempt to make any precise statement on this head, but I imagine that the sort of unit with which the geologist deals is 100 million years, and that he would not consider any estimate involving from one to twenty of such units as unreasonable.

Mellard Reade has attempted to determine geological time by certain arguments as to the rate of denudation of limestone rocks, and arrives at the conclusion that geological history is comprised in something less than 600 million years*. The uncertainty of this estimate is wide, and I imagine that geologists in general would not lay much stress on it.

Joly has employed a somewhat similar, but probably less risky, method of determination†. When the earth was still hot, all the water of the globe must have existed in the form of steam, and when the surface cooled that steam must have condensed as fresh water. Rain then washed the continents and carried down detritus and soluble matter to the seas. Common salt is the most widely diffused of all such soluble matter, and its transit

* *Chemical Denudation in relation to Geological Time*, Bogue, London, 1879; or *Proc. Roy. Soc.*, Vol. xxviii. (1879), p. 281.

† "An Estimate of the Geological Age of the Earth," *Trans. Roy. Dublin Soc.*, Vol. vii., Series iii., 1902, pp. 23—66.

to the sea is an irreversible process, because the evaporation of the sea only carries back to the land fresh water in the form of rain. It seems certain, then, that the saltness of the sea is due to the washing of the land throughout geological time.

Rough estimates may be formed of the amount of river water which reaches the sea in a year, and the measured saltness of rivers furnishes a knowledge of the amount of salt which is thus carried to the sea. A closer estimate may be formed of the total amount of salt in the sea. On dividing the total amount of salt by the annual transport Joly arrives at the quotient of about 100 millions, and thence concludes that geological history has occupied 100 million years. I will not pause to consider the several doubts and difficulties which arise in the working out of this theory. The uncertainties involved must clearly be considerable, yet it seems the best of all the purely geological arguments whence we derive numerical estimates of geological time. On the whole I should say that pure geology points to some period intermediate between 50 and 1,000 millions of years, but the upper limit is more doubtful than the lower. Thus far we do not find anything which renders the tidal theory of evolution untenable.

But the physicists have formed estimates in other ways which, until recently, seemed to demand in the most imperative manner a far lower scale of time. According to all theories of cosmogony, the sun is a star which became heated in the process of its condensation from a condition of wide dispersion. When a meteoric stone falls into the sun the arrest of its previous motion gives rise to heat, just as the blow of a horse's shoe on a stone makes a spark. The fall of countless meteoric stones, or the condensation of a rarefied gas, was supposed to be the sole cause of the sun's high temperature.

Since the mass of the sun is known, the total amount of the heat generated in it, in whatever mode it was formed, can be estimated with a considerable amount of precision. The heat received at the earth from the sun can also be measured with some accuracy, and hence it is a mere matter of calculation to determine how much heat the sun sends out in a year. The total heat which can have been generated in the sun divided by the annual output gives a quotient of about 20 millions. Hence it seemed to be imperatively necessary that the whole history of the solar system should be comprised within some 20 millions of years.

This argument, which is due to Helmholtz, appeared to be absolutely crushing, and for the last forty years the physicists have been accustomed to tell the geologists that they must moderate their claims. But for myself I have always believed that the geologists were more nearly correct than the physicists, notwithstanding the fact that appearances were so strongly against them.

And now, at length, relief has come to the strained relations between the two parties, for the recent marvellous discoveries in physics show that concentration of matter is not the only source from which the sun may draw its heat.

Radium is a substance which is perhaps millions of times more powerful than dynamite. Thus it is estimated that an ounce of radium would contain enough power to raise 10,000 tons a mile above the earth's surface. Another way of stating the same estimate is this: the energy needed to tow a ship of 12,000 tons a distance of six thousand sea miles at 15 knots is contained in 22 ounces of radium. The *Saxon* probably burns three or four thousand tons of coal on a voyage of approximately the same length. Again, M. and Mme Curie have proved that radium actually gives out heat *, and it has been calculated that a small proportion of radium in the sun would suffice to explain its present radiation. Other lines of argument tend in the same direction †.

Now we know that the earth contains radio-active materials, and it is safe to assume that it forms in some degree a sample of the materials of the solar system. Hence it is almost certain that the sun is radio-active also; and besides it is not improbable that an element with so heavy an atom as radium would gravitate more abundantly to the central condensation than to the outlying planets. In this case the sun should contain a larger proportion of radio-active material than the earth.

This branch of science is as yet but in its infancy, but we already see how unsafe it is to dogmatise on the potentialities of matter.

It appears, then, that the physical argument is not susceptible of a greater degree of certainty than that of the geologists, and the scale of geological time remains in great measure unknown.

I have now ended my discussion of the solar system, and must pass on to the wider fields of the stellar universe.

Only a few thousand stars are visible with the unaided eye, but photography has revealed an inconceivably vast multitude of stars and nebulæ, and every improvement in that art seems to disclose yet more and more. About twenty years ago the number of photographic objects in the heavens was roughly estimated at about 170 millions, and some ten years later it had increased to about 400 millions. Although Newcomb, in his recent book on *The Stars* refrains even from conjecturing any definite number,

---

* Lord Kelvin has estimated the age of the earth from the rate of increase of temperature underground. But the force of his argument seems to be entirely destroyed by this result. See a letter by R. J. Strutt, *Nature*, December 21, 1905.

† See W. E. Wilson, *Nature*, July 9, 1903; and G. H. Darwin, *Nature*, Sept. 24, 1903.

yet I suppose that the enormous number of 400 million must now be far below the mark, and photography still grows better year by year. It seems useless to consider whether the number of stars has any limit, for infinite number, space, and time transcend our powers of comprehension. We must then make a virtue of necessity, and confine our attention to such more limited views as seem within our powers.

A celestial photograph looks at first like a dark sheet of paper splashed with whitewash, but further examination shows that there is some degree of method in the arrangement of the white spots. It may be observed that the stars in many places are arranged in lines and sweeping trains, and chains of stars, arranged in roughly parallel curves, seem to be drawn round some centre. A surface splashed at hazard might present apparent evidence of system in a few instances, but the frequency of the occurrence in the heavens renders the hypothesis of mere chance altogether incredible.

Thus there is order of some sort in the heavens, and, although no reason can be assigned for the observed arrangement in any particular case, yet it is possible to obtain general ideas as to the succession of events in stellar evolution.

Besides the stars there are numerous streaks, wisps, and agglomerations of nebulosity, whose light we know to emanate from gas. Spots of intenser light are observed in less brilliant regions; clusters of stars are sometimes imbedded in nebulosity, while in other cases each individual star of a cluster stands out clear by itself. These and other observations force on us the conviction that the wispy clouds represent the earliest stage of development, the more condensed nebulæ a later stage, and the stars themselves the last stage. This view is in agreement with the nebular hypothesis of Laplace, and we may fairly conjecture that the chains and lines of stars represent pre-existing streaks of nebulosity.

As a star cools it must change, and the changes which it undergoes constitute its life-history, hence the history of a star presents an analogy with the life of an individual animal. Now, the object which I have had in view has been to trace types or species in the physical world through their transformations into other types. Accordingly it falls somewhat outside the scope of this address to consider the constitution and history of an individual star, interesting although those questions are. I may, however, mention that the constitution of gaseous stars was first discussed from the theoretical side by Lane, and subsequently more completely by Ritter. On the observational side the spectroscope has proved to be a powerful instrument in analysing the constitutions of the stars, and in assigning to them their respective stages of development.

If we are correct in believing that stars are condensations of matter originally more widely spread, a certain space surrounding each star must have been cleared of nebulosity in the course of its formation. Much thought has been devoted to the determination of the distribution of the stars in space, and although the results are lacking in precision, yet it has been found possible to arrive at a rough determination of the average distance from star to star. It has been concluded, from investigations into which I cannot enter, that if we draw a sphere round the sun with a radius of twenty million millions of miles*, it will contain no other star; if the radius were twice as great the sphere might perhaps contain one other star; a sphere with a radius of sixty million millions of miles will contain about four stars. This serves to give some idea of the extraordinary sparseness of the average stellar population; but there are probably in the heavens urban and rural districts, as on earth, where the stars may be either more or less crowded. The stars are moving relatively to one another with speeds which are enormous, as estimated by terrestrial standards, but the distances which separate us from them are so immense that it needs refined observation to detect and measure the movements.

Change is obviously in progress everywhere, as well in each individual nebula and star as in the positions of these bodies relatively to one another. But we are unable even to form conjectures as to the tendency of the evolution which is going on. This being so, we cannot expect, by considering the distribution of stars and nebulæ, to find many illustrations of the general laws of evolution which I have attempted to explain; accordingly I must confine myself to the few cases where we at least fancy ourselves able to form ideas as to the stages by which the present conditions have been reached.

Up to a few years ago there was no evidence that the law of gravitation extended to the stars, and even now there is nothing to prove the transmission of gravity from star to star. But in the neighbourhood of many stars the existence of gravity is now as clearly demonstrated as within the solar system itself. The telescope has disclosed the double character of a large number of stars, and the relative motions of the pairs of companions have been observed with the same assiduity as that of the planets. When the relative orbit of a pair of binary or double stars is examined, it is found that the motion conforms exactly to those laws of Kepler which prove that the planets circle round the sun under the action of solar gravitation. The success of the hypothesis of stellar gravitation has been so complete that astronomers have not hesitated to explain the anomalous motion of a seemingly single star by the existence of a dark companion; and it is interesting to know that the more powerful telescopes of recent times have disclosed, in at least two cases, a faintly luminous companion in the position which had been assigned to it by theory.

* This is the distance at which the earth's distance from the sun would appear to be 1″.

By an extension of the same argument, certain variations in the spectra of a considerable number of stars have been pronounced to prove them each to be really double, although in general the pair may be so distant that they will probably always remain single to our sight. Lastly, the variability in the light of other apparently single stars has proved them to be really double. A pair of stars may partially or wholly cover one another as they revolve in their orbit, and the light of the seemingly single star will then be eclipsed, just as a lighthouse winks when the light is periodically hidden by a revolving shutter. Exact measurements of the character of the variability in the light have rendered it possible not only to determine the nature of the orbit described, but even to discover the figures and densities of the two components which are fused together by the enormous distance of our point of view. This is a branch of astronomy to which much careful observation and skilful analysis has been devoted; and I am glad to mention that Alexander Roberts, one of the most eminent of the astronomers who have considered the nature of variable stars, is a resident in South Africa.

I must not, however, allow you to suppose that the theory of eclipses will serve to explain the variability for all stars, for there are undoubtedly others whose periodicity must be explained by something in their internal constitution.

The periods of double stars are extremely various, and naturally those of short period have been the first noted; in times to come others with longer and longer periods will certainly be discovered. A leading characteristic of all these double stars is that the two companions do not differ enormously in mass from one another. In this respect these systems present a strongly marked contrast with that of the sun, attended as it is by relatively insignificant planets.

In the earlier part of my address I showed how theory indicates that a rotating fluid body will as it cools separate into two detached masses. Mathematicians have not yet been able to carry their analysis far enough to determine the relative magnitudes of the two parts, but as far as we can see the results point to the birth of a satellite whose mass is a considerable fraction of that of its parent. Accordingly See (who devotes his attention largely to the astronomy of double stars), Roberts, and others consider that what they have observed in the heavens is in agreement with the indications of theory. It thus appears that there is reason to hold that double stars have been generated by the division of primitive and more diffused single stars.

But if this theory is correct we should expect the orbit of a double star to be approximately circular; yet this is so far from being the case that the eccentricity of the orbits of many double stars exceeds by far any of the eccentricities in the solar system. Now See has pointed out that when two bodies of not very unequal masses revolve round one another in close proximity the conditions are such as to make tidal friction as efficient as possible in transforming the orbit. Hence we seem to see in tidal friction a cause which may have sufficed not only to separate the two component stars from one another, but also to render the orbit eccentric.

I have thought it best to deal very briefly with stellar astronomy, in spite of the importance of the subject, because the direction of the changes in progress is in general too vague to admit of the formation of profitable theories.

We have seen that it is possible to trace the solar system back to a primitive nebula with some degree of confidence, and that there is reason to believe that the stars in general have originated in the same manner. But such primitive nebulæ stand in as much need of explanation as their stellar offspring. Thus, even if we grant the exact truth of these theories, the advance towards an explanation of the universe remains miserably slight. Man is but a microscopic being relatively to astronomical space, and he lives on a puny planet circling round a star of inferior rank. Does it not then seem as futile to imagine that he can discover the origin and tendency of the universe as to expect a housefly to instruct us as to the theory of the motions of the planets? And yet, so long as he shall last, he will pursue his search, and will no doubt discover many wonderful things which are still hidden. We may indeed be amazed at all that man has been able to find out, but the immeasurable magnitude of the undiscovered will throughout all time remain to humble his pride. Our children's children will still be gazing and marvelling at the starry heavens, but the riddle will never be read.

APPENDIX

27.

MARRIAGES BETWEEN FIRST COUSINS IN ENGLAND AND THEIR EFFECTS.

[*Journal of the Statistical Society*, Vol. XXXVIII. (1875), pp. 153—182.]

TABLE OF CONTENTS.

I. *The Proportion of First Cousin Marriages to all Marriages.*

IT is well known that when the Census Act, 1871, was passing through the House of Commons, an attempt was made by Sir J. Lubbock, Dr Playfair, and others, to have a question inserted with respect to the prevalence of cousin marriages, under the idea that when we were in possession of such statistics we should be able to arrive at a satisfactory conclusion as to whether these marriages are, as has been suspected, deleterious to the bodily and mental constitution of the offspring. It is unfortunately equally well known that the proposal was rejected, amidst the scornful laughter of the House, on the ground that the idle curiosity of philosophers was not to be satisfied.

It was urged, that when we had these statistics it would be possible to discover by inquiries in asylums, whether the percentage of the offspring of consanguineous marriages amongst the diseased was greater than that in the healthy population, and thus to settle the question as to the injuriousness of such marriages. The difficulty of this subsequent part of the inquiry was, I fear, much underrated by those who advocated the introduction of these questions into the census. It may possibly have been right to reject the proposal on the ground that every additional question diminishes the trustworthiness of the answers to the rest, but in any case the tone taken by many members of the House shows how little they are per-meated with the idea of the importance of inheritance to the human race.

In the summer of 1873 the idea occurred to me that it might be, in some measure, possible to fill up this hiatus in our national statistics. In looking through the marriages announced in the *Pall Mall Gazette*, I noticed one between persons of the same surname; now, as the number of surnames in England is very large, it occurred to me that the number of such marriages would afford a clue to the number of first cousin marriages.

In order to estimate what proportion of such marriages should be attributed to mere chance, I obtained the Registrar-General's *Annual Report* for 1853, where the frequency of the various surnames is given. I here find the following table, p. xviii. :—

Number of persons whose names were registered	Number of different surnames occurring in the whole Register	Number of different surnames to every 100 persons, *i.e.*, $\frac{275,405}{100} \times 11\cdot9 = 32,818$	Number of persons to one surname, *i.e.*, $8\cdot4 \times 32,818 = 275,405$
275,405	32,818	11·9	8·4

The fifty commonest names embraced 18 per cent. of all the population. It appears that one in 73 is a Smith, one in 76 a Jones, one in 115 a Williams, one in 148 a Taylor, one in 162 a Davies, one in 174 a Brown, and the last in the list is one Griffiths in 529. Now it is clear that in one marriage in 73 one of the parties will be a Smith, and if there were no cause which tended to make persons of the same surname marry, there would be one in 73^2 or 5,329 marriages, in which both parties were Smiths. Therefore the probability of a Smith-Smith marriage, *due to mere chance*, is $\frac{1}{5329}$; similarly the chance of a Jones-Jones, a Davies-Davies and a Griffiths-Griffiths marriage would be $\frac{1}{76^2}$, $\frac{1}{162^2}$ and $\frac{1}{529^2}$, respectively. And the sum of fifty such fractions would give the probability of a *chance* marriage, between persons of the same surname, who owned one of these fifty commonest names. The sum of these fifty fractions I find to be 0·0009207, or 0·9207 per thousand. It might however be urged, that if we were to take more than fifty of the common names, this proportion would be found to be much increased. I therefore drew a horizontal straight line, and at equal distances along it I erected ordinates proportional to $\frac{1}{73^2}$, $\frac{1}{76^2}$, $\cdots$ $\frac{1}{529^2}$. As I found that the names decreased in value very gradually, I, as a fact, omitted every other name; but had I taken every name the result would have been sensibly the same. The upper ends of these twenty-five ordinates were found to lie in a curve of great regularity, remarkably like a rectangular hyperbola, of which my horizontal straight line was one asymptote; and the ordinate corresponding to Griffiths was exceedingly short. Observing the great regularity of the curve, I continued it beyond the fiftieth surname by eye, until it sensibly coincided with the asymptote, at a point about where the hundred and twenty-fifth name would have stood, and then cut out the whole (drawn on thick paper), and weighed the part corresponding to the fifty surnames, and the conjectural part. The conjectural addition was found to weigh rather more than one-tenth of the other part $\left(i.e.\ \dfrac{124}{920}\right)$; and as the chance of same-name marriages is proportional to the areas cut out, I think I may venture confidently to assert that, in England and Wales, about one marriage in a thousand takes place in which the parties are of the same surname, and have been uninfluenced by any relationship between them bringing them together. Now, it will appear presently that far more than one marriage in a thousand is between persons of the same surname; and, as I do not profess to have attained results of an accuracy comparable to 0·1 per cent., I am entitled to say that same-name marriages, when they take place, are due to the consanguinity of the parties. If it permitted such accuracy, the method pursued would however include a compensation for this disturbing cause.

With the help of an assistant the marriages announced in the *Pall Mall Gazette* in the years 1869–72, and part of 1873, were counted, and were found to be 18,528. Out of these 232 were between persons of the same surname, that is 1·25 per cent. were same-name marriages. The same marriage is occasionally announced twice over, but as there can be no reason to suppose that this course has been pursued oftener or seldomer with same-name marriages than with others, the result will not be vitiated thereby. In order to utilize this result, it now became necessary to determine—

(1) What proportion of this 1·25 per cent. were marriages between first cousins.

(2) What proportion marriages between first cousins of the same surname bear to those between first cousins of different surnames.

If these two points could be discovered, the percentage of first-cousin marriages *in the upper classes* could be at once determined. I have endeavoured to find out these proportions in several ways.

An assistant was employed to count the marriages of the *men* in the pedigrees of the English and Irish families occupying about 700 pages of Burke's *Landed Gentry*, marking every case where the marriage was "same-name." I then tried in every such case to discover, from a consideration of the pedigree, whether the marriage had been between first cousins. I found that in a certain number of cases I was unable to discover this. The total number of pedigrees in the 700 pages was about 1,300; and of these I had to exclude 71, thinking that by only including family trees where I could discover the relationship of the parties, I should not obtain an unfair selection of the whole. The marriages of the men alone were included, because, had I included those of the women, many marriages would have been counted twice over,—once in the pedigree under consideration, and again in that of the husband. In this way, then, I found that out of 9,549 marriages given by Burke 72 were same-name first cousin marriages, and 72 were same-name marriages not between first cousins. This gives the percentage of same-name marriages as 1·5 (not strikingly different from the 1·25 deduced from the *Pall Mall Gazette*), and of this percentage 0·75 is to be attributed to first cousin marriages.

I further collected in the same way 1,989 marriages from the *English and Irish Peerage*, and of these 18 were same-name first cousin marriages, or 0·91 per cent. The number of same-name marriages not being first cousin marriages was not however compared in this case. It will be observed, that the proportion is nearly 0·2 per cent. higher than with the *Landed Gentry*, and as the nobility are known to marry much *inter se*, this was perhaps to be expected; however, 2,000 is too small a number on which to base a conclusion on this head with safety. The Peerage and Burke combined give 90 out of 11,538, or 0·78 per cent., of same-name first cousin marriages.

The next step was to send out a very large number of circulars (about 800) to members of the upper middle and upper classes, in which I requested each person to give me the names of any members of the following classes, who married their first cousins; viz. (1) the uncles, aunts, father and mother of the person; (2) the brothers, sisters and the person himself; (3) the first cousins of the person. I further asked for the names of any persons in the above classes, who contracted same-name marriages *not* with first cousins. I confined my questions to near relations, because, had the more distant ones been included, a risk was run of getting a selected set of marriages,—a risk which I am inclined to suspect was not avoided, as will hereafter appear.

In about 300 of the circulars I further asked for the total number of marriages contracted by the persons included in the Classes 1, 2 and 3. Care was taken to exclude, as far as possible, those persons who had cousins in common, so that each answer should embrace a fresh field. I must here return my thanks to the many persons who so kindly filled in and returned the circulars.

The following result was obtained :—

TABLE A.

Same-name First Cousin Marriages	Different-name First Cousin Marriages	Same-name *not* First Cousin Marriages
66	182	29

From 181 circulars returned, in which the total number of marriages in each class was given, the following was the result :—

TABLE B.

Total number of marriages	Total number of First Cousin Marriages	Percentage of First Cousin Marriages	Percentage of same-name marriages, whether Cousin or not Cousin
3,663	125	3·41	1·38 *

* Compare this with 1·25 deduced from *Pall Mall Gazette*.

Persons having no cousin marriages to fill in were asked to return the circular blank, in those cases where the total number of marriages was not asked for. Of such blank returns, together with those where the total number of marriages was not given, 207 came back to me. From Table (B) it will be seen that $3,663 \div 181$, or $20\cdot2$ marriages were recorded in each return; to judge, therefore, of the congruity of the 207 blank returns with the others, I impute to each of these 207 circulars, 20 marriages, and therefore add 4,140 marriages to the 3,663; as a grand total, with this conjectural addition, the following is the result :—

TABLE C.

Total number of Marriages	Total number of First Cousin Marriages	Percentage of First Cousin Marriages
7,803	248	3·18

It is thus seen that the 207 returns are tolerably congruent with the 181 returns in Table (B); for 3·18 differs but slightly from 3·41.

From Table (A) it is seen that there were 182 different-name cousin marriages to 66 same-name cousin marriages; *i.e.* for every same-name cousin marriage there were $2\frac{3}{4}$ different-name cousin marriages.

And again, there were 66 same-name cousin marriages to 29 same-name-not-cousin marriages; that is rather more than two to one. The last result disagrees so much with that obtained from Burke and the Peerage, where the proportion was, as above stated, found to be as 1 to 1, that I am inclined to suspect that I had either a run of luck against me, or more probably that a considerable number of marriages between persons of the same surname, not being first cousins, escaped the notice of my correspondents. This latter belief is somewhat confirmed by what follows. If, however, I combine the results obtained from Burke with those from my circulars, I obtain the following :—

$$\frac{\text{Same-name cousin marriages}}{\text{All same-name marriages}} = \frac{142}{249} = \cdot57$$

And in default of anything more satisfactory I am compelled to accept this result as the first of my two requisite factors.

As to the second factor, the proportion $2\frac{3}{4} : 1$ for different-name cousin marriages to same-name cousin marriages is, I fear, also unsatisfactory. But before entering on this point I will indicate the sources of error in my returns :—

(1) The sensitiveness of persons in answering the question in cases where there are cousin marriages, particularly when any ill results may have accrued.

(2) The non-return by persons who had no such marriages to fill in, and who would say, "I have no information, what is the use of returning this?* "

---

* The circulars were ready stamped for return, which would induce many to return them by saving trouble.

(3) The ignorance of persons of the marriages of their relations. This ignorance would be more likely to affect the returns of different-name marriages than of same-name ones. I feel convinced that this has operated to some extent, as will be seen hereafter.

(4) In the cases of same-name marriages, persons would be more likely to know of the marriages between first cousins, than of other such marriages. The discrepancy between Burke and my circulars leads me to believe that this too has operated.

I have been much surprised to find how very little people know of the marriages of their relations, even so close as those comprised in my three classes. As it is clear that the marriages contracted by a man's uncles and aunts and by his brothers and sisters, would be less likely to escape his notice than would those contracted by his first cousins, I made an analysis of my circulars, including only the first two classes, viz. (1) uncles, aunts, father and mother; (2) brothers and sisters and the person himself, with the following result :—

TABLE D.

Same-name First Cousin Marriages	Different-name First Cousin Marriages	Same-name *not* First Cousin Marriages
42	121	21

And from the returns where the total number of marriages was required, the following is the result :—

TABLE E.

Total number of Marriages	Total number of First Cousin Marriages	Percentage of First Cousin Marriages
1,929	81	4·2

It appears then, that

$$\frac{\text{Same-name cousin marriages}}{\text{Different-name cousin marriages}} = \frac{1}{3} \text{ nearly}$$

and

$$\frac{\text{Same-name cousin marriages}}{\text{All same-name marriages}} = \frac{2}{3}$$

And these results I take to be more trustworthy than those given above, but I think that even here many different-name first cousin marriages and same-name-not-cousin marriages have escaped notice, and that the indirect method, to which I now proceed, is on the whole more reliable.

It is possible to discover the proportion between the same-name and different-name marriages in an entirely different way, and this I have tried to do. A man's first cousins may be divided into four groups, viz. the children of (*a*) his father's brothers, (b) his father's sisters, (c) his mother's brothers, (d) his mother's sisters. Of these four groups only (a) will in general bear the same surname as the person himself. On the average, the number of marriageable daughters in each family of each of the four groups will be the same. Were the four groups then equally numerous, we might expect that the same-name would bear to the different-name marriages the proportion of one to three. Since however a man cannot marry his sisters, this cannot hold good; for the classes (a) and (d) are clearly on the average smaller than (b) and (c), and the proportion we wish to discover is $\dfrac{(a)}{(b) + (c) + (d)}$, which must evidently be less than $\dfrac{1}{3}$. To take a numerical example: A's father is one of 3 brothers, who married and have children, and A's father had 2 sisters, who married and have children: A's mother had 1 brother, who married and has children, and was one of 5 sisters, who married and have children. Then clearly the class

> (a) consists of 2 families.
>
> (b)   „   2   „
>
> (c)   „   1 family.
>
> (d)   „   4 families.

So that the above fraction becomes $\dfrac{2}{2 + 1 + 4} = \dfrac{2}{7}$. In this case we may conclude that if A marries a first cousin, it is 5 to 2 that he will marry one of a different surname. In another case the numbers might have been different, and therefore the fraction and the betting also different. And what we wish to discover is the *average* value of this fraction. But, for the various members of a large community, there will be a very large number of such fractions, and some will occur more frequently than others; so that in finding this average value, each fraction should have its proper weight assigned to it. In order to assign the weight to, say, the above fraction $\frac{2}{7}$, we must take a thousand families and find in how many of them there were 3 sons and 2 daughters who married and had children, and in how many there were 1 son and 5 daughters who married and had children. Having sufficiently indicated how the required proportion may depend on probabilities, I may state that I sent out a number of circulars to members of the upper middle, and upper classes, and obtained and classified statistics with respect to 283 families. The following table gives the results, excepting that I have supposed that I had collected 1,000 families, that is, the numbers given in the table are the actual numbers multiplied by $\dfrac{1,000}{283}$.

N.B. (a, b) *means a family in which there are* (a) *sons who married and had children; and* (b) *daughters who married and had children. Only such families are included as have, so to speak, done marrying.*

—	0,1	0,2	0,3	0,4	0,5	0,6	—
	82	39	21	7	0	4	
1,0	1,1	1,2	1,3	1,4	1,5	1,6	—
92	117	99	29	14	7	0	
2,0	2,1	2,2	2,3	2,4	2,5	2,6	—
64	78	43	46	11	14	0	
3,0	3,1	3,2	3,3	3,4	3,5	3,6	3,7
39	32	32	14	4	7	0	4
4,0	4,1	4,2	4,3	4,4	4,5	4,6	—
7	28	14	4	7	11	0	
5,0	5,1	5,2	5,3	5,4	5,5	5,6	—
0	0	14	0	4	0	0	
6,0	6,1	6,2	6,3	6,4	6,5	6,6	—
0	7	0	0	0	4	0	

As the number (283) of families collected is so small, the proportion of the rarer order of families will be of course incorrect, thus there are no families of the form (0,5), whilst there are four of the form (3,7). Any small error in these rarer orders of families will have but an infinitesimal effect on my results. I treated the question in four different ways. It might be supposed that a man, who had five families of first cousins in relation to himself, would be five times as likely to marry a first cousin as a man who had only one such family, or again it might be supposed that he would be only equally likely. The truth, however, will certainly lie between these suppositions. The question, when treated from this point of view, leads to the result that $\frac{\text{same-name cousin marriages}}{\text{different-name cousin marriages}}$ is greater than $\frac{1}{4 \cdot 44}$ and less than $\frac{1}{4 \cdot 12}$. So that the true proportion would be about $\frac{1}{4\frac{1}{4}}$.

The two other methods are founded on the same grouping of families, and depend on the fact that my class (a) will on the average be equal in number to class (d); and class (b) to class (c), and all that is necessary is to find what value should be assigned to the ratio (a) or (d) : (b) or (c). It would be tedious to indicate the precise method employed, but suffice it to say that after a correction for the greater prevalence of the second marriages of men than of women, the result comes out that $\dfrac{\text{same-name cousin marriages}}{\text{different-name cousin marriages}}$ is greater than $\dfrac{1}{4\cdot23}$ and less than $\dfrac{1}{4\cdot14}$, so that the proportion would be really about $\dfrac{1}{4\frac{1}{6}}$; a result which differs but very slightly from that given by the two other methods.

The amount of arithmetical labour was so great that I was obliged in the first two methods to rank all families of a higher order than $(3,3)$ as $(3,3)$, or a family $(5,1)$ as being the same as a family of the form $(3,1)$; in the two latter methods I was able to go as high as $(4,4)$. These higher orders of families are of very rare occurrence, and thus the reduction of all families to those of lower orders would not materially affect the results, but as far as it goes it would make the above fractions too small.

I think on the whole it may be asserted that the same-name first-cousin marriages are to the different-name first-cousin marriages as 1 to 4. It may perhaps be worth mentioning that a second grouping of families from Burke's *Landed Gentry* led to almost identical results, notwithstanding the bias introduced by the fact that the eldest sons have a constant premium on marriage.

It appears to me on the whole that this latter result is considerably more reliable than that from my circulars, and this as before stated I can only explain on the supposition that many different-name marriages have escaped notice. The whole is very perplexing, and may perhaps be held to make all my results value-less. My final result for the two required factors then is that

$$\frac{\text{Same-name first-cousin marriages}}{\text{All same-name marriages}} = \cdot57$$

and
$$\frac{\text{Same-name first-cousin marriages}}{\text{Different-name first-cousin marriages}} = \frac{1}{4}$$

If this be applied to the percentage $1\cdot25$ of the *Pall Mall Gazette*, we get $3\cdot54$ or $3\frac{1}{2}$ per cent. as the proportion of first cousin marriages to all marriages in the middle classes. If it be applied to the peerage we get $4\frac{1}{2}$ per cent., and for the landed gentry $3\frac{3}{4}$ per cent., and for both combined $3\frac{9}{10}$ per cent.—To sum up, the direct statistical method gives from $3\frac{1}{5}$ to $3\frac{2}{5}$ per cent., or including only the classes (1) and (2), comprising uncles, aunts, brothers and sisters, $4\frac{1}{5}$ per cent.; the indirect method $3\frac{1}{2}$ per cent.; and the partly indirect and partly statistical, founded on the peerage and Burke, gives $3\frac{9}{10}$. There is, however, some reason to suppose that the proportion is really higher amongst the landed classes. There is a serious discrepancy between the direct and indirect methods as to the proportion of same-name and different-name marriages, which goes far to invalidate the results.

Whether, however, these proportions are actually correct or not, there can be little doubt that if the area taken is large enough, the percentage of first cousin marriages in any class is proportional to the percentage of same-name marriages; so that if the latter is, say, only half the former, the cousin marriages are also only half. I therefore obtained from the General Registry of Marriages at Somerset

House a return of the proportion of same-name marriages in 1872 in various districts, namely (1) London, (2) large towns, viz. Bradford, Leeds, Manchester, Portsmouth, Southampton, Exeter, Plymouth, Birmingham, Witney, Banbury, Northampton, Wellingborough, Peterborough, Bedford, and (3) Agricultural districts of Hampshire, Devonshire, Middlesex, Herts, Bucks, Oxon, Northampton, Huntingdon, Bedford, and Cambridge, &c. I must take this opportunity of returning my warm thanks to the superintendent of the statistical department, Dr Farr, for the very great kindness both he and Mr N. A. Humphreys, of the General Registry Office, have shown in helping me in this inquiry by every way in their power. The following table, in which the third column is introduced for the sake of comparison with the statistics from the *Pall Mall Gazette*, gives the results :—

	Number of marriages registered	Percentage of same-name Marriages	Ratio to the number (1·25) from *Pall Mall Gazette*	Percentage of First Cousin Marriages as deduced by previous method
1. London metropolitan districts	33,155	0·55	$\frac{1}{2}$	$1\frac{1}{2}$
2. Urban districts ...	22,346	0·71	$\frac{7}{12}$	2
3. Rural „ ...	13,391	0·79	$\frac{2}{3}$	$2\frac{1}{4}$

The numbers in most of the towns and counties, taken individually, were too small to give any trustworthy results.

It thus appears that in London, comprising all classes, the cousin marriages are about half what they are in the upper middle class, that is probably $1\frac{1}{2}$ per cent. In urban districts they are about $\frac{7}{12}$ths of what they are in the upper middle classes, that is, probably 2 per cent. In rural districts they are about two-thirds of what they are in the upper middle classes, that is, probably $2\frac{1}{4}$ per cent. In the middle and upper middle class or in the landed gentry probably $3\frac{1}{2}$ per cent. In the aristocracy probably $4\frac{1}{2}$ per cent. This is in accordance with what might have been expected *à priori*: for the aristocracy hold together very much, the landed gentry slightly less, the business class again less. And beginning from the other end, London is an enormous community, recruited from every part of England; the large towns form communities, only one degree less heterogeneous; and the country is still less heterogeneous. I am, however, somewhat surprised at finding the proportion in the rural districts so small, for one would imagine that agricultural labourers would hold together very closely*.

---

* I may mention that Mr Clement Wedgwood made very careful inquiries for me concerning 149 marriages of skilled artisans in the potteries, and did not find a single case of first cousin marriage, and only three where there was any kind of relationship between the husband and wife. He was further assured that such marriages never take place amongst them. It may be worth giving the following table of consanguineous marriages from *Italia Economica nel* 1873, kindly sent to me by Dr Farr.

Persons accustomed to deal with statistics will be able to judge better than myself what degree of reliance is to be placed on the previous results. My own *impression* is that there is not an error of 1 per cent. in asserting that amongst the aristocracy the proportion of first cousin marriages to all marriages is $4\frac{1}{2}$ per cent., and that for the upper middle classes, and the urban and rural districts, the error in the percentages is somewhat less, and lastly for London decidedly less. But this is an impression that I hardly know how to justify, and I therefore leave an ample field for adverse criticism.

II. *Inquiries in Asylums.*

I now pass on to the second part of my inquiry, namely, the endeavour to discover, by collecting statistics in asylums, whether first cousin marriages are injurious or not.

The method I intended to pursue was as follows : to get the superintendents of asylums to ask each one of the patients under their charge, either personally or through their subordinates, the question, "Were your father and mother first cousins or not ? " In the case of the insane, I thought, in my ignorance, that those who had charge of them would have so intimate a knowledge of the character of each individual case, as to be able to sift those whose answers could be depended on from those who were quite untrustworthy. In this it appears that I was mistaken, as will be shown by the remarks sent me by the various gentlemen who so kindly took up this inquiry. I cannot help thinking, however, that they undervalue the statistics which they have collected for me. I must take this opportunity to return my warm thanks to all the gentlemen mentioned below for the immense pains they have been at in collecting these results. I could hardly have believed that so many men, much occupied by their business, could have shown a stranger so much kindness, more especially as many of them seemed convinced that their labours were almost in vain. To Dr W. Lauder Lindsay, Dr Crichton Browne, Dr Maudsley, and Dr Scott, I must return my

(continued on p. 568)

Consanguineous Marriages Contracted in Italy from 1868–70 inclusive.

Marriages	1868–70	Per Annum	Per 100	Percentage to all Marriages
Between brothers-in-law and sisters-in-law	2,392	797	33·27	0·413
Uncles and nieces............................	292	97	4·05	0·050
Aunts and nephews	50	17	0·70	0·009
Cousins	4,455	1,485	61·98	0·769
Total	7,189	2,396	100·0	1·24

It must be borne in mind that in Roman Catholic countries a dispensation is requisite to permit the marriage of cousins.

English and Welsh Asylums	Doctors	Number of Patients		Answers to "were parents First Cousins?"	
		Males	Females	Males	Females
1. West Riding, Wakefield (lunatics and idiots)	Crichton Browne	700	707	337	318
2. Hanwell (lunatic)............	Rayner	166	214	110	145
3. Warneford, Oxford (lunatic)	Byewater Ward...	30	29	20	
4. Mickleover, Derby (lunatic)	Murray Lindsay	174	190	99	99
5. Metropolitan District, Caterham (lunatic)	Adam	877	1,038	434	126
6. Glamorgan County (lunatic)	Yellowlees	254	238	102	116
7. Chester County (lunatic)...	Lawrence	About 450		115	110
8. County Lunatic, Snenton, Nottingham	Phillimore	184	206	97	103
9. Grove Hall, Bow	Mickle	427	—	181	—
10. Hatton, Warwick	Oscar Woods......	537		258	
11. Earlswood, Surrey (idiot)...	Grabham	—		1,388	
12. Broadmoor Criminal (lunatic)	Orange	370		150	
Totals for England and Wales	—	8,170 very nearly		4,308	
SCOTCH ASYLUMS					
1. Montrose (lunatic)	Howden	179	227	49	92
2. Crichton Royal Institution, Dumfries	Gilchrist............	87	59	31	20
3. Southern Counties, Dumfries	Anderson	178	140	108	92
4. Murray Royal Institution, Perth	Lauder Lindsay...	42	38	28	16
5. Perth District, Murthly ...	McIntosh	99	130	37	41
Totals	—	585	594	253	261
		1,179		514	
IRISH ASYLUMS					
1. Maryborough..................	Through Dr Courtenay	217		—	
2. Limerick District............	Courtenay	434		—	
Totals	—	651		—	

Untrustworthy, or unable to answer		Offspring of First Cousins		Observations
Males	Females	Males	Females	
363	389	14	17	Examination conducted with great care; cases of doubt excluded. Almost all who gave answers were lunatic and not idiotic
56	169	2 or 1	1	Only those are given as trustworthy where the history of the patient could be ascertained. Amongst the males there were twelve cases of doubtful consanguinity, but whether first cousins or not, is not stated
39		0		Patients of the farmer and tradesmen class
75	91	2	2	Dr Lindsay thinks these statistics worth little
433	912	8	12	Statistics very imperfect; trustworthiness of answers uncertain. The total number of patients is overstated by nine, but whether males or females I know not
152	122	5	4	Statistics worth little. Of those who did not answer, 137 were ignorant, and 137 incapable
about 225		1	2	Patients of the labouring class
87	103	2 or 5	2 or 4	Statistics to be little depended on
246		8	—	Patients old soldiers
277		9 or 8		Patients labourers and artisans. The offspring of first cousins belonged to seven families. Examination conducted with great care
—		53		Facts derived from parents, and therefore tolerably trustworthy
220		2		Dr Orange places little reliance on these results
about 3,860		149 or 142		Between 3·46 and 3·29 per cent. of the patients who answered said they were offspring of first cousin marriages
130	135	2	6	Dr Howden thinks the inquiry useless. No inquiry was made of the idiots in this asylum
56	39	2	2	
70	48	4	4	
14	22	4	0	Dr Lindsay thinks the results *very* doubtful. The failure to get answers was due to incapacity and refusal
62	89	1	2	Patients paupers
332	333	13	14	5·25 per cent. of the patients who answered said that they were offspring of first cousin marriages
665		27		
—		2		Patients agricultural labourers
—		3		Twenty patients of better class; the rest labourers
—		5		No information as to numbers who failed to answer. Dr C. considers these statistics of little value. Roman Catholics do not marry first cousins. 0·77 per cent. of *all* the patients say they are offspring of first cousin marriages

especial thanks for the really extraordinary vigour with which they took up the subject, and gave me every help in their power.

The tables on pp. 566—7 give the results collected from lunatic and idiot asylums in England and Wales, Scotland and Ireland.   Besides the results tabulated above, Dr Wilkie Burman, of the Wilts County Asylum, Devizes, informed me that he could collect no statistics worth giving ; Dr Bacon, of the Fulbourn Asylum, Cambridge, whilst kindly offering to persevere, expressed his conviction of the uselessness of the attempt ; Dr Shuttleworth, of the Royal Albert Asylum, Lancaster, estimated that, out of 200 patients, 5 per cent. were the offspring of first cousins ; and Dr Clouston, of the Royal Edinburgh Asylum, tells me that, out of 750 patients, two *said* that they were offspring of first cousin marriages, but most could not answer.

The columns of observations show how very unsatisfactory the collectors consider these results.   From various circumstances, it appears that the results from Earlswood, Hatton, and the West Riding Asylums are considerably more trustworthy than the others.

Including, then, only these three asylums, it appears that, out of 2,301 patients, 90 or 91 were offspring of first cousins, that is 3·9 per cent.   The fact that this agrees pretty closely with the 3·4 per cent. deduced from the whole table, leads me to think that the trustworthiness of the results collected has been under-estimated by the collectors themselves.

At Hanwell, where also there were some circumstances leading one to believe in tolerable accuracy, the percentage is very small, and this agrees well with what I should have been led to expect, from the small percentage of cousin marriages I found in London, by the methods of the first part of this paper.   It is to be observed, however, that there were twelve cases reported of doubtful *consanguinity*.

It will be seen that the percentage of offspring of first cousin marriages is so nearly that of such marriages in the general population, that one can only draw the negative conclusion that, as far as insanity and idiocy go, no evil *has been shown* to accrue from consanguineous marriages.

From the high percentage (5¼) of offspring of first cousin marriages in the Scotch asylums, I should be led to believe that such marriages are more frequent in Scotland than in England and Wales, and from the mountainous nature of the country this was perhaps to be expected.

The methods of the first part of this paper throw no light on the question as far as concerns Scotland.

From the two Irish asylums no results whatever can be deduced.

But whatever the value of these statistics may be, the opinion of prominent medical men, who have had especial advantages of observation, and are many of them also men of science, cannot be without interest.

Dr Crichton Browne writes to me that the investigation was impossible in the case of idiots, except through the medium of the parents.   "It has always seemed to me that the great danger attending such marriages consists in the intensification of the morbid constitutional tendencies, which they favour. Hereditary diseases and cachexiæ are much more likely to be shared by cousins than by persons who are in no way related...(and these) are transmitted with more than double intensity when they are common to both parents....They seem to be the square or cube of the combined volume....Even healthy temperaments, when common to both parents, often come out as decided cachexiæ in the children." He adds that persons of similar temperaments ought not to intermarry. Elsewhere he tells me that he did not at first make sufficient allowance for the ignorance "and stupidity of my patients." In such an investigation, congenital effects, he says, should be distinguished from the acquired. I fear, however, that I must leave this to some hands more skilful than mine.

Dr Rayner, of Hanwell, says that amongst the fishermen of Whitstable there is much intermarriage. The results seem to show that the prevalent diathesis is developed, whether it be strumous, rheumatic, or otherwise. He says that it was very difficult to discover the facts from his patients.

Dr Howden, of Montrose, says : "As regards insanity, my own impression is, that unless there exists a hereditary predisposition the marriage of cousins has *no effect* in producing it....Neither in insanity nor in any other abnormal propensity do two plus two produce four; there is always another factor at work neutralising intensification and bringing things back to the normal." Dr Howden thus disagrees with Dr Crichton Browne, who, I take it, would maintain that, in insanity, two plus two make more, and not less, than four.

Dr Lauder Lindsay is of opinion that the ill effects of cousin marriage, including insanity, are much less than represented. He quotes "Stonehenge" (Mr J. H. Walsh), "On the Dog" (*Field* newspaper, p. 188, 1859), to the effect that in-and-in breeding sometimes reduces dogs "to a state of idiocy and delicacy of constitution, which has rendered them quite useless...full of excitability...with a want of mental capacity." He also urges the "impossibility" of obtaining trustworthy answers from the patients themselves; and even the results of personal inquiries from the nearest relatives of the patients would be liable to much error. Several of my correspondents expressed a belief that consanguinity of parents was more potent in producing idiocy than insanity. The results from Earlswood do not seem, however, to confirm this, and here the results sent seemed peculiarly trustworthy.

I had intended to pursue my inquiries in hospitals and asylums for other diseases, but the attempt which I made with respect to deaf-mutes has shown me that the difficulties which arise are so great, that it is almost useless to persevere in this course any further. I will now give the results which I have collected.

The first return relates to the College for the Blind at Worcester. The results were communicated through the kindness of the Rev. Robert Blair and Mr S. S. Foster. The college is small, and only 20 cases are recorded, and particulars of each case were sent. Of these 20, the offspring of first cousins were one, and of second cousins a case of two brothers. Of the 20 cases, 2 were due to accidents. Thus, out of 17 families, there was one offspring of first cousins.

Dr Scott, of Exeter, has informed me that out of 241 families, in which there were children born deaf and dumb, there were 7 cases of first cousin marriage. In three or four of these families there were more than one child so afflicted.

Dr Scott also kindly offered to place me in communication with the superintendents of a number of institutions for the deaf and dumb, and having availed myself of his kindness, I have collected the following answers.

Mr Arthur Hopper, of the Deaf and Dumb School near Birmingham, conducted an inquiry with the utmost care. He tells me that out of 122 pupils, he has received information about the parentage of all but 9. The 113 pupils, whose parentage is known, belonged to 109 families; of these 113, there were deaf from accident or disease 37, and of 10 the cause of deafness was unknown. Of these 10 pupils and the 66 congenitally deaf, not one was the offspring of a *consanguineous* marriage. Of the 37 who became deaf from disease, one was the offspring of first cousins. I am not informed whether the cases, where several were deaf in a family, belonged to the congenital cases, but it is almost certain to be so, and in any case I will assume (as the most unfavourable assumption) that it is so. Thus, out of 62 congenitally deaf families, not one was the offspring of even a consanguineous marriage. If we were to assume the 10 other cases to be cases of congenital deafness, it would be, not one in 72 congenitally deaf families was the offspring of a consanguineous marriage.

Mr Patterson, of the Manchester School for Deaf-Mutes, kindly informs me that his 130 pupils belonged to 123 families. Concerning 8 of these families no information could be obtained; in 67 such families the deaf-mutism resulted from disease; in 63 it was congenital; and only one family was the offspring of first cousins.

Mr Neill, of the Northern Counties Institution, at Newcastle-on-Tyne, says, "350 have been admitted into this institution, and I do not think more than 6 of the parents were cousins. In one family whose parents were cousins there were 4 deaf-mutes."

I have thus accurate information with respect to 366 families (*i.e.* 241 + 62 + 63), and out of these 8 were offspring of first cousins; that is to say, nearly 2·2 per cent. were offspring of first cousins. And, including the 350 cases at Newcastle, the percentage is $\frac{1400}{716}$, or 1·9 per cent. It is curious to notice that I deduced $2\frac{1}{4}$ per cent. as the proportion of first cousin marriages in urban districts, other than London. Thus as far as these meagre results go, no evil in the direction of deaf-mutism would appear to arise from first cousin marriages. The failure to collect more statistics of this kind does not arise from any inability to get at the best sources of information; on the contrary, I have on all hands received the kindest assurances of willingness to help me.

Mr David Buxton, of the Liverpool School, says the mode of investigation is simply impracticable; but he has sent me several pamphlets on the subject, his own excellent paper amongst the number*.

---

* Since my paper has been in print, Mr Buxton has sent me additional statistics, and from these I make some extracts.

Mr William Sleight, of the Brighton School, tells me that the children know nothing, and the parents are unwilling to communicate the fact inquired after, and says, " As far as I have been able to ascertain, about 7 per cent. of born deaf children are the offspring of parents who were cousins." (Query, first cousins ?)

Mr Patterson also writes to me that he is of opinion that, "though the result of the marriage of near relatives may not be seen in the deafness of their immediate offspring, yet the result is a deterioration of the constitution of the offspring, which may show itself in deafness in a few generations."

Mr Neill, who has been engaged in the tuition of the deaf and dumb for forty years, thinks the cases of offspring of cousins so afflicted are fewer than is supposed. He also gives me facts showing how strongly heritable congenital deafness is where both parents are deaf-mutes ; marriages are, moreover, by no means uncommon between pupils of these institutions.

To sum up the results of the whole investigation : It seems probable that in England, among the aristocracy and gentry, about 4 per cent. of all marriages are between first cousins ; in the country and smaller towns between 2 and 3 per cent. ; and in London perhaps as few as 1½ per cent. Probably 3 per cent. is a superior limit for the whole population. Turning to lunatic and idiot asylums, probably between 3 and 4 per cent. of the patients are offspring of first cousins. Taking into account the uncertainty of my methods of finding the proportion of such marriages in the general population, the percentage of such offspring in asylums is not greater than that in the general population, to such an extent as to enable one to say positively that the marriage of first cousins has any effect in the

---

Mr Buxton himself collected the following cases twenty years ago.

Twenty-six families, offspring of first-cousin marriages, gave 54 deaf-mutes, 1 deaf of one ear, 1 semi-mute, and 1 idiotic. 17 of these families contained 85 children, of whom 42 were afflicted and the remaining 43 sound.

In another family each of the parents had had families by previous marriages, but neither of these previous families were affected by deaf-mutism.

Mr Buxton has also sent me a number of extracts from the reports of American Institutions for the Deaf and Dumb ; they refer in general to consanguineous marriages, but the following refer to first-cousin marriages.

In Illinois, out of 893 children, 42 (or 4·7 per cent.) were offspring of first cousins.

Out of 36 children who entered the Pennsylvanian Institution in 1872, one was the offspring of first cousins.

Mr Buxton also quotes a return made to the French Academy. A *résumé* of M. Boudin's paper is given in the *Comptes Rendus*, Tom. LIV., 1862. The statement is as follows :—

Two per cent. of all French marriages are consanguineous. Deaf-mute offspring of consanguineous marriages equal, at Lyons, 25 per cent. of the congenitally afflicted, at Paris 28 per cent., and at Bordeaux 30 per cent. " Taking the ordinary risk of deaf-mute offspring as 1, there are 18 such births from unions of first cousins, 37 from those of uncles and nieces, and 70 from those of nephews and aunts. Healthy parents, if related in blood, may have deaf-mute children, while deaf and dumb parents not so related rarely have any."

The paper is given *in extenso* in the *Annales d'Hygiène Publique*, Tom. XVIII., pp. 5—82.

production of insanity or idiocy, although it might still be shown, by more accurate methods of research, that it is so. With respect to deaf-mutes, the proportion of offspring of first cousin marriages is precisely the same as the proportion of such marriages for the large towns and the country, and therefore there is no evidence whatever of any ill results accruing to the offspring in consequence of the cousinship of their parents.

III. *Literature on the Subject.*

For the sake of any persons who may desire to make investigations in this subject at any future time, I will append a short sketch of what I know of the literature on the subject. I cannot, however, pretend to have studied previous writings at all deeply.

To the best of my belief, the most thorough investigation ever made is contained in some papers* "On Blood Relationship in Marriage," by Dr Arthur Mitchell, a Deputy Commissioner in Lunacy for Scotland.

In 1860, Dr Mitchell collected the histories of 45 consanguineous marriages, and amongst these found 8 cases of no evil results, 8 cases of sterility, and 29 cases of evil resulting to the offspring. He feels sure, however, that these cases do not present the rule, but that they were really selected from their striking nature, although the observers had doubtless no intention of making any such selection; an equal number of marriages, where no kinship existed between the parties, might easily be collected, presenting a yet sadder picture. He observes that some families seem to have a tendency to cousin marriages, and I have noticed the same thing in my investigation. He points out that the most satisfactory mode of investigation appears to be—

1st. Take a large number of instances of any defect, and ascertain how many are the offspring of kinsmen; then compare the result with the proportion of cousin marriages to other marriages in the same community. This is, in fact, the method I have tried to pursue.

2nd. Take a locality and collect the family history of every marriage there, and compare results. Such an inquiry, if sufficiently wide to be accurate, is almost beyond the power of a private individual.

In applying the former of these methods, Dr Mitchell states that certain Scotch counties have an aggregate population of 716,210; and that he investigated the cases of 711 idiots from these counties. Of 84 of these the parentage was unknown, and yet of the whole 711, 98, or 13·6 (query 13·8) per cent. were *shown* to be offspring of blood relations. Marriages of blood relations are notoriously not 13·6 per cent. of all marriages, but Dr Mitchell thinks that it may be regarded as certain that such a ratio is about ten times the reality, or that the actual percentage of consanguineous marriages is about 1·3 per cent. I here venture to differ from him and I should not be greatly surprised if the marriages of *first* cousins alone in Scotland were as many as 4 per cent. Dr Mitchell has

* *Edinburgh Medical Journal*, March, April, and June, 1865.

elsewhere shown that illegitimacy tends to produce defective children ; deducting, therefore, the illegitimate children (of whom, of course, the parentage was unknown) from the above 711 idiots, the percentage of the offspring of consanguineous marriages rises to 18·9 per cent.

The 98 cases of blood relations, whose offspring were idiots, were first cousins in 42 cases ; second cousins in 35 cases ; third cousins in 21 cases. It is probable, he says, that more second and third cousins marry than first cousins, so that these statistics show that the nearer the alliance the greater the danger.

Out of 177 insane persons from the counties of Ross and Wigtown, he found 23 per cent. were offspring of first, second, or third cousins ; or, including those about whose parentage no information could be obtained, out of 260 patients 16 per cent. were such offspring.

The influence of these marriages, he says, in producing insanity is clear. It appears also that its influence is more felt in producing imbecility and idiocy, than in insanity acquired late in life. It does not, of course, follow that blood relationship of parents is the *cause* of mental weakness in the children.

A valuable collection of references will be found in this paper, and I have myself consulted some of the originals, but as I am unable to improve on Dr Mitchell's abstract, I will only give the outline of what he says, and refer readers for more accurate details to the original sources.

In 1846, Dr Howe showed that in Massachusetts, the parentage of 359 idiots, out of 574 cases, was ascertained, and that there were 17 cases certainly, and probably 3 more, of offspring of consanguineous marriages ; that is, about 5 per cent.

Again the reports of the Commissioners on idiocy to the General Assembly of Connecticut showed that in 1856, out of 310 cases, consanguinity of parents was the probable cause of idiocy in 20 cases, or nearly 7 per cent. of the whole number. But Dr Mitchell finds that the question as to consanguinity of parents was only answered in 160 cases, so that we ought to count the 20 cases as $12\frac{1}{2}$ per cent. Of these 20 cases, 12 were the offspring of first cousins.

With respect to deaf-mutism, Mr Buxton, of the Liverpool Institution for the Deaf and Dumb, in a paper in the *Medicochirurgical Journal* (January, 1859), says that he found one deaf-mute in ten to be the offspring of first cousins.

Dr Peet, of the New York Institution, gives the same proportion, and adds that in that part of the United States hardly one family in fifty is the offspring of a cousin marriage.

Dr Mitchell himself instituted an inquiry into institutions for deaf-mutes, and found that in English institutions, out of pupils representing 323 families, 15 families were the offspring of consanguineous marriages ; and that in Scotland the corresponding numbers were 181 and 9. Making a deduction of 25 per cent. for cases of acquired mutism, Dr Mitchell finds that 1 case in 17 of congenital mutism is the offspring of a consanguineous marriage. The numbers, however, collected by me are larger than these, and show a very different result, unless indeed the marriage of more distant cousins is more fruitful of this evil than those of first cousins—a very unlikely result.

Dr Peet says in his *Thirty-fifth Annual Report*, in analysing Sir C. Wilde's *Status of Disease in Ireland**, that it appears that in Ireland about 1 in 16 of deaf-mutes were offspring of first, second, or third cousins.

From the Irish statistics, Dr Mitchell deduces the result that deaf-mutism, as it appears among the children of cousins, seems to be more congenital than in marriages between persons not akin.

Dr Mitchell carried out his second plan by collecting family histories, as complete as possible, of the whole populations in the islands of St Kilda, Scalpay, the parish of Berneray in Lewis, and some small fishing villages on the south-east coast of Scotland. In all these places he had been led to suppose that cousin marriages were frequent. It should be mentioned that in every case the frequency was found less than was supposed, although it sometimes rose as high as one marriage in four. These districts were chosen because the populations are much isolated. Very full details of the status of disease are given; but Dr Mitchell was disappointed in the degree of accuracy to which he was able to attain. He sums up the general result of his second method as follows :—" The facts which I have detailed appear to show a great unsteadiness in the character of this influence (consanguineous marriages). Sometimes we seem to find little or no proof that it is an evil influence. At other times this bloodship in the parentage appears to be the origin of much injury to the offspring. More frequently still the facts admit of various interpretations, and are not very clear or satisfactory in their teaching. It is of importance, however, to know that these differences or seeming differences may occur, and to learn that it is necessary to widen the field of observation, and carefully to inquire into all those circumstances by which it is quite clear the results may be, and often are, exaggerated, modified, or concealed....If taken *as a whole* and fairly interpreted, it appears to me that they lead to the same conclusion as that drawn from the first line of inquiry, viz. that consanguinity in parentage tends to injure the offspring."

Dr Mitchell came to the conclusion that, under favourable conditions of life, the apparent ill effects were frequently almost nil, whilst if the children were ill-fed, badly housed and clothed, the evil might become very marked. This is in striking accordance with some unpublished experiments of my father, Mr Charles Darwin, on the in-and-in breeding of plants; for he has found that in-bred plants, when allowed enough space and good soil, frequently show little or no deterioration, whilst when placed in competition with another plant, they frequently perish or are much stunted.

The general conclusions, drawn from the whole of Dr Mitchell's investigation, may be shortly stated as follows :

1st.   Consanguinity of parents is injurious to the offspring.

---

* *Blue Book*, containing Report of the Census (for Ireland), 1864.

2nd. Where the children seem to escape, the injury may show itself in the grandchildren.

3rd. In many isolated cases, and even groups of cases, no injurious result can be detected.

4th. These unions influence idiocy and imbecility more than the forms of insanity acquired later in life.

5th. The frequency of these unions in Scotland (although not so great as supposed) somewhat increases the amount of idiocy there.

All who are interested in the subject should certainly refer to the originals of these papers.

It will be observed that my investigation, so far as it is worth anything, tends to invalidate Dr Mitchell's results; but perhaps the apparent invalidation is due to the fact that a large majority of Englishmen live under what are on the whole very favourable circumstances.

The next paper to which I will refer is by Signore Paolo Mantegazza, Professor at Pavia, and is entitled "Studj sui Matrimonj Consanguinei*."   The Professor begins with an interesting historical sketch of the legislation against consanguineous marriages, which has obtained in the various ages of history.   He says it would be useless to repeat all that has been written on the subject, and refers to the works of Chipault†, Reich‡, his own work§, and to Dr Mitchell's paper above referred to.   He gives a list of fifty-seven authors who have opposed these marriages, and of fifteen who have defended them, but says that we ought not to lay stress on the great inequality of these numbers.   He also gives a long list of experiments made on the in-breeding of animals.

He says that in 1863 an inquiry was made by the Government of France as to the proportion of consanguineous marriages, but that he failed to obtain (at least at that time) an inquiry of a like nature in Italy.   The professor therefore gives the cases of 512 consanguineous marriages (inclusive of Mitchell's) from all countries collected by him; and of these he found 409 cases of bad results, and 103 with no ill results.   It cannot, however, he observes, be asserted from this that it is 4 to 1 that the result of such a marriage will be ill.   They are selected cases, for they naturally caught the attention of observers.   Many of the evils recorded in the list are doubtless quite independent of consanguinity.

The only fact which can be safely deduced from these numbers is the effect in producing sterility; and he finds that 46, or 8·9 per cent., of the marriages were sterile.

The results, which he deduces from his consideration of the cases, may be shortly summed up as follows:

* Milan, Gaetano Brigola, 1868, price 75 centimes.
† *Étude sur les mariages consanguins*, Paris, 1863.
‡ *Geschichte, Natur, &c.*, Cassel, 1864.
§ *Elementi d'Igiene*, Milan, 1868, p. 437.

1. Consanguineous marriages are, on the whole, more unfavourable to the offspring than others.

2. The proportion of 4 : 1, deduced from his table, is not a correct view of the case.

3. The injury arises from the multiplication of pathological germs of the same nature.

4. This influence alone would weaken the offspring of relations. This is confirmed by the frequency of sterility, and of miscarriages, and the appearance of diseases new to the family.

5. The best proved results of these unions are, failure of conception, abortive conception and miscarriage, monstrosities, disposition to nervous complaints, arrested mental development, scrofulous and tubercular diathesis, lowered vitality, high rate of mortality, especially amongst infants, dysmenorrhœa, small generative power, pigmental retinitis.

6. The nearer the kinship, the greater the danger.

7. He gives the obvious conclusions as to choice in marriage.

8. It is tolerably probable that the danger is greater in cases of uterine kinship : first, because more evil or good is heritable from the mother; and secondly, because "we are entirely sons of our mother, but are not equally so, nor always, the sons of our father." This last sentence can hardly refer to conjugal infidelity; and yet, if it does not, how does the second cause differ from the first ?

Amongst the references given in this paper is one to "a very recent work" by Dr Loubrieu on "Deaf-mutism," and others to Dr Down's "Marriages of Consanguinity," &c. (*London Hospital Report III*, pp. 224 and 236), to Saint-Lager's *Études sur les causes du Crétinisme, &c.*, Paris, 1867, and to *Studj sui sordomuti e rendiconto degli Istituti, &c.*, Milan, 1864.

I have to thank Mr Buxton, of the Liverpool School for Deaf-Mutes, for sending me a pamphlet by Mr J. Scott Hutton*. At the Halifax (Canada) School Mr J. Scott Hutton found that out of 54 families (with 100 children), with respect to which information could be obtained, 15 families (with 37 children) were offspring of first cousins. He sums up by saying :—"Thus out of 110 deaf-mute children of whom we have definite information, 56 are the offspring of cousins...an expressive example of the melancholy consequences flowing from cousin-marriages." It should be added that there are two apparent discrepancies between the figures given in this part of the paper. The statistics of other countries are not so striking. "In England, the deaf and dumb in marriages within the limits of consanguinity are in the proportion of 6 per cent., in France 25 to 30 per cent., in Kentucky 20 per cent., in Illinois 12 per cent., and in Ontario 28 per cent." No authority is given for these figures, nor is the phrase "limit of consanguinity" defined.

* *American Annals of the Deaf and Dumb*, January, 1869. Washington, D.C., pp. 15—17.

Mr Buxton's admirable paper on " Deaf-mutism " has been already referred to in the sketch of Dr Mitchell's paper.

Sir W. Wilde, in an Appendix to his *Aural Surgery*, gives a very complete account of the history, of the tuition of deaf-mutes, and of the causes which produce the disease. He says consanguinity may be looked on as paramount. "Many conjectures have been offered on the subject, but the question has been set at rest by the results of the Irish census." This appendix embodies the results of the inquiry carried out in the Irish census, and is referred to in Dr Mitchell's papers.

So general a consent as to the ill effects of cousin-marriages must certainly have far greater weight than my purely negative results. But it strikes me that in no case has the investigation been free from flaws, for in no case has it been really determined what is the proportion of consanguineous marriages in the whole population. The exceedingly various estimates which different people have given me of the frequency of cousin-marriages (from 10 per cent. down to 1 in 1,000, if my memory serves me right), lead me to believe that general impressions on this point are almost valueless. Every observer is biased by the frequency or rarity of such marriages amongst his immediate surroundings.

My paper is far from giving anything like a satisfactory solution of the question as to the effects of consanguineous marriages, but it does, I think, show that the assertion that this question has already been set at rest, cannot be substantiated.

The subject still demands attention, and I hope that my endeavour may lead more competent investigators to take it up from some other side.

Marriages between Cousins in Relation to Infertility and a High Death-rate amongst the Offspring.

Professor Mantegazza states in his paper † that he may conclude with tolerable safety, from his collection of 512 cases of consanguineous marriage, that consanguinity tends to cause sterility ; for he found that between 8 and 9 per cent. of the recorded marriages were sterile. It is not clear, however, how he is entitled to draw this conclusion, unless he knows what is the proportion of sterile marriages in the general population, and he admits that he has no statistics on this point. M. Boudin, who wrote at an earlier date, is of the same opinion, and considers, further, that even where sterility does not afflict the consanguineous marriage itself, it is apt to affect the offspring ‡. Dr Balley is also of opinion that the ill effects of such marriages are liable to appear in the second generation §.

Since reading my paper on "Cousin Marriages in England," on the 16th of March, before this Society, a method has occurred to me of settling these points pretty satisfactorily. This method is by a comparison between the fertility of the marriages of first cousins and of the marriages of their offspring, as recorded in the

* Sent in subsequent to the reading of the paper.
† *Studj Sui Matrimonj Consanguinei.* Milan, 1868.
‡ *Annales d'Hygiène Publique*, Tom. xviii., pp. 5—82.
§ *Comptes Rendus*, Tom. lvi., p. 135.

pedigrees in Burke's *Landed Gentry* and the *Peerage*, with the fertility of marriages between persons not akin.

I had already got a large number of marriages marked as being between first cousins, and I accordingly proceeded to count the number of children arising therefrom. The marriages made within the twenty years immediately preceding the publication of those works were excluded; so that only complete families were counted. It soon became evident that the lists of the daughters were very incomplete, and that the daughters were perhaps sometimes omitted altogether; the sons dying in infancy are also frequently omitted (especially in the *Landed Gentry*); and when such occurred I excluded them. I think that the lists of the sons surviving infancy are, however, pretty complete, and any incompleteness will clearly affect the record of marriages between persons not akin as much as it does the first cousin marriages. The comparison to be made must, therefore, be only between the numbers of sons. I shall use the words *sterile* or *infertile* to mean the absence of children surviving infancy. The number of daughters recorded will be given, so as to show the extent of incompleteness.

In this manner 116 families, offspring of first cousins, were collected. In all but 12 of them the marriages were between children of brothers. In 11 of the 116 it is merely stated that there was issue of the marriage, and in 8 others there is no information as to whether there was issue or not. I found in a subsequent inquiry, by cross references to other pedigrees, that where there was no information there was nevertheless often a family; so that the absence of information is no indication of sterility, and indeed is perhaps some slight indication of fertility, because the family is omitted in order to economise space, and d. s. p. (*decessit sine prole*) is frequently added where there *was* no issue. In this case, however, cross references were of no avail, because the family would be recorded in the pedigree under consideration or not at all. The absence of information is here then a slightly greater indication of sterility than in my later inquiry, where it is no indication at all.

The cases where issue was recorded may clearly be disregarded in making the comparison, since they might be matched by similar cases amongst the non-consanguineous marriages.

Subtracting, then, the 11 recorded cases of issue and the 8 cases of no information, we are left with 97 families; these gave 202 sons and 153 daughters. It is probable that about 200 daughters should have been recorded. Now 202 sons to 97 marriages is at the rate of 2·07 sons to each marriage; or, supposing the 8 cases of doubt to have been all sterile, we get 105 marriages as giving 202 sons, that is at the rate of 1·92 sons to each marriage.

Thus the average number of sons who survive infancy, arising from a marriage of first cousins amongst the gentry of England, is between 1·92 and 2·07.

The next step was to collect the non-consanguineous marriages. In order to secure myself from bias, I opened my book by chance and counted all the marriages in the pedigree which fell under my eye. I then did the same in another place, and so on. In this way 217 families arising from persons not akin were collected, and found to give 416 sons and 340 daughters. Here, as before, the daughters are deficient, and about 420 daughters ought probably to have been given. Now 416 sons to 217 marriages is at the rate of 1·91 sons to each marriage. Thus the average number of sons who survive infancy, arising from non-consanguineous marriages, is 1·91.

The balance of fertility is therefore slightly on the side of the cousins, but the small difference is probably due to chance.

In order to feel greater confidence in this result, a second method of analysis was carried out. If cousin marriages tend to cause sterility, they probably tend to cause partial sterility. Now amongst the 97 cousin marriages, 14 were sterile (in the sense defined), and amongst the 217 non-consanguineous marriages 33 were sterile. Thus we have 83 fertile cousin marriages and 184 fertile non-consanguineous marriages; the former gave 202 sons, the latter 416 sons. It will be observed that this course entitles me to disregard the 8 cases of "no information" before referred to, for if they were sterile they are to be subtracted *ex hypothesi*, and if there was issue, they could be matched by similar cases amongst the non-consanguineous. Thus fertile first cousin marriages produce sons at the rate of 2·43 sons to each marriage, and fertile non-consanguineous marriages produce sons at the rate of 2·26 sons to each marriage.

Therefore the analysis leads to a similar slight balance in favour of the fertility of the first cousins, just as did the former one.

I offer the following suggestion as a possible explanation of the greater fertility of the cousins, although mere chance is the more probable cause of the difference. Marriages between first cousins will be more apt to take place where there is a large group of persons who bear that relationship to one another. In such families fertility will be hereditary; hence it is possible that the comparison is to some extent being effected between abnormally fertile families, and those in which fertility is only normal.

The next point to investigate is as to whether the offspring of first cousin marriages are themselves affected by sterility.

To test this, recourse was again had to the *Peerage* and *Landed Gentry*, and 136 marriages of the offspring of first cousins were collected. Concerning 29 of these no information could be obtained, and, for the reasons before assigned, these may be set aside. Of the 107 remaining marriages, it is recorded that 14 had issue. Subtracting these, we are left with 93 marriages, and these gave 180 sons and 157 daughters. It should be mentioned that some few of the marriages were recent, so that the families would be not quite complete in these cases. Now 93 marriages giving 180 sons is at the rate of 1·93 sons to each marriage.

Again, 16 of these marriages were sterile, so that 77 fertile marriages gave 180 sons, that is at the rate of 2·34 sons to each marriage. If these two numbers, viz. 1·93 and 2·34, be compared with the corresponding numbers, viz. 1·91 and 2·26, for the non-consanguineous marriages, it is clear that there is again no evidence of want of fertility in the offspring of first cousin marriages.

The results with respect to fertility may be summed up in the following table :—

Parentage	Average number of sons to each marriage	Percentage of sterile marriages *	Average number of sons to each fertile marriage
Not consanguineous..............	1·91	15·9	2·26
Parents first cousins .... }	between 2·07 and 1·92	between 14·7 and 20·9 }	2·43
One parent the offspring of a marriage between first cousins	1·93	17·2	2·34

The comparison may be best effected by means of the numbers in the last column. The figures in the second column are not of much value, since in some cases it was difficult to decide whether the entry should be made as being a case of "no information" or of sterility.

The comparison of the figures in the first and last columns shows, without much room for doubt, that the alleged infertility of consanguineous marriages, whether direct or indirect, cannot be substantiated.

I now pass on to the question of the youthful death-rate.

It has been stated by M. Boudin and others that the offspring of consanguineous marriages suffer from an excessively high rate of infant mortality. I have tried to put this to the proof as follows :—

I recurred to the families in the *Peerage* which were offspring of first cousins, and marked every case where it is recorded that a son or daughter died in infancy or youth. Where the age of the child was mentioned, ten years was taken as the standard of youth. Burke's *Landed Gentry* was of no avail in this inquiry, because I found that children dying in infancy were never, or very rarely, mentioned therein.

From the *Peerage* I could only obtain 37 fertile first cousin marriages; in two of these there were no children surviving youth. The 37 gave 86 sons, who survived infancy, 15 children (boys and girls) who died in infancy or youth, and 4 more as to whom the period of death was doubtful. Besides this, it is stated of one family that "all died young except one daughter." Now in the previous part of this paper it is shown that the average number of sons to a fertile first cousin marriage is nearly 2½; so that it may not be unreasonable to credit this family with 4 infants who died.

On this supposition we should have 37 fertile marriages of first cousins giving 86 sons, who survived, and between 23 and 19 boys and girls, who died early. Reducing these numbers to percentages, I find that—

One hundred fertile marriages of first cousins would give from 51 to 62 children who die young, and that for every 100 sons, offspring of first cousins,

* Sterility means absence of children surviving infancy.

who survive youth, there are from 22 to 27 boys and girls (their brothers and sisters) who die early.

These numbers cannot be used as giving the actual infant death-rate, on account of the imperfections in the pedigrees in the *Peerage*, but they may be used in a comparison with other statistics deduced from the same source.

Now 89 fertile non-consanguineous marriages (collected by chance from the *Peerage*) gave 197 sons, and 44 sons and daughters who died young. Reducing these numbers to percentages as before, I find—

That 100 fertile non-consanguineous marriages would give 49 children who die young, and that for every 100 sons, offspring of fertile non-consanguineous marriages, who survive infancy, there are 22 boys and girls (their brothers and sisters) who die early.

The numbers to be compared are therefore 51 or 62 with 49, and 22 or 27 with 22.

These are merely two different ways of consulting the facts, and it appears that both methods give some evidence of a slightly lowered vitality amongst the off-spring of first cousins.

Thirty-seven cases form, however, far too small a total on which to base satisfactory statistics. The numbers thus collected are far scantier than those collected by others, but as far as I am aware this is the only occasion in which the method of collection has been one in which the unconscious bias of the collector could not operate. In all these inquiries I was ignorant as to whither the figures were tending until I came to add up the totals.

This last inquiry is, I fear, worth but little, but so far as it goes it tends to invalidate the alleged excessively high death-rate amongst the offspring of cousins, whilst there remains a shade of evidence that the death-rate is higher than amongst the families of non-consanguineous marriages.

28.

NOTE ON THE MARRIAGES OF FIRST COUSINS.

[Journal of the Statistical Society, Vol. XXXVIII. (1875),
pp. 344—348.]

AFTER I had read my paper on this subject in March last* before the
Statistical Society, Mr Arthur Browning (a Fellow of the Society) suggested
to me another method of determining whether cousin marriages were injurious
or not. This method was to discover whether the proportion of offspring of first
cousins, amongst persons distinctly above the average, either physically or mentally,
was less or greater than the general proportion given by my paper for persons in a
similar rank of life.

Mr Browning and I agreed to carry out this scheme together; but we thought
it would be well to delay extensive operations until we saw what success was
attainable in a more limited inquiry. The results are so very unequal to our
expectations that we do not intend to proceed further. The statistics are,
however, of some interest as far as they go.

The boating eights, who race at Oxford and Cambridge in May, are a picked
body of athletic men. There are twenty boats at Oxford, and thirty at Cambridge,
in the "first and second divisions"; and their crews are 400 men, exclusive of
coxswains. We accordingly sent circulars to the stroke-oars of these fifty boats,
during their preparatory training, begging them to ask the members of their crews
whether their parents were first cousins or not. Where there were several brothers
rowing in the eight, they were only to be counted as one case; and cases of refusal
to answer were also to be marked. We received answers from nineteen Oxford
crews, and from eighteen at Cambridge. Three or four men appear not to have
been asked, probably on account of their absence at the time that the circular was
being filled up; and there were two cases of two brothers rowing in the same boat,
but they were not offspring of first cousins. We here beg leave to return our
warm thanks to the gentlemen who so kindly answered the queries.

* [Paper 27.]

Besides these answers, the circular addressed to the stroke of the second boat of Corpus College, Cambridge, came back to us with a carefully falsified return; it was by mere chance that I was able to detect the fraud.

One member of a crew was accidentally disabled, but we have thought it proper to include him, as well as his substitute; he is a son of first cousins.

Altogether the parentage of 290 men belonging to different families was ascertained, and of these seven were found to be offspring of first cousins, and one man refused to answer the query. The result is therefore that 2·41 or 2·75 per cent. (according as we exclude or include the case of refusal) of boating men are offspring of first cousins. The proportion of first cousin marriages to all marriages, amongst the same class of society, was determined at 3 to 3½ per cent. in my former paper. Thus these numbers appear, to some extent, to justify the belief that offspring of first cousins are deficient physically, whilst at the same time they negative the views of alarmist writers on this subject. But taking into consideration the smallness of the number 291, and the uncertainty of my previous methods, the indication is very slight.

The next step was to send circulars to masters at sixty-five of the principal schools for the upper and middle classes in England. We begged them to put the circulars before the School Natural History Club, or else into the hands of any boy who would be likely to take an interest in the investigation. The collector of statistics was asked to form a list of the best cricketers, football players, and other athletes, such list not to comprise more than 20 per cent. of the whole school; and only one of several brothers was to be entered therein. Each of the boys on the list was then to be asked whether his parents were first cousins or not, and the answers to be returned to us.

Returns were, however, only received from six schools. The work was in most cases undertaken by the masters themselves. We here beg leave to thank all the collectors for their great kindness.

The following table gives the number of boys from whom the selection was made, and the numbers on the selected lists :—

School	Number of boys	Selected athletes	Percentage of selected list compared to the whole number of boys
Rugby	117 over 16 years of age	34	19·9
Sherborne	243	39	16·0
Lancing	145	15	10·3
Taunton	130	18	13·8
Giggleswick	120	24	20·0
Bury St Edmunds	64	13	20·3
Total	819	143	16·4

At Rugby and Sherborne the standard of athleticism is high, as also at Lancing and Taunton, where only about one boy in ten was taken from the whole school. At Bury St Edmunds it would be rather low, but at Giggleswick allowance was made for the ages of the boys, so that the 20 per cent. was distributed over the whole school.

Out of the 143 athletes, one was the offspring of first cousins, a sturdy boy in the highest class of his school; and three either did not know, or refused to answer the query.

These figures are thus almost nugatory, for we have from one to four offspring of first cousins amongst 143 boys, that is to say from 0·7 to 2·8 per cent. Combining the boating statistics with these we get from eight to twelve sons of first cousins amongst 434 athletes, that is to say from 1·84 to 2·76 per cent.

I take the higher number, 2·76, as probably more near the truth than the lower one. The same remarks as those made on the results of the boating inquiry are therefore applicable to the whole.

The following observation of Mr Browning, with respect to longevity in children of consanguineous parents, is perhaps worth giving.

He is a director and the honorary secretary of the French Protestant "Hospice," where forty old women and twenty old men, descendants of French refugees, find a comfortable home. They are seldom admitted much under 70, and their average age is 77; three or four are over 90. They were questioned as to whether their parents were first cousins. Out of thirty-seven women, four were absent and four were ignorant as to the fact; out of the remaining twenty-nine, one was the daughter of first cousins. Out of twenty men, three were absent and one was ignorant; of the remaining sixteen, none were offspring of first cousins. Thus, out of fifty very aged persons, one was the offspring of first cousins, and five were uncertain as to the fact. The steward, a man of about 40, also a descendant of French Protestant refugees, had married his first cousin. These people are in the fifth or sixth generation from the original refugees. In the earlier generations there would, doubtless, have been much intermarriage amongst them, but Mr Browning says that they now have almost entirely lost their French characteristics, and are merged in the general population. If, however, there is *any* class feeling remaining, cousin marriages would be, doubtless, more prevalent amongst them than elsewhere.

With respect to intellectual powers, I happen to know that amongst the sixty Fellows of one of our larger colleges at Cambridge there are three sons of first cousins, and there may be more; the tenure of a fellowship betokens, at least, great power of acquiring knowledge.

Since March last, Mr Huth's work on *The Marriage of Near Kin* has appeared, and I find therein some confirmation of my own results as to the prevalence of cousin marriages in England.

It appears (p. 120) that M. Dally examined the registers of the *mairie* of the eighth district in Paris, and found that out of 10,567 marriages celebrated between

1853 and 1862, 141 were between first cousins, eight between uncles and nieces, and one between a nephew and aunt—total, 150 consanguineous marriages within the above degree. "(These numbers may vary from 146 to 152, on account of three figures which are uncertain.) These numbers give us a proportion of 1·4 per cent., and it appears to me (*i.e.* M. Dally) impossible to admit otherwise than this—that in a district of Paris which is inhabited by foreigners, showing a considerable floating population, there are many less marriages between cousins than in the midst of small towns, and in the country."

Now, it will be remembered that I estimated the proportion of first cousin marriages in London by a totally different method, at $1\frac{1}{2}$ per cent., which lies very close to 1·4 per cent. ; and it would be likely that the proportion of consanguineous marriages in two such immense towns as London and Paris would be nearly the same. M. Dally further considers himself authorised (from the context, I presume by M. Legoyt, the chief of the Statistical Department of France) to say that M. Boudin's estimate of 0·9 per cent. for marriages over the whole of France within the above degrees, is between three and four times too small; according to M. Legoyt, therefore, the proportion for the whole of France lies between about $2\frac{1}{2}$ and $3\frac{1}{4}$ per cent. This estimate may be compared with my results of 3 to $3\frac{1}{2}$ per cent. for the upper classes, 2 per cent. for the larger towns, and $2\frac{1}{4}$ per cent. for the country, and, as far as it goes, it tends to confirm my figures. I should certainly expect that the equal division of property under the *Code Napoléon* would tend to promote first cousin marriages, as the family property would be thereby kept together. On the other hand, the Roman Catholic Church discourages these marriages; yet it is stated (p. 209) that legal dispensations are only requisite for marriages between uncles and nieces, and between nephews and aunts, and not for those between first cousins, so that the discouragement would not be likely to be very efficacious.

It cannot be doubted that M. Boudin's estimate of 2 per cent. for consanguineous marriages within the degree of second cousins is very far too low for France; probably 5 to 8 per cent. would be nearer the mark.

It is stated (Huth, p. 212) that the Irish Census Commissioners found that in 1871 6·7 per cent. of the parents of deaf-mutes were cousins within the sixth degree; in 1861 6·9 were cousins within the fourth degree; and in 1851 4·9 were cousins within the third degree. These figures have been taken to show the appalling injury resulting from consanguineous marriages; if, however, M. Legoyt's estimate for France may be taken as even nearly accurate, and may be extended to Ireland (also chiefly Roman Catholic), these figures would rather show that the evil has been exaggerated. Altogether, considering my own results in combination with these figures, the safest verdict seems to be that the charge against consanguineous marriages on this head is not proven.

In a short criticism of my paper in the *Spectator*, it was objected that the women of a family keep up intercourse much more than the men; this reminds me of the old jingle :

"Your son is your son until he's a wife,
Your daughter's your daughter all her life."

And there is probably some truth in the criticism. Now from this cause different name first cousin marriages should be slightly more frequent than they otherwise would be. On the other hand, the mere identity of surnames between two families doubtless tends to keep them together.

But granting the soundness of the objection, the only effect is, that I have under-estimated the extent of first cousin marriage, and it is so much the harder for those, who hold extreme views as to the ills of these marriages, to prove their case.

The *Spectator*, however, takes no notice of the fact that my indirect method, partly indirect method, and purely statistical method, all point to approximately the same result.

Mr Huth says: "We have absolutely no basis from which to start a statistical inquiry as to the effect of consanguineous marriage on the offspring." If this is the case, the value of my own imperfect estimate is enhanced.

INDEX TO VOLUME IV.

CAMBRIDGE: PRINTED BY JOHN CLAY, M.A. AT THE UNIVERSITY PRESS.

Printed in the United States
By Bookmasters